Fundamentals of Water Security

Fundamentals of Water Security

Quantity, Quality, and Equity in a Changing Climate

Jim F. Chamberlain
University of Oklahoma, United States of America

David A. Sabatini
University of Oklahoma, United States of America

WILEY

This edition first published 2022

Registered Office
John Wiley & Sons, Inc., 111 River Street, Hoboken, NJ 07030, USA

Editorial Office
John Wiley & Sons, Inc., 111 River Street, Hoboken, NJ 07030, USA

For details of our global editorial offices, customer services, and more information about Wiley products visit us at www.wiley.com.

Wiley also publishes its books in a variety of electronic formats and by print-on-demand. Some content that appears in standard print versions of this book may not be available in other formats.

Library of Congress Cataloging-in-Publication Data applied for:

Paperback ISBN: 9781119824640

Cover images: Courtesy of Jim F. Chamberlain; Used with Permission from Sniffin the Wind Photography, Fred R. Bowman; © Pixel Embargo/Shutterstock
Cover design by Red Owl Graphics; Wiley

Set in 9.5/12.5pt STIXTwoText by Straive, Chennai, India

SKY10035252_071322

The authors dedicate this work to:

My students who inspire me with their passion for the world's most disadvantaged peoples, and my brother-in-law, Frank Zemanek, who paddled with me many peaceful and beautiful waters.

(Jim)

My wife - Frances Sabatini - and my children and their families - Caleb, Kirbie and Conrad Sabatini, and Peggy and Hunter McDonald - for their love, encouragement, and endearing support.

(Dave)

Contents

Preface

The world is rapidly changing. Burgeoning populations along with ever-increasing standards of living, especially for those in emerging countries, increase the *demand* and stress on our water resources. What is not increasing, however, is the *supply*, the total amount of water in earth's biosphere, water that is integral to all standards of living. This water is needed not only for human consumption and growing the food we eat but also for generating electricity, for sustaining health and healing illnesses, for maintaining ecosystem services, and for making various other products and services that comprise life's modern amenities. No less importantly, water has been, and continues to be, used for transportation, recreation, cultural well-being, and religious ceremony. Water, thus, is valuable from physical, cultural, and spiritual vantage points. While the total amount of water remains the same, the amount that is available and useful at any given time and place, for human and ecological flourishing, is transitory between its various stocks and flows.

Water security is the modern term that describes a state of having sufficient quantity and quality of water necessary to equitably meet both immediate and long-term human and ecological needs. As such, water security spans multiple environmental science and engineering disciplines, including groundwater and surface-water hydrology, geography, geology, water resources, soil science, ecology, atmospheric science, chemistry, biology, and health science, and also the disciplines of social and political science.

This volume is an introductory volume directed toward both the academic student and the practitioner. As a textbook, this volume could be used for the education of upper-level undergraduate or entry-level graduate students who already have some background in chemistry, biology, and mathematics. For the practitioner, this volume serves as an introduction to the diverse field of water security, perhaps serving as a supplement to critical hands-on training and experience.

In the **Introduction (Part I)** section, we begin by defining water security as existing at the water quantity–quality–equity nexus along with a brief overview of historical water challenges and insecurity and the manner in which affected communities adapted. The second section of the book – **The Context of Water Security (Part II)** – offers basic principles of water security under the threefold headings of water quantity, quality, and equity. Here, the student is given a basic understanding of hydrology, watershed management, aquatic chemistry and biology, and the social dynamics of water access and distribution. The final chapter in this section is devoted to climate change, which impacts all three of these principles. The third section – **Competing Uses of Water and Threats to Security (Part III)** – delves deeper into the diverse uses of water – for food, energy, industry, and ecosystems – and security threats that exist within, and at the nexus of, these uses.

The fourth section, entitled **Sustainable Responses and Solutions (Part IV)**, presents modern attempts to correct and/or mitigate the challenges of global water stress in pursuit of water security. The final section – **Resilience, Economics, and Ethics (Part V)** – invites the reader to consider various approaches to water planning as well as the fundamentals of water resource economics, critically important to water security. Finally, we propose an ethic of water for the modern pilgrim.

Case studies and examples are given throughout the text in order to both illustrate the principles and to create a sense of solidarity with sisters and brothers throughout the world. Regions in both developed and developing countries are now, or will soon be, experiencing water stress, and so we include issues and examples from a range of global contexts. We also highlight seven persons who have been working in various aspects of water security in sections called **The Practice of Water Security** following several chapters. These women and men are winners of the University of Oklahoma International Water Prize, awarded biennially from 2009 to 2022. Finally, most chapters culminate in a "Foundations" section, which provides a more quantitative probe into the science of water security. Depending on the background and orientation of the reader or student, this section may be considered optional.

The study of water security can be both sobering and optimistic. Because water is most often experienced as a local resource, there are "haves" and "have-nots" in the world of water security. But there are also signs of a brighter future – from a greater application of old and new technologies, a growing appreciation for the benefits and vulnerability of our water resources. and a desire for water cooperation and innovation among diverse entities.

Acknowledgments

This book began with material developed for two University of Oklahoma courses – an on-campus course "Introduction to Water" (taught by Chamberlain) and a new "Water Security" course (taught by Sabatini) developed as part of our online Master's program in *Hydrology and Water Security*. We would like to acknowledge those students who provided helpful input into both courses based on their particular knowledge and experience. We especially thank Christian Newkirk, Heath Orcutt and Shannon Mathers for their assistance in the tedious work of producing figures as well as gaining permission for the use of existing figures. Their contributions are felt – and seen – throughout this book. We would also like to acknowledge the School of Civil Engineering and Environmental Science (CEES), the Gallogly College of Engineering, and the University of Oklahoma administration for their support of our efforts, which are an outgrowth of the OU WaTER Center, the Hydrology and Water Security program, and the larger water enterprise on campus. Without their support, and invaluable interactions with our esteemed colleagues, this volume would not have been possible.

We want to thank the many academic colleagues from across the US that provided valuable input into the content of the above courses and the outline of this book. Their input was invaluable. We also want to thank water experts that provided detailed reviews of specific chapters – Drs Emma Colven, Randy Kolar, R. David Lamie, Mark Shafer, Aondover Tarhule, Evan Tromble, and Paul Weckler – and a very special thanks to Dr. Michael Campana, affectionately known as Aquadoc (Doctor Water), for reviewing the book draft in its entirety.

Finally, we give thanks to our generous and passionate colleagues, the Directors of the OU WaTER Center – Drs Yang Hong, Robert Knox, Robert Nairn, Jason Vogel and, especially, Randy Kolar, who continue to lead us and inspire us to do great work on behalf of the common good.

About the Companion Website

This book is accompanied by a companion website.

www.wiley.com/go/chamberlain/fundamentalsofwatersecurity

This website includes solutions to selected end-of-chapter problems.

Part I

Introduction

1

Introduction to Water Security

In this chapter, readers are introduced to the concept and importance of global water security. Many of the United Nations Sustainable Development Goals address or depend upon water security. Several water-security definitions are presented with a brief discussion on the focus and limitations of each. We then offer a network of considerations that form the context for water security – water quantity, quality, and social equity. In the context of these three components, water security is found at the nexus where water is of suitable quality for the user's purpose(s), is of sufficient quantity for the user's purpose(s), and is equitably available to users, regardless of age, gender, social, or economic status. And all of this happens in the modern setting of a changing climate. The chapter then discusses various ways in which water security is measured, using metrics that are quantifiable and useful for comparison. The analysis of water security can be at watershed, household (local), regional, or global level for bases of comparison.

Learning Objectives

Upon completion of this chapter, the student will be able to:

1. Understand the concept of global water security with its various facets and components.
2. Articulate a working definition of water security.
3. Discuss the ways in which many of the United Nations Sustainable Development Goals, either directly or indirectly, overlap with water security.
4. Quantify water security using several commonly used metrics.
5. Understand the various scales of water security – from local to global.
6. *Utilize the various units of measurements of water quantity and quality.

1.1 Introduction

What do we think of when we hear the term **"water security"**? We might immediately think of water that is needed to survive, to live. This is the water that we drink to sustain our bodily functions. Water is necessary for all biochemical processes, as reflected by the fact that our bodies are composed of 60% water (Ford 2016). But water is also needed to grow food, from

Fundamentals of Water Security: Quantity, Quality, and Equity in a Changing Climate, First Edition.
Jim F. Chamberlain and David A. Sabatini.

Companion website: www.wiley.com/go/chamberlain/fundamentalsofwatersecurity

cabbage to cow and everything in between. We might next think of the other ways that we use water in the home – for cooking, cleaning, bathing, and watering plants and landscape. Water security then would involve the availability of enough water to accomplish these very personal tasks upon which we all depend.

But as our horizons expand a bit, we might also realize that populations historically tended to settle near bodies of water, along rivers, lakes, seas, springs (fed by groundwater), and coastlines. This early development reflects the fact that water is necessary for sustaining agriculture, for transporting goods, for power generation, and for supplying the industrial sector that undergirds a nation's economy. Thus, water security is not only a local phenomenon but also a regional and global one. Whole populations can be affected by drought, by flooding, and by widespread contamination that threatens the quality of a large water body. Whole segments of a population may also lack access to water of sufficient quality and quantity. This lack of access may be because they cannot afford to purchase water or because they are unfairly prevented from accessing a water source. These examples foretell the inverse of water security, that is, the *in*security that comes with an imbalance of water quantity, a degradation of water quality, or water inequity that results in an inability to access the water necessary to sustain life.

These initial thoughts lead us to realize that *water security* is a consequence of scale, focus, and function. The *scale* of water security may be local (household), regional (watershed, aquifer), national, transboundary (within or between countries or other jurisdictions), or global. The *focus* of water security may be on the quantity, quality, or equitable access (equity) to water. The *function* of water security may be on basic human health and wellbeing, economic progress, ecosystems' functioning, or any combination of the above. These aspects are captured in the four editorial foci for the journal *Water Security* (Lall et al. 2017):

- Shortage (water quantity)
- Flooding (water quantity)
- Governance (water equity)
- Health and sanitation (water quality)

Likewise, a threat to water security (i.e. the threat of water *in*security) will also be faced, analyzed, and mitigated according to water quantity, quality, and equity, which are all intertwined. The lack of quality, quantity, or equity represents an uneven risk to people, the economy, and/or the environment. And so, a drought that threatens the Ethiopian teff harvest may affect rural people to a greater extent than city dwellers who have a more diversified diet. The construction of a Tennessee dam that generates hydroelectric power for local benefit may upset aquatic ecosystems downstream in neighboring states. The Bangladeshi reliance on groundwater wells to avoid cholera-impacted surface water may expose unsuspecting villagers to water tainted with arsenic. In addition, nearly all of these threats will be impacted by a climate that is warming gradually and consistently.

In this introductory chapter, we describe both the measures and usefulness of water security as a basis for study and analysis. Subsequent chapters will zoom in on various foci within the areas of water quantity, quality, equity, and climate change while looking at water security at a number of scales.

1.2 Sustainable Development Goals (SDGs)

Adopted in 2015, the United Nations **Sustainable Development Goals** (SDGs) set forth 17 targets (endpoints) designed to achieve a better and more sustainable future for all peoples (United Nations 2015) (Figure 1.1). One of these goals specifically addresses "Clean Water and Sanitation" (SDG 6). The goal of SDG 6 is to "Ensure availability and sustainable management of water and sanitation for all." This goal will be achieved by, among other things, reducing water pollution, increasing water-use efficiency, protecting ecosystems and watersheds, and strengthening proper water management (Table 1.1).

Other SDGs are also intertwined with water security (United Nations 2015). Table 1.1 illustrates the connections between water and several of the SDGs. Water is needed to grow food and put an end to hunger (SDG 2). Clean water is needed for the elimination of waterborne diseases (SDG 3). Children must be regularly healthy in order to attain a quality education (SDG 4). The need for women and girls to fetch water at long distances is a threat to their safety and ability to attend school along with the boys. Such is a deterrent to gender equity (SDG 5). Water-related disasters result in economic losses that are much harder on the poor and vulnerable in urban settings (SDG 11). Water safety entails the proper management of chemicals that might be released into the hydrosphere (SDG 12), and the cooperation of international bodies is required to protect terrestrial and inland freshwater ecosystems and their services (SDG 15). In order to meet these goals, stakeholders will need to use the basic tools of water security, including knowledge of hydrology, integrated

Figure 1.1 Seventeen Sustainable Development Goals (SDGs) agreed upon by member states of the United Nations, ushered in on 1 January 2016 (Martin 2015). Source: With permission from United Nations.

Table 1.1 Sustainable development goals (SDGs) and targets that are directly and indirectly relevant to water security.

<table>
<tr><th colspan="2">SDGs DIRECTLY RELEVANT TO WATER SECURITY</th></tr>
<tr><td colspan="2">SDG 6: “Ensure availability and sustainable management of water and sanitation for all”

6.1 Achieve universal and equitable access to safe and affordable drinking water for all.
6.2 Achieve access to adequate and equitable sanitation and hygiene for all.
6.3 Improve water quality by reducing pollution.
6.4 Increase water-use efficiency across all sectors and reduce the number of people suffering from water scarcity.
6.5 Implement integrated water resources management at all levels.
6.6 Protect and restore water-related ecosystems.
6.A Expand international cooperation and capacity-building support to developing countries.
6.B Strengthen the participation of local communities for improving water and sanitation management.</td></tr>
<tr><th colspan="2">SDGs INDIRECTLY RELEVANT TO WATER SECURITY</th></tr>
<tr><td>SDG 2: “End hunger, achieve food security and improved nutrition and promote sustainable agriculture.”

2.1 End hunger and ensure access by all people, in particular the poor and people in vulnerable situations … to safe, nutritious and sufficient food all year round.
2.2 End all forms of malnutrition … and address the nutritional needs of adolescent girls, pregnant and lactating women and older persons.
2.3 Double the agricultural productivity and incomes of small-scale food producers, in particular women, indigenous peoples, family farmers, pastoralists and fishers.</td><td>SDG 3: “Ensure healthy lives and promote well-being for all at all ages”

3.3 End the epidemics of AIDS, tuberculosis, malaria, and neglected tropical diseases.
3.9 Substantially reduce the number of deaths and illnesses from hazardous chemicals and air, water, and soil pollution and contamination.</td></tr>
<tr><td>SDG 4: “Ensure inclusive and equitable quality education and promote lifelong learning opportunities for all.”

4.5 Eliminate gender disparities in education and ensure equal access to all levels of education and vocational training for the vulnerable, including persons with disabilities, indigenous peoples and children in vulnerable situations.</td><td>SDG 5: “Achieve gender equality and empower all women and girls.”

5.2 Eliminate all forms of violence against all women and girls in the public and private spheres.</td></tr>
<tr><td>SDG 11: “Make cities and human settlements inclusive, safe, resilient and sustainable.”

11.5 Significantly reduce the number of deaths and the number of people affected … decrease the direct economic losses relative to global gross domestic product caused by disasters, including water-related disasters, with a focus on protecting the poor and people in vulnerable situations.</td><td>SDG 12: “Ensure sustainable consumption and production patterns.”

12.4 Achieve environmentally sound management of chemicals and all wastes throughout their life cycle … and significantly reduce their release to air, water and soil to minimize their adverse impacts on human health and the environment.</td></tr>
</table>

Table 1.1 (Continued)

SDGs INDIRECTLY RELEVANT TO WATER SECURITY	
SDG 15: "Protect, restore and promote sustainable use of terrestrial ecosystems, sustainably manage forests, combat desertification, and halt and reverse land degradation and halt biodiversity loss." 15.1 Ensure conservation, restoration and sustainable use of terrestrial and inland freshwater ecosystems and their services, in particular forests, wetlands, mountains, and drylands, in line with obligations under international agreements; 15.8 significantly reduce the impact of invasive alien species on land and water ecosystems, and control or eradicate the priority species.	

Source: United Nations (2015).

water management, and the food–energy–water nexus, to achieve the future envisioned by the SDGs.

1.3 Definitions of Water Security

The above discussion refers to water security without yet providing a definition of water security. The Global Water Partnership (GWP) has offered a very straightforward goal of water security, in which "every person has access to enough safe water at an affordable cost to lead a clean, healthy and productive life, while ensuring that the natural environment is protected and enhanced" (Lankford et al. 2013). Water quantity ("enough") and quality ("safe") are explicit in this definition while equity is implicit, using affordability as a surrogate.

Another practical definition is similar: "Water security is a condition in which there is a sufficient quantity of water, at a fair price, and at a quality necessary to meet short and long term human needs to protect their health, safety, welfare, and productive capacity at the local, regional, state, and national levels." (Kaplowitz and Witter 2002; Lankford et al. 2013). This definition reminds us that there are various levels to be considered, from local to national, and even international.

The United Nations gives an even more comprehensive definition of water security – "The capacity of a population to safeguard sustainable access to adequate quantities of and acceptable quality water for sustaining livelihoods, human well-being, and socio-economic development, for ensuring protection against water-borne pollution and water related disasters, and for preserving ecosystems in a climate of peace (equity) and political stability." (United Nations 2013). This definition is robust because it encompasses water quantity, quality, and equity while also acknowledging the threats that water brings in the form of pollution and disasters, such as flooding. It also presents several of water security's outcomes, including human well-being, socioeconomic development, ecosystem preservation, and peace and political stability.

Figure 1.2 is a word cloud made from nine definitions of "water security" found in the literature. This word cloud highlights several key themes – quality, access, availability, acceptability, and ecosystem. Based on the above, and for the purposes of this text, we generate an operative definition of water security that is a slight modification to one given by Grey and Sadoff (Grey and Sadoff 2007).

Figure 1.2 Word cloud made from nine definitions of water security. Source: Author original (using Appelgren 1997; GWP 2000; Kaplowitz and Witter 2002; WHO 2003; Xia et al. 2006; Grey and Sadoff 2007; Calow et al. 2010; ADB 2011; Norman et al. 2011).

Water security is the equitable availability
of a suitable quantity and quality of water for health and well-being,
with an acceptable level of water-related risks
to people, environment, and economies.

This definition encompasses scale, focus, and function while also acknowledging the reality that water itself brings risk to communities and the environment in which we live. The quality needed will depend upon the use of the water, and equity includes both access and affordability, which varies across nations and peoples.

1.4 Water Security at the Nexus of Quantity, Quality, and Equity

As is now becoming evident, water security develops at the nexus of the appropriate balance of water quantity, quality, and equity (Figure 1.3). This balance includes the minimization of unacceptable risks due to an overabundance (flooding) or lack (drought) of water, natural or manmade water pollution, and physical, societal, or political limitations to water access.

With regard to water, equity can be defined as the just and appropriate accessibility to sufficient water resources across gender, socioeconomic, spatial, and generational differences. For example, water resources are *not* equitable when:

- Women bear an inordinate burden of household water management, thereby limiting their education and development, when such a burden can be corrected

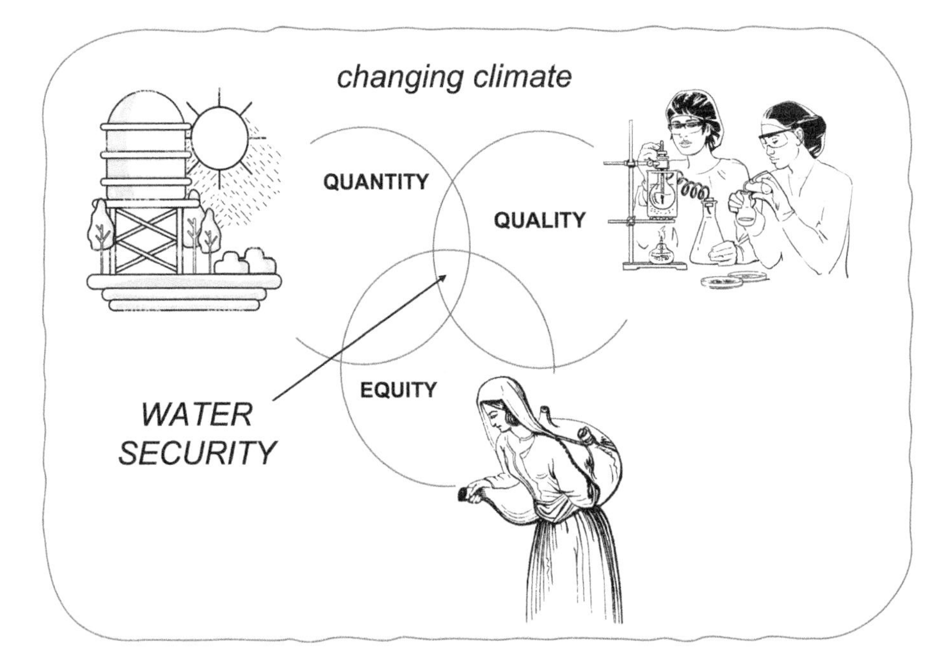

Figure 1.3 Water security is found at the nexus of the appropriate water quantity, quality for purpose, and equitable access, all in the context of a changing climate. Source: JFC author original.

- The price of water is beyond the reach of a typical household
- Rural villagers are excluded from water provision and management when such exclusion can be corrected
- An upstream entity uses an inordinate amount of water resulting in a lack of water security to an entity downstream
- An upstream entity releases a waste stream resulting in a degradation in water quality to an entity downstream
- The current generation is consuming fresh water at a rate that places a heavy burden on successive generations, as in the case of aquifer depletion.

A simple example here can be used to illustrate the water-security nexus of water quantity, quality, and equity. **Managed aquifer recharge** (MAR) in India can be accomplished by constructing small check dams across rivers or streams. In addition to storing water underground, where evaporation is limited or nonexistent, or replenishing the existing water in the subsurface rocks and sediments (aquifer), this practice has also been shown to improve water quality by reducing salinity levels as well as fluoride and arsenic concentrations in the groundwater. The improvement in quantity and quality makes this water more available for drinking or irrigation. But the practice is not as widespread as one might think given its obvious benefits. Stakeholders tend to be both wary of the intentions of central government (which funds or subsidizes the capital costs) and concerned about future maintenance issues and the control of quality of water entering the aquifer (Gunda et al. 2019). Water rights can also be an issue. Water security happens when local political and social conditions are favorable and supportive.

The benefits of water security are illustrated in Figure 1.4. The "bottom billion" are citizens of the 50 poorest nations, representing a population of about one billion, most of whom live

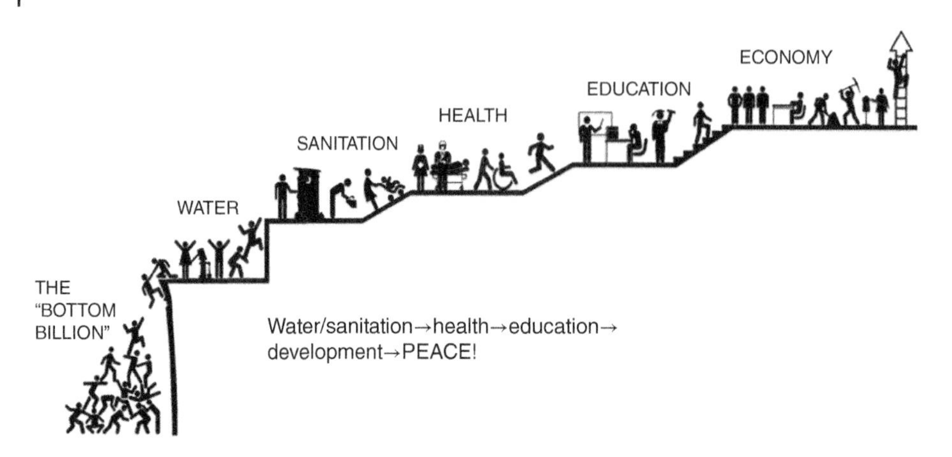

Figure 1.4 The "bottom billion" can reach the final goals of education and a firm economy only through climbing the prior "rungs" of water, sanitation, and health. Source: DAS author original; adapted from Living Water International.

on less than $1 a day (Collier 2008). Each rung of the ladder of development enables them to reach an acquired benefit: water and sanitation lead to better health, better health improves the chances of acquiring a good education, and a good education allows one to contribute more fully to a stable economy while achieving a higher standard of living (e.g. food, access to medical care). The final outcome of this development ladder is hopefully a more stable society, a more peaceful world.

1.5 Metrics for Quantification and Comparison

In order to both evaluate water security and compare across populations, certain quantifiable measures have been found useful. This section describes some of the more common metrics.

- *Water stress and scarcity*. One common measure (and the simplest to calculate) is based on population and renewable water resources available on an annual basis. Using this measure, **water stress** is defined as water supply at or below 1700 m^3 of available water per person per year and **water scarcity** is supply at or below 1000 m^3 per person per year. *Absolute water scarcity* is applied to countries with less than 500 m^3/cap-year or roughly 1400 l/day per person, for all uses. By this definition, 49 countries are water stressed, 9 of which experience water scarcity and 21 experience absolute water scarcity (Table 1.2). By 2025, 1.8 billion people could be living in absolute water scarcity (United Nations 2013; Ford 2016).
- *SDG water withdrawal intensity (WWI)*. The SDG 6.4.2 (UN SDGs) uses a slightly different indicator of water stress. This indicator tracks the portion (%) of freshwater being withdrawn by all economic activities, compared to the total actual renewable freshwater resources (TARWR) available when flow needed for ecosystem services is safeguarded. As an equation,

$$\mathrm{WWI}\,(\%) = 100 \times \frac{\text{Total freshwater withdrawal}}{[\text{TARWR–Environmental flow requirements}]} \tag{1.1}$$

Table 1.2 Examples of countries that are categorized as experiencing water stress, water scarcity, or absolute water scarcity.

Absolute water scarcity (< 500 m^3/cap)	Water scarcity (500–999 m^3/cap)	Water stress (1000–1699 m^3/cap)
Algeria	Morocco	Poland
Libya	Egypt	Nigeria
Saudi Arabia	Sudan	Ethiopia
China (northeast regions)	Kenya	India
	South Africa	Pakistan
	Morocco	Belgium
	Burkina Faso	Czech Republic

Source: Black (2016).

In this equation, the numerator includes water withdrawals by all economic activities, with a focus on agriculture, manufacturing, electricity, and water collection, treatment, and supply [(UN-Water 2021), AQUASTAT database]. The denominator includes an accounting of surface and groundwater inflows, surface runoff, and treaty obligations and environmental flow as needed (United Nations 2020).

This metric provides an estimate of pressure by all economic activities on the country's renewable freshwater resources, directly responding to the environmental component of the target – "to ensure *sustainable* withdrawals and supply of freshwater." For example, Egypt had a WWI level of 117% in 2017, suggesting that it had overdrawn its freshwater resources by 17% (Figure 1.5). Algeria recorded a WWI level of 138% whereas the countries of India and South Africa had WWIs of 66 and 62%. Using this metric, many nations are withdrawing freshwater at a rate that is not sustainable.

- *Water poverty index (WPI).* A third metric is the WPI, which produces a composite variable composed of five components that can be quantified: (i) access to water (Access); (ii) water resource quantity, quality, and variability (Resource); (iii) water uses by sector (domestic, food, productive purposes) (Use); (iv) capacity for water management (Capacity); and (v) environmental aspects (Environment) (Sullivan et al. 2003; Mason 2013). Each of the five components is weighted by experts and used for evaluating specific sites or regions. The components may, of course, be weighted equally, and a component may be omitted due to lack of data. The WPI can be used to compare communities, and those with a low index may be prioritized for attention.

$$\mathrm{WPI} = \frac{w_{\mathrm{A}}A + w_{\mathrm{R}}R + w_{\mathrm{U}}U + w_{\mathrm{C}}C + w_{\mathrm{E}}E}{w_{\mathrm{A}} + w_{\mathrm{R}} + w_{\mathrm{U}} + w_{\mathrm{C}} + w_{\mathrm{E}}}$$

where "*A*, *R*, ..." are the quantities assigned to each component: access, resource, use, capacity, and environment, respectively, and "w_{A}, w_{R}, ..." are the weights given to access, resource, use, capacity, and environment, respectively.

For example, a comparison of WPI was made for a large rural town in each of the three countries of South Africa, Tanzania, and Sri Lanka. In this case, each of the five components was given equal weight. The resulting WPIs are given as 43.1 (South Africa),

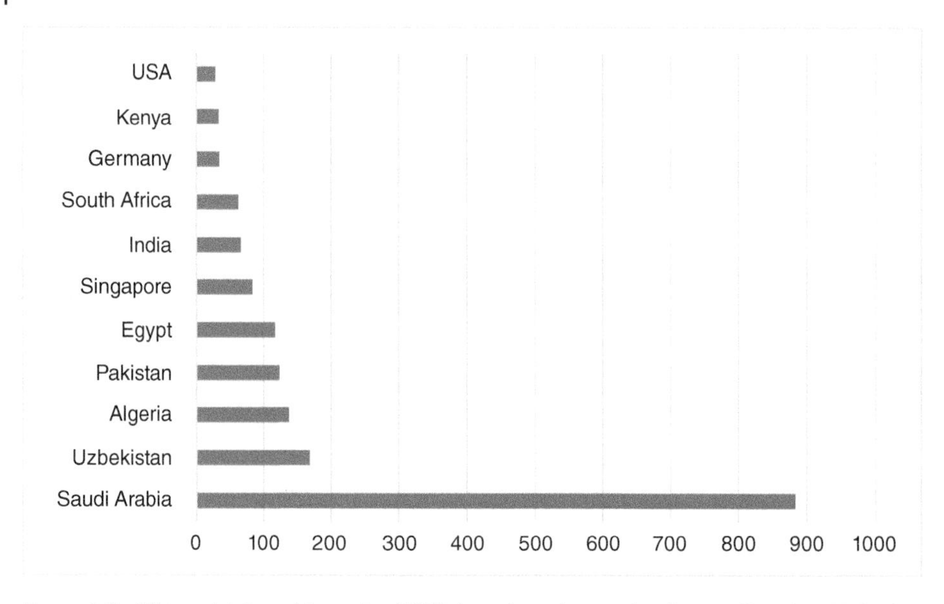

Figure 1.5 Water withdrawal intensity (WWI) for selected countries. Source: Based on United Nations (2020).

52.8 (Tanzania), and 46.4 (Sri Lanka). Figure 1.6 illustrates the comparison for three rural towns. The village in Tanzania scored high for "Environment," reflecting an awareness and wise use of natural resources and little degradation in the relevant watershed(s). The South African village scored high in surface water and groundwater availability and quality ("Resources"). Finally, the Sri Lankan village scored the highest in the "Use" category, which illustrates a more sustainable balance among domestic, livestock, agricultural, and industrial users.

- *Aqueduct water risk atlas.* The World Resources Institute (WRI) calculates water risk using water withdrawals, availability, and groundwater levels [see Dormido (2019) for more details]. Figure 1.7 highlights the countries that show extremely high risk (ExHR). The number indicates a country ranking, with Qatar (#1) as the country with the highest level of water risk. Note that this map does not reveal the regions within a country that could vary by risk. In addition, less obvious are the numbers of peoples that fall into the highest risk category. India has a population of 1.4 billion, more than all the other ExHR countries combined. Pakistan and Iran together have 287 million residents and the remaining ExHR countries comprise 102 million people.

Each of these metrics has its own advantages and limitations. Many of them depend upon accurate current data, which are often difficult to obtain. While care must be taken to make comparisons with these qualifications in mind, these metrics do have the advantage of providing a rough quantitative basis for prioritization and decision-making. They also alert us to the global nature of water security.

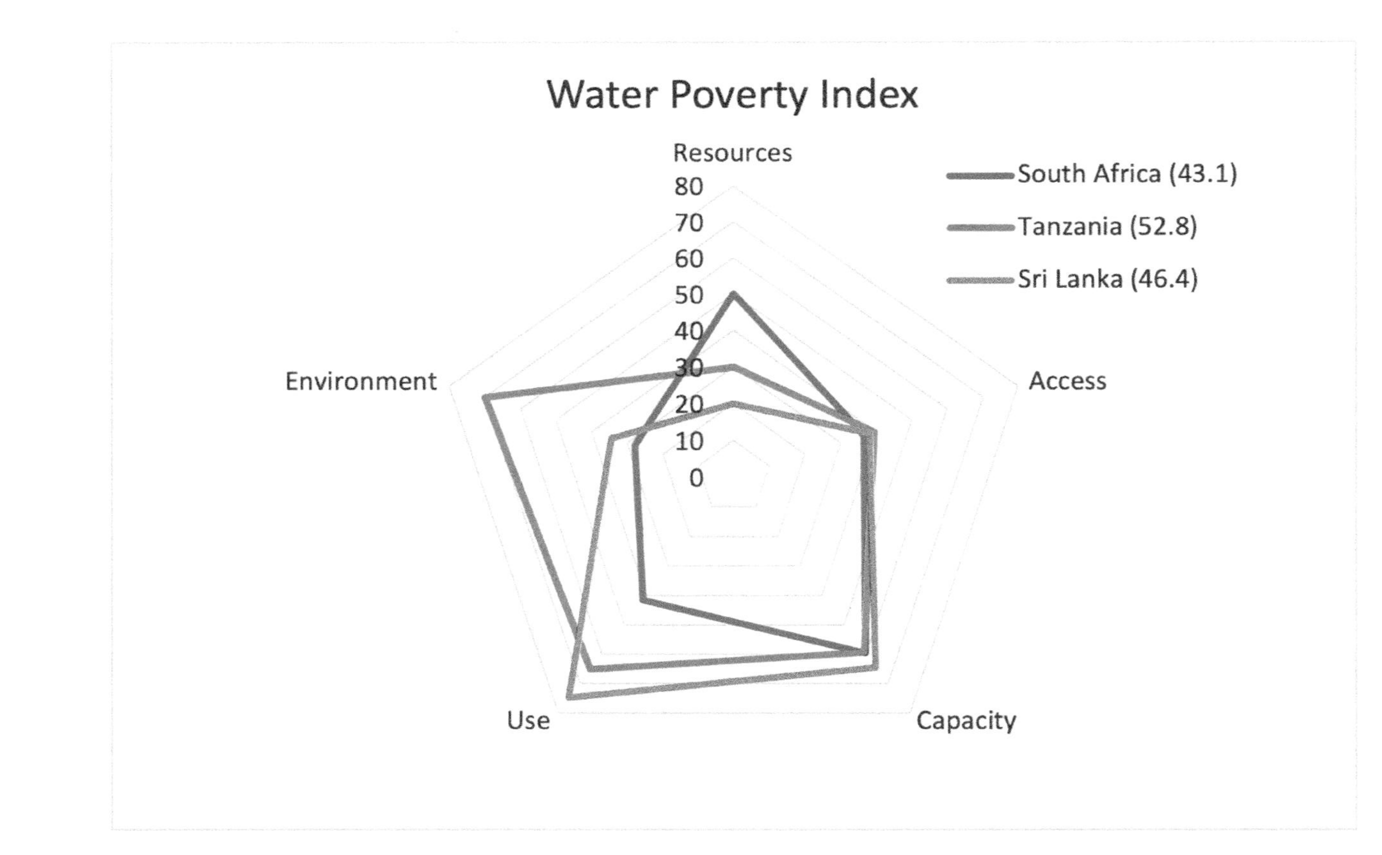

Figure 1.6 Water poverty index (WPI) determination for three countries, rural setting. A region with a higher score is considered more water-secure. Source: Original adapted from Sullivan et al. (2003).

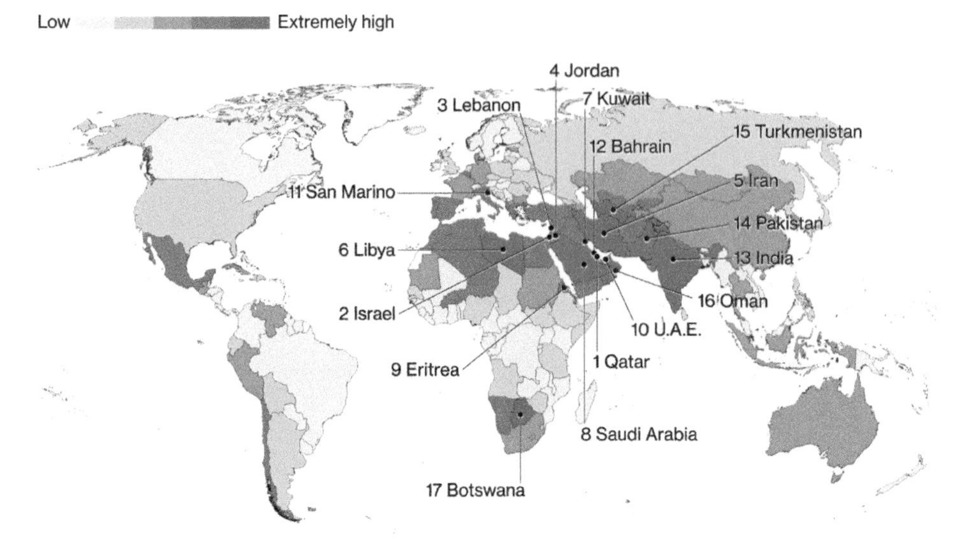

Figure 1.7 Baseline water-stress analysis using data on water withdrawal, and available water and groundwater for the year 2013. Source: World Resources Institute's Aqueduct Water Risk Atlas (Dormido 2019)/with permission from World Resource Institute.

1.6 Watershed Protection – Integrated Water Resources Management

Rain and subsequent runoff do not follow the political boundaries of a state or nation any more than birds do as they migrate south each Fall from their North American habitats. A watershed is an areal extent of land in which all precipitation falling on the land is channeled to a single outfall. That outfall might be a river, reservoir, bay, or ocean. Activities and land use within the watershed may have an effect on the quantity and quality of water leaving the watershed, whereas activities and land use outside of the watershed will not. Thus, the watershed provides a natural boundary for resource management within.

In the simplest of terms, **integrated water resources management (IWRM)** is the process of managing human activities (including land use) and natural resources on a watershed basis (Figure 1.8). This process takes into consideration the needs of the local environment and ecological species, shown in the red circle. It also includes societal needs, shown in the light brown circle, including consideration of both water quality and quantity. At the same time, proper management allows for both short-term and long-term economic benefits (blue circle), as long as these are sustainable and not damaging to the other two sectors. One can immediately see that a larger, more diverse watershed is sure to result in the need for a more complex IWRM scheme, as it will include many more stakeholders, users, and types of water demands.

The growth of urban areas is the most consistent global phenomenon in our lifetime and will be for some time yet to come. Impervious cover is increased by the proliferation of buildings, roads and highways, parking lots, and housing. Rain can no longer infiltrate into the soil but is forced to run off at a much faster rate with a higher peak flow. Not only is the quantity of runoff altered, but also the quality of the runoff, as the stormwater now contains oils, greases, metals, bacteria, and other suspended solids and particulates. IWRM has become a vehicle for responding to such a phenomenon.

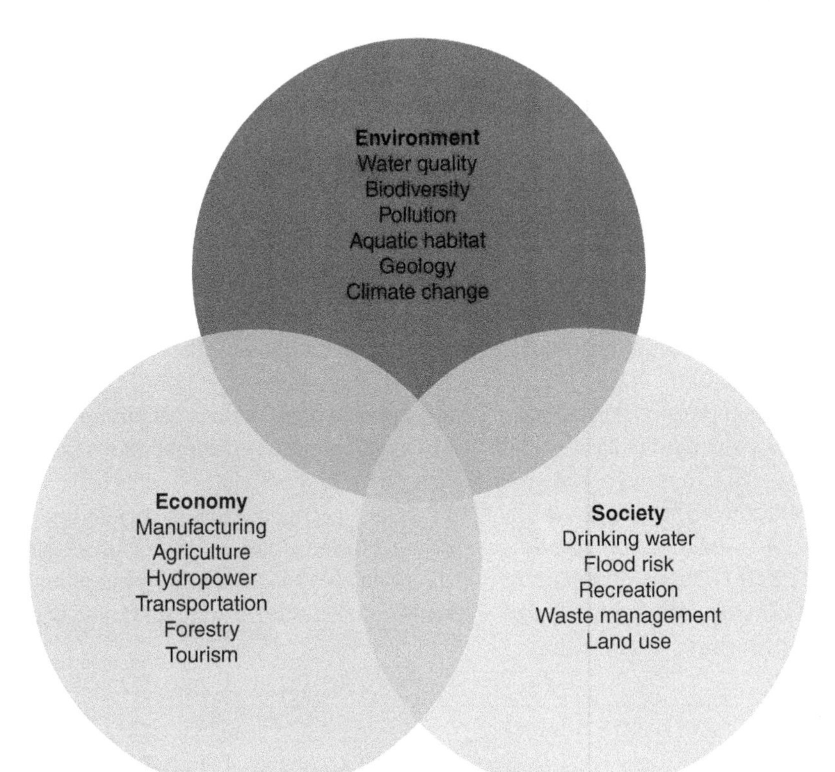

Figure 1.8 Integrated water resources management (IWRM) considers all aspects of environment, society, and economy on a watershed basis. Source: JFC author original.

Ideally, IWRM is a process of decision-making that results in a plan (or strategy) that:

- Is designed to meet both long- and/or short-term targets of water quality, quantity, or equity;
- Contains explicit management options for addressing the stated problem(s);
- Is developed with the support of relevant stakeholders; and
- Has broad public acceptance.

The IWRM process may take many months or years with several iterations in order to come to an agreed-upon conclusion. Even when a true consensus is not reached, the process has both increased awareness of the real or potential threats to water security and has built the community's capacity for decision-making (Heathcote 2009). This process will be described in more detail in a subsequent chapter.

1.7 Levels of Study – Local to Global

The intersection of water supply and demand can happen at different geographic scales. A river can receive its water from within the surface boundaries of its **watershed** (basin) and its subsurface flow, which might extend far beyond that basin. Both the basin and the subsurface flow can cross international (or other jurisdictional) borders. For example, the Rio Tijuana (Tijuana River) is a 120-mile intermittent river that drains in a northwesterly course through northern Mexico and into the US state of California (Figure 1.9). Its waters flow through a salt marsh and a national estuarine reserve before reaching the Pacific Ocean. The marsh is home to hundreds of species of birds and is the largest coastal wetland in California. Beach and ocean health are crucial to the recreation economy just south of San Diego. However, periodic flooding from winter rains frequently overwhelms a small wastewater treatment plant and carries untreated sewage into the marshlands. Sustainable management of this international river system depends upon cooperation of stakeholders speaking different languages on both sides of the border. But the resulting decisions have an impact on many levels – ecosystem services of the marsh, diversity of wetlands species and coastal birds, and socioeconomic benefits of tourism to the local human community (Tijuana River NERR 2010).

A second example is more global in nature. The rising global population is becoming both more urbanized and more middle-class. With this rise comes an increased demand for a more "Westernized" diet, rich in animal-based protein, high-fat dairy products, and high-fructose corn syrup. The water footprint (discussed in a later chapter) of such a diet is substantially higher than a grain-based diet and results in a greater burden upon water resources for agriculture (Black 2016). The ramifications of this dilemma entail local solutions but global implications in the trade of water that is embedded within food products that are traded and transported across the globe.

Figure 1.9 The Tijuana River collects water in an arid watershed that straddles two international jurisdictions (Wright and Vela 2005). Source: With permission from San Diego State University Press.

Both examples of water insecurity are complicated by climate change in which drier areas become drier and wetter areas become wetter. The threats to local ecosystems and local agriculture are exacerbated in a warmer global climate, especially when resources for adaptation are limited. This book will address these and many more challenges that face water security.

1.8 Conclusion

In two very ancient texts, water's potential and peril are featured as a subplot to the main narrative. The Yahwist account of creation in Genesis says that "a river rises in Eden to water the garden. Beyond there it divides and becomes four branches." (Gen. 2:10-14). These four rivers are named as the Pishon, Gihon, Tigris, and Euphrates rivers, ancient names for important rivers of modern-day Iraq. These four tributaries flow out to all the known peoples of the time, watering their own gardens and providing the basis for life and culture.

Written even earlier, probably in the 1300s BCE, the first great epic of literature was the story of Gilgamesh, the priest-king of Uruk. In this epic, the city of Shurrupak along the Euphrates river is growing quickly and too noisy for the impatient god, Enlil. He and his fellow gods agree to wipe out all the mortals with a great flood. "The flood and wind lasted six days and six nights, flattening the land. On the seventh day, the storm was pounding like a woman in labor." (*Epic*, Tablet XI). The story was very likely based on a storm of immense magnitude.

Thus, early in our history, water featured prominently in the Fertile Crescent, both in its life-giving richness and in its death-dealing power. Water security, then, has long been integral to the narrative of human flourishing, an anonymous current that becomes part of the rationale for human settlements, movements, and conquests. This volume will help us explore both the qualitative and quantitative aspects of water security, with a focus on the water quantity–quality–equity nexus and with an ample supply of global applications and implications for water management. All these are addressed within the broader context of climate change. In the next chapter, we will survey some important incidents in history in which this narrative was disrupted resulting in human suffering or death, all because of a loss or lack of water security.

*Foundations: Units of Measurement in Water Security

Having thus delved into the world of water security, with its many facets and subcomponents, we end this chapter, as we do most chapters, with a section entitled "Foundations." The purpose of this section is to provide quantitative tools that will be useful in the practice of water security.

Water *quantity* is usually expressed in terms of volume and flow (volume/time) using volume units of gallons, liters, or cubic meters. In hydrology, one also encounters units that describe rainfall volume over a watershed area, such as acre-feet (ac-ft) or hectare-millimeters (ha-mm). Volume units are used for water that is stationary, in lakes, ponds, reservoirs, and tanks, for example. Flow units describe water that is moving in streams, channels, pipelines, and ditches. Groundwater is sometimes described in both volume units to express storage and flow units to reflect its movement. Table 1.3 gives some simple factors for conversion between units.

Water *quality* is often described in terms of concentrations of dissolved substances, expressed in units of mass/volume or mass/mass. These substances may be beneficial to life, such as dissolved oxygen, or harmful at high levels, such as arsenic or hexavalent chromium. The concentration of substances is important as it drives not only toxicity but also the fate and transport of chemicals within the water and between environmental media, such as air and water, or soil and water.

Concentrations are most often expressed in mg/l, understood as the "mass of the substance" divided by the "volume of solution." Since we are working mostly with dilute environmental solutions, the denominator is almost always assumed to be water with a density of ~1000 g/l. To express a concentration unit as mass/mass, we most often use parts per million (ppm_m) or parts per billion (ppb_m). Thus, a concentration of calcium in water may be expressed in two ways:

$$\frac{25\text{ mg of calcium}}{\text{liter of water}} = 25\,{}^{\text{mg}}/_{\text{l}} \times \frac{1\text{ l}}{1000\text{ g}} \times \frac{1\text{ g}}{1000\text{ mg}} = \frac{25}{10^6} = 25\text{ ppm}_\text{m}$$

Table 1.3 Commonly used units for area, volume, and flow of water.

Unit A	Multiply A by ___ to obtain B:	Unit B
AREA		
acre	43 560	square feet
hectare	10 000	square meters
hectare	2.47	acres
VOLUME		
gallons	3.8	liters
cubic meters	1000	liters
cubic meters	35.3	cubic feet
cubic feet	28.3	liters
acre-in	10.3	hectare-mm
acre-ft	325 851	gallons
FLOW		
gpm	3.8	lpm
cfs	1700	lpm
cms	35.3	cfs
MGD	3785	m^3/day

When the gram molecular weight of a substance is known, this concentration can also be expressed in *molarity*, or moles/liter:

$$\frac{25\text{ mg of calcium}}{\text{liter of water}} = 25\,{}^{\text{mg}}\!/_{\text{l}} \times \frac{1\text{ mole Ca}}{40\text{ g Ca}} \times \frac{1\text{ g}}{1000\text{ mg}} = 0.00625\text{ moles Ca/l}$$

Sometimes, it is important for us to know the concentration of a *common constituent* in several compounds. For example, nitrogen may be present in the form of nitrate (NO_3^-), ammonia (NH_3), or ammonium (NH_4^+). We can use simple stoichiometry (ratio of chemical components) and the molecular weights to determine the concentration of the constituent of concern. Consider a water sample with two nitrogen species – 30 mg/l of ammonia and 5 mg/l of nitrate (Mihelcic and Zimmerman 2014). We want to know the partial contribution of N to each species and for the total solution:

$$\frac{30\text{ mg NH}_3}{\text{liter}} \times \frac{\text{mole NH}_3}{17\text{ grams}} \times \frac{1\text{ mole N}}{1\text{ mole NH}_3} \times \frac{14\text{ g}}{\text{mole N}} = \frac{24.7\text{ mg NH}_3 - \text{N}}{\text{liter}}$$

$$\frac{5\text{ mg NO}_3^-}{\text{liter}} \times \frac{\text{mole NO}_3^-}{62\text{ g}} \times \frac{1\text{ mole N}}{1\text{ mole NO}_3^-} \times \frac{14\text{ g}}{\text{mole N}} = \frac{1.1\text{ mg NO}_3^- - \text{N}}{\text{liter}}$$

The total contribution of nitrogen (N) to this solution is $24.7 + 1.1 = 25.8$ mg/l.

Another aspect that illustrates the importance of nitrogen units is the drinking water standard for nitrate. We may see this standard listed as 10 mg/l in one place but 45 mg/l in another place. This seeming inconsistency can be rectified by realizing that 45 mg/l as nitrate is the same as 10 mg/l of nitrate *as nitrogen*:

$$\frac{45\text{ mg NO}_3^-}{\text{liter}} \times \frac{\text{mole NO}_3^-}{62\text{ g}} \times \frac{1\text{ mole N}}{1\text{ mole NO}_3^-} \times \frac{14\text{ g}}{\text{mole N}} = \frac{10\text{ mg NO}_3^- - \text{N}}{\text{liter}}$$

Thus, when a value of nitrate is reported, it is important to know if it is reported as nitrate (NO_3) or as nitrate-nitrogen (NO_3-N) so we know whether to compare it to 10 mg/l or 45 mg/l as the standard. For example, 30 mg/l of nitrate exceeds the standard if it is 30 mg/l of NO_3-N but is below the standard if it is 30 mg/l of NO_3.

We do well to use different units in different situations. It is human nature that we are more comfortable working with numbers in a certain range of magnitude. For example, we sometimes express salinity in ppt (thousands) or ppm. Seawater, for example, has a salinity of 35 ppt or 35 000 ppm. We would not use ppb for seawater as this would be noted as 35 000 000 ppb. Conversely, we would use ppb for the arsenic standard (10 ppb) versus ppm (0.01 ppm) or ppt (0.00001 ppt). We have a preference to work with numbers closer to one versus numbers much greater or less than one and pick our units accordingly. Likewise, the unit of ac-ft is convenient for the volume of a lake (e.g. Lake Thunderbird in Norman, OK, has an average volume of 120 000 ac-ft) as that is a more manageable number than gallons (3.9×10^9 gal). So, while some units may seem strange to us (ppt or acre-ft), they allow us to express numbers in an order of magnitude that is more comfortable to us.

As we proceed through our study of water security, we will see many examples of water volumes and chemical concentrations in water. A potentially harmful chemical, such as arsenic or chromium VI, may not be easily seen (Figure 1.10), but it is always important from the perspective of water safety and security.

Figure 1.10 Chemical concentrations in a dilute solution of tap water may not always be visible, but they are important for water security. Source: New Africa/Adobe Stock.

End-of-Chapter Questions/Problems

1.1 Compare and contrast the two definitions of water security given by the GWP and the United Nations on the basis of scope, incorporation of threats, and goals/outcomes.

1.2 Use an Internet search (e.g. the FAO-Aquastat database) to answer the following questions for the country of Angola (latest data available):

a. What is the total surface water produced internally? (m^3/year)
b. What is the total groundwater produced internally? (m^3/year)
c. What is the overlap between surface water and groundwater? (m^3/year)
d. What is meant by the "overlap"?
e. How is the total IRWR (internal renewable water resources) calculated? Give equation in word form.

1.3 Use the FAO-Aquastat database to compare the percentages of freshwater withdrawals between Australia and the United Kingdom.

a. What is percent (%) withdrawals for agricultural, municipal, and industrial sectors in Australia?
b. What is percent (%) withdrawals for agricultural, municipal, and industrial sectors in the United Kingdom?
c. Comment on the differences between these two nations.

1.4 Using the latest census figures and the FAO-Aquastat database, calculate the water-stress level (m^3/cap-year) for the following nations:

a. Vietnam
b. Iraq
c. Peru
d. Spain
e. Comment on the differences (or similarities) between these nations.

f. Why is this indicator perhaps a poor water-stress indicator for large nations such as Peru and Spain?

1.5 Groundwater is a very valuable stored resource as it usually does not undergo evaporation and remains in place until tapped by the human community. How much of the internal renewable water resources is groundwater (%) in each of the following nations?
a. Chile
b. Haiti
c. Canada
d. Venezuela
e. Belgium
f. How might this single metric be a good indicator of water security? What are the constraints upon accessibility for this stored resource?

1.6 The Chesapeake Bay watershed is an important, large, and complex watershed. Use the Internet to retrieve a management plan for some portion or all of this watershed and describe at least three specific management strategies that are recommended to protect the watershed.

1.7 Find a watershed-protection plan for a watershed near where you live.
a. What are the major challenges in this watershed?
b. What are the specific management strategies that are being recommended?

1.8 Conduct an Internet search for information on the Tijuana River Watershed. How does social equity play a role in the management of this watershed?

1.9 A 2-l water sample was analyzed and found to contain 40 mg of sulfur, whose molecular weight is 32 g/mole.
a. What is the concentration of sulfur in mg/l?
b. What is the concentration of sulfur in ppm_m?
c. What is the concentration of sulfur in moles/l?

1.10 Three different water samples were analyzed and found to contain the following concentrations of nitrate and nitrite. Convert these concentrations to nitrate-N and nitrite-N basis, sum them, and see if the samples exceed the combined nitrate+nitrite standard of 10 mg/l as nitrogen.
a. 10 mg/l of nitrate/30 mg/l of nitrite
b. 20 mg/l of nitrate/20 mg/l of nitrite
c. 30 mg/l of nitrate/10 mg/l of nitrite

Further Reading

Black, M. (2016). *The Atlas of Water: Mapping the World's Most Critical Resource*, 3e. University of California Press.

Gleick, P. (1993). *Water in Crisis: A Guide to the World's Fresh Water Resources*, 1e. New York: Oxford University Press.

Gleick, P. and Iceland, C. (2018). *Water, Security and Conflict | World Resources Institute*. World Resources Institute https://www.wri.org/publication/water-security-and-conflict.

Lankford, B., Bakker, K., Zeitoun, M., and Conway, D. (2013). *Water Security: Principle, Perspectives and Practices*. Earthscan.

References

ADB (2011). *Asian Water Development Outlook 2011*. Manila: Asian Development Bank.

Appelgren, B. (1997). "Keynote Paper-Management of Water Scarcity:National Water Policy in Relation to Regional Development Cooperation." In . Cairo: FAO.

Black, M. (2016). *The Atlas of Water: Mapping the World's Most Critical Resource*, 3e. University of California Press.

Calow, R.C., MacDonald, A.M., Nicol, A.L., and Robins, N.S. (2010). Ground water security and drought in Africa: linking availability, access, and demand. *Ground Water* 48 (2): 246–256.

Collier, P. (2008). *The Bottom Billion: Why the Poorest Countries Are Failing and What Can be Done about it*. USA: Oxford University Press.

Dormido, H. (2019). "These Countries Are the Most at Risk from a Water Crisis." *Bloomberg.Com*, 2019. https://www.bloomberg.com/graphics/2019-countries-facing-water-crisis.

Ford, T. (2016). Chapter 16: Water and health. In: *Environmental Health: From Global to Local*, 3e (ed. H. Frumkin), 413–450. John Wiley & Sons.

Grey, D. and Sadoff, C.W. (2007). "Sink or swim? Water security for growth and development." *Water Policy* 9 (6): 545–71. doi:https://doi.org/10.2166/wp.2007.021.

Gunda, T., Hess, D., Hornberger, G.M., and Worland, S. (2019). "Water security in practice: the quantity-quality-society nexus." *Water Security* 6 (March): 100022. doi:https://doi.org/10.1016/j.wasec.2018.100022.

GWP (2000). *Towards Water Security: A Framework for Action*. Stockholm: Global Water Partnership.

Heathcote, I.W. (2009). *Integrated Watershed Management: Principles and Practice*, 2e. Hoboken, N.J: Wiley.

Kaplowitz, M.D. and Witter, S.G. (2002). Identifying water security issues at the local level: the case of Michigan's Red Cedar River. *Water International* 27 (3): 379–386.

Lall, U., Davis, J., Scott, C., Merz, B., and Lundqvist, J. (2017). "Pursuing water security." *Water Security* 1 (July): 1–2. doi:https://doi.org/10.1016/j.wasec.2017.07.002.

Lankford, B., Bakker, K., Zeitoun, M., and Conway, D. (2013). *Water Security: Principle, Perspectives and Practices*. Earthscan.

Martin (2015). "Sustainable Development Goals Launch in 2016." *United Nations Sustainable Development* (blog). December 30, 2015. https://www.un.org/sustainabledevelopment/blog/2015/12/sustainable-development-goals-kick-off-with-start-of-new-year.

Mason, N. (2013). Easy as 1,2,3? Political and technical considerations for designing water security indicators. In: *Water Security: Principles, Perspectives and Practices* (ed. B.A. Lankford), 357. Routledge.

Mihelcic, J.R. and Zimmerman, J. (2014). *Environmental Engineering: Fundamentals, Sustainability, Design*, 2e. Wiley.

Norman, E.S., Bakker, K., and Dunn, G. (2011). Recent developments in Canadian water policy: an emerging water security paradigm. *Canadian Water Resources Journal* 36 (1): 53–66.

Sullivan, C.A., Meigh, J.R., and Giacomello, A.M. (2003). "The water poverty index: development and application at the community scale." *Natural Resources Forum* 27 (3): 189–99. doi:https://doi.org/10.1111/1477-8947.00054.

Tijuana River NERR (2010). "Tijuana River National Estuarine Research Reserve – Comprehensive Management Plan." Tijuana River National Estuarine Research Reserve.

United Nations (2013). "Water Security and the Global Water Agenda." UN-Water. 2013. https://www.unwater.org/publications/water-security-global-water-agenda.

United Nations (2015). "Water and the Sustainable Development Goals (SDGs) | 2015 UN-Water Annual International Zaragoza Conference. Water and Sustainable Development: From Vision to Action. 15-17 January 2015." 2015. https://www.un.org/waterforlifedecade/waterandsustainabledevelopment2015/open_working_group_sdg.shtml.

United Nations (2020). "6.4.2 Water Stress | Sustainable Development Goals | Food and Agriculture Organization of the United Nations." 2020. http://www.fao.org/sustainable-development-goals/indicators/642/en.

UN-Water (2021). "Indicator 6.4.2 – Water Stress." Sdg6monitoring. N.D. https://www.sdg6monitoring.org/indicators/target-64/indicators642.

WHO (2003). *WHD (World Health Day) Brochure, Part IV: The Priorities and Solutions for Creating Healthy Places.* World Health Organization.

Wright, R.D. and Vela, R. (2005). *Tijuana River Watershed Atlas.* IRSC and SCSU Press.

Xia, J., Zhang, L., Liu, C., and Yu, J. (2006). Towards better water security in North China. *Water Resources Management* 21 (1): 233–247.

2

Historical Examples of Water Insecurity

In this chapter, readers are introduced to past examples of the lack of water security, delineated by a lack of sufficient quantity of water of acceptable quality to meet the needs of all members of society. These instances of water insecurity sometimes resulted in widespread social disruption, morbidity, and death. In many cases, the affected populations responded with technological solutions, such as the aqueducts from ancient Rome or provisional wastewater reuse in modern-day Texas. In some cases, the insecurity led to the creation of new public health practices or institutions, such as in the cases of London and Chicago. The global proliferation of dams has led to conflict over who reaps the benefits, and who pays the costs, for such major hydrologic alterations. While emerging economies are especially vulnerable to water insecurity, developed nations are also vulnerable, although they have many more resources with which to respond.

Learning Objectives

Upon completion of this chapter, the student will be able to:

1. Describe several historical instances in which there was insufficient quantity of water, such as periods of drought, and the resulting social effects.
2. Describe several historical instances in which there was water of dangerous quality, such as contaminated with biological pathogens, and the social effects of the degraded water.
3. Describe historical instances in which one segment of society was excluded from water of sufficient quantity and quality, producing inequities of human development.
4. Describe a modern instance in which climate change is a significant factor in water insecurity.
5. *Understand the basic concepts of mass balance and reaction kinetics as it relates to population growth and contaminant reduction.

2.1 Introduction

In the ancient world civilizations needed water for survival, causing new settlements to locate near a dependable water source, such as a spring or river. But they also formed population clusters based on additional factors, such as topography, economy, and security.

Fundamentals of Water Security: Quantity, Quality, and Equity in a Changing Climate, First Edition.
Jim F. Chamberlain and David A. Sabatini.

Companion website: www.wiley.com/go/chamberlain/fundamentalsofwatersecurity

Ancient Rome was initially settled on Palatine Hill, which was easily defended. London was established on the Thames River, which was deep enough for ocean-going ships that could deliver valuable goods and supplies. Paris was founded on the Seine River, valued by the Franks for its many ports and wharves for easy loading and unloading of ships. Chicago began as a trading post, situated conveniently between Lake Michigan and the Mississippi River, thereby allowing ready access to both water supply and water-based transportation.

Early civilizations quickly realized that water for growing food and household use would have to be provided in one way or another – by local access or by transport in from sources both near and far. When sufficient water was not available, the result was often both water and food insecurity. Further, any water source worth using was also worth defending. Around the year 700 BCE, the Israelite King Hezekiah was determined to defend Jerusalem and its inhabitants against the Assyrian ruler, Sennacherib. The city's main water supply, Gihon Spring, was located outside the city walls and made the city vulnerable to a fatal siege from the enemy. Hezekiah set two groups of men digging tunnels toward each other, one from the spring and one from a well inside the city walls, in order to divert the water inside the protective fortress. "When Hezekiah saw that Sennacherib was coming with the intention of attacking Jerusalem, he took the advice of his princes and warriors to stop the waters of the springs outside the city; they promised their help. A large force was gathered and stopped all the springs and also the stream running nearby. For they said, 'Why should the kings of Assyria come and find an abundance of water?'" (2 Chronicles 32:2-4). If it had not been for the forethought of their king, the many inhabitants of Jerusalem would have been forced into famine and thirst, the results of water insecurity.

In this chapter, we offer several examples of past and present periods of water insecurity. Each of these can be classified as a lapse of adequate water quantity, water quality, or equitable access, or some combination of the above. In more recent decades, climate change has become an additional factor in water insecurity. These are snapshots in time, but they combine to create a historical collage that provides inspiration and incentive (we hope!) for diving into the remainder of the book.

2.2 Ancient Rome – A City of Fountains and Aqueducts (Water Quantity)

Rome's legendary aqueducts were expansive but not original. The Etruscan peoples of northern Italy had already been constructing tunnels to drain wetlands, to irrigate crops, and to provide additional water to settlements between the Arno and Tiber Rivers. The earliest Roman aqueduct was built in 312 BCE and was called the Aqua Appia after its patrician booster, Appius Claudius Caecus. The 20-km channel stayed underground most of the way into the city so as to protect it from malicious enemies. Over time, many additional aqueducts were constructed (Figure 2.1). The Aqua Anio Vetus (269 BCE) also stayed underground for its 69 km in length. The Aqua Marcia (140 BCE) was the first to convey water aboveground, reaching 91 km from the Anio Valley into the city proper. This aqueduct supplied water at the rate of over 49 000 gal/day and was a contributing factor for the city's stunning growth to a city of one million inhabitants around the year 100 CE (Fagan 2011). From the span of 300 BCE to 300 CE, the Romans built 11 aqueducts to bring supplemental water into the city.

These water structures were not built to provide drinking water or sanitation, as most Roman houses had wells and cisterns. Most of this supplemental water was used in

Figure 2.1 Crossing of two Roman aqueducts in Via Latina, photograph of a painting ca. 1936 by M. Zeno Diemer, ca. 1910/1935. Source: Credit: Wellcome Collection. Attribution 4.0 International (CC BY 4.0).

public baths, places of conversation and confraternity, located on a ground or upper floor. Ingeniously and effectively, the excess water was used to carry away human waste from toilets installed in a floor one level beneath the baths. This was done to control odors rather than disease as the feces-to-victim pathway would not be understood for over a millennium. As there were no shutoff valves for this system, an extremely high water usage rate resulted.

In 572 CE, the invading Goths laid siege to the city and essentially cut off the aqueduct water supply. Once labeled one of the Seven Wonders of the World, the grandiose baths were rendered useless and their complexes abandoned. The loss of the aqueducts resulted in lack of quantity for hygiene and adequate sanitation using the proven technique of sewage removal.

2.3 London – Cholera and the Theory of Miasma (Water Quality)

In the mid-nineteenth century, London was a booming Victorian metropolis of two and a half million people, all living in a 30-mile circumference (Johnson 2007). Although a central waste-collection system was long in the future, humanure waste recycling was already happening. Night-soil collection, bone picking, and other scavenging trades developed among people who, by industry or despair, learned how to eke out a living from the waste of others. In addition, millions of microbial scavengers were at work below city streets and alleyways, converting solid waste into methane gas with its putrid odor, which characterized the city's bloom. Where disease was present, the prevalent theory was in favor of "fouled air" as the cause, a theory known as the miasma theory of disease.

Water was a necessity of life, but it was also the carrier of disease. This fact was not yet appreciated when residents began dying of a cholera epidemic in 1854 that would claim 500 lives in 10 days. The *Vibrio cholerae* bacteria, likely arriving as an unknown passenger on a ship from the East, found in London an ideal atmosphere to replicate and move from host to host. An infected person would expunge large quantities of cholera-rich waste from their bowels. Because water sources were shared and poorly protected, the bacteria survived long enough to find another host, thereby perpetuating the disease cycle.

The physician John Snow lived on the edge of the Soho district, the epicenter of the epidemic that encompassed a five-block neighborhood. Reading accounts of previous incidents of cholera and speaking with chemists, Snow understood that the disease was contagious but that it was not correlated merely by proximity. That is, cholera was not contagious in the same way that the flu or smallpox were contagious. Cholera's victims had somehow ingested the poison. Snow surveyed the neighborhood and interviewed families, door to door, performing what would later come to be known as an epidemiologic study. His focus on the Broad Street well as the source was confirmed when he interviewed a widow who lived outside of the neighborhood but preferred the well's water because of its taste. She died three days after receiving her last jug of water from that well (Hrudey and Hrudey 2010). Snow convinced the local board to remove the pump handle from the well, serving to curb the spread of the disease, which had already begun to decline as those with the means had fled the area (Figure 2.2).

This account of a cholera epidemic instructs us in two competing roles of water in disease. While water can be the primary pathway for spread of a disease, such as cholera, typhoid, or dysentery, untainted water is also critical to recovery from the disease that kills primarily through dehydration. As we have said, the human body's systemic functions are all dependent upon water. The cholera bacteria disrupt the small intestine's ability to absorb and process water. Since the body excretes more water than it absorbs, the blood becomes more viscous, and the organs gradually begin to fail. This awful death can be forestalled by ample quantities of clean water with sufficient electrolytes to redress the blood imbalance. In the case of London in 1854, the number of mortality cases decreased not only because of the removal of the Broad Street pump handle, preventing usage of contaminated water

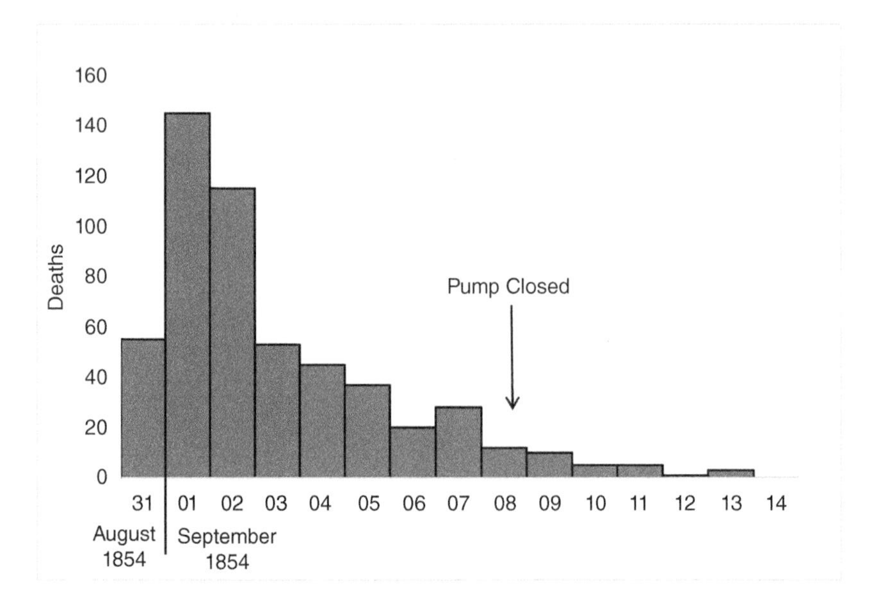

Figure 2.2 Number of cholera mortality cases in Soho District of London, 1854, prior to and following the removal of the Broad Street water-pump handle. Source: Adapted from Hrudey and Hrudey (2010).

Figure 2.3 The coauthor (Sabatini) next to the commemorative pump at the original site of the Broad Street well, London (notice the missing pump handle). Source: DAS author.

(Figure 2.3), but also by the forced return to cleaner sources of water in other areas of the Soho district.

Thus, in this case, water quality was at the root of intense water insecurity and danger to human health.

2.4 Chicago – Typhoid and the Great Reversal (Water Quality)

In roughly this same time period, another of the world's great cities, Chicago, was also victim to waterborne disease epidemics – not only cholera but also typhoid fever. Typhoid, like cholera, is an infectious enteric disease spread by a bacterium found in contaminated water. Prior to major infrastructure improvements in Chicago, typhoid fever outbreaks resulted in death rate as high as 150 deaths per 100 000 people (Figure 2.4). At this time, sewage from the Chicago River flowed into Lake Michigan, which, unfortunately, was also the city's source of drinking water. When the water supply intake was moved two miles farther out from shore, the rates of infection decreased significantly, as seen in Figure 2.4. Disasters, such as the great Chicago fire, and natural trends (such as low lake levels) would increase the infection rates, as did technological disruptions, such as pump failures. All of these served to increase the human exposure to pathogens, thereby increasing the death rate.

But Chicago gradually adopted more heroic measures. In the year 1900, the Chicago Sanitary and Ship Canal was completed, reversing the flow of the Chicago River away from Lake Michigan (Salzman 2017). Sewer lines, which had previously been built on top of the ground to allow for gravity flow, now emptied into the river that carried the effluent downstream away from the city's source of drinking water. This achievement, along with the adoption of chlorination in 1913, managed to reduce the incidences of typhoid to near zero by the year 1940.

The Chicago example illustrates the importance of providing **multiple barriers of protection** to safeguard drinking water quality (USEPA 2002), as well as the role of both risk prevention and risk management. Steps to reverse the flow of sewage and to extend the drinking water intake farther into the lake were steps of risk prevention. The addition of chlorination was a risk management step. The Chicago water tower, built in 1869, was built with a tall standpipe to hold water for firefighting and control surges of water pressure. The tower was one of a few buildings that survived the Great Chicago Fire of 1871. An architectural wonder, today it also stands as a marvelous icon to resilience and risk management and the importance of water security (Figure 2.5).

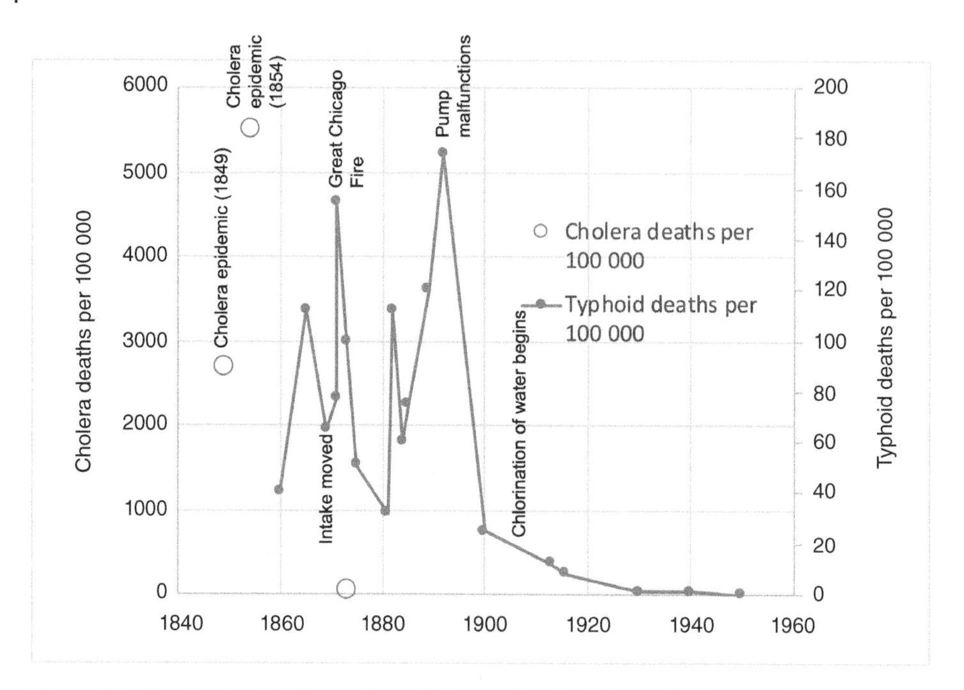

Figure 2.4 Death rates in Chicago from cholera and typhoid fever, 1840–1960. Source: Adapted from Chicago Public Library (2007) and Hrudey and Hrudey (2010).

Figure 2.5 The Chicago water tower, an architectural treasure and a symbol of resilience and risk management against water insecurity. Source: Photograph provided by author (DAS).

2.5 The Blue Nile River – Basin Disputes and Competing Needs (Water Quantity/Equity)

The countries of North Africa and the Middle East include arid and semiarid nations whose modern development has been shaped by the politics of oil. But as one author explains: "If oil built modern Middle East society, water holds the key to its future." (Solomon 2010). Nowhere is this more evident than in the fragile water-driven relationship between Egypt and Ethiopia. Egypt's agricultural economy has always depended on the nutrient-rich floodwaters of the Nile River, which flow northward from upstream African nations. Cheap oil has allowed the deserts of Egypt to bloom with food crops to feed the populous cities of Cairo and Alexandria. Originating in Ethiopian highlands, the Blue Nile is the largest of the two major tributaries, contributing 85% of the Nile's overall flow. By contrast, the White Nile, which contributes 15% of the overall flow, has its headwaters in Uganda, with the Blue and White Nile confluence in Sudan.

Egypt has always claimed priority to the majority of the Nile's flow, citing historical prior usage, and was able to sign an agreement with Sudan in 1959 allowing that country to receive 1/4 of the flow while Egypt would receive the remaining 3/4. But the rains of the highlands were unpredictable, and Egypt had the resources to complete the immense Aswan Dam in 1978 to store two years' worth of flow while protecting its burgeoning downstream population from the vagaries of flooding and drought. The natural flow of the Nile, as well as its delivery of nutrient-rich silt, was forever disturbed and the Upper Nile was rendered a managed irrigation canal.

The irony of this development was that its success became an incentive for other similar projects, including in Ethiopia! The Grand Ethiopian Renaissance Dam (GERD) has been under construction since 2011 (Figure 2.6), and, when filled, will more than double Ethiopia's current electricity production and potentially earn hundreds of millions of export dollars for the country (Walsh et al. 2018). Talks among the three relevant nations (Ethiopia, Egypt, Sudan) have been ongoing but have failed to reach agreement as to how this dam should be managed. The waters of the Blue Nile are waters that observe no national boundaries and represent a real challenge in stakeholder cooperation.

When filled, Ethiopia's new dam will both provide water for irrigation and house the largest hydroelectric power plant in Africa. Thus, this project is a matter not only of water quantity but also of water equity. About 65% of Ethiopians lack access to electricity, a vital component of modern human development (Mutahi 2020). The human development index (HDI) is a measure of a long and healthy life, access to a good education, and a decent standard of living. Egypt has a higher HDI of 0.7 (scale of 0–1) compared with Ethiopia (0.46) and Sudan (0.5) (all values from year 2017) (Roser 2019). It is important to remember that water security is the provision of safe and adequate water to all peoples in need, regardless of their place on the development ladder.

2.6 Bangladesh – Arsenic and Well-Intentioned Groundwater Wells (Water Quality/Quantity/Equity)

Bangladesh is the eighth most populous county in the world and one of the highest in population density. Its major rivers and streams are consistently contaminated with human and

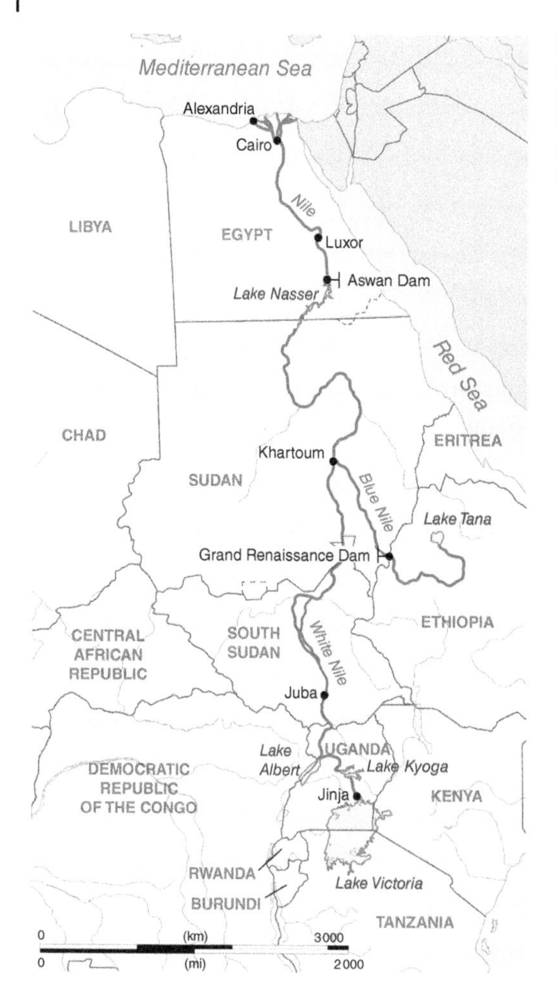

Figure 2.6 The Nile River watershed with the two largest dams, Aswan Dam and Grand Renaissance Dam. Source: Hel-hama/ Wikimedia Commons, https://commons.wikimedia.org/wiki/File:River_Nile_map.svg/ CC BY-SA 4.0.

animal feces, and enteric infectious diseases, such as cholera, have plagued its citizens for centuries. Shifting from surface water to shallow tubewells for drinking water was heavily promoted in the 1970s, resulting in millions of rural villagers now having access to water that was much safer from microbial pathogens. But this water was later discovered to be high in naturally occurring (geogenic) arsenic, causing arsenicosis and death from resulting cancers. In 2009, about 20 million and 45 million people were found to be exposed to concentrations above the national standard of 50 μg/l and the World Health Organization's guideline value of 10 μg/l, respectively (Flanagan et al. 2012) (Figure 2.7). This has been correlated with 19 000–24 000 adult deaths annually. In times of drought or during the dry season, poor rural villagers often have no choice but to drink from wells tainted with arsenic.

The majority of Bangladesh's land is in a low-lying delta for three major rivers, called the Ganges–Brahmaputra delta, the world's largest river delta. Because of its low elevation, the delta and much of the country are extremely prone to flooding. With elevated global average air temperatures due to climate change, the country is extremely vulnerable to both storm surges from increased cyclone intensity and sea-level rise. It has been estimated that 10% of its land area would be inundated if the sea were to rise by only 1 m in depth. Major infrastructure barriers are expensive and perhaps too costly for a country in which about 1/3 of the population still live below the poverty line (IFAD 2020). Months later, these same areas

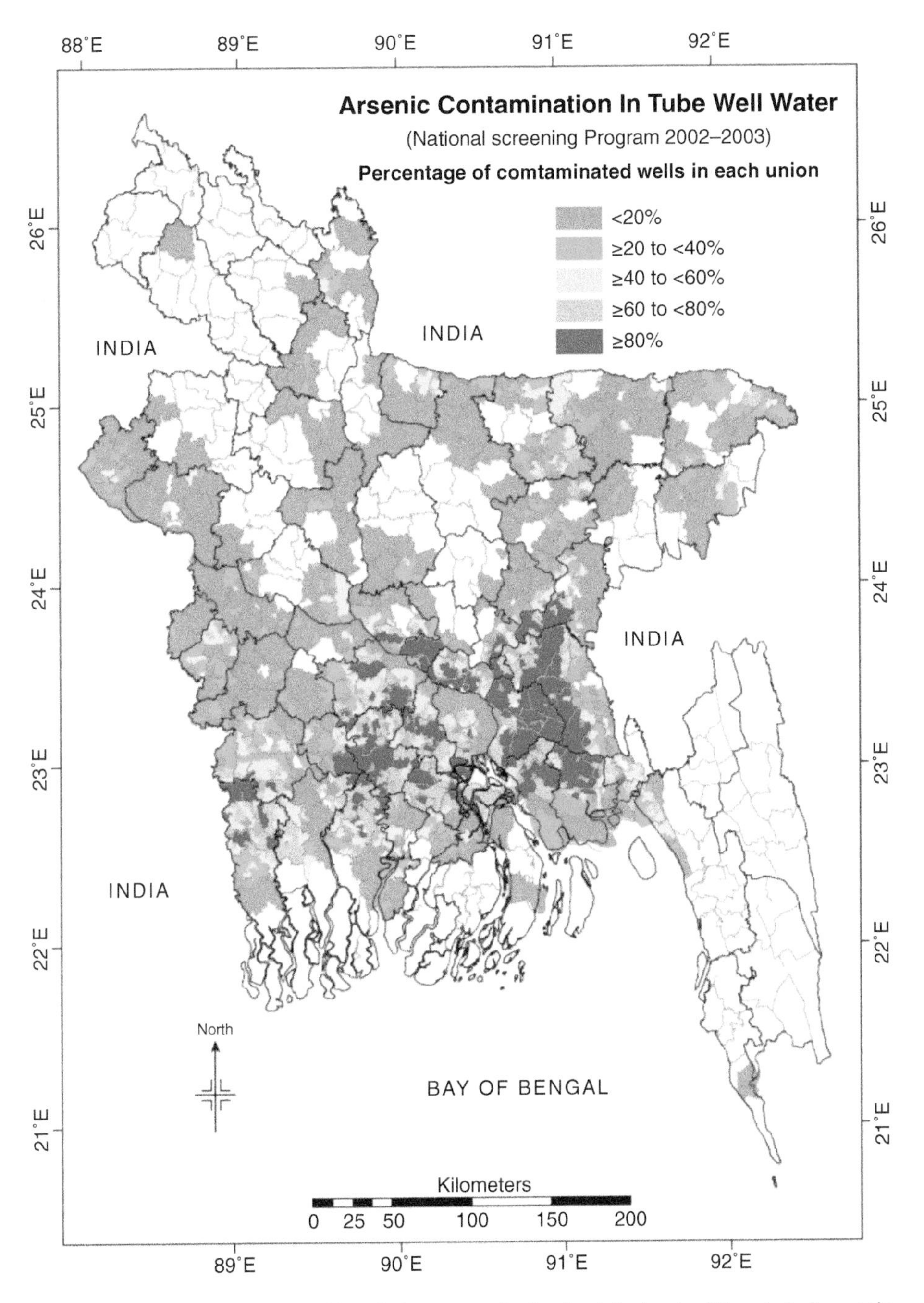

Figure 2.7 A high percentage of boreholes across the fertile delta lands of Bangladesh reveals arsenic contamination above acceptable levels (Gordon et al. 2004). Source: BAMWSP.

can be subject to severe drought. Thus, in addition to water-quality issues, Bangladesh is also subject to severe water-quantity challenges. And the rural poor are most vulnerable, leading to water-equity issues as well.

2.7 Flooding, Drought, and Dam Construction – Senegal, El Salvador, and Texas (Equity/Climate Change)

For centuries, farmers, fishermen, and herders in *Senegal* have depended upon the annual floods over the banks of the Senegal River (Figure 2.8). Farmers have learned how to symbiotically farm the sandy uplands during the brief rainy season and then plant the floodplain (lowlands) after the floodwaters have receded, a practice known as *recession farming*. Livestock, too, followed the seasonal shift of prime pasture land, and fish were abundant with as many as 30 000 tons caught annually (Loucks and Van Beek 2018). Two dams constructed in Mali have eliminated this rich soil-soaking event, accelerating desertification and exacerbating food insecurity for the 500 000–800 000 people who depend on it. Dam construction was intended to bring prosperity by producing hydropower, increasing irrigation water, and making the river navigable all year round. Unfortunately, the people most affected are also the most politically powerless and have either fled as refugees, become sharecroppers under commercial farming enterprises, or suffered from malnutrition due to inadequate food sources. Water insecurity and inequity clearly lead to poverty and food insecurity.

El Salvador villagers living in the Bajo Lempa region had experienced flooding of the Rio Lempa before. But this day in the year 2009 was different. After a few hours of rising water,

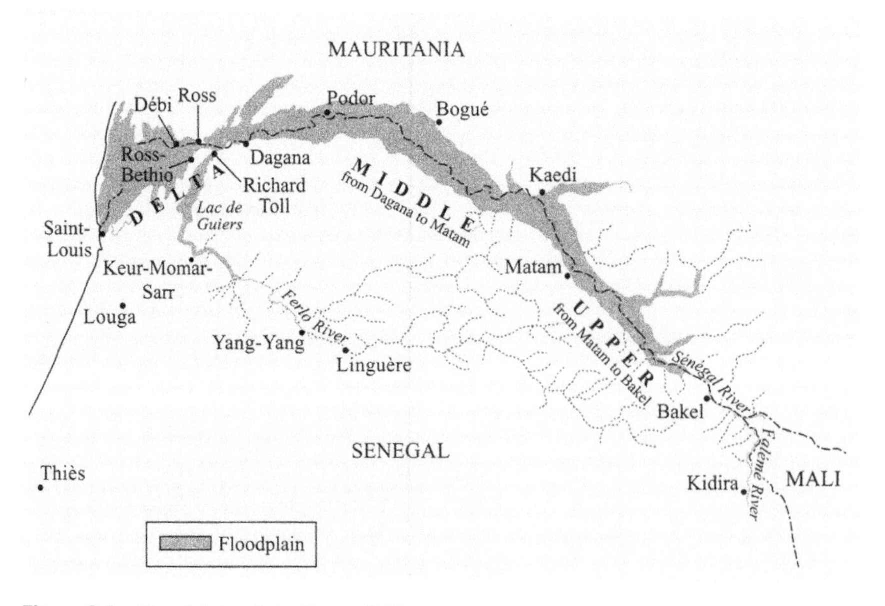

Figure 2.8 Floodplain of the Senegal River, stretching along both banks of the river from Mali to the coastal city of Saint-Louis (DeGeorges and Reilly 2006). Source: UN photos/with permission from Cambridge University Press.

the water temperature began to turn icy cold (Figure 2.9). The villagers recognized that this water had been released from the dam upstream at the Quince de Septiembre reservoir. The Rio Lempa originates in Guatemala, flows through Honduras, and enters the Pacific Ocean on the southwestern coast of El Salvador. It has a watershed area of over 10 000 km^2, about half the size of the entire country. The three reservoirs and associated hydroelectric power plants provide 60% of the country's electricity. The reservoirs are managed for hydropower generation and not for flood control. Consequently, most flooding is associated with releases from these reservoirs (USACE 1998). The villagers in the downstream portion of the watershed have learned to adapt to this enormous challenge by starting their own early warning system using their community radio station, called Mangrove Radio (Aguirre 2013). Using this technological tool, residents were warned to take shelter in buildings that have raised flooring and structural solidity. In upland regions, torrential rains cause deaths by mudslide, exacerbated by the country's extreme deforestation (Taylor 2005). But in the Bajo Lempa (Lower Lempa) region, lives were saved because of the ingenuity of the local villagers who had taken water security into their own hands.

On 30 January 2012, the water wells for Spicewood Beach, *Texas*, began to go dry. Clean water soon began to be trucked into this town in a wealthy state in the richest country in the world. The city of 1100 people sits on a peninsula that juts into Lake Travis just northwest of Austin, a lake that had fallen to just 37% of its capacity. The irony of having no water was not lost on the residents of this city surrounded by a lake and named after a beach. Stage 4 rationing was enforced (Figure 2.10), which allowed only water usage for cooking, cleaning, drinking, and watering house foundations, a necessity in a geology of shrinking and swelling clays. The same two-year drought (2011–2012) in Texas forced other communities to promptly begin wastewater-reuse (water-reclamation) projects. Wichita Falls and Big Spring, Texas, both instituted direct potable reuse after additional advanced treatment of wastewater. This drought shows that even developed nations are susceptible to instances of water insecurity, and the solutions are often familiar ones to those living in places of frequent water stress.

The water insecurity in these last two cases – El Salvador and Texas – arises in large part due to a climate that is changing more quickly than populations can adjust and for which they can plan. As air temperature rises due to increased concentrations of greenhouse gases in the atmosphere, dry regions become drier and wet regions become wetter. These points will be discussed in greater detail in a following chapter (Chapter 6).

Figure 2.9 Flooded streets in the Bajo Lempa region of El Salvador due to a low-pressure system from the Pacific Ocean in 2009. Photo credit: EcoViva.

Figure 2.10 Water-warning signs posted in Spicewood Beach, Texas (January 2012). Stage 4 restrictions ban all outdoor watering and permit only the "essential use" of water – cooking, cleaning, and drinking (Henry 2012). Source: Photograph courtesy of author (DAS).

2.8 Conclusion

In this chapter, we have described several instances of water insecurity due to droughts or flooding, inferior water quality, and/or lack of equitable water access. History has proven that even developed nations are susceptible to periods of water stress or degradation, although the undeveloped nations often experience more severe consequences as they lack adequate resources to respond quickly enough to protect their citizens. All nations are more vulnerable to water insecurity due to a climate that is changing faster than communities can prepare to respond. Some areas, such as El Salvador, will likely receive more extreme rainfall events. Other areas will be subject to more frequent droughts, such as those that have occurred in central Texas.

In subsequent chapters, we will delve more deeply into the three facets of water security – quantity, quality, and equity – in order to understand both the effects and possible solutions to each of these cases. All of these facets are viewed in the context of a rapidly changing climate.

*Foundations: Water-Mass Balance and Kinetics/Exponential Growth and Decay

We conclude this chapter with a discussion of mass (water) balance and kinetics. Mass balance concepts allow us to conduct a water balance (e.g. input minus output in a lake determines whether the lake is filling or emptying) as well as a contaminant balance (e.g. the mass or concentration of a contaminant increasing or decreasing in a lake). Kinetics, the science of time rates of change, is important to understanding future water demand as well as the predicted contaminant concentration under natural or treatment conditions (e.g. natural growth in pathogens versus decreased pathogen concentration due to disinfection).

A simple way of considering mass balance is to think of the money balance in a bank account. To evaluate the balance of money in an account, one needs to consider the inputs (deposits) and the outputs (withdrawals) as well as any regular fees and, under ideal conditions, growth (interest). The money balance at any future time will equal to the starting balance plus increases (deposits and growth) minus decreases (withdrawals and fees):

$$\text{Future Balance} = \text{Current balance} + \text{deposits} + \text{interest} - \text{withdrawals} - \text{fees} \quad (2.1)$$

Changing from money to water, one can consider the water level in a sink or bathtub. Assuming there is a certain amount of water in the sink to begin with, the future amount of water (level in the sink) will equal the starting amount plus inflows minus outflows. Inflows could be from the faucet and from your grandson's water pistol, while outflows could be down the drain or water your grandson splashes out of the sink. Thus, in a word equation, this would be:

$$\begin{aligned}\text{Future Water Volume} = {} & \text{Current water volume} - \text{inputs (faucet, water pistol)} \\ & - \text{outputs (drain, water splashing out)} \quad (2.2)\end{aligned}$$

Converting Eq. (2.2) from word form to equation form results in Eq. (2.3):

$$(\text{Vol})_{\text{future}} = (\text{Vol})_{\text{current}} + (\text{Vol})_{\text{faucet}} + (\text{Vol})_{\text{water pistol}} - (\text{Vol})_{\text{drain}} - (\text{Vol})_{\text{splashed out}} \quad (2.3)$$

If we greatly increase the surface area of our sink to the size of a lake, we can consider the inputs as river inflow and precipitation on the lake and the outputs to be flow out of the dam, evaporation, groundwater infiltration, and, in the case of a drinking water supply reservoir, water withdrawn for human consumption. Eq. (2.3) now becomes Eq. (2.4):

$$\begin{aligned}(\text{Vol})_{\text{future}} = {} & (\text{Vol})_{\text{current}} + (\text{Vol})_{\text{river}} + (\text{Vol})_{\text{precipitation}} - (\text{Vol})_{\text{dam}} - (\text{Vol})_{\text{infiltration}} \\ & - (\text{Vol})_{\text{evaporation}} - (\text{Vol})_{\text{drinking water}} \quad (2.4)\end{aligned}$$

If any of these terms is zero (or negligible), then the equation can be simplified by dropping those terms out of the equation. If the inflows equal the outflows, then the water volume (and thus water level) remains constant and the system is said to be at steady state or equilibrium. While water is still flowing in and out of the system, there is no net change over time. In other words, $(\text{Vol})_{\text{future}} = (\text{Vol})_{\text{current}}$.

We can use a local example to illustrate this concept. East of Norman, OK (home of the University of Oklahoma), is Lake Thunderbird. The lake was constructed between 1962 and 1965 for the primary purpose of providing municipal water to the nearby communities of Del City, Midwest City, and Norman. The lake is named for the Native American legend of the Thunderbird, a supernatural bird of power and strength. Fed by a watershed area of 257 square miles, Lake Thunderbird has an average surface area of 6000 acres and an average depth of 20 ft leading to an average volume of 120 000 acre-ft (ac-ft). During a significant rainfall event the inflow will exceed outflow, infiltration, evaporation and drinking water withdrawal, and the lake water volume, and water level, will increase. During an extended drought, the inflow decreases dramatically compared to drinking water withdrawal, evaporation and infiltration, and the lake water volume, and level, will decrease.

Now let us shift our thought from volume of water to mass of a contaminant. While here we focus on contaminant mass, it is easy to convert from mass to concentration by dividing by the volume of water. The contaminant of interest could be a pathogen, an inorganic compound (e.g. nitrogen or phosphorous) or an organic compound (e.g. pesticides). We can consider the mass (M) of contaminant in our lake at some future time as equal to the initial mass of the contaminant plus mass in the inflow minus mass in the outflow minus any mass lost due to transformations (e.g. degradation) plus any mass produced by transformations (e.g. nitrogen transformation from ammonia to nitrate if nitrate is what we are tracking):

$$(\mathrm{M})_{\mathrm{future}} = (\mathrm{M})_{\mathrm{initial}} + (\mathrm{M})_{\mathrm{inflow}} - (\mathrm{M})_{\mathrm{outflow}} - (\mathrm{M})_{\mathrm{degradation}} + (\mathrm{M})_{\mathrm{produced}} \tag{2.5}$$

Speaking of transformations, this leads us to our next topic of discussion – kinetics or the rate of increase or decrease in a population (of people or pathogens) or mass of a contaminant. You may recall from an introductory chemistry course that the discussion of equilibrium reactions informs us as to whether or not a given reaction will occur. Equilibrium considerations do not shed light on how quickly, or slowly, such a reaction will occur – this is the topic of kinetics.

We will begin by focusing on kinetics as the change in mass (or concentration) with respect to time. For a chemical reaction, we often consider the rate to be a function of the chemical concentration. In words, we would say that the change in concentration (ΔC) with respect to time (Δt) is proportional to the chemical concentration (C); by introducing a proportionality constant (K), we have Eq. (2.6). The minus sign on the right-hand side of Eq. (2.6) indicates that the concentration is decreasing over time (e.g. biodegradation). If we replace the minus sign with a positive sign, then the concentration would be increasing over time (the chemical is being produced). As an asider, change from C to P (population), and maintain a positive sign, the equation can be used to describe increasing population:

$$\frac{\Delta C}{\Delta t} = -K\,C \tag{2.6}$$

If our time increment (Δt) becomes infinitesimally small, we convert to the differential form of the equation as shown in Eq. (2.7). Since the expression is proportional to the concentration raised to the power of one (as opposed, for example, to the power of two, or concentration squared), we say that the reaction is first order with respect to concentration:

$$\frac{dC}{dt} = -K\,C \tag{2.7}$$

If we rearrange Eq. (2.7) by moving C to the left and dt to the right of the equal sign, we end up with Eq. (2.8), where K is the rate constant. As can be seen from Eq. (2.8), larger values of K indicate a greater decrease in concentration (dC) with time or a faster reaction rate:

$$\frac{dC}{C} = -K\,dt \tag{2.8}$$

From calculus, we know that we can integrate Eq. (2.7) from C_o to C over the time period ranging from 0 to t and end up with Eq. (2.9):

$$\ln\,C - \ln\,C_o = \ln\,(C/C_o) = -K\,t \tag{2.9}$$

By taking the exponent of both sides of Eq. (2.9), we get Eq. (2.10), where exp is exponent or e raised to the power of exponent ($\ln x$) = x, and the exponent of $-Kt$ is the same as e raised to the power of $-Kt$:

$$\exp\,(\ln\,C/C_o) = C/C_o = \exp(-Kt) = e^{-Kt} \tag{2.10}$$

Equation (2.10) is the well-known exponential decay function, probably best known as it describes radioactive decay. If we remove the minus sign and change from C to P

(population), this equation can be used to describe exponential growth. Eq. (2.10) can be used to determine K for a given set of data (chemical loss, population growth) and, once the K value is known, can be used to predict future chemical concentrations (or populations, P) based on the initial concentration (population); Eqs. (2.11a) and (2.11b) show the predictive form of Eq. (2.10) for C (decreasing concentration) and P (increasing population – notice positive sign in exponent for P):

$$C = C_o\, e^{-Kt} \tag{2.11a}$$

$$P = P_o\, e^{+Kt} \tag{2.11b}$$

For radioactive decay, the term half-life is often used to describe the rate of decay. The half-life is just another way of expressing the rate constant K, as can be seen by introducing $C = 1/2\, C_o$ (the final concentration equals 1/2 the initial concentration) and $t = t_{1/2}$ (the time to achieve $C = 1/2\, C_o$) into Eq. (2.10):

$$0.5C_o/C_o = 0.5 = \exp(-K\, t_{1/2}) \tag{2.12}$$

By taking the ln of both sides of Eq. (2.12) and rearranging we get Eq. (2.13), which shows that the half-life is inversely proportional to the rate constant K; i.e., as K gets larger (the reaction is faster), the half-life gets smaller (it takes less time for the concentration to be reduced to 1/2 of its initial concentration). While K is the more convenient term for use in the predictive equation (Eqs. 2.11a and 2.11b), $t_{1/2}$ is often more meaningful; a K value of 0.0069/days has less tangible meaning than a half-life of 100 days (it would take 100 days for the concentration to decrease to ½ of its initial concentration, 200 days to reduce to 1/4 300 days to reduce to 1/8, etc., – see Figure 2.11):

$$t_{1/2} = -\ln 0.5/K = 0.69/K \tag{2.13}$$

When considering population growth, we would replace 0.5 C_o with 2 P_o (twice the initial population) on the left-hand side of Eq. (2.12) and $t_{1/2}$ with t_2 as well as remove the negative sign on the right-hand side of the equation. In this way, we arrive at Eq. (2.14):

$$t_2 = \ln 2/K = 0.69/K \tag{2.14}$$

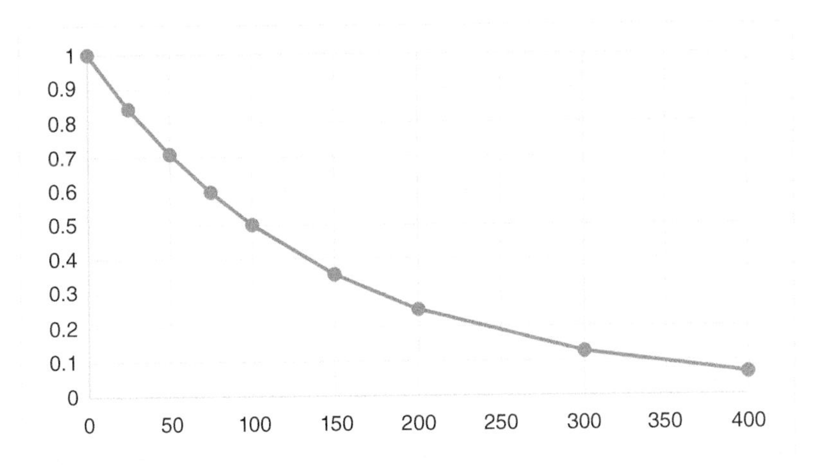

Figure 2.11 Exponential decay plot of C/C_o versus time (hours) with half-life of 100 days ($C/C_o = 0.5$ at 100 days). Source: Original graph – DAS author.

So, if the K value for population growth is 0.069/years (notice units of years versus days above), the t_2 would be 10 years – it would take 10 years for the population of the city to double.

We now use the country of Nigeria as an example. In 2018, the population was 196 million inhabitants and the growth rate has been steady for the last two decades at around 2.6% per year. Given that a 2.6% growth rate means that the population next year will be 1.026 times greater than this year ($P/P_0 = 1.026$), Eq. (2.11b) can be used to solve for k as follows:

$$^{P}/_{P_0} = 1.026 = e^{kt} \tag{2.15}$$

Since $t = 1$ year, and taking the ln of both sides of the equation:

$$\text{Ln}\, 1.026 = k = 0.026\,\text{yr}^{-1} \tag{2.16}$$

We thus see that the annual growth rate, usually expressed as a percent (e.g. 2.6%), when expressed in decimal form (0.026), approximates the K value for population growth in the equation above.

Knowing the value of K (0.026/years), Eq. (2.11b) can now be used in a predictive mode. For example, we can predict how many years it will take for the population of Nigeria to double ($P/P_0 = 2$). Using the formula, it is determined to be 27 years. Thus, the population of Nigeria is predicted to double from 196 million in the year 2018 to 392 million in 2045, thereby surpassing the United States to become the world's third most populous country.

A general rule of thumb is that doubling time can be estimated by dividing 70 by the growth rate expressed as a percent (the "rule of 70"). Thus, for growth rates of 2.6 and 10%,

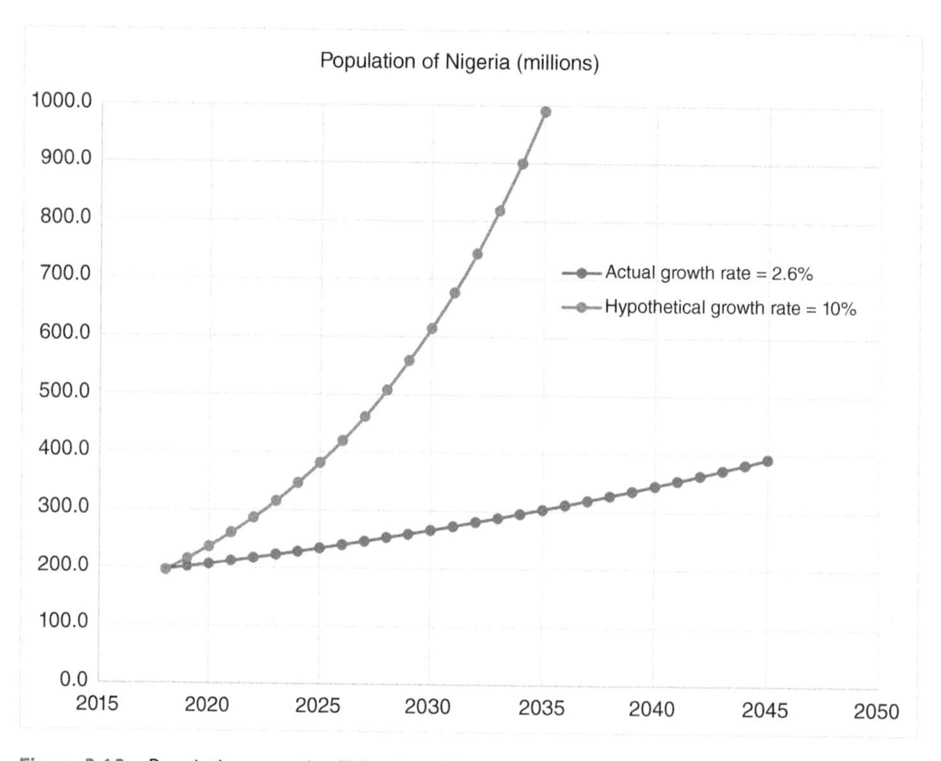

Figure 2.12 Population growth of Nigeria, with the actual 2.6% growth rate per annum and a hypothetical growth rate of 10%. Note that the actual population is predicted to double in the year 2045 as compared to the year 2018. Source: JFC author original.

the population would be predicted to double in 27 (70/2.6) and 7 years (70/10), respectively. These two growth rates are shown in Figure 2.12 for illustration purposes, along with quantitative predictions based on Eq. (2.11b).

The concepts of mass balance and kinetics are important to topics of water quantity (water balance) and water quality (pathogen and contaminant concentrations) as well as population growth, all of which impact water security.

End-of-Chapter Questions/Problems

2.1 The Pacific Institute publishes a chronology of world water conflicts. It can be found here:
http://www.worldwater.org/water-conflict
Use the chronology to answer the following questions.

a. In the nineteenth-century time period, 1800–1899, what was the nature of a conflict over waters in the Nile region?
b. In this same period, describe the conflicts that occurred in the states of Indiana and New Hampshire regarding disputes over water bodies.
c. In the period, 2010 till present, describe two water conflicts that involve food security, e.g., water for irrigation.

2.2 Using an online population database (such as the United Nations Population Projections 2019 Database), determine the population rate constant from 2010 to 2020 for two nations, Uganda and Chile. Based on the result, determine which country's population is increasing at a faster rate.

2.3 The following values provided in the table below were determined for a reservoir with a duration of one year. The reservoir must release 75 cfs to protect the downstream ecosystems, which accounts for 54 446 ac-ft over the duration of a year.
Using Eq. (2.4) and the values provided in the table below,

a. Determine how much water is available to a downstream community while also maintaining the health of downstream ecosystems and a reservoir volume of at least 100 000 ac-ft.
b. Based on this value, determine if it is sufficient to meet the needs of the capital of El Salvador, San Salvador, or if it would be better suited for a city the size of San Miguel (also in El Salvador).
c. If it is not sufficient, use the categories of water stress provided in Chapter 1, Table 1.2 to determine the level of water scarcity.

Source	Volume (ac-ft)
Current volume	150 417
River inflow	321 368
Precipitation	31 252
Groundwater infiltration	31 289
Evapotranspiration	33 068
Ecosystem flow	54 446

2.4 Compare and contrast two events of water insecurity based on water quality: the Flint Water Crisis and the London Cholera outbreak.

2.5 Research the infamous "Day Zero" that occurred in Cape Town, South Africa, in 2018 and identify four to five ways in which the city prolonged their water supply to avoid complete depletion.

2.6 Using the concepts of half-life as discussed in the chapter, determine:
   a. If the annual growth rate of a population is 1.6%, how long would it take for the population to double?
   b. If the half-life of BTEX is 693 days, how long would it take for the concentration to decrease from 200 to 5 μg/l?

2.7 Using the estimated half-life "rule of thumb" as explained in the text, compare the calculated doubling time from 2.6a to the estimated value.

2.8 In the Navajo Nation in the southwestern United States, roughly 15% of the population does not have access to a water distribution network (US EPA). Therefore, many residents are forced to gather water from unregulated water sources. Research the most prevalent concerns in regard to water quality, quantity, and equity.

2.9 Research the history of the Hoover Dam construction and the current status of Lake Mead. Using the volume budget analysis, explain what the original water allocations failed to consider and how this now poses a threat to water security.

2.10 According to the EPA, an acceptable amount of phosphorous within a water body is not to exceed 10–40 μg/l. The volume of a small pond is 200 m^3. The total phosphorous in the lake is 7 μg/l. In a given day, a nearby farm loads 450 mg of phosphorous into the lake. A stream leaving the lake removes 200 mg of phosphorous from the lake per day. Within the lake, phosphorous transformations are negligible for the duration of one day. Determine the total phosphorous (mg) in the lake after a one-day cycle and if it exceeds the EPA guideline.

Further Reading

Fagan, B.M. (2011). *Elixir: A Human History of Water*. London: Bloomsbury UK.

Hrudey, S.E. and Hrudey, E.J. (2010). *Safe Drinking Water: Lessons from Recent Outbreaks in Affluent Nations*, 1e. London, UK: IWA Publishing.

Solomon, S. (2010). *Water: The Epic Struggle for Wealth, Power, and Civilization*, 1e. New York: Harper.

References

Aguirre, M. (2013). "Re-Launching Mangrove Radio." EcoViva. 2013. https://ecoviva.org/re-launching-mangrove-radio.

Chicago Public Library (2007). "Chicago: 1900 Flow of Chicago River Reversed." March 7, 2007. https://web.archive.org/web/20070307091435/http://www.chipublib.org/004chicago/timeline/riverflow.html.

DeGeorges, A. and Reilly, B.K. (2006). "Dams and Large Scale Irrigation on the Senegal River: Impacts on Man and the Environment." UNDP. http://hdr.undp.org/en/content/dams-and-large-scale-irrigation-senegal-river.

Fagan, B.M. (2011). *Elixir: A Human History of Water*. London: Bloomsbury UK.

Flanagan, S.V., Johnston, R.B., and Zheng, Y. (2012). "Arsenic in tube well water in Bangladesh: health and economic impacts and implications for arsenic mitigation." *Bulletin of the World Health Organization* 90 (11): 839–46. doi:https://doi.org/10.2471/BLT.11.101253.

Gordon, B., Mackay, R., and Rehfuess, E. (2004). *Inheriting the World: The Atlas of Children's Health and the Environment*. World Health Organization http://www.who.int/ceh/publications/atlas/en.

Henry, T. (2012). "During Texas Drought, Will Spicewood Beach Be the First to Run Dry?" *StateImpact Texas* (blog). 2012. https://stateimpact.npr.org/texas/2012/01/26/during-texas-drought-will-spicewood-beach-be-the-first-town-to-run-dry.

Hrudey, S.E. and Hrudey, E.J. (2010). *Safe Drinking Water: Lessons from Recent Outbreaks in Affluent Nations*, 1e. London, UK: IWA Publishing.

IFAD (2020). "Bangladesh." IFAD. 2020. https://www.ifad.org/en/web/operations/country/id/bangladesh.

Johnson, S. (2007). *The Ghost Map: The Story of London's Most Terrifying Epidemic--and how it Changed Science, Cities, and the Modern World*. Riverhead Trade.

Loucks, D.P. and Van Beek, E. (2018). *Water Resource Systems Planning and Management: An Introduction to Methods, Models, and Applications*. Springer.

Mutahi, B. (2020). "Egypt-Ethiopia Row: The Trouble over a Giant Nile Dam." *BBC News*, 2020, sec. Africa. https://www.bbc.com/news/world-africa-50328647.

Roser, M. (2019). "Human Development Index (HDI)." *Our World in Data*. https://ourworldindata.org/human-development-index.

Salzman, J. (2017). *Drinking Water: A History*. Overlook Duckworth.

Solomon, S. (2010). *Water: The Epic Struggle for Wealth, Power, and Civilization*, 1e. New York: Harper.

Taylor, J. (2005). "El Salvador Flood Disaster Worsened by Deforestation." The Independent. 2005. www.independent.co.uk/environment/el-salvador-flood-disaster-worsened-by-deforestation-5348221.html.

USACE (1998). *Water Resources Assessment of El Salvador*. US Army Corps of Engineers.

USEPA (2002). *Consider the Source: A Pocket Guide to Protecting your Drinking Water*. US Environmental Protection Agency.

Walsh, D., Sengupta, S., and Ahmed, H. (2018). "As a Dam Rises in Ethiopia, Its Manager Is Found Dead." *The New York Times*, July 26, 2018, sec. World. https://www.nytimes.com/2018/07/26/world/africa/ethiopia-dam-manager.html.

2a The Practice of Water Security: Public Health Impacts of Clean Water

The World Health Organization (WHO) estimates that about 1/10 of the global disease burden can be attributed to unsafe water, lack of sanitation, inadequate hygiene, and poor management of water resources (Prüss-Üstün and World Health Organization 2008). While there are many options for interventions, these need to be evaluated as to their efficacy and cost–benefit value. Ideal practitioners for this assessment are those who know the science of both water and medicine.

Steve Luby is a medical doctor … a researcher … and a public health practitioner (Figure 2a.1). After receiving a bachelor's degree in philosophy at Creighton University, he earned his medical degree and completed a residency in internal medicine. He studied epidemiology and preventative medicine at the Centers for Disease Control (CDC) in Atlanta where he also worked as a medical epidemiologist. After first living and working in Pakistan, Steve moved his family to Bangladesh, where for eight years he served as Director of the Centre for Communicable Diseases at the International Centre for Diarrheal Diseases Research (ICDDR) while he and his wife raised three children.

Figure 2a.1 Stephen Luby stays active in global clean water testing and interventions. Source: Photo: Stanford Center for Innovation in Global Health.

Drawing upon these varied experiences, Steve can navigate easily through the overlapping worlds of medicine, public health, academia, and developing country programs with the goal of improving human health and water security. As a scientist, Steve has been a lead researcher in the areas of water treatment, water supply, and household hygiene in developing countries (Figure 2a.2). He has dutifully assessed the health impacts of:

- Household water treatments, e.g. PuR sachets and Safe Water Systems (SWS),
- Handwashing interventions, and
- Concurrent multiple interventions.

He has detailed the usefulness of microbiological indicators as well as lack of effectiveness of some standard interventions, including the shock chlorination of wells.

Figure 2a.2 Dr. Stephen Luby has spent many years in developing countries studying the impacts of sanitation and hygiene interventions. Source: Photo: Ruggles (2015).

Diarrhea and respiratory infections kill more than 3.5 million children aged five years and under, with most of them in low-income neighborhoods of developing countries. From a study funded by Procter & Gamble, Luby discovered that the availability of soap, and instructions on when to use it, reduced diarrhea incidences by half in poor Karachi, Pakistan, neighborhoods. In Bangladesh, he evaluated the effectiveness of a combined flocculent disinfection product in reducing the body burden of arsenic among village residents drinking arsenic-contaminated water. Also in Bangladesh, Luby worked with a mother to nudge other mothers to local clinics to acquire diphtheria, tetanus, and other vaccinations for their children. Luby was convinced that low-tech solutions used faithfully would, over time, solve even the most insurmountable challenges of water and health. According to his friends and coworkers, he is patient, he listens to people, and he does not let a little thing like a flood or worker strike stop him! One of Steve's favorite sayings is: "If you want peace, work for public health."

Stephen Luby is the recipient of the 2009 OU International Water Prize

References

Prüss-Üstün, A. and World Health Organization (2008). *Safer Water, Better Health Costs, Benefits and Sustainability of Interventions to Protect and Promote Health.* Geneva: World Health Organization http://whqlibdoc.who.int/publications/2008/9789241596435_eng.pdf.

Ruggles, R. (2015). "Dr. Stephen Luby, an Omaha Native, Takes Pragmatic Approach to Global Public Health Research." *Omaha World-Herald*, 2015. https://omaha.com/livewellnebraska/health/dr-stephen-luby-an-omaha-native-takes-pragmatic-approach-to-global-public-health-research/article_2d5f626e-39d3-5032-92f5-c31f282605ec.html.

Part II

The Context of Water Security

3

The Context of Water Security – The Quantity of Water

The objective of this chapter is to provide a broad overview of the supply of and demand on the world's water resources. Proper understanding of water security depends upon a knowledge of the hydrologic cycle, including both stocks (reservoirs) of water and the movement of water (fluxes) between stocks. Further, the parameters of water supply and water demand will vary according to region (spatial) and time (temporal) differences. Human influence is also considered as part of the overall hydrologic system. Specific examples will highlight the potentially severe consequences of drought and flooding events. Finally, interbasin water transfers occur in many regions of the planet as communities search outside their watershed to obtain the desired water quantity for their specific needs.

Learning Objectives

Upon completion of this chapter, the student will be able to:

1. Understand the various components that comprise water's natural hydrologic cycle.
2. Give rough comparisons of the quantity of water that is present in fluxes and bound up in global water stocks or reservoirs.
3. Explain how human activities have and will continue to alter the hydrologic cycle and the global water budget.
4. Understand the range of water demand versus supply ratios that exist across regions and nations.
5. Understand and appreciate the severe risks from flooding and drought that are present in all communities, especially the most vulnerable.
6. Be able to give estimates of and compare relative water demands from various sectors, including domestic, agriculture, energy, and industry uses.
7. *Be able to perform simple hydrologic calculations regarding peak flow, storm runoff volume, and the sizing of a detention basin.

3.1 Introduction

The study of quantity of water is, in large part, the study of hydrology. Just as geology is the study of Earth (from "geo" = earth, "logos" = speech), hydrology is the study of water – its movements, cycles, and reservoirs (Figure 3.1). Water falls from the atmosphere

Fundamentals of Water Security: Quantity, Quality, and Equity in a Changing Climate, First Edition.
Jim F. Chamberlain and David A. Sabatini.

Companion website: www.wiley.com/go/chamberlain/fundamentalsofwatersecurity

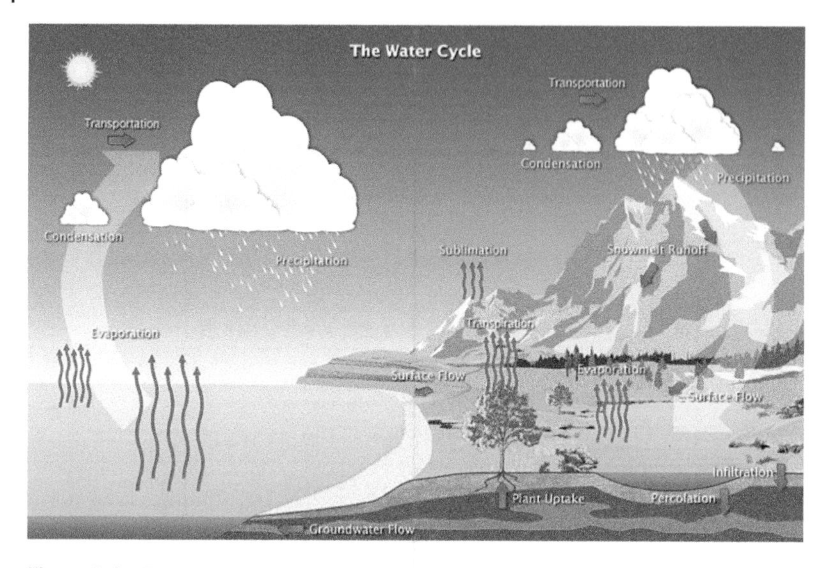

Figure 3.1 The hydrologic cycle, including both saltwater and freshwater stocks and fluxes (NASA 2021).

as precipitation, landing on solid ground or in a surface water body – e.g. ocean, lake, or river. If it falls on the ground, it may percolate (infiltrate) through the soil and into an underground reservoir where it then becomes groundwater. Or it may be taken up by plants who use the water to transport vital nutrients up into stems and leaves. The water may also run off the land as overland flow, where it may eventually enter a surface water body, such as a stream channel. (Note that the term "stream" may refer to a water body as large as the Mississippi River or as small as a backyard creek.)

At any point along its path, water may be transferred back into the atmosphere by either direct evaporation or transpiration through plant leaves. A portion of the water vapor then condenses back into water droplets, forming clouds, and eventual precipitation. The hydrologic cycle is thus complete. The water does not travel alone but carries sediment particles, valuable nutrients, and harmful contaminants that make up the water's overall quality.

Hydrology is the study, in space and time, of these various stocks (reservoirs) and flows (fluxes) of Earth's water. All living things – plants, animals, and people – are at the mercy of these stocks and flows. A *water budget* can be used to describe the annual flow of water from one compartment to the other. Not surprisingly, the largest fluxes are from precipitation that falls on the ocean and returns via evaporation from the ocean. But the combined evaporation from surface water bodies and soil and transpiration from vegetation is also significant and is roughly 65% of the total precipitation that falls on the land (Schlesinger 2018).

The hydrologic cycle as a biophysical phenomenon can no longer be studied as a system independent of human influence and perturbations (Vogel et al. 2015). Human activity can drastically impact the evaporation flux from soil and vegetation through land use change, for example the replacement of wooded lands with asphalt streets and parking lots.

Other human activities, such as withdrawal and storage of groundwater, discharge of wastewater flows to surface water bodies, and desalination projects, also affect natural stocks and flows.

In addition, climate change is a global imbalance that has a local impact. Major disruptions from climate change, such as drought, flooding, and damage to water quality, can result in either chronic or acute harm to living species. For example, climate change models predict that Sweden will experience an increase of up to 4 °C and up to 40% more precipitation. These conditions would increase the nitrogen loading in streams, lakes, and estuaries, a consequence of atmospheric deposition and nitrogen leaching from the soil (Arheimer 2005).

With some exceptions, Earth's living organisms need water that is "fresh," that is, having low concentrations of dissolved salts. And so, with regard to water security, we may ask the following:

- How much of the water on the Earth is freshwater versus saltwater?
- Of this freshwater, how much is available (e.g. not frozen)?
- Of the available freshwater, how much is surface water versus groundwater?

These questions are answered pictorially in Figure 3.2.

From the figure, we can conclude that ~2.5% of all water on earth is freshwater. About 30% of freshwater is available (not frozen), and of this, less than 1% is readily available as surface water. Any additional water must be extracted from the ground. So, while we live on a "blue planet," only a small fraction of the blue is easily available freshwater capable of sustaining life.

In the recent past, hydrologic design of channels, culverts, bridges, and reservoirs has relied on some assurance that natural systems operate within a predictable range of variability, a characteristic known as **stationarity** (Milly et al. 2008). Given enough historical data for the design time period, a water analyst can develop a probability of occurrence for any hydrologic

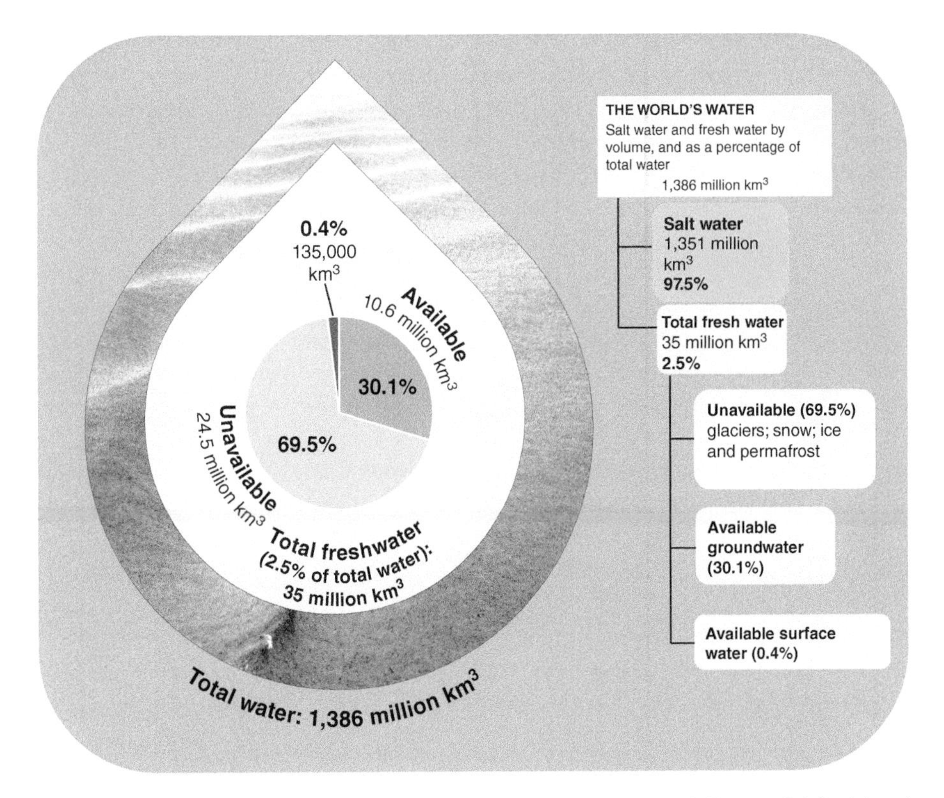

Figure 3.2 Distribution of water in the world (graphic is not to scale). Source: Original drawing adapted from USGS (2021).

variable, such as streamflow or rainfall event. This value can then be used to assess risk or to design effective water structures. For example, the size of a 100-year storm event (a storm that has a probability of occurrence equal to 1/100, or 1%, in any given year) in a watershed upstream of the city of Houston, Texas, can be used to size stormwater culverts, drainage channels, spillways, reservoirs, and heights of stream overpasses, as appropriate.

With increased anthropogenic land-use alteration and the acceleration of climate change, however, the predicted precipitation characteristics and stream flow for a given design storm return interval may not be accurately characterized based on historical data. The impacts of land-use change can be modeled as deterministic processes; that is, the change in runoff can be directly correlated with a change in land use (Lins 2012). For example, the change of one acre from grassland to concrete parking lot will result in an increase in runoff that can be reasonably predicted. However, the size of the rainfall event has a range of variability that can no longer easily be determined. Such a state of **nonstationarity** will likely require the collaboration of statisticians, climate scientists, and water resource engineers to make more accurate predictions of future hydrologic variations.

The purpose of this chapter is to introduce the challenges to water security that come with too little or too much water (i.e. droughts and flooding) and to describe how water varies across the globe in both supply and demand. A water balance budget can be used to quantify this supply and demand using the elements of the water cycle already described.

3.2 Distribution of Global Water Resources

Much of the inequity in water security is a function of two interlocking factors: *hydrologic distribution* (spatial and temporal) and *population preference*. First, Earth is composed of at least four zones of aridity, based on the amount of water lost to the atmosphere relative to the total rainfall that falls in that zone (Sombroek and Sene 1993). This results in a very uneven natural *hydrologic distribution* of water resources. In hyperarid and arid zones, such as the African Sahara, the Atacama Desert of Chile, and Baja California, the utter lack of water has constrained human settlements and only specialized life has adapted and evolved. Semiarid and humid regions are much more amenable to life forms and historically populations have tended toward these regions where life is more water sustainable. A single state or country can include all levels of aridity as seen in Australia where the largest populations are in the coastal, humid regions along the east coast and the inland arid areas with thin soils are sparsely inhabited.

The temporal distribution of water occurs both *intra-annually* (within the year) and *inter-annually* (from one year to the next). The variation within a year is often articulated as a "rainy" season and a "dry" season. This is illustrated in Figure 3.3 which shows rainfall variation within the year for Bangladesh. Figure 3.3a shows a clear regional variation of rainfall within the country, from 4300 mm/year in the east to 1300 mm/year in the west. Figure 3.3b shows a rainy season from April to September while November to March constitutes the dry season. Areas prone to flooding in the rainy season may be in severe drought a few months later. Human activities, the planting of crops, and the care of animals all revolve around the knowledge and expectation of this hydrologic variation, which cannot be captured by an average annual rainfall number. Water infrastructure –rainwater-collection vessels, reservoirs, and aquifers – are used in an effort to help dampen the extremes in water supply from both intra-annual and inter-annual variability (Gaupp et al. 2015).

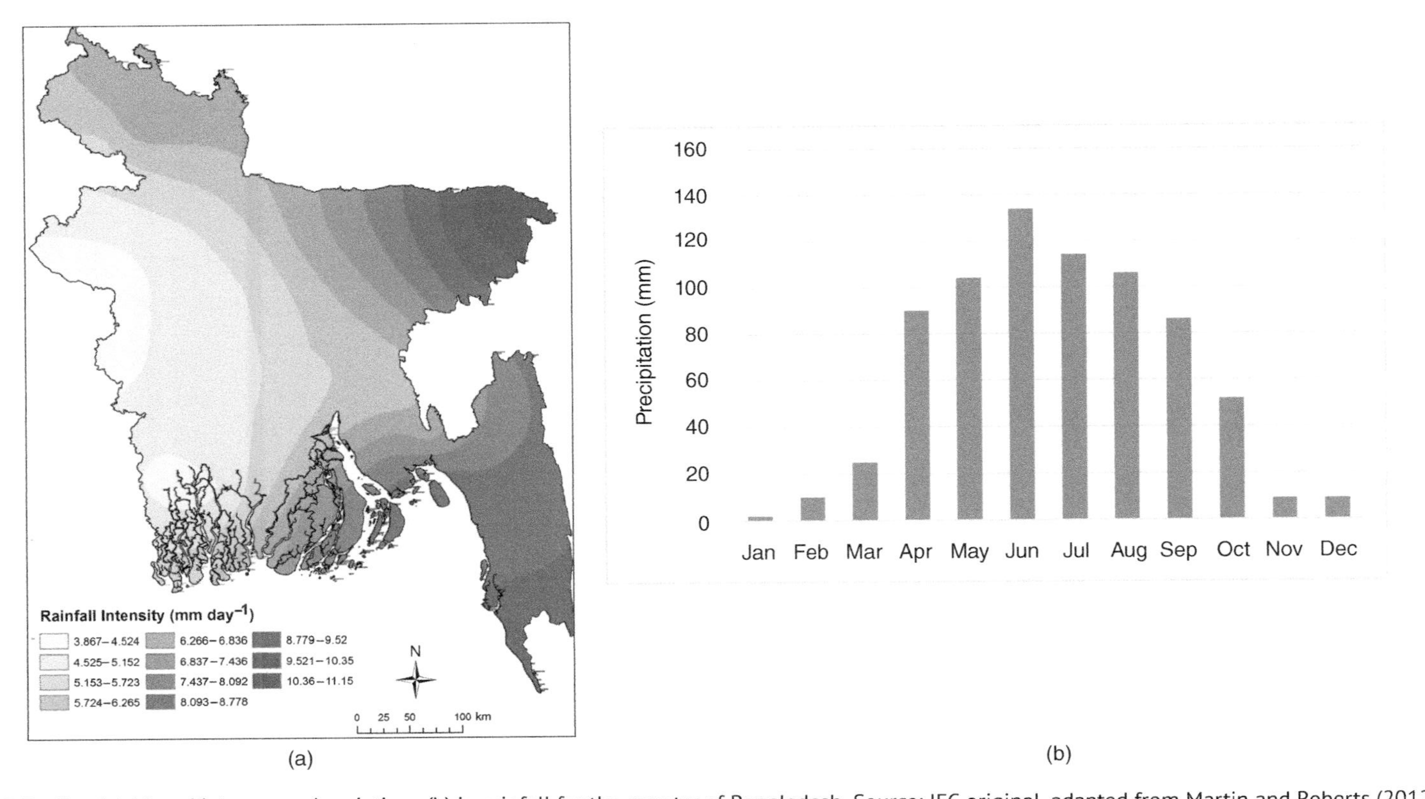

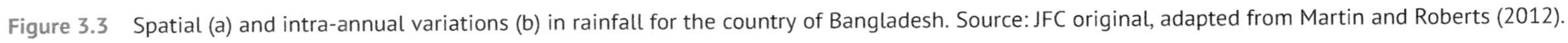

Figure 3.3 Spatial (a) and intra-annual variations (b) in rainfall for the country of Bangladesh. Source: JFC original, adapted from Martin and Roberts (2012).

The inter-annual hydrologic variation is more difficult to predict and can usually only be estimated based on recent and historical trends. For example, Figure 3.4 illustrates variations in the average annual rainfall for the state of Oklahoma from 1895 to 2015 centered around the long-term statewide average of 39 in/year. Areas shaded green lie above the long-term average and correspond to wetter time periods while areas shaded brown correspond to dryer time periods. Notables are two major dry spells in the 1930s and 1950s – both a decade long. The 1930s is the famous "Dust Bowl" period, so named because the land was so dry that fine soil particles were scattered widely by the wind. Fortunately, the 1950s were not as catastrophic due to improved farming and water resource management practices. On the other side, the wet years encompassing the 1980s and 1990s have the tendency to erase all memories of drought and, if flooding occurs, may even prompt some residents to suggest lowering reservoir storage levels to better absorb future rainfall and mitigate flooding, not realizing the catastrophic effect this would have in the event of an ensuing drought. Thus, such long-term historical data are invaluable to informing management of reservoirs and water resources.

The trends and variability shown in Figure 3.4 have the potential to be further exacerbated by ongoing climate change. Climate change is impacting both types of rainfall variation (intra- and inter-annual) and, along with population growth, is a critical factor influencing future water security.

With the advent of industrialized farming practices, however, decisions based on *population preference* increasingly served to override water availability. Man-made irrigation opened up land not previously inhabitable based on rain-fed agriculture. When given the options, human families have tended to settle at intersections of trade or transportation routes, in regions of economic prosperity, in places of natural beauty, or where there is an abundance of affordable land. Water insecurity then results in places where hydrologic distribution and population preferences are not aligned with one another. Table 3.1 shows significant variability in water security for the countries listed using water withdrawals as a percentage of available water as a barometer. For example, as a nation overall, Canadians withdraw

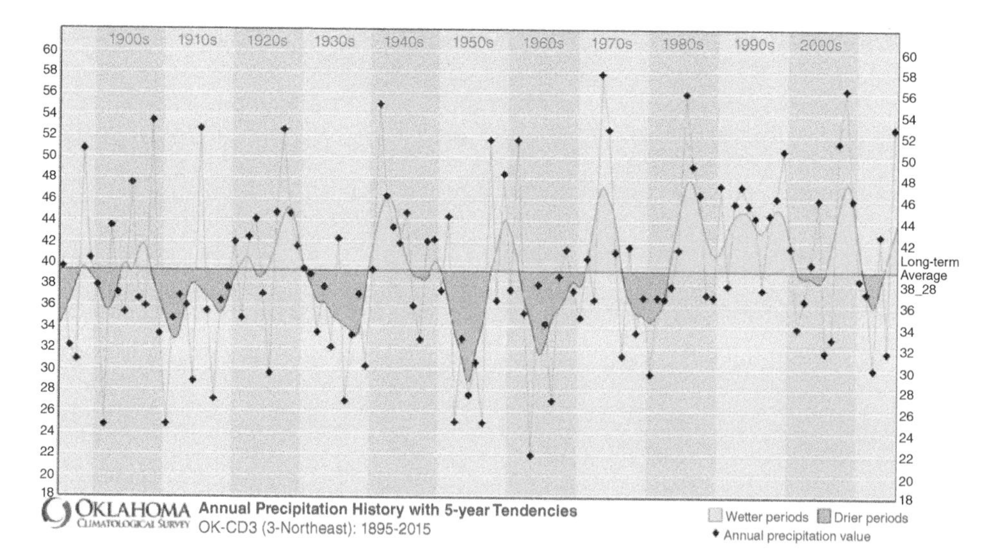

Figure 3.4 Average annual precipitation history for Oklahoma showing variability about the long-term average of 34 in/year. Source: OCS (2020) (Public document state of Oklahoma)/with permission from Board of Regents of the University of Oklahoma.

Table 3.1 Water supply and demand, by country.

Country	Internal water available (m^3/capita/year)	Withdrawal as % of available H_2O
Canada	109 510	1
USA	10 060	19
China	2520	16
India	2270	21
Tunisia	490	53
Libya	170	374
Egypt	40	97

Source: Clarke (1991)/with permission from Taylor & Francis.

a tiny fraction of their available water and thus enjoy an enviable level of water security, even though certain regions of Canada may lack sufficient water (and the cost of moving water long distances is often prohibitive). In contrast, Libyans' water withdrawals far exceed their water supply by withdrawing 95% of their water from a deep, nonrenewable (fossil water) underground reservoir (aquifer) that spans the political boundaries of four countries in northeastern Africa (Dahab and El Sayed 2001). It is not difficult to see that withdrawal data as percent of available water can be useful in determining potential areas of future water insecurity for a nation and potential conflict among neighboring nations.

Hydrologic challenges are often met with a combination of mechanized irrigation and, unfortunately, wishful thinking. The western half of the Great Plains in the United States was settled partly in response to government incentives that gave a settler cheap land as reward for settling parts of the country considered by many to be uninhabitable. It is estimated that about 1.2 billion people live in areas of the world that are water-challenged due to geography alone (Black 2016). Brazil receives the bulk of its water in the Amazonian region, but a fifth of the population is in cities along the northeast coast that receives only 2% of the country's rainfall. China has 19% of the world's population but only 5% of water (Black 2016). And this is further exacerbated by the fact that the majority of fresh water exists in the southern part of China while the majority of the population resides in the northern portion of the country, as discussed later in this chapter.

3.3 Human-Impacted Water Budget

The water availability for a watershed can be estimated by a simple water budget (Figure 3.5). In a given watershed/basin, the change in storage is equal to the total water entering minus the water leaving the watershed boundaries:

$$\text{Change in storage } (\Delta S) = \text{Water in} - \text{water out}$$

$$\begin{aligned} \Delta S &= P + GW_{\text{in}} - (ET + Q + GW_{\text{out}}) \\ &= P - (ET + Q + \Delta GW) \end{aligned} \tag{3.1}$$

In Eq. (3.1), P is precipitation, Q is streamflow, ΔGW is change in groundwater seepage ($GW_{\text{out}} - GW_{\text{in}}$), and ΔS is the change in storage. Evapotranspiration (ET) is a combination of evaporation from water surfaces and transpiration through the leaves and stems of plants. Units are given in volume per time period, such as cubic meters per year (m^3/year).

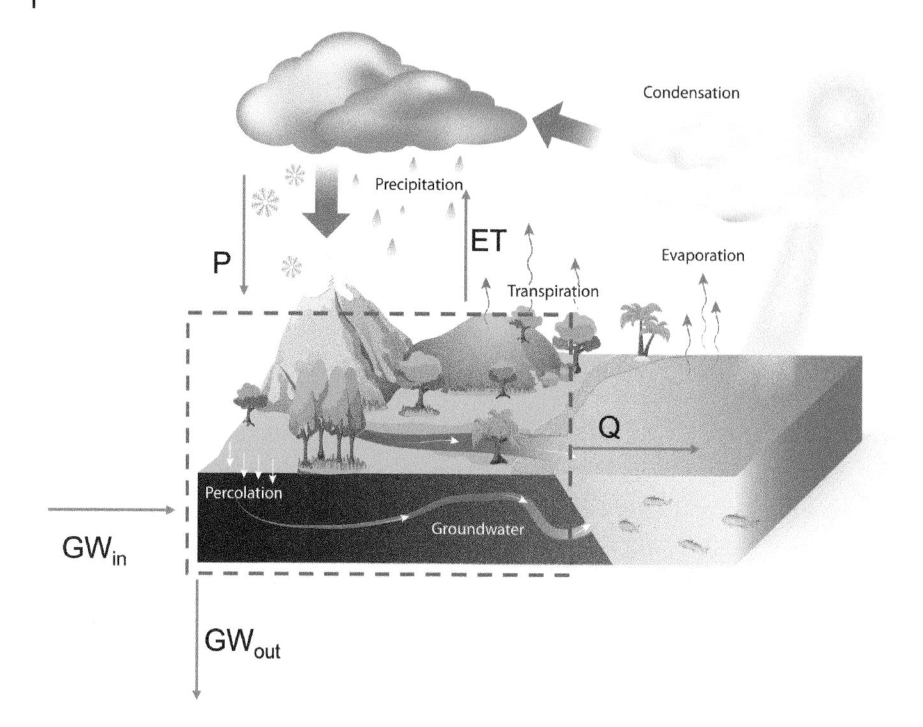

Figure 3.5 Watershed boundaries show both inputs (precipitation, groundwater inflow) and outputs (evaporation, transpiration, streamflow, groundwater outflow). Source: JFC.

Quantification of most of the parameters is relatively straightforward. Precipitation, P, can be measured with a rain gauge and streamflow, Q, can be observed with in-stream flow monitoring. The change in groundwater seepage (ΔGW) is estimated using observation wells but may be considered relatively insignificant on some time scales of interest. ΔS can be estimated based on the change in groundwater and storage volumes.

ET is more difficult to assess but it is still extremely important for two reasons. First, the *ET* component is a significant portion of the overall budget. In sunny, arid regions such as Arizona, the *ET* component may be more than 95% of the annual precipitation (Brooks et al. 2012). In these regions, the ET/P ratio is very close to 1, and the availability of surface water (represented by Q) is small. Second, this factor is considerably affected by land-use change, which either increases or decreases the original (natural) rate of ET.

If we now include the human factor, we must add terms for human withdrawals (or export from the basin) as well as human additions (return flows or import into the basin) (Vogel et al. 2015). The human withdrawals are water that is consumed for irrigation, municipal water needs, or industrial usage. The flow additions may come from once-through usage – such as cooling tower or wastewater treatment plants – or from **interbasin transfers** (IBTs).

Neglecting the change in groundwater storage (ΔGW, GW_{in}, GW_{out}), our overall equation then becomes:

$$\Delta S = P + H_{in} - (ET + Q + H_{out}) \tag{3.2}$$

where

H_{in} = human return flows or imports to the basin
H_{out} = human withdrawals for local use or export from basin

Table 3.2 Water system hydrologic data for two water systems.

System	R_T	H_{in}	Sum of inflows	Q_{out}	H_{out}	Sum of outflows	del S = Inf – Out	H_{in} as % of net inflow	H_{out} as % of net outflow
Surface water system (all units in mm/year, except as noted)									
Yellow River (China) 1998–2000 (area = 742 000 km²)	72	11	83	9.2	67	76.2	6.8	13%	88%
Aquifer system (all units in mm/year, except as noted)									
Central Valley, California (USA) 1961–1977 (area = 51 800 km²)	48	223	271	7.1	283	290.1	−19.1	82%	98%

Source: Adapted from Vogel et al. (2015).

Table 3.2 presents hydrologic data for two water regimes as an illustration of human influence on the local hydrology. The total natural recharge for a surface basin, R_T, can be expressed for a surface water basin as "$P - ET$." The total recharge for an aquifer includes precipitation, surface water recharge, and adjacent aquifer recharge. Likewise, the natural discharge out of a basin can be expressed as surface water flow, Q_{out}, or as total aquifer discharge to surface waters, adjacent aquifers, and ET.

The Yellow River is the second longest river in Asia (3395 miles) and is one of the most important hydrologic regimes in China. The climate is arid, with an *ET* to P ratio of 85%. Irrigation withdrawals from the river increased by ninefold over the last half of the twentieth century. As Table 3.2 shows, human withdrawals from the basin were six times as great as human input or returns over the two-year period of study. In addition, during drought periods the human withdrawals (H_{out}) may exceed the natural recharge, R_T. This imbalance can seriously deplete the water regime and can result in water stress and aquatic ecosystem degradation due to reduced annual or seasonal stream flows.

The highly productive agricultural region of the Central Valley in California relies on imported snowmelt from the Sierra Nevada range as well as pumping from an underlying aquifer. From Table 3.2, note the large percentage of human inflow and outflow to the local subsurface (aquifer) hydrology. The change in storage was net negative over the 16-year study period, and the recycling of irrigation return flows brings higher salinity and other water-quality degradation (Vogel et al. 2015).

3.4 Flooding and Overabundance of Water

Two types of meteorological events are representative of the overabundance of water for exposed populations (Black 2016). Prolonged heavy rainfall or above-normal seasonal snow melt can result in the overflowing of streambanks with loss to land, animals, and human lives. This might be considered an *outward* force of flooding. Major storms can also cause the sea to move inland, breaking shore seawalls and causing intense flooding in an *inward* land trajectory. Both are becoming more frequent with a changing climate.

Several examples may be given to illustrate the devastation of flooding:

- Over 2.5 million people were displaced and 2.4 million acres of cropland affected during heavy monsoon rains in the Punjab region of India (2014).
- On Easter weekend, floods in Durban, South Africa, resulted in at least 70 deaths caused by collapsed buildings, mudslides, and sinkholes (2019).
- After a year of severe drought, seven East African countries received excessive rains over a three-month period, causing the deaths of more than 500 people. In addition to flooding and massive landslides, several dams were breached and failed (2018).
- Hurricane Harvey inundated hundreds of thousands of homes in Houston, Texas, displacing more than 30 000 people and prompting more than 17 000 rescues (2017).

Flooding is exacerbated when the land is altered for development. Trees lessen the danger of erosion by breaking the force of rainfall and by holding slopes firm with their root systems. When slopes are rendered treeless and without other vegetation, significant amounts of silt and pollutants wash into the stream channel and catastrophic landslides become more prevalent. In cities, conversion of natural terrain to rooftops and impermeable surfaces increases the chance of flash flooding while urban contaminants are also washed downstream affecting local ecologies. Temporary catchments, such as detention ponds and rain gardens, can serve to reduce the peak storm flow and nutrient loading while permeable pavements and other low impact development (LID) practices can increase infiltration and reduce runoff. These will be discussed in a later chapter (Chapter 13).

3.5 Droughts and Water Scarcity

Equally dangerous is the ever-present threat of drought in many regions of the world. Drylands are tropical and temperate areas with an aridity index (AI) of less than 0.65. AI is a ratio of precipitation to the potential (maximum) *ET* for a given region (Cherlet 2019). AI values lower than 1 indicate an annual moisture deficit, and an area with AI <0.65 is an area in which annual mean potential *ET* is at least 1.5 times greater than annual mean precipitation. Drylands range from dry subhumid lands (such as Mozambique) to hyperarid lands (deserts, such as much of Egypt and Libya). Climate-change models predict that dryland areas will become drier and less able to support the 1.5 billion people that are currently living in these fragile zones (Putnam and Broecker 2017). Figure 3.6 shows a map illustrating the distribution of these aridity zones for Africa, Europe, Asia, and Australia. The Sahara Desert region of Africa stands out boldly on this map (light beige), serving as the geographic barrier that separates north African nations from Sub-Saharan African countries with their vastly different ethnicities and cultures.

Each continent has had its share of droughts, with accompanying human misery:

- California survived a multiyear drought (2011–2014), which sparked widespread wildfires. This was followed by an El Nino flooding event whose effects were exacerbated by the loss of millions of trees, increasing the destruction done by landslides (Hanak 2016).
- Central America suffered a drought in 2015 at the mercy of an El Nino event that caused flooding elsewhere. More than two million people found themselves in need of immediate food assistance, health care, and nutritional support. This region of vulnerable countries

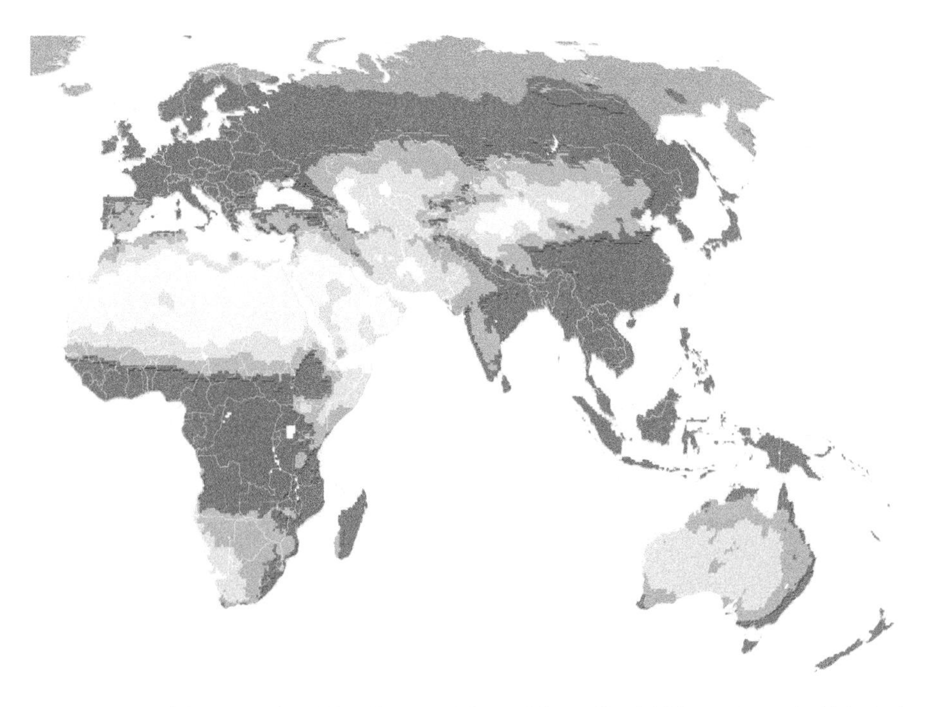

Figure 3.6 Aridity zones for Africa, Europe, Asia, and Australia. Arid (brown), semiarid (beige), and hyperarid (light beige) regions clearly demarcate areas of potential water scarcity. Source: JRC (2018). (Content owned by the EU on this website is licensed under the Creative Commons Attribution 4.0 International [CC BY 4.0] license.)

that are the most prone to drought or extreme precipitation – Guatemala, El Salvador, Honduras, and Nicaragua – is now known as the Dry Corridor (FAO 2017).

- The 2011 drought in East Africa caused a severe food shortage across the region as livestock died and crops failed. Tens of thousands of Somalians fled into Kenya and Ethiopia where they faced malnutrition and disease in crowded migrant camps (HuffPost 2011).
- The year 2019 was the driest in Australia's recorded history. Water for rural residents in New South Wales was carted in by tanker truck and motorcycle. Nonpotable water supply from a nearby lake had been depleted and rainwater tanks had run dry. Aboriginal families struggled to keep alive their cultural practices of hospitality with lack of water (Readfern 2020).
- The 2010–2011 drought in China was the worst to hit the country in 60 years with devastating effects on wheat production. The Hubei Lake in central China dried up to 1/8 of its normal surface area and 1/5 its usual depth, while over four million people faced a drinking water shortage (Watts 2011).

Not all water scarcity can be blamed on natural causes alone. For centuries, Iraqi agriculture relied on water from the Tigris and Euphrates Rivers, which flow through upstream Turkey and Syria. Recent dam construction and accompanying water losses due to evaporation in Turkey and Syria have resulted in a 90 and 47% reduction of Euphrates and Tigris river flow into Iraq, respectively, hampering Iraqi agriculture (Al-Ansari and Knutsson 2011). Built water infrastructure has both local and regional consequences, which must be considered in water security regarding drought.

3.6 Primary Demands on US and Global Water Supply

Water use may be defined as either consumed or withdrawn but not consumed. **Consumptive use** is the removal of water from the environment for a good or service without returning the water to the local surface or groundwater source. For example, when the US Geological Survey (USGS) reports human withdrawals (demand) of fresh water, it is reporting the *consumptive* use of water, water that is withdrawn and not returned to the land-water system. This is water that is either evaporated to the atmosphere, consumed through drinking or other use, or incorporated into a product. **Nonconsumptive use** of water (also called *use by withdrawal*) is discharged back into the environment after usage. This usage includes power plant cooling water or greywater and sewage water from a household that is sent to a wastewater-treatment plant, treated, and then returned to a natural water body. As we have said, this is water that has been withdrawn but not consumed. There is also *non-withdrawal* or "in-stream" uses of water, such as water used for hydropower generation, recreation, navigation, and ecosystem protection and services. These uses will typically demand a minimum quantity of flow and/or depth in order to provide their many benefits and services. The study of water security encompasses all of these types of uses as they are critical to understanding the overall picture of supply and demand.

From a global perspective, much more water is withdrawn by agriculture than any other sector (69%) (Figure 3.7). Industry follows, withdrawing 19% of the water, while domestic household use accounts for 12% of the total. In the uses by agriculture, much of the water is consumed and thus lost to the local hydrology. Industry may consume the water or withdraw and return, for example, by cooling a processing unit. More importantly, the annual water withdrawal globally has been increasing since 1900 at an average rate of about 30 km^3/year. The increases were greatest in the middle decades of the twentieth century (3.0–4.8%/year), but the rate of increased withdrawals has slowed to 0.4% from 2000 to 2010 (FAO 2020).

In the United States, agriculture is a major user of ground and surface water, accounting for approximately 80% of the nation's consumptive water use and over 90% in many Western States. When all uses are considered, thermoelectric power is the biggest user by withdrawal

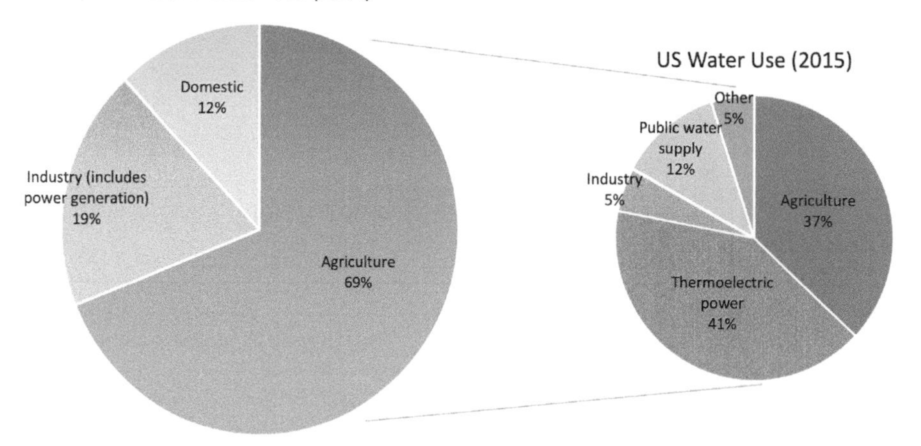

Figure 3.7 Freshwater withdrawals by sector – world water use (Gleick et al. 2018) and US water use (Dieter 2018). Source: Original drawing.

(41.5%), while agriculture is second (37%). Public water supply accounts for only 12% of total freshwater withdrawals (Figure 3.7).

Total water use in the United States, including both freshwater and saline water, has declined since 1980 and in 2010 was at its lowest level in 40 years, at 490 km^3/year. In spite of a growing population, all major water sectors – thermoelectric power generation, agriculture, and municipal/industrial – have shown increased efficiencies and lower water-use intensities (Gleick et al. 2018). In the case of thermoelectric generation, water-intensive once-through cooling has largely been replaced by recirculating or dry cooling. Irrigated agriculture has expanded into more acreage but is using more sprinkler and micro-irrigation methods while reducing flood irrigation. Reductions in municipal/industrial usage have been driven by (i) a shift from water-intensive manufacturing to a service-oriented economy and (ii) greater gains in water-efficiency improvements in response to federal and state standards and incentives. For example, the US EPA's WaterSense program helps consumers pick appliances with better water efficiencies (Gleick et al. 2018).

3.7 Interbasin Transfers (IBTs) of Water

The human-devised transfers of water across watershed boundaries are often called IBTs. These transfers employ various types of water infrastructure – canals, pipelines, and ditches – and are used to distribute water resources according to supply and demand (Dickson and Dzombak 2017). As previously discussed, IBTs were prevalent in ancient Rome when aqueducts carried water across hills and mountains to meet the needs and desires of population centers. These transfers, from one watershed to another, were propelled by gravity, often at very gradual grades, and often from mountain springs (Fagan 2011).

Since the dawn of the Industrial Revolution, water transfers were aided by mechanical pumping in addition to gravity flow through open channels. An example is the Central Arizona Project (CAP), which delivers about 1.5 million acre-feet of water from the lower Colorado River to Central and Southern Arizona every year. Beginning at Lake Havasu, the water is pumped uphill more than 2900 ft in elevation. This complex system is a 336-mile-long system of aqueducts, pipelines, tunnels, and pumping plants, bringing water to more than 80% of the state's population. Along the way, the open canal loses about 16 000 acre-feet of water a year to evaporation, or about 1% of its flow. Because of the enormous energy needed to pump water, the CAP is the largest single user of electricity in the state of Arizona (CAP 2020).

Most of the IBTs in the United States are clustered in the states of Texas, Florida, Southern California, and North Carolina (Dickson and Dzombak 2017). The imported water is used for irrigation, generation of electricity, or municipal water supply. The need for IBTs is not necessarily correlated with climate – note the clusters in both humid and dry climates. Instead, the clusters coincide with areas of high-density population and water-rich agriculture. The number of IBTs in the United States has grown from 256 in 1985–1986 to over 2100 in 2017 (Dickson and Dzombak 2017). The city of Los Angeles, California, operates the largest IBT in the world, moving 2.3 billion gallons of water per day over 240 miles to meet the needs of 13.2 million people (Postel 2017).

Globally, the most significant IBT is the South–North Water Transfer Project in China, designed to transfer 44.8 billion cubic meters of water annually from the Yangtze River in the south to the more arid and populous region in the north (Figure 3.8). The Yangtze basin is a remote, hilly watershed with a typical summer monsoon season that regularly produces

Figure 3.8 The South–North Water Diversion Project of China delivers water from the humid south to the arid north of the country. Source: Qimingfei/Wikimedia Commons.

flooding. The river water is collected or intercepted at three locations (one planned and two in service) for its travel northward via open channel canals and pipelines. Today, when residents in Beijing turn on their tap water, 2/3 of their water is coming from this prolific basin (The Economist 2018).

IBTs are one solution to a growing demand for water in places where water supply is not sufficient. In addition to political disputes over water rights, the practice could result in negative environmental consequences. The exporting basin could retain less flow for maintenance of wetlands and aquatic species, some of which may be endangered. Later, we will consider other solutions for the geographic imbalance between water supply and water demand.

3.8 Conclusion

Water, as opposed to many other necessities in life, has no substitute. Even water of poor quality can sustain human and ecological life for a time, albeit with challenges and limitations. Thus, the quantity of available water is of primary concern to hydrologists, environmental scientists, and water planners in all global areas but especially in regions of recurring drought.

Although there is ample water available on the planet for all of Earth's needs, only a small fraction is freshwater that is easily available. In addition, this water is not distributed evenly across the planet. Thus, large populations have settled in areas that are either prone to water scarcity or are dependent upon a water supply that has been artificially augmented, such as via an IBT. Such an augmentation has the potential for malfunction and for conflict. Water security, in this case, is uncertain and unsustainable.

In this chapter, we have introduced the threat of water quantity – droughts and flooding. In the following chapter, we will consider the correlative problem of inferior water quality. In subsequent chapters, we present responses to both of these threats to water security.

*Foundations: Relevant Design Calculations in Hydrology

The ability to measure stormwater flow and runoff is an essential tool for determining the expected peak and total quantity of flow resulting from a rain event. These calculations will affect both groundwater recharge and water quality as runoff inevitably carries unwanted contaminants in its flow. In this section, we consider methods for determining:

- Design storm,
- Time of concentration,
- Peak stormflow,
- Runoff volume, and
- Detention basin design.

Design storm. A design storm is the storm of a specific intensity, duration, and frequency for a location of interest (Figure 3.9). For example, a six-hour, 100-year storm in Bangkok has an intensity of 34 mm/hour. This would suggest that this rainfall event would drop ~204 mm of precipitation during the storm's duration.

Water infrastructure, such as culverts, bridges, and reservoirs, may be designed to carry or contain a storm of specified design. The information about a design storm can be used to determine peak flowrate and total volume of runoff. Design storm information for the United States and much of the world can be found at NOAA's Hydrometeorological Design Studies Center website (NOAA 2020).

Time of concentration. The time of concentration, t_c, is the time required for a raindrop falling at the farthest hydraulic location in the watershed to reach the outlet (Figure 3.10). The parameter is not subject to direct measurement but is conceptualized as the time from the beginning of the rainfall event to the point when the entire watershed is contributing to runoff at the outlet.

One can use a topographic map (giving slope and distance) along with knowledge of the land use to estimate the time of concentration. Travel time in any watercourse is given by

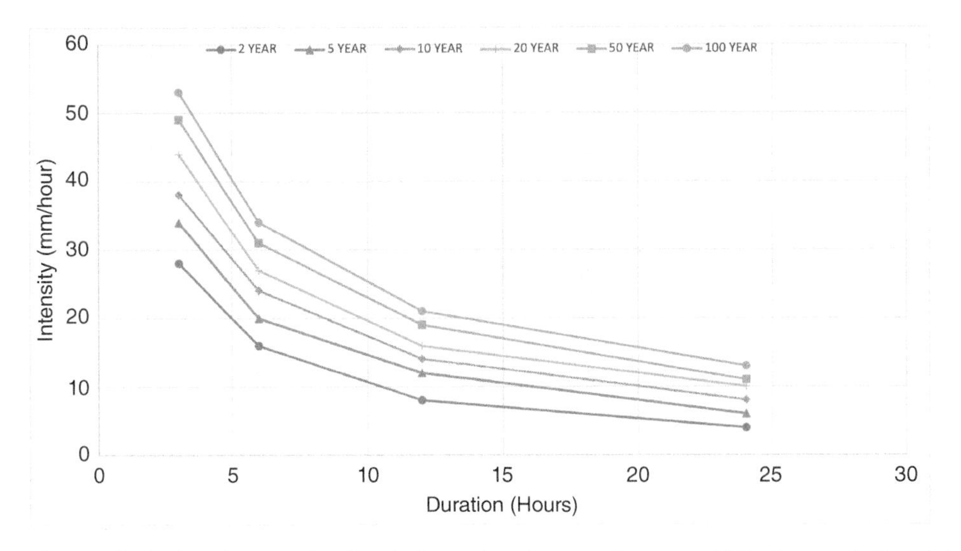

Figure 3.9 Series of curves showing the intensity-duration-frequency (IDF) of storms in Bangkok, Thailand. Source: Original drawing, from Shrestha (2013)/with permission from Ashish Shrestha.

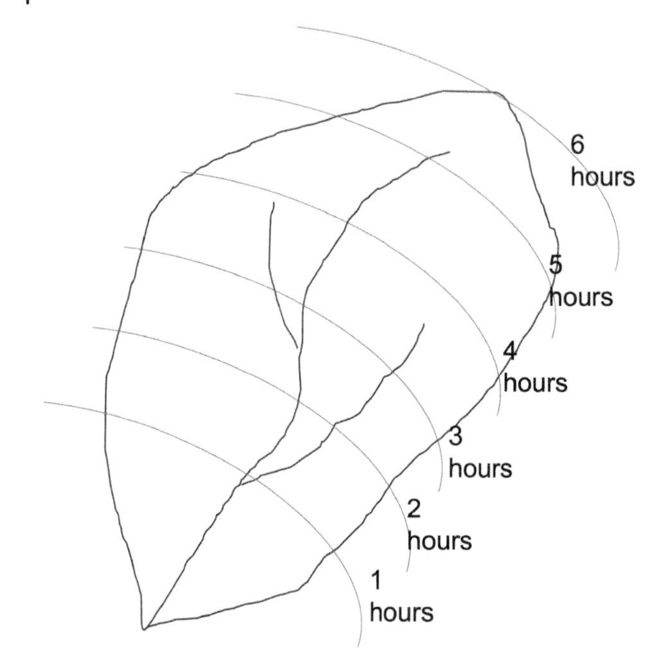

Figure 3.10 Conceptual presentation of the time of concentration for a watershed. In this pictorial, the t_c would be about six hours. Source: JFC.

length divided by velocity, L/v. Thus,

$$t_c = \frac{L}{v} = \frac{L}{3600kS^{0.5}} \tag{3.3}$$

where

t_c = time of concentration (hours)
L = length of longest channel (ft)
k = coefficient, a function of hydraulic radius and roughness of channel
S = slope of channel (%)

Table 3.3 gives representative values for the k coefficient. Note how quickly the k value, and thus the velocity, increases as the flow moves from across rangeland to grassy channel to paved channel flow.

Peak stormflow. The rational equation was developed nearly a century ago for estimating peak flow rates for small urban watersheds (<50 acres). The method contains a few implicit assumptions:

- Rainfall intensity is uniform over the drainage basin during the duration of the rainfall.
- Maximum runoff rate occurs when the rainfall lasts as long or longer than the time of concentration (Schaefer 2019).

The peak flow rate, Q, is given as:

$$Q = CiA \tag{3.4}$$

where

Q = peak flow rate (cfs)
C = runoff coefficient, dimensionless

Table 3.3 Representative values of k coefficient for use in determining the time of concentration, t_c.

Watershed land use	k
Forest:	
Dense underbrush	0.7
Light underbrush	1.4
Grass:	
Bermuda grass	1.0
Short grass	2.1
Rangeland	1.3
Grassed waterway	15.7
Paved area (sheet flow)	20.8
Paved gutter	46.3

Source: Wurbs and James (2001).

i = rainfall intensity (inches/hour)
A = watershed area (acres)

The value for area, A, is straightforward. The value used for rainfall intensity, i, is determined based on the design storm recurrence interval and a duration equal to the time of concentration (t_c) of the watershed. For example, the analysis might be for a six-hour storm event that occurs once every 10 years. The runoff coefficient, C, is a function of soil type, topography, vegetation, and land use. Attempts have been made to refine this parameter more accurately, but the range of values can be quite large (Table 3.4).

One can see from the table that developed areas – with increased pavement and solid structures – will have higher C values and thus higher peak runoff rates. All else being equal, the rational method can be used to compare peak runoff for predevelopment and postdevelopment conditions.

Runoff volume. A common method of estimating watershed runoff is found in the Natural Resources Conservation Service (NRCS) curve number (CN) method:

$$Q = \frac{(P - 0.2S)^2}{(P + 0.8S)} \tag{3.5}$$

where

Q = excess runoff (inches)
P = precipitation (inches)
S = potential maximum retention after runoff begins; a function of the CN:

$$S = \frac{1000}{CN} - 10$$

The CN is a single parameter (similar to runoff coefficient, C, above) that characterizes the watershed according to soil type, topography, vegetation, land use, and antecedent moisture conditions (AMC) – dry (I), average (II), and heavy rainfall within the previous five days (III) (Wurbs and James 2001). Soil groups fall into four hydrologic soil groups depending on their runoff and infiltration characteristics. CN values can be found in tables online, including in the National Engineering Handbook (NRCS 2004).

Table 3.4 Example values of runoff coefficient, C, for agricultural and urban areas.

Type of drainage area	Runoff coefficient, C
Woodland:	Sandy loam/silt loam/clay
Flat (0–5% slope)	0.10/0.30/0.40
Rolling (5–10% slope)	0.25/0.35/0.50
Hilly (10–30% slope)	0.30/0.50/0.60
Pasture:	Sandy loam/silt loam/clay
Flat	0.10/0.30/0.40
Rolling	0.16/0.36/0.55
Hilly	0.22/0.42/0.60
Lawns:	
Sandy soil, average, 2–7%	0.10–0.15
Heavy soil, average, 2–7%	0.18–0.22
Business:	
Downtown areas	0.70–0.95
Neighborhood areas	0.50–0.70
Residential:	
Single-family areas	0.40–0.60
Apartment dwelling areas	0.50–0.70
Industrial:	
Light areas	0.50–0.80
Heavy areas	0.60–0.90
Parks, cemeteries	0.10–0.25
Playgrounds	0.20–0.35
Unimproved areas	0.10–0.30
Streets:	
Asphaltic	0.70–0.95
Concrete	0.80–0.95
Brick	0.70–0.85
Drives and walks	0.75–0.85
Roofs	0.75–0.95

Source: NRCS (2019).

Detention basin design. The tools discussed previously can be used to estimate the required detention volume for a storage basin in order to reduce the peak flow rate from a 100-year design storm to an allowable rate over a specific duration.

Let us assume we have a watershed of 20 acres and our postdevelopment C value (Table 3.4) is 0.90. We have determined that our time of concentration, t_c, is 10 minutes. We also know that the desired maximum outflow rate out of the basin is 8.2 cfs, which is usually some percentage of the predevelopment outflow rate.

Table 3.5 Detention basin design using the modified rational method for a 100-year design storm of various storm durations.

A	B	C	D	E	F	G
Storm duration (min)	Storm intensity (in/hour)	Inflow rate (cfs)	Runoff volume (cf)	Desired outflow rate (cfs)	Outflow volume (cf)	Storage volume (cf)
10	6.8	122.40	73 440	8.2	4920	68 520
15	5.73	103.14	92 826	8.2	7380	85 446
30	4.39	79.02	142 236	8.2	14 760	127 476
60	3.02	54.36	195 696	8.2	29 520	166 176
120	1.9	34.20	246 240	8.2	59 040	187 200
180	1.42	25.56	276 048	8.2	88 560	187 488
360	0.925	16.65	359 640	8.2	177 120	182 520
720	0.592	10.66	460 339	8.2	354 240	106 099

Source: Modified from Schaefer (2019).

The steps are described with references to columns in the results table (Table 3.5).
Steps:

1) Obtain storm duration and intensity data from NOAA website (Columns A, B). Note that intensity decreases with storm duration.
2) Calculate the peak inflow rate for each storm duration (Column C). Use the equation above, $Q = CiA$.
3) Calculate the runoff volume in cubic feet (cf) by multiplying storm duration (Column A) by the inflow rate (Column C) by 60 seconds/minute (Column D).
4) The desired outflow rate is a constant (Column E).
5) Calculate the outflow volume (cf) by multiplying the storm duration (Column A) by the desired outflow rate (Column E) by 60 seconds/minute (Column F).
6) The storage volume is simply the runoff volume (Column D) minus the outflow volume (Column F). The results are given in Column G. For example, the required storage volume for a 100-year storm of two-hour duration would be 246 240–59 040 = 187 200 cubic feet, or roughly 1 acre at an average water depth of 4 ft.

End-of-Chapter Questions/Problems

3.1 In your own words, what is the difference between freshwater withdrawal and consumptive use of freshwater? Give two examples of each.

3.2 Visit the USGS website for the National Water-Use Science Project. Find the Focus Area Studies of the National Water Census.

a. Pick one study of interest.
b. Describe the basin – area, population, activities, etc.
c. Describe the “Water Use” that happens in this watershed of study.

3.3 Use the internet to define the term "available precipitation."
   a. To what value does this correspond in Table 3.1?
   b. Why is this a more useful indicator of water availability as opposed to rain-gauge precipitation?

3.4 Use the US EIA (Energy Information Agency) website to describe the estimated thermoelectric water withdrawals in the United States for the coming years up to 2050. What impact will these withdrawals have on regions of the United States that are predicted to see reduced rainfall in the same years?

3.5 What are some challenges in classifying IBTs? Why has not a national inventory in the United States been conducted since the 1980s?

3.6 What is the hydrologic unit code (HUC) system of watershed classification? Describe the four levels of classification.

3.7 The rational equation, $Q = CiA$, assumes English units – cfs, resulting from in/hour and acres. Using dimensional analysis, show why there need not be a conversion factor when using these units.

3.8 A developer wants to develop a 20-acre site that is currently forested land with 8% slope. The proposed site will have 50% impervious surfaces with the balance of land remaining in forest. Calculate the peak flowrate for pre- and postdevelopment conditions for three design storms of the following frequency and intensities:
   a. 2-year (2.27 in/hour)
   b. 10-year (3.18 in/hour)
   c. 100-year (4.39 in/hour)

3.9 Use the rational method to solve the following problem:
A new 10-acre subdivision in an urban setting has the following characteristics: 1 acre = playground, 2 acres = streets (asphaltic), and 7 acres = pasture (hilly). The soil is heavy clay, and the t_c is 20 minutes. Find the instantaneous 100-year frequency peak discharge for design of a stormwater conveyance channel in this area near Asheville, North Carolina.

3.10 Complete the table that gives information for one important interbasin transfer of water involving each of the following countries or regions. One example is given (Table 3.6).

Table 3.6 Four locations of interbasin transfers of water.

Country/Region	Length/Type of transfer	Quantity (MGD)	Primary purpose(s)
Arizona: Central Arizona project	336 miles/open channel canal from Colorado River to central Arizona	~1400 MGD	Agriculture; water for the city of Phoenix
Florida			
South Africa			
Australia			

Further Reading

Black, M. (2016). *The Atlas of Water: Mapping the World's Most Critical Resource*, 3e. University of California Press.
Gleick, P.H., Cohen, M., Cooley, H. et al. (2018). *The World's Water Volume 9: The Report on Freshwater Resources*. CreateSpace Independent Publishing Platform.
Wurbs, R. and James, W. (2001). *Water Resources Engineering*, 1e. Upper Saddle River, NJ: Pearson.

References

Al-Ansari, N. and Knutsson, S. (2011). Toward prudent Management of Water Resources in Iraq. *Journal of Advanced Science and Engineering Research* 1 (January): 53–67.
Arheimer, B. (2005). Climate change impact on water quality: model results from Southern Sweden - ProQuest. *AMBIO: A Journal of the Human Environment* 34 (7): 559–566.
Black, M. (2016). *The Atlas of Water: Mapping the World's Most Critical Resource*, 3e. University of California Press.
Brooks, K.N., Ffolliott, P.F., and Magner, J.A. (2012). *Hydrology and the Management of Watersheds*, 4e. Ames, Iowa: Wiley-Blackwell.
CAP (2020). "Central Arizona Project: Background & History." https://www.cap-az.com/about-us/background.
Cherlet, M. (2019). "Patterns of Aridity." World Atlas of Desertification. https://wad.jrc.ec.europa.eu/patternsaridity.
Clarke, R. (1991). *Water: The International Crisis*. Taylor & Francis https://www.amazon.com/gp/product/B00AXHD6WO/ref=dbs_a_def_rwt_hsch_vapi_taft_p1_i8.
Dahab, K.A. and El Sayed, E.A. (2001). "A Study of Hydrgeological Conditions of the Nubian Sandstone Aquifer in the Area between Abu Simbel and Toschka, Western Desert, Egypt." In.
Dickson, K.E. and Dzombak, D.A. (2017). Inventory of interbasin transfers in the United States. *JAWRA Journal of the American Water Resources Association* 53 (5): 1121–1132. https://doi.org/10.1111/1752-1688.12561.
Dieter, C.A. (2018). "Estimated Use of Water in the United States in 2015." Circular 1441. USGS. https://pubs.er.usgs.gov/publication/cir1441.
Fagan, B.M. (2011). *Elixir: A Human History of Water*. London: Bloomsbury UK.
FAO (2017). "Chronology of the Dry Corridor: The Impetus for Resilience in Central America | Agronoticias: Agriculture News from Latin America and the Caribbean | Food and Agriculture Organization of the United Nations." http://www.fao.org/in-action/agronoticias/detail/en/c/1024539.
FAO (2020). "AQUASTAT Global Information System on Water and Agriculture." http://www.fao.org/aquastat/en/overview/methodology/water-use.
Gaupp, F., Hall, J., and Dadson, S. (2015). The role of storage capacity in coping with intra- and inter-annual water variability in large river basins. *Environmental Research Letters* 10 (12): 125001. https://doi.org/10.1088/1748-9326/10/12/125001.
Gleick, P.H., Cohen, M., Cooley, H. et al. (2018). *The World's Water Volume 9: The Report on Freshwater Resources*. CreateSpace Independent Publishing Platform.
Hanak, E. (2016). "California's Latest Drought - Public Policy Institute of California." https://www.ppic.org/publication/californias-latest-drought.
HuffPost (2011). "Somalia Food Crisis One Of Biggest In Decades: U.S. State Department Official | HuffPost." https://www.huffpost.com/entry/somalia-food-crisis_n_899811.

JRC (2018). "World Atlas of Desertification, from the European Commission's Joint Research Centre." European Commission. https://wad.jrc.ec.europa.eu/patternsaridity.

Lins, H.F. (2012). *A Note on Stationarity and Nonstationarity*. CHy Advisory Working Group, World Meteorological Organization.

Martin, O.E. and Roberts, T.M. (ed.) (2012). *Rainfall: Behavior, Forecasting and Distribution*. UK ed. Hauppauge N.Y: Nova Science Pub Inc.

Milly, P.C.D., Betancourt, J., Falkenmark, M. et al. (2008). Stationarity is dead: whither water management? *Science* 319 (5863): 573–574. https://doi.org/10.1126/science.1151915.

NASA (2021). "Hydrologic Cycle | Precipitation Education." https://gpm.nasa.gov/education/water-cycle/hydrologic-cycle.

NOAA, National Oceanic and Atmospheric Administration (2020). *Hydrometeorological Design Studies Center*. NOAA's National Weather Service https://www.weather.gov/owp/hdsc.

NRCS (2004). Chapter 9: Hydrologic soil-cover complexes. In: *National Engineering Handbook*, 20. NRCS.

NRCS (2019). *Hydrology Training Series*. US NRCS.

OCS (2020). "Oklahoma Climatological Survey | Precipitation History - Annual, Statewide." http://climate.ok.gov/index.php/climate/climate_trends/precipitation_history_annual_statewide/CD00/prcp/Annual.

Postel, S. (2017). *Replenish: The Virtuous Cycle of Water and Prosperity*. https://smile.amazon.com/Replenish-Virtuous-Cycle-Water-Prosperity/dp/1642830100/ref=sr_1_1?dchild=1&keywords=Replenish+Postel&qid=1593794678&s=books&sr=1-1.

Putnam, A.E. and Broecker, W.S. (2017). Human-induced changes in the distribution of rainfall. *Science Advances* 3 (5): e1600871. https://doi.org/10.1126/sciadv.1600871.

Readfern, G. (2020). "http://Https://Www.Theguardian.Com/Australia-News/Audio/2020/Feb/17/the-Unequal-Cost-of-the-Drought."

Schaefer, L. (2019). "Stormwater Calculations." Presented at the SWMDR Training Day 1, NJDEP Division of Water Quality. https://www.njstormwater.org/pdf/SMDR_Stormwater_Calculations_Slides.pdf.

Schlesinger, B. (2018). "The Fate of Rainfall". https://blogs.nicholas.duke.edu/citizenscientist/the-fate-of-rainfall/

Shrestha, A. (2013). "Impact of Climate Change on Urban Flooding in Sukhumvit Area of Bangkok." https://doi.org/10.13140/RG.2.2.27839.00160.

Sombroek, W. and Sene, E.H. (1993). *Land Degradation in Arid, Semi-Arid and Dry Sub-Humid Areas: Rainfed and Irrigated Lands, Rangelands and Woodlands*. Rome: FAO http://www.fao.org/3/x5308e/x5308e02.htm.

The Economist (2018). "A Massive Diversion - China Has Built the World's Largest Water-Diversion Project | China | The Economist." https://www.economist.com/china/2018/04/05/china-has-built-the-worlds-largest-water-diversion-project.

USGS (2021). "Where Is Earth's Water?". https://www.usgs.gov/special-topic/water-science-school/science/where-earths-water?qt-science_center_objects=0#qt-science_center_objects.

Vogel, R.M., Lall, U., Cai, X. et al. (2015). Hydrology: the interdisciplinary science of water. *Water Resources Research* 51 (6): 4409–4430. https://doi.org/10.1002/2015WR017049.

Watts, J. (2011). "China Bids to Ease Drought with $1bn Emergency Water Aid | Environment | The Guardian." *The Guardian*. https://www.theguardian.com/environment/2011/feb/11/china-drought-emergency-water-aid.

Wurbs, R. and James, W. (2001). *Water Resources Engineering*, 1e. Upper Saddle River, NJ: Pearson.

4

The Context of Water Security – The Quality of Water

The objective of this chapter is to describe the various aspects of water's quality using physical, biological, and chemical characteristics. In a resource-constrained world, water security relies on the proper quality of water for its intended use. Water obtained from natural sources may contain harmful substances. These may fall under various descriptor categories of geogenic or anthropogenic origin, point or nonpoint source, and organic or inorganic contaminants. Water-quality targets may be based on health outcomes, contaminant concentrations, method effectiveness, or technology selection. Epidemiological tools of study are used to describe populations at risk, and analytical tools employ modern techniques of analysis. An overview of water quality sets the background for discussion of water socioeconomic equity in the following chapter.

Learning Objectives

Upon completion of this chapter, the student will be able to:

1. Understand the most important contributors to water quality – physical, biological, and chemical components.
2. Identify commonly used classifications of water based on source and composition.
3. Understand the various water-quality targets that can be used to assess water quality and its potential threat to human or ecological health.
4. Use the tools of epidemiology to detect and quantify outbreaks of water-based diseases on a local and global scale.
5. Understand the various techniques used to track and monitor water quality for the public good.
6. *Perform basic water-quality calculations regarding phase partitioning and disinfection kinetics.

4.1 Introduction

As we have seen, water is essential for life. But it can also be life-threatening. Moreover, the dangerous elements that it carries are often invisible to the naked eye. At its most elementary

Fundamentals of Water Security: Quantity, Quality, and Equity in a Changing Climate, First Edition.
Jim F. Chamberlain and David A. Sabatini.

Companion website: www.wiley.com/go/chamberlain/fundamentalsofwatersecurity

level, water is a polar molecule of two elements, hydrogen and oxygen. Thus, it is sometimes called the "universal solvent" because of its ability to attract and dissolve both ionic (charged) and nonionic substances. These dissolved substances become part of the water's nature and affect the water's quality – its ability to be lifesaving or its potential to be life-threatening.

Even when an ample *quantity* of water is available, the *quality* of water may not be suitable for its intended use. Water that is tainted with human pathogens may be dangerous as a drinking water source at any concentration but acceptable for use in a cooling tower or for irrigation of a golf course when present at low concentrations. Water that is rich in nutrients may cause harmful algal blooms in a surface water body but may actually be beneficial, acting like a fertilizer, when applied to an agricultural field. We speak of water as being "fit for purpose" when it has the necessary quality that makes it favorable, or at least suitable, for its intended use.

When we examine a beaker of glass of clear water, we expect to see a colorless, odorless liquid that is safe. But even a water of such an appearance may contain invisible contaminants that are dangerous, even deadly, for human consumption. Dissolved metals, bacteria, and pesticides are just some of the many contaminants that may be present in an otherwise pristine glass of water. This chapter examines water from the standpoint of its quality, that is, its overall descriptive characteristics, both seen and unseen.

4.2 Characteristics of Water

Rather than producing mandatory standards, the World Health Organization (WHO) promulgates water-quality guidelines that are used as the basis for regulation and standards on a global scale. However, the US EPA has established both mandatory **standards** and suggested **guidelines** to govern public water supply. Maximum contaminant levels (MCLs) are the maximum contaminant concentrations allowed in drinking water. These are in reference to primary drinking water standards regarding chemical and biological characteristics described below. These standards are set as minimum criteria to be met, but individual US states may adopt more stringent standards. The secondary maximum contaminant levels (SMCLs) are not mandatory standards but are guidelines that protect against undesirable taste, color, and odor in drinking water (US EPA 2015b).

These standards and guidelines are based on the various aspects of physical, chemical, and biological water quality as described here.

4.2.1 Physical Aspects

The physical characteristics of water affect its taste, odor, color, temperature, cloudiness (turbidity), and general aesthetic appearance, those qualities that can be assessed by the human senses. Natural groundwater will generally be of high physical quality, although it may have a high dissolved solids content or exude from a high-sulfur zone giving the water an odor of rotten eggs. Surface freshwaters can carry natural organic materials that affect color, odor, and taste, while seawater or brackish water is distasteful due to its dissolved salt content. These physical aspects of water are undesirable and impact the aesthetics of the water, but they do not pose the health risk of other contaminants and are thus regulated by the US EPA secondary drinking water standards, which are further described in Table 4.1.

Table 4.1 EPA's secondary drinking water standards with corresponding secondary maximum contaminant levels (SMCLs) for selected contaminants.

Contaminant	SMCL	Noticeable effects above the secondary MCL
Aluminum	0.05–0.2 mg/l	Colored water
Chloride	250 mg/l	Salty taste
Color	15 color units	Visible tint
Iron	0.3 mg/l	Rusty color; sediment; metallic taste; reddish or orange staining
Odor	3 TON (threshold odor number)	"Rotten-egg," musty or chemical smell
Sulfate	250 mg/l	Salty taste
Total Dissolved Solids (TDS)	500 mg/l	Hardness; deposits; colored water; staining; salty taste
Zinc	5 mg/l	metallic taste

Source: US EPA (2015b).

Interestingly, at lower concentration levels certain compounds in Table 4.1 are desirable while at higher levels they become a nuisance. For example, the stressed body needs the electrolytes (e.g. sodium, potassium, chloride) provided by sports drinks such as Gatorade®, while at higher levels these dissolved solids become a nuisance (does anyone want to drink saltwater?) and potentially dangerous. Low levels of iron can have health benefits (e.g. mitigation of anemia) while at higher levels iron becomes a nuisance by rust-coloring white shirts, tablecloths, and the inside surfaces of bathtubs. One of the authors (DAS) recognizes the irony of touring an iron-removal water-treatment plant in Memphis, TN, while reflecting that his wife was slightly anemic and was at the time on an iron-supplement diet. Thus, secondary standards address "aesthetic" or personal preference issues while primary standards, discussed below, address human health issues.

4.2.2 Chemical Characteristics

The chemical characteristics of interest are often a function of the source but are important for allocation to an appropriate use of the water. Dissolved metals and organic contaminants, such as pesticides, may present a health danger in the use of water for human consumption, and elevated salt contents may prevent the water from being used in crop irrigation. Health effects at levels above national standards may vary from skin and bone damage to kidney disease and death (Table 4.2).

4.2.3 Biological Aspects

The presence of biological pathogens in drinking water can have an enormous impact on global citizens, in terms of both mortality (deaths) and morbidity (illness and disease). Approximately 829 000 people are estimated to die each year from diarrhea as a result of unsafe drinking water, sanitation, and hand hygiene, mostly children under the age of five years and primarily in developing countries who still suffer from lack of universal access to safe water and sanitation (WHO 2021a). For comparison, in the single year 2012, over

Table 4.2 A selection of US EPA national primary drinking water standards with corresponding maximum contaminant levels (MCLs), sources, and health effects.

Contaminant	MCL	Sources	Potential health effects
Inorganic chemicals			
Arsenic	10 μg/l	Erosion of natural deposits; runoff from orchards, runoff from glass and electronics production wastes	Skin damage or problems with circulatory systems, and may have increased risk of getting cancer
Chromium	0.1 mg/l	Discharge from steel and pulp mills; erosion of natural deposits	Allergic dermatitis
Fluoride	4.0 mg/l	Water additive that promotes strong teeth; erosion of natural deposits; discharge from fertilizer and aluminum factories	Bone disease (pain and tenderness of the bones); children may get mottled teeth
Lead	15 μg/l	Corrosion of household plumbing systems; erosion of natural deposits	Infants and children: Delays in physical or mental development; children could show slight deficits in attention span and learning abilities Adults: Kidney problems; high blood pressure
Mercury (inorganic)	2 μg/l	Erosion of natural deposits; discharge from refineries and factories; runoff from landfills and croplands	Kidney damage
Nitrate (measured as nitrogen)	10 mg/l	Runoff from fertilizer use; leaking from septic tanks, sewage; erosion of natural deposits	Infants below the age of six months who drink water containing nitrate in excess of the MCL could become seriously ill and, if untreated, may die. Symptoms include shortness of breath and blue-baby syndrome.
Organic chemicals			
Alachlor	0.002 mg/l	Runoff from herbicide used on row crops	Eye, liver, kidney, or spleen problems; anemia; increased risk of cancer
Atrazine	0.003 mg/l	Runoff from herbicide used on row crops	Cardiovascular system or reproductive problems
Glyphosate	0.7 mg/l	Runoff from herbicide use	Kidney problems; reproductive difficulties
Polychlorinated biphenyls (PCBs)	0.0005 mg/l	Runoff from landfills; discharge of waste chemicals	Skin changes; thymus gland problems; immune deficiencies; reproductive or nervous system difficulties; increased risk of cancer
Toluene	1.0 mg/l	Discharge from petroleum factories	Nervous system, kidney, or liver problems

Table 4.2 (Continued)

Contaminant	MCL	Sources	Potential health effects
Trichloroethylene	0.005 mg/l	Discharge from metal degreasing sites and other factories	Liver problems; increased risk of cancer
Radionuclides			
Radium 226 / 228 (combined)	5 pCi/l	Erosion of natural deposits	Increased risk of cancer
Uranium	30 μg/l	Erosion of natural deposits	Increased risk of cancer, kidney toxicity

Note: Complete standards may be found at US EPA (2015a).

367 000 deaths in Sub-Saharan Africa were attributed to diarrhea from inadequate water, sanitation, or hygiene (WaSH), as compared to only 11 500 similar deaths in the United States for the same year (Black 2016). Equally disturbing are the untold number of work and school days lost due to intestinal illness. All of these instances of mortality and morbidity are preventable with clean drinking water and proper sanitation.

In the United States, the water-quality goal – a maximum contaminant level goal (MCLG) – is to have no biological pathogens present in the drinking water supply. As Table 4.3 shows, if an MCLG cannot be reached, a specified treatment technique (TT) is required to reduce the level of a pathogen in drinking water as low as possible.

Biological pathogens that may be transmitted through water (waterborne) can be categorized into three main classes, which are – in order of increasing size and complexity – viruses, bacteria, and protozoa. *Viruses* are the smallest and simplest in this group, ranking in size from 20 to 100 nm (0.1 μm). They must infect and inhabit a living cell to reproduce and are moderately susceptible to chlorination. *Bacteria* are more complex, being able to survive on their own as single-celled organisms, and are larger than viruses, being on the order of 0.3–1.0 μm in size. They exist in virtually every ecosystem, living in the intestines of warm-blooded animals to aid in digestion, and of the three classes are most susceptible to chlorination. Many bacteria are critical to human and animal life, with only a fraction being pathogenic – classic examples of pathogenic bacteria include those that cause typhoid fever and cholera. *Protozoa* are the most complex and largest of the three classes, being 2–50 μm in size, and are sometimes referred to as parasites. Protozoa are able to form cysts and oocysts; their protective coating makes them more resilient against drying, temperature changes, etc. Protozoa are the least susceptible to chlorination, but are highly susceptible to UV disinfection, as are viruses and bacteria. *Giardia* and *cryptosporidium* are examples of pathogenic protozoa that can occur even in pristine mountain streams due to beaver, deer, and muskrat activity (Hrudey and Hrudey 2010).

Having introduced these three categories of waterborne pathogens, we will highlight four specific waterborne pathogens of concern. The first is *Escherichia coli*, a vital bacterial component of the intestinal flora of warm-blooded animals. While the vast majority of *E. coli* strains are nonpathogenic, several strains are pathogenic and can cause profuse, watery diarrhea, cramping, and vomiting. The incubation time (time between exposure and onset of symptoms) is one to three days and the infective dose (level above which exposure can lead

Table 4.3 A selection of US EPA national primary drinking water standards as pertaining to pathogenic microorganisms.

Contaminant	MCLG (mg/l)	MCL or TT (mg/l)	Sources of contaminant in drinking water	Potential health effects from long-term exposure
Cryptosporidium	Zero	TT	Human and animal fecal waste	Gastrointestinal illness (such as diarrhea, vomiting, cramps)
Heterotrophic plate count (HPC)	n/a	TT	HPC measures a range of bacteria that are naturally present in the environment; this would include bacteria that cause cholera and typhoid fever	No health effects; an analytic method used to measure the variety of bacteria. The lower the concentration of bacteria in drinking water, the better maintained the water system is.
Giardia lamblia	Zero	TT	Human and animal fecal waste	Gastrointestinal illness (such as diarrhea, vomiting, cramps)
Legionella	Zero	TT	Found naturally in water; multiplies in heating systems	Legionnaire's disease, a type of pneumonia
Total coliforms (including fecal coliform and *E. coli)*	Zero	No more than 5.0% samples total coliform-positive in a month	Coliforms are naturally present in the environment; as well as feces; fecal coliforms and *E. coli* only come from human and animal fecal waste	Not a health threat in itself; it is used to indicate whether other potentially harmful bacteria may be present
Viruses (enteric)	Zero	TT	Human and animal fecal waste	Gastrointestinal illness (such as diarrhea, vomiting, cramps)

"MCLG" = maximum contaminant level goal, "TT" = treatment technique
Source: US EPA (2015a).

to infection and symptoms) is 10^6–10^{10} organisms, depending on the health of the receptor (person exposed) and other environmental factors. Exposure levels below the infective dose generally do not cause an infection (Hrudey and Hrudey 2010).

The second waterborne pathogen is also a bacterium – *Salmonella typhi*, the agent responsible for typhoid fever. While it is less pervasive than in the nineteenth and early twentieth centuries, typhoid fever still exists today, with a global toll of 11–20 million cases and up to 161 000 deaths annually (WHO 2021c). *S. typhi* has an incubation time of 12–36 hours and an infective dose of 10^4–10^9 organisms, smaller than for *E. coli*.

Another bacterium of great global concern is *Vibrio cholerae*, the cause of the acute diarrheal infection known as cholera. The bacteria are found in water polluted by human feces; therefore, cholera rates are a measure of the inequities of sanitation and human development. Every year there are still 1.3–4.0 million cases of cholera, with deaths due to cholera in the range of 21 000–143 000 annually (WHO 2020). Not all people infected with cholera will show symptoms, but the bacteria can be present in the host's feces for up to 10 days after infection. Sufficient quantities of clean drinking water can be used to treat and subsume an epidemic of cholera in the event of a breakout.

Finally, *cryptosporidium* is a protozoan that was first recognized in the late 1900s, with humans, cattle, birds, and reptiles as potential hosts (including beavers, deer, muskrats, as mentioned above). Symptoms include watery diarrhea, cramping, abdominal pain, and fatigue, with an incubation time of 1–12 days and an infective dose of as few as 30 oocysts. Symptoms typically last 30 days – assuming good health, diet, and medical care available for recovery. Those victims who lack the above, or especially the elderly and immunodeficient, are highly vulnerable and the disease may be fatal (Hrudey and Hrudey 2010).

Having discussed these pathogens and potential morbidity or mortality, the obvious question is how to prevent, or at least limit, the degree of exposure to these pathogens. This leads us to a discussion of the well-known F-diagram, as illustrated in Figure 4.1. This diagram effectively illustrates potential pathways from pathogen-laden feces to potential future victims. Pathways include fluids (e.g. groundwater or surface water contaminated by human or animal feces), fingers (that come into direct contact with feces or indirect contact by touching something previously contacted by someone else), fields (e.g. crops fertilized with animal or human waste that has not been sterilized), or flies (flies coming into contact with feces and then landing on food). These are all potential pathways for pathogen transport from infected-person-feces to a future victim, and the transmission repeats itself again and again, unless the cycle is somehow broken. Coauthor Sabatini (DAS) recalls a visit with his personal physician after being infected with *cryptosporidium*. In this instance, the doctor used the F-diagram to assure his wife that, because of good sanitation and hygiene, she was safe from being infected! In addition to potential pathways, Figure 4.1 also demonstrates protective barriers to break the cycle, showing how water purification (clean water), adequate sanitation, and hygiene practices can stop the spread of the disease. Notice that these three lifesaving barriers – water, sanitation and hygiene – make up the acronym WaSH that is used to refer to this sector in the health field. Thus, while Figure 4.1 appears to be a fairly simple diagram, the concepts imbedded in this figure are critically important, providing a roadmap to saving countless lives!

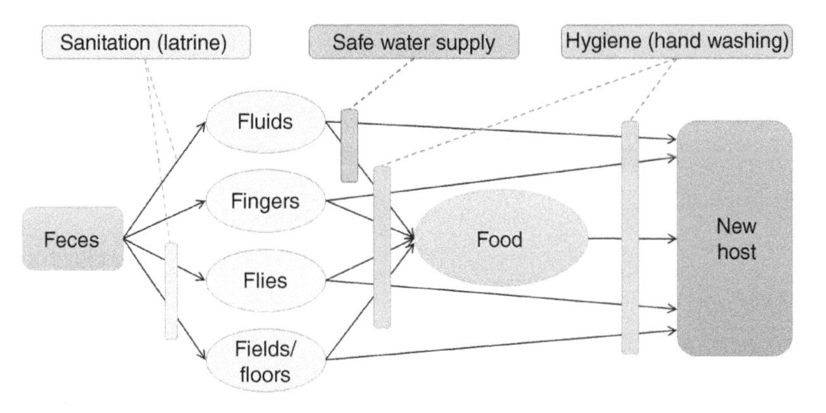

Figure 4.1 The F-diagram, illustrating pathways of pathogen transfer from feces to future victim. Source: Original drawing, adapted from Cairncross and Feachem (2011).

4.3 Geogenic and Anthropogenic Classifications

In addition to describing characteristics of water, it is often helpful to discuss water quality in terms of major, often binary, classifications. Two broad categories – geogenic and anthropogenic – are based on source, while further classifications are based on level of diffusion and composition.

Waterborne contaminants can be divided into *geogenic* and *anthropogenic*, that is, those that originate from the earth (*geo*) and those that can be attributed to the actions of humans (*anthro*). Geogenic contaminants, such as arsenic and fluoride, exist in natural *background* levels that may or may not be harmful to humans based on the concentration and levels of exposure. Anthropogenic contaminants, such as lead and mercury, nitrates from fertilizers, pesticides, and biological pathogens from human waste are likewise site-specific but can be traced to human behavior and human–water interactions. In both cases, actions are necessary to eliminate the contaminant source or to mitigate the risk.

4.3.1 Geogenic Compounds

When either surface water or groundwater is used as a drinking water source, arsenic and fluoride are the two geogenic contaminants that pose the greatest health burden (Howard and Bartram 2003). These two dissolved contaminants may be present in drinking water with no discernible effect on the taste or appearance of the water. Rural villagers in developing countries, such as Bangladesh and Ethiopia, are often faced with the difficult choice of ingesting tubewell water that may have elevated arsenic or fluoride versus water from a pond or stream that is likely infested with biological pathogens capable of causing intestinal diseases. The presence of either may occur without visual detection. Because the chronic effects of both arsenic and fluoride are often seen only after years of consumption, villagers may well choose the tubewell water, clear in appearance, and risk the associated danger (Figure 4.2). Conversely, they may choose the surface water based on convenience alone,

Figure 4.2 This household hand pump is supplying water from a Bangladeshi tubewell (borehole) tainted with arsenic. See discussion of Bangladesh in Chapter 2. Source: Photo courtesy of DAS.

without factoring in the health consequence. The burden for them is "Which water is less unsafe to drink?" although the degree of this burden is beyond what most people in affluent nations can fully comprehend.

Arsenic and fluoride both have their own prevalent geology in which they are naturally found. However, when both contaminants occur simultaneously in groundwater research has shown a correlation with volcanic eruptions, geothermal currents, and/or mining activities (Alarcón-Herrera et al. 2013). Figure 4.3 shows modeled predictions of the probability of geogenic arsenic globally. The well-known contaminated regions of southeast Asia (parts of India and Bangladesh) correspond well with the model (Amini et al. 2008a). The vast majority of people exposed to high levels of As (> 10 μg/l) are using water from alluvial aquifers (Ravenscroft 2006), that is, shallow aquifers that are influenced by surface river waters, such as the middle Yellow River in China and the Rift Valley of Ethiopia. Other types of aquifers in which they are found include geological strata in which bacteria-induced reducing conditions or pH conditions liberate arsenic from minerals with which it is bound. A classic example is iron oxide minerals, in which reducing conditions dissolve the iron oxide and associated arsenic while elevated pH liberates arsenic from the mineral surfaces that remain intact (Ravenscroft 2006).

While one could die in a few hours from acute levels of arsenic poisoning, those concentrations are beyond normal groundwater levels and the more common health effect is mortality and/or morbidity by *chronic arsenic poisoning,* which occurs over a longer period of time (months, years, or decades, depending on arsenic levels, dietary, antecedent health conditions, etc.). Historians suggest that Napoleon Bonaparte died of chronic arsenic poisoning during his exile on the island of Saint Helena, a theory supported by arsenic concentrations in his hair measured at 13 times the normal level (Whorton 2011). Thus, below we examine the health effects resulting from lifetime consumption of drinking water tainted with toxic levels of arsenic. Early health effects of arsenic are manifested as black spots on the upper body, bronchitis and loss of bodily sensation (Figure 4.4). In more serious cases, high levels of exposure result in swollen legs, renal malfunction, and cancer (Salzman 2017). The lethality of arsenic is seen clearly in the units of its drinking water standard. Whereas many chemical contaminants have standards in the range of parts per million (mg/l), the arsenic standard is 1000 times smaller, in the range of parts per billion (μg/l). The United States established an arsenic standard of 50 ppb in 1942, decades before the EPA was established. Based on

Figure 4.3 Modeled global probability of geogenic arsenic contamination in groundwater. Source: Amini et al. (2008a)/with permission from American Chemical Society.

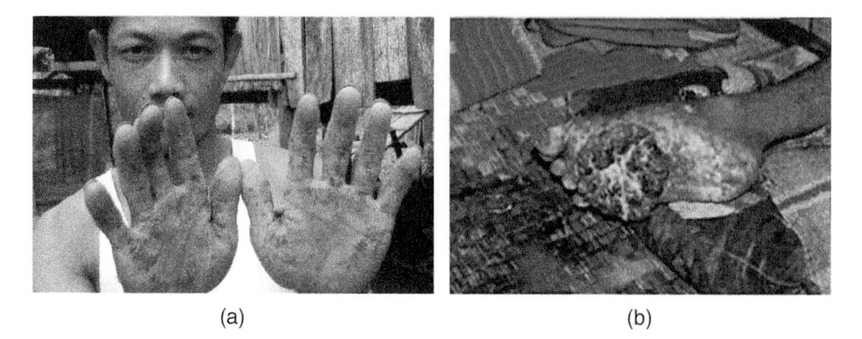

Figure 4.4 Images of devastating effects of consumption of arsenic in drinking water, Cambodia – mild skin keratosis (a) and severe foot lesions (b). Source: Photos courtesy of Marc Hall.

increasing evidence of the harmful effects at even lower levels from around the world, a more stringent standard of 10 ppb was adopted in 2001, matching the guideline set by the WHO (WHO 2011).

A second prevalent geogenic element is fluoride, an element that occurs in both surface waters and groundwater in geologies that facilitate fluoride mineral dissolution (Apambire et al. 1997) (Figure 4.5). In the case of fluoride, toxicity is very much a function of the dose. In much of the developed world, fluoride has traditionally been added to public drinking water supplies as a supplement for reducing cavities and dental decay. The range for this addition is usually 0.5–1.5 mg/l, although the practice is losing favor as there are other sources of fluoride in a typical modern diet (Freeze and Lehr 2009).

In other parts of the world, endemic levels of fluoride in the range of 1.5–5 mg/l can cause mottled teeth, while skeletal fluorosis is a debilitating condition that occurs over a lifetime of drinking water with fluoride concentrations in excess of 5 mg/l (Figure 4.6). With this condition, agricultural laborers may find it difficult to raise their hands up to eye level and above. This makes manual labor very difficult in regions where such labor is required for making a living or for raising sufficient food for a household.

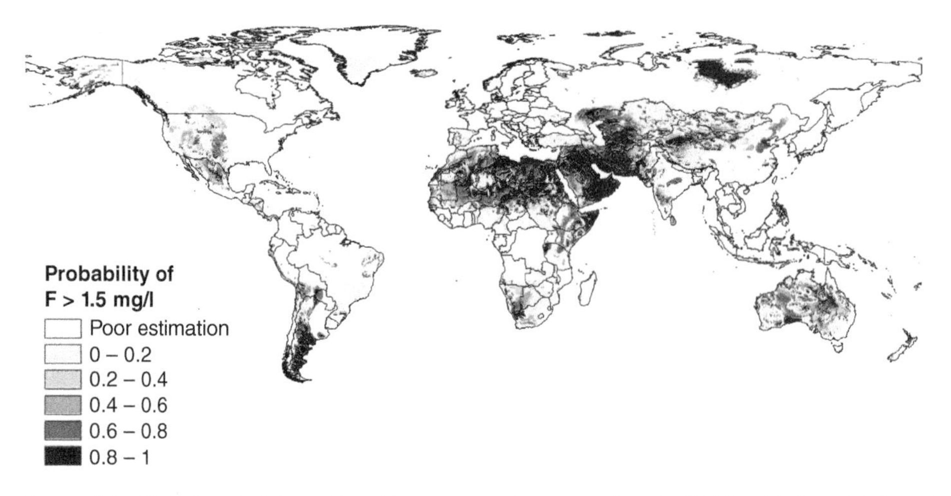

Figure 4.5 Global probability of fluoride concentration exceeding the WHO guideline of 1.5 mg/l in drinking water. Source: Amini et al. (2008b)/with permission from American Chemical Society.

Figure 4.6 Images of effects of consumption of excess fluoride in drinking water, Ethiopia. (a) Dental fluorosis, woman with author JFC; (b) Skeletal fluorosis in India. Sources: (a) Courtesy of author JFC; (b) Adrian Page/Alamy Stock Photo.

4.3.2 Anthropogenic Sources of Contamination

Many human activities – whether in energy production, mining, agriculture, industry, or domestic households – withdraw freshwater and return it to the environment in a damaged form by carrying unwanted contaminants. These contamination sources may be categorized in several ways:

- Point and nonpoint sources
- Organic and inorganic compounds
- Human and animal excreta

Point and nonpoint sources. Point sources are sources of pollution that enter the environment via a single point of entry, such as a pipe carrying industrial waste to a stream. Both the flow and concentration of point sources are more easily measured and thus more easily regulated. In the United States, the US EPA administers a permit program that monitors and regulates the discharge of pollution from these sources (US EPA 2010).

Nonpoint sources are more diffuse, such as stormwater flowing through city streets or rainfall runoff spreading across a farmer's fields. Because of their diffuse nature, these latter sources are more difficult both to measure and to control. As a result of both point and nonpoint pollution, over half of China's groundwater and surface water sources (lakes and reservoirs) are too polluted to be used for anything other than agriculture or industrial use (Figure 4.7).

Organic and inorganic compounds. A second categorization is between *organic* (carbon-containing) and *inorganic* contamination (Figure 4.8). One source of *organic* chemicals is

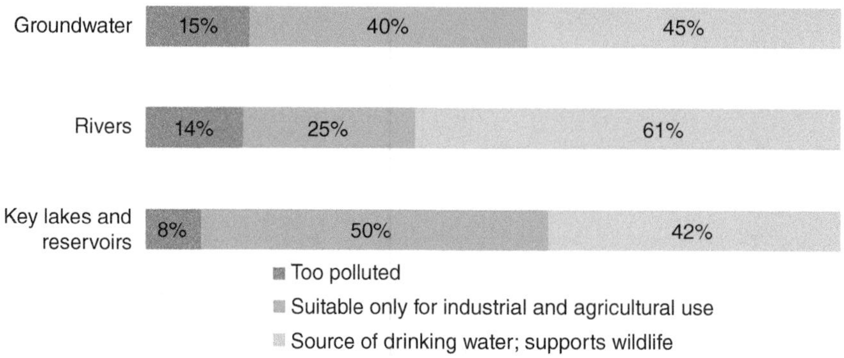

Groundwater 15% 40% 45%

Rivers 14% 25% 61%

Key lakes and reservoirs 8% 50% 42%

Too polluted
Suitable only for industrial and agricultural use
Source of drinking water; supports wildlife

Figure 4.7 The state of China's water supply is showing gradual signs of improvement, but less than half of lakes and groundwater aquifers are suitable for drinking water and the support of wildlife. Source: Original, modified from Black (2016).

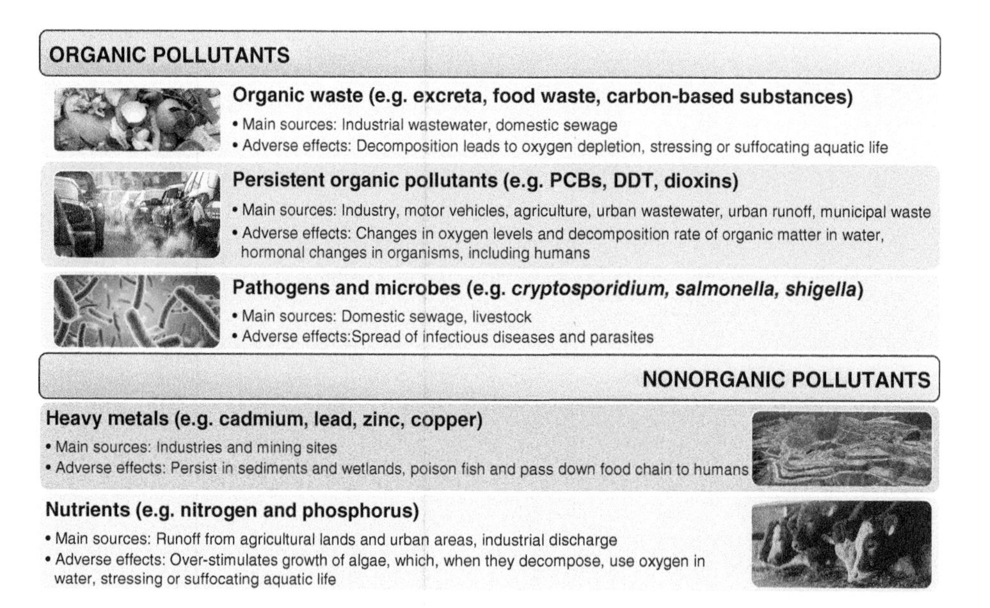

Figure 4.8 Organic and nonorganic pollutants that arise from human activities. Source: Modified from Black (2016).

domestic sewage, which, compared to industrial wastewater, is fairly similar from location to location. Domestic sewage biodegrades naturally over time by aquatic bacteria. Unfortunately, such biodegradation, if concentrated enough, can overwhelm natural aquatic systems and deplete dissolved oxygen in the receiving waters. Another source of organic chemicals is industrial wastewater, the composition of which is highly dependent on the industry. A slaughterhouse will have a concentrated but fairly biodegradable wastewater, while a pharmaceutical or pesticide industry will discharge a more complex and varied wastewater with varying degrees of biodegradability.

Inorganic contaminants include heavy metals (e.g. cadmium, zinc, copper, lead) and nutrients (e.g. nitrogen, phosphorous). Heavy metals enter waterways from mining activities and industrial discharges (e.g. car or aircraft manufacturing, metal plating operations) while nitrogen and phosphorous enter waterways through industrial discharges (point) and fertilizer application (nonpoint) sources, potentially impacting the aquatic ecosystem as well as human health (see Figure 4.8).

Global food production as currently practiced uses 70–90% of locally available freshwater and returns much of this water to nearby streams and rivers, adding excess *nutrients* (nitrogen and phosphorus) that can result in downstream algal blooms. When these aquatic organisms die, they become abundant food for bacteria, which also consume the dissolved oxygen in the water body, creating anoxic conditions and fish kills in sensitive areas, such as the Mississippi Gulf coast region. Globally, nitrogen effluents are projected to grow from 6 to 14 million tonnes between the years 2000 and 2030 (Black 2016). In addition, approximately 60% of global ecosystem services are being degraded or used unsustainably. These services include fresh water, capture fisheries, air and water purification, the regulation of regional and local climate, and the control of pests (Millennium Ecosystem Assessment [Program] 2005).

Human and animal excreta. Finally, *human and animal excreta* are a dominant threat to water quality across the world, as 2.6 billion people still live without improved sanitation, defined as that which ensures the hygienic separation of human excreta from human contact (WHO 2021b). It includes connection to a piped sewer system, septic tank, or pit latrines. Further, up to 90% of the world's wastewater flows untreated into densely populated coastal zones, augmenting the marine dead zones of fish kills (Corcoran and GRID--Arendal 2010). As a result, these untreated wastewater flows become a threat to food security for those who depend on fishing for their food and livelihood. In addition, polluted streams disrupt and destroy the diverse ecosystems.

What are the consequences of water-quality degradation? In addition to human health impacts, severe environmental impacts can also occur. For example, freshwater ecologies are especially abundant in diverse populations but are also sensitive to water quality. These require an inflow of clean rainwater to maintain these habitats (Black 2016). Changes in water quality (e.g. temperature, salinity, chemical composition including those toxic to the biota) that alter the delicate ecosystem equilibrium can threaten and, ultimately destroy, a native ecosystem. Likewise, water-quality changes that promote the presence of invasive plants or animals can threaten an ecosystem. Further, dam and reservoir construction, as well as other water projects, can greatly alter the indigenous ecosystem.

Figure 4.9 shows a sample of freshwater endangered species, their status and trends based on International Union for Conservation of Nature (IUCN) assessments. The importance of these ecosystems cannot be overstated, both as part of the global natural system and for their benefit to mankind. These benefits will be discussed in greater detail in a later chapter.

4.4 Water-Quality Targets

In order to determine the need for and effectiveness of water-quality interventions, suitable water-quality targets must be established and measured for compliance. The WHO gives the following four main classes of health-based targets (WHO 2017):

1. Health outcome targets

ENDANGERED SPECIES	Percent accessed as threatened (IUCN)
Waterbirds More than 800 species of bird (ducks, geese, herons, etc.) depend on wetlands for part of their cycle. Migratory birds travel thousands of kilometers, using wetlands as vital stopping points. Populations of more than half of these birds are declining.	17%
Water-dependent mammals A range of mammals, including species of dolphin, otter, seal, shrew, and hippopotamus, live or feed in fresh water. Of those assessed, over a third are threatened.	38%
Freshwater amphibians Species of amphibians that are associated with flowing water have a higher likelihood of being threatened than those that use still water.	26%
Freshwater turtles Of the 229 species assessed, 32 are considered critically endangered. Around 3/4 of Asia's species are threatened due to the high value of their medicinal qualities.	59%
Crocodiles Of the 23 species that inhabit marshes, swamps, rivers, lagoons and estuaries, 7 are considered critically endangered.	48%
Freshwater fish A high proportion are considered threatened with extinction. This is often prompted by the introduction of non-native species.	33%
Freshwater crayfish and crabs Many freshwater crayfish and crabs are threatened, not only by pollution and loss of habitat, but also by invasive species that actively predate them, or out-compete them for food.	31%

Figure 4.9 List of endangered species showing the importance of water quality in maintaining diverse, healthy ecosystems. Source: Modified from Black (2016).

2. Water-quality targets
3. Performance metrics
4. Specific technology targets

These are further described in Table 4.4.

One commonly used *health outcome target* is the **disability-adjusted life years (DALYs)**, defined as the number of years lost due to illness and/or death:

$$\text{DALYs} = \text{YLL (years of life lost)} + \text{YLD (years lived with a disability or illness)}$$

This metric thus includes both acute (YLL) and chronic (YLD) impacts of a disease and is thus a combined measure of both mortality and morbidity. The higher the DALY, the greater the overall burden from that disease. A severity factor for the particular symptoms of the disease is used in calculating the years lost due to disability (YLDs). For example, an infection with the *Cryptosporidium* pathogen (*C. parvum*) causes watery diarrhea. Based on data from the Global Burden of Disease (GBD) project, the severity factor of watery diarrhea is 0.067 (as compared to death = 1) and the average duration is seven days in the case of a *C. parvum* infection (Havelaar and Melse 2003). In addition, the infection causes death in very young children in 0.015% of cases. Thus, in the case of *Cryptosporidium*,

$$\text{YLL (per 1000)} = 1000 \times 1 \times 0.9\ \text{years} \times 0.00015 = 0.135\text{YLD (per 1000)}$$

$$\text{YLD (per 1000)} = 1000 \times 7\ \text{days}/365\ \text{da/year} \times 0.067 = 1.34\text{DALY (per 1000)}$$

$$\text{DALY (per 1000)} = \text{YLL} + \text{YLD} = 1.47$$

Table 4.4 Nature, application, and examples of four health-based targets for water quality.

Type of target	Nature of T = target	Typical applications	Example(s)
Health outcome	Defined tolerable burden of disease	High-level policy target	No greater than 10^{-6} (1 per million) DALYs per person per year
Water-quality guideline	Guideline values – based on individual chemical risk assessments	Chemical hazards	Arsenic, 10 µg/l Fluoride, 1.5 mg/l
Performance metric	Specified removal of hazards, based on quantitative microbial risk assessment (QMRA) or chemical guideline values	Microbial hazards (log-reductions) Chemical hazards (percentage removals)	10-log removal of biological pathogens
Technology target	Defined technology(ies) – set at national level	Control of microbial and chemical hazards	Best available technology (BAT)

"DALY" = disability-adjusted life year
Source: Modified from WHO (2017).

Table 4.5 shows a comparison of various waterborne diseases using the metric of DALYs. The table shows that *E. coli* is a bigger killer than *C. parvum* and that rotavirus is much more prevalent and dangerous in low-income countries.

DALYs provide a metric of comparison for use in determining priorities for health interventions and assistance. In the WHO guidelines, a tolerable burden of disease corresponds to a single excess case of cancer per 100 000 people who are drinking water at the water-quality target over a 70-year period (WHO 2017). Use of the DALY is limited, however, due to large gaps in the knowledge needed for its calculation.

Water-quality guideline values are more direct descriptors and are based on risk assessments of the health effects from exposure to chemicals in water. These guidelines are based on a no-observed-adverse-effect level of the contaminant and include other assumptions, such as toxicity of the chemical, daily ingestion rate, body weight, and time period of ingestion. National targets may differ appreciably from the guideline values, since many chemicals are present as well in other sources, such as food, air, and consumer products, such as lead pipes

Table 4.5 Summary of disease burden for various drinking-water contaminants.

	Disease burden per 1000 cases		
Contaminant	YLD	YLL	DALY
C. parvum	1.34	0.13	1.47
E. coli O157	13.8	40.9	54.7
Rotavirus – high-income countries	2.0	12	14
Rotavirus – low-income countries	2.2	480	482

Source: Adapted from Havelaar and Melse (2003).

or paint. Thus, if there is another source of a chemical, then the guidelines may be adjusted lower for this chemical in drinking water.

Performance metrics are often used for microbial hazards in distribution systems. These targets assist in the selection of treatment options and control measures that prevent the passage of pathogens, such as bacteria, viruses, and protozoa. The targets are normally expressed in terms of log reductions. For example, a 3-log reduction results in a final count that is $1/10^3$ ($= 10^{-3}$) as large as the original count. Reduction also implies inactivation of the microorganism, so that a 3-log inactivation rate means that 99.9% of the pathogens are rendered inactive.

Technology targets are recommendations of specific technologies to be used in certain circumstances. Often the recommendation is the best available technology (BAT), the most effective technology based on assessments of source water type (e.g. surface water or groundwater) and water quality. An example might be the required use of a chlorine-disinfection system that leaves a residual in the distribution system. These technology targets can be used to address water with either microbial or chemical hazards.

All of these health-based targets may be used in the formation of water safety plans, which will be discussed in a later chapter (Chapter 14).

4.5 Epidemiological Tools of Study

"On warm summer nights, the heat in their cramped London flat was unbearable, and the father would often wake after midnight and send one of the boys out to fetch some cool well water to combat the sweltering air . . . The tailor and his 12-yr old boy had been struck in the first hours of the outbreak. By Saturday, both were dead." This passage from *The Ghost Map* describes the speed with which cholera killed its victims, as a result of simply drinking cool well water (Johnson 2007). The root cause of this tragic epidemic was discovered by the London physician, John Snow, considered now to be the "Father of Epidemiology" and discoverer of the waterborne disease pathway.

Officially, **epidemiology** is the study of health outcomes (positive and negative) in specified populations to apply this knowledge for the betterment of public health. In the case of water, epidemiology pays detailed attention to the occurrences of waterborne diseases in affected populations to determine prevalence of disease or, positively, the efficacy of interventions.

We use here a case study to introduce several of the most common terms and concepts used in epidemiological studies. Walkerton, Ontario, is a peaceful farming community of 4800 residents, nestled between Lake Huron and Lake Erie. The town normally relied on cool, clear, pleasant-tasting groundwater from a group of wells that received very small doses of chlorine to meet regulations (Morris 2008). In the year 2000, a series of mishaps culminated in a perfect storm of public water supply contamination. A construction project during a heavy rainstorm coincided with an uninstalled chlorinator on one of the city's wells. This well pumped water from an aquifer that was hydrologically connected, especially during extreme rainfall events, to the fields of a farmer who used horse and cow manure to fertilize his freshly planted rows of corn. An *E. coli* outbreak resulted in 2300 serious illnesses and 7 deaths. Under normal circumstances one can speak of a low *prevalence* of *E. coli* outbreaks in the general population, that is, a low regular rate of occurrence. If we assume that the prevalence rate in Canada is similar to the United States, then the annual infection rate is ~8 per 10 000

(0.08%) (CNN 2020). When the outbreak occurred in May of 2020, there was an *incidence rate* of 2300 cases per 4800 residents, or a rate of nearly 480 per 1000 per month! Of the 2300 cases who contracted the illness, seven died for a *case-fatality ratio* of 7/2300 = 0.3%. These three measures of occurrence – prevalence, incidence rate, and case-fatality ratio – can be used to compare the severity of waterborne outbreaks in an affected population.

Most diseases are present to some extent in a range of populations and exhibit a prevalence rate, a kind of background measure. If an outbreak is established and is well above the prevalence rate, then we might say that an *epidemic* is occurring. An example is the 2010 cholera epidemic in Haiti that killed over 8000 people. If the epidemic is spread over several regions and nations, then it becomes a *pandemic*, as in the pandemic of coronavirus of 2020–2022.

Epidemiological studies seek to establish a causal relationship between two factors, e.g. between illness and a water contaminant. The study begins with evidence of a *correlation* between the determinant and the outcome, taking into account potential *confounding factors*. Further analysis will confirm whether or not there is cause and effect (*causation*) due to the presence of the determinant. In the case of Walkerton, it initially seems that the effect of *E. coli* illness was from drinking the City's public water, since all those who presented admitted doing so. But could there have been something in their diet that also might have been a cause, for example, tainted hamburger meat? When all other potential confounding factors are ruled out, the epidemiologist can be fairly certain of causation, especially if there is scientific evidence to support it, such as the chain of evidence from the manure to the tainted well.

The case we have just discussed can be termed an *observational study*, since we were observing actual cause and effects. But we can also establish an *experimental study* in which we determine the beneficial effects of an intervention, such as handwashing to reduce the rates of diarrhea. One study was set up to determine whether "nudging" will help children in Bangladesh practice more frequent handwashing after using the toilet (Grover et al. 2018). A series of colorful footprints was painted on the pavement to nudge children toward the hand-washing station (Figure 4.10). The results showed that both nudging and hygiene

Figure 4.10 Colorful footprints painted on the exits from latrines help to "nudge" children toward a hand-washing station in Bangladesh. Source: Grover et al. (2018). Wiley pub – Tropical Medicine and International Health journal.

education were effective in increasing hand-washing rates by as much as triple the rate before intervention.

Epidemiological tools are important in the study of water security as they provide (i) a basis for quantifying and describing the extent of a disease within a population and (ii) a method for assessing the impact of an intervention.

4.6 Analytical Tools of Water Quality

Advanced analytical tools are increasingly being used to monitor the presence of water contaminants. When used to identify the "water criminals" in environmental litigation, the investigation is called *pollution forensics*. And when the goal is public health and the mined data are from wastewater, the science can be called *wastewater-based epidemiology (WBE)*.

Pollution forensics. Some of the more traditional tools of laboratory analysis involve gas chromatography and are based on the volatilization of wastewater components with subsequent partitioning of these volatile compounds from an inert carrier gas to a solid absorbent in the separation column. The variation in partitioning characteristics along with adjustments to the oven temperature causes the various components in the waste stream to be "chromatographically separated" – by travelling through the column at different rates – and thus exiting the column and being detected at different times. This technique can be combined with mass spectrometry, which identifies an organic compound according to its mass-to-charge ratio (mass spectrum), and the overall technique – gas chromatography/mass spectrometry (GC/MS) – is very accurate for separating and identifying compounds. For compounds with limited volatility (i.e. very high boiling points), high-performance liquid chromatography (HPLC) can separate compounds without the need to change the compound to a gaseous phase; in this case, the carrier fluid is an organic phase rather than an inert gas and the partitioning occurs into a hydrophobic-modified solid phase (Manahan 2009). These methods are becoming more and more important as the desire for detection of emerging contaminants – e.g. pharmaceuticals, endocrine disruptors, polybrominated compounds, and personal care products – becomes more urgent.

Wastewater-based epidemiology. The detection and tracing of a virus or bacteria can be costly and time-consuming when limited to random population sampling and laboratory testing. WBE is the analysis of wastewater for purposes of tracking an emerging pollutant or monitoring the public health of a community in a noninvasive manner. This kind of analysis has been used as detection tool for illicit drug use in Europe for many decades (Daughton 2021).

The coronavirus pandemic of 2020–2022 spread faster than viral testing stations (using nasal swabs) could be established in areas of great need. Alternatively, chemical data could be mined from wastewater treatment plants, which are operating year-round and are present in all sewered populations. The cost savings come in the two-step process whereby WBE is used to identify hot spots, followed by clinical testing in these affected areas. The quantification of the coronavirus in wastewater is a function of wastewater temperature and the decay rate (half-life) of the virus biomarker.

Standard hydraulic modeling – e.g. an EPA stormwater model such as Storm Water Management Model (SWMM) – can be used to calculate in-sewer residence time (from household to outfall) based on volumetric flow rates and velocities. It would be most important to note how much non-sewer contributions (such as stormwater) are included in the overall flow. With a detection limit of 10 genomes per milliliter of wastewater, this method can theoretically detect 1 infected person out of a group of 100 (1:100) to 1 out of 2 million (1:2M) healthy individuals (Hart and Halden 2020).

Thus, WBE provides a potentially time-saving and cost-effective approach to monitoring an infectious disease in any population with sewered wastewater systems. The analysis can detect the RNA (ribonucleic acid) of a targeted biomarker even from samples of patients who are asymptomatic or paucisymptomatic (showing few symptoms), as was demonstrated in the early weeks of the SARS-CoV-2 outbreak in Italy (La Rosa et al. 2020). The monitoring of individual sewer sub-basins can be used to reduce the travel time of the virus, identify geographical hot zones, and allocate viral testing to sub-populations more efficiently.

Water quality affects water security in many aspects – its necessity as a component of health and hygiene, its ability to serve an economic and societal purpose, and its potential to carry harmful constituents as a threat to public welfare. In this chapter, we have introduced the main concepts of water quality, which will be built upon in later chapters.

4.7 Conclusion

Water has been described as a universal solvent since it can carry more dissolved substances than any other liquid on earth (USGS 2021). Wherever water travels then, it can pick up substances in its path, substances that are either beneficial or harmful to a user at the end of its journey. Some of these substances are critical enough to be measured and monitored and perhaps even removed using water-treatment methods (to be discussed later).

The quality of water can be described in many ways based on its source, its state (liquid, solid, gas), and its physical, chemical, and biological constituents. A complete understanding of water quality will allow for its accurate assessment regarding usability for defined purposes. These purposes may be for the production of food, energy, or products, for ecological sustainability or for human health and well-being. As we shall see, these various uses depend on varying levels of water quality.

Water security depends on the three foci of appropriate water quantity, quality, and equity. Chapter 5 will consider this last focus area – the equitable distribution of water to all.

*Foundations: Water-Quality Calculations – Equilibrium Reactions, Phase Partitioning, and Disinfection Kinetics

Concepts important to water quality include equilibrium reactions (e.g. chemical and biological transformation reactions), phase partitioning (e.g. migration of chemicals from water to air or solid phases), and disinfection kinetics (e.g. rate of pathogen disinfection by chlorine). Each of these concepts will be described below.

Equilibrium reactions. Chemical transformation can occur in the environment via chemical and/or biological reactions. We will discuss two such reactions here: nitrification (transformation of ammonia to nitrite to nitrate) and biotransformation of organic matter (e.g. carbohydrates to carbon dioxide and water).

Nitrification is an important environmental process as it transforms ammonia to nitrate, which, at elevated levels, can cause methemoglobinemia, otherwise known as blue baby syndrome. Nitrates can change hemoglobin to methemoglobin in the body, thereby decreasing the ability of blood to carry oxygen, a condition that can be fatal in infants. This potential danger has resulted in a nitrate drinking water standard of 10 mg/L NO_3-N (see Table 4.2). Fertilizer is a potential source of nitrate in water, either as ammonium or as ammonium nitrate. The ammonium (NH_4^+) dissociates to ammonia (NH_3) in water, which is then transformed to nitrite (NO_2-) and nitrate (NO_3-) as a two-step process shown in Eqs. (4.1, 4.2):

$$NH_3 + O_2 \rightarrow NO_2\text{-} + 3H^+ \tag{4.1}$$

$$NO_2\text{-} + H_2O \rightarrow NO_3\text{-} + 2H^+ \tag{4.2}$$

Since the reaction of nitrite to nitrate is often rapid, we seldom see nitrite in the environment.

From Eqs. (4.1) and (4.2), for a given concentration of ammonia, we can estimate the nitrate concentration that will result. For example, we can answer the following question: Given nitrification of 20 mg/l of ammonia, how much nitrate will be formed? Looking at Eqs. (4.1) and (4.2), we observe that 1 mole of ammonia produces 1 mole of nitrite and 1 mole of nitrite produces 1 mole of nitrate. Thus, if we convert 20 mg/l of ammonia into moles/l, and given that we have the same number of moles of nitrate, we can convert from moles/l to mg/l of nitrate, as shown below:

$$\begin{aligned} &20\text{ mg/L of }NH_3 \times 1\text{ mole }NH_3/17\text{ g }NH_3 \times 1\text{ g}/10^3\text{ mg} = 1.17 \times 10^{-3}\text{ moles }NH_3 \\ &\quad \times 1\text{ mole }NO_3\text{-}/1\text{ mole }NH_3 \times 62\text{ g }NO_3\text{-/mole} \times 10^3\text{ mg/g} \\ &= 72.9\text{ mg/L }NO_3\text{-} \end{aligned}$$

Or, expressed on a nitrogen basis:

$$72.9\text{ mg/L }NO_3\text{-} \times 14\text{ mg N}/62\text{ mg }NO_3\text{-} = 16.5\text{ mg/L }NO_3\text{-N}$$

Thus, 20 mg/l of NH_3 generates 16.5 mg/l NO_3-N, which is greater than the 10 mg/l NO_3-N drinking water standard. Thus, methemoglobinemia is an environmental concern related to excessive use of ammonia-based fertilizers. (Another concern, eutrophication, will be discussed in Chapter 10).

Phase partitioning. We now transition to considering phase-partitioning reactions, in which a compound of interest migrates from one phase (e.g. water) into another phase (e.g. gas), or vice versa, based on its relative affinity between the two phases (i.e. desire to be in one or the other phase). We will here consider partitioning between the following phases: liquid–gas, liquid–liquid, and liquid–solid, as depicted in Figure 4.11b. Earlier we mentioned gas–water partitioning when we talked about oxygen partitioning from the air into water as microbial degradation depleted the oxygen in the river. This case is depicted in Figure 4.11c where the gas phase has an abundance of oxygen, whereas the water phase has a smaller concentration of oxygen at equilibrium. When this equilibrium is perturbed (e.g. when oxygen is depleted by sugar biodegradation), oxygen transfers from the air to the water in an attempt to reestablish the equilibrium ratio illustrated in Figure 4.11c.

Thus, Figure 4.11c depicts a case in which the compound (e.g. oxygen) has a limited solubility in water. If we look at partitioning of another gas, carbon dioxide, from air to water, we find higher levels of carbon dioxide in the water phase (>300 ppm versus <10 ppm for oxygen) while having lower levels of carbon dioxide in the air (0.35%) than oxygen (21%), and thus our outcome shifts toward the case illustrated in Figure 4.11a. To carbonate soda, the soda water phase is exposed to concentrated carbon dioxide under pressure, thereby achieving a much higher level of carbon dioxide in cola than in tap water in equilibrium with air.

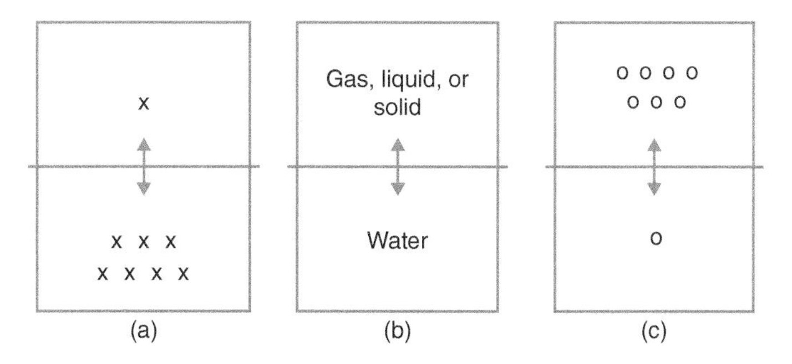

Figure 4.11 Schematic depicting a compound partitioning (b) between water and a second phase – gas, liquid, or solid; (a) preference for the water phase; (c) preference for the gas, liquid, or solid phase.

In this case we can consider that the soda is supersaturated with carbon dioxide, causing the soda to "fizz" when opened to the atmosphere. The excess carbon dioxide slowly partitions out of the cola until it equilibrates with the carbon dioxide in the atmosphere, but by this time the cola carbon dioxide concentration is sufficiently low that we refer to the cola as "flat" – or decarbonated.

Looking at Figure 4.11, we observe that the partitioning (distribution) of a compound between water and gas equilibrates to a common point (ratio) for a given chemical – while the relative concentrations of the compound may vary in the two phases, the ratio will remain constant. We describe this relationship as Henry's Law and the ratio as Henry's constant, as shown in Eq. (4.3):

$$P_{\mathrm{A}} = H_{\mathrm{A}} \times C_{\mathrm{A}} \quad \text{or} \quad H_{\mathrm{A}} = P_{\mathrm{A}}/C_{\mathrm{A}} \tag{4.3}$$

where H_{A} is Henry's constant for compound A, P_{A} is the concentration of A in the gaseous phase, and C_{A} is the concentration of compound A in the water phase. Thus, if the concentration in the gaseous phase becomes 10 times higher, the concentration in the water phase will also be 10 times higher since Henry's constant is independent of the concentration of A. Thus, relatively higher values of Henry's constant correspond to more of the compound residing in the gas phase (corresponding to Figure 4.11c), while relatively lower values of Henry's constant indicate that more of the chemical will be in the water phase (corresponding to Figure 4.11a). Table 4.6 shows some typical values of Henry's constant for several compounds of interest (as well as several other parameters discussed below). The larger the Henry's constant the greater the potential to partition out of water into air; thus, benzene and PCE are more volatile while phenanthrene, lindane, and DDT (dichlorodiphenyltrichloroethane) are less volatile and prefer to remain in the water phase.

The Henry's constants in Table 4.6 are actually temperature dependent. With increasing temperature, the activity (thermal motion) of a compound increases in water – this increased energy gives the compound a greater propensity to escape water and go into the air phase (much like heating water produces more water molecules with increased activity and an increased propensity to escape the water phase and evaporate). Thus, with increasing temperature the Henry's constant for a compound increases, and vice versa – at lower temperatures, the compound has a lower propensity to escape the water phase. This temperature dependence helps explain why rivers and lakes tend to have a lower dissolved oxygen level in the summer and a higher dissolved oxygen level in the winter. This also explains why we sometimes notice air bubbles escaping our hot water coming out of the kitchen faucet – the

Table 4.6 Compilation of partitioning coefficients for select chemicals of environmental interest.

Chemical	H_A	Log K_{ow}	Log K_{oc}[a]	R_f[b]
Benzene	2.2×10^{-1}	2.12	1.91	1.3
DDT	2×10^{-3}	6.19	5.98	3900
Lindane	1.3×10^{-4}	3.7	3.49	14
Phenanthrene	1.6×10^{-3}	4.52	4.31	85
Tetrachloroethene (PCE)	6.2×10^{-2}	2.6	2.39	2.0

a) Using Eq. (4.10).
b) Using Eqs. (4.10, 4.11) with an $f_{oc} = 0.001$, $n = 0.4$, and $\rho_b = 1.6\ \text{g/cm}^3$.
Note: All units are dimensionless.
Source: Knox et al. (1993). Copyright © 1992. Reproduced by permission of Taylor and Francis Group, LLC, a division of Informa plc.

elevated temperature in the enclosed hot water tank causes the water to be supersaturated with oxygen, with the supersaturation "released" when exposed to the atmosphere. Finally, this explains why carbonation of soda is conducted at lower temperatures – the soda can "hold" more carbon dioxide at lower temperatures.

Let us now consider a two-phase system of water and octanol, an organic compound with chemical formula $CH_3(CH_2)_7OH$. Given that octanol is less dense than water, Figure 4.11 remains accurate with water being on the bottom of a mixture and octanol being on the top. Eq. (4.3) now becomes Eq. (4.4) with K_{ow} being the octanol–water partition coefficient. Just as Henry's constant describes partitioning between water and air, K_{ow} describes partitioning between water and an organic phase, in this case octanol. Thus, higher values of K_{ow} correspond to great partitioning into octanol and, by analogy, into other organic phases. From Table 4.6 we observe that DDT, phenanthrene, and lindane partition into octanol to the greater extent (are more hydrophobic) with benzene and PCE preferring the water phase:

$$C_{\text{octanol}} = K_{ow} \times C_{\text{water}} \quad \text{or} \quad K_{ow} = C_{\text{octanol}}/C_{\text{water}} \tag{4.4}$$

Considering contaminant partitioning into soil organic matter (aquifer organic content) is an example of the third two-phase partitioning scenario – water to solid. In the simplest case, this can be described by the expressions in Eq. (4.5), where K_d is the linear sorption coefficient (distribution of the contaminant between the soil and the water):

$$C_{\text{soil}} = K_d \times C_{\text{water}} \quad \text{or} \quad K_d = C_{\text{soil}}/C_{\text{water}} \tag{4.5}$$

As this soil–water partitioning is analogous to the octanol–water partitioning, relationships have been developed to predict K_d from the contaminant's K_{ow} and the soil/aquifer organic content (f_{oc}); Eq. (4.6) illustrates one such relationship (Knox et al. 1993). Thus, contaminants with higher K_{ow} values and soils with higher f_{oc} values experience higher levels of contaminant adsorption (partitioning of the contaminant from the water into the soil) and thus slower migration through the soil/aquifer (referred to as retardation factor – R_f as shown in Eq. (4.7), where ρ_b is the soil/aquifer bulk density and n is the porosity) (Knox et al. 1993). The larger the retardation factor, the slower the contaminant migration is relative to the groundwater (a retardation factor of 3 means that the contaminant moves 1/3 as fast as the groundwater or will take three times longer than the groundwater to show up in a down gradient well). So if a well is 100 m downgradient of a release and groundwater is flowing at 0.5 m/day, it will take 200 days for a nonsorbing contaminant to migrate with the ground water to the well and 600 days for a contaminant with a retardation factor of 3 to migrate to the well. Table 4.6 shows Log K_{oc} values and R_f values for the chemicals assuming an aquifer organic content of 0.1% ($f_{oc} = 0.001$). Thus, we see that DDT and phenanthrene will be least mobile (move slowest relative to the groundwater) in the aquifer system:

$$K_d = K_{oc} \times f_{oc} \text{ and } \log K_{oc} = \log K_{ow} - 0.21 \tag{4.6}$$

$$R_f = 1 + \frac{\rho_b}{n} K_d \tag{4.7}$$

Disinfection kinetics. Disinfection is the process of inactivation and/or destruction of microorganisms in a contaminated water supply. Disinfection kinetics has been found to be a first-order reaction, following Chick's Law (1908) as shown in Eq. (4.8), where N is the concentration of microorganisms, K is the first-order rate constant, and t is time. Integrating from time = 0 to t and from N_o to N gives Eq. (4.9). Taking the exponent of both sides gives Eq. (4.10):

$$dN/dt = -K\,N \tag{4.8}$$

$$\ln (N/N_o) = -K\,t \tag{4.9}$$

$$N = N_o \exp(-K\,t) \tag{4.10}$$

Eq. (4.10) is the common expression of microorganism decay as a function of decay rate and time. Knowing the K value, it is possible to estimate how much time is required to go from N_o to N. For example, let us assume a desired 99.9% rate of disinfection, i.e. out of 1000 organisms present initially, we want a maximum of 1 remaining after disinfection. Thus, $N_o = 1000$, $N = 1$ and let us assume a value of $K = 0.4$/minute. Plugging these values into Eq. (4.13) and solving for t gives us $t = 17.3$ minutes. Thus, 17.3 minutes should allow 99.9% efficiency in disinfection. In a water-treatment facility, the chlorine contact tank is typically designed for 30 minutes, which is consistent with, and provides a safety factor to, the 17.3 minutes calculated above.

We now note that this contact time (17.3 minutes) achieves a three order of magnitude (or log-3) reduction in N, that is, N decreases from 1000 (10^3) to 1 (10^0) microorganisms. If we double this time (34.6 minutes), we expect a similar reduction again, or six orders of magnitude reduction (much as when we doubled the half-life we saw the concentration went from one-half to one-half of one-half or one-fourth of the initial concentration). If we plug 34.6 minutes and 0.4/minute into Eq. (4.14), we find that $N_o/N = 10^6$, which means if we start with 10^6 microorganisms we end up with a six order of magnitude (6 log) reduction in microorganisms or an N value of 1.

While the above equations suggest that decay kinetics are only a function of N, the addition of a disinfectant dose, C, to the equation will reflect the impact of both N and C on microorganism decay. In this case, disinfection is a function of N (number of microorganisms) and C (disinfectant concentration) (Eq. (4.11)):

$$dN/dt = -K\,C\,N \tag{4.11}$$

Water-treatment plants are not only required to meet a reaction time requirement (e.g. 30 minutes as mentioned above) but also a Ct (concentration times reaction time) requirement. This parameter, Ct, accounts for the fact that the disinfectant concentration may vary over the reaction time and often requires a tracer test to establish the actual reaction time rather than relying on the theoretical detention time (takes into account nonidealities in the flow through the reactor). If UV disinfection is used in place of chlorine, then the parameter becomes It, where I is the UV lamp intensity. This approach is based on the same principle discussed above but utilizing a more comprehensive reaction expression.

In this Foundations section, we have thus looked at equilibrium reactions, phase partitioning, and disinfection kinetics, all important principles when considering water quality as an integral component of water security.

End-of-Chapter Questions/Problems

4.1 Why are color, odor, sulfate, and zinc all considered secondary standards?

4.2 Nitrate contamination in drinking water is said to cause "blue-baby syndrome." Use online resources to answer the following. Document your references used.
a. What does the chemical do in a person's body?
b. What is the reason behind the name "blue-baby syndrome"?

4.3 Public concern has recently focused on glyphosate, the active ingredient in a popular pesticide called "Roundup" and a suspected carcinogen. Find a recent study on the health effects of Roundup.
a. What are the findings of this study?
b. According to the study, what further research needs to be done?

4.4 You are the public water manager for a medium-sized city and you would like to test the integrity of the distribution system, a system of pipes that carry water to your many customers. But you only have budget to conduct one test for a pathogenic organism.
a. Which organism(s) would you test for? Why?
b. Where (at what locations) would you do your testing? Why?

4.5 Describe and compare three methods that can be used to remove arsenic from drinking water. Which of these would be most suitable for use in a developing country? Why?

4.6 Describe and compare three methods that can be used to remove fluoride from drinking water. Which of these would be most suitable for use in a developing country? Why?

4.7 Typhoid fever was once a prevalent waterborne disease in the United States.
a. What are the causes and symptoms of typhoid fever? How does it spread?
b. Discuss the incident of "Typhoid Mary" in the early twentieth century. How did she easily infect so many people?
c. Compare modern-day India and the United States regarding prevalence and mortality rate of typhoid fever.

4.8 The island nation of Haiti suffered a severe cholera outbreak in the year 2010.
a. What are the causes and symptoms of cholera? How does it spread?

b. Describe the devastation that was caused by cholera in Haiti during this outbreak. How many people died? How many fell ill?
c. What is the best way to respond to an outbreak such as happened in Haiti?

4.9 Many studies have been done to determine the best health interventions to prevent diarrhea in developing countries.
a. Find a study that compares two or more WaSH (water, sanitation, hygiene) interventions in a developing country context.
b. Which method(s) are the most effective? Use the F-diagram to discuss the results of the study.

4.10 If an NH_3 concentration of 35 mg/l transforms into NO_3, what is the resulting concentration expressed both as NO_3 and as NO_3-N?

4.11 For an aquifer with an $f_{oc} = 0.002$, $n = 0.4$, and $\rho_b = 1.6\ g/cm^3$,
a. What are the corresponding R_f values for (i) benzene and (ii) lindane?
b. For a downgradient well 100 m away, and given a ground water velocity of 0.2 m/day, how long (days) would it take for the following compounds to appear in the well (neglecting dispersion)?
   i. Nonsorbing compound (travels at the velocity of the ground water)
   ii. Benzene
   iii. Lindane

4.12 Using Chick's law (first-order degradation) with a K value of 0.23/min and an $N_0 = 500$:
a. Calculate and plot N at the following times (minutes): 1, 2, 3, 5, 10, and 15
b. How much time would be required to reach $N = 0.5$ (99.9% kill)?

Further Reading

Carlson, G. (2022). *Human Health and the Climate Crisis*, 1e. Burlington, MA: Jones & Bartlett Learning.

Hrudey, S.E. and Hrudey, E.J. (2010). *Safe Drinking Water: Lessons from Recent Outbreaks in Affluent Nations*, 1e. London, UK: IWA Publishing.

Salzman, J. (2017). *Drinking Water: A History*. Overlook Duckworth: Revised and Updated edition.

World Health Organization (WHO) (2017). *Guidelines for Drinking-Water Quality, 4th Edition, Incorporating the 1st Addendum*. Geneva, Switzerland: World Health Organization https://www.who.int/publications-detail-redirect/9789241549950.

References

Alarcón-Herrera, M.T., Bundschuh, J., Nath, B. et al. (2013). Co-occurrence of arsenic and fluoride in groundwater of semi-arid regions in Latin America: genesis, mobility and remediation. *Journal of Hazardous Materials* 262 (November): 960–969. https://doi.org/10.1016/j.jhazmat.2012.08.005.

Amini, M., Abbaspour, K.C., Berg, M. et al. (2008a). Statistical modeling of global Geogenic arsenic contamination in groundwater. *Environmental Science & Technology* 42 (10): 3669–3675.

Amini, M., Mueller, K., Abbaspour, K.C. et al. (2008b). Statistical modeling of global geogenic fluoride contamination in groundwaters. *Environmental Science & Technology* 42 (10): 3662–3668. https://doi.org/10.1021/es071958y.

Apambire, W.B., Boyle, D.R., and Michel, F.A. (1997). Geochemistry, genesis, and health implications of fluoriferous groundwaters in the upper regions of Ghana. *Environmental Geology* 33 (1): 13–24. https://doi.org/10.1007/s002540050221.

Black, M. (2016). *The Atlas of Water: Mapping the World's Most Critical Resource*, 3e. University of California Press.

Cairncross, S. and Feachem, R. (2011). *Environmental Health Engineering in the Tropics: An Introductory Text*, 3e. Earthscan Publications Ltd.

CNN (2020). "E. Coli Outbreaks Fast Facts - CNN." 2020. https://www.cnn.com/2013/06/28/health/e-coli-outbreaks-fast-facts/index.html.

Corcoran, E. and GRID--Arendal (2010). *Sick Water? The Central Role of Wastewater Management in Sustainable Development: A Rapid Response Assessment*. Arendal, Norway: UNEP/GRID-Arendal.

Daughton, C. (2021). "Wastewater-Based Epidemiology May Pay off for Covid-19." *STAT* (blog). January 7, 2021. https://www.statnews.com/2021/01/07/wastewater-based-epidemiology-20-year-journey-pay-off-for-covid-19.

Freeze, R.A. and Lehr, J.H. (2009). *The Fluoride Wars: How a Modest Public Health Measure Became America's Longest Running Political Melodrama*, 1e. Hoboken, N.J: Wiley.

Grover, E., Hossain, M.K., Uddin, S. et al. (2018). Comparing the behavioural impact of a nudge-based handwashing intervention to high-intensity hygiene education: a cluster-randomised trial in rural Bangladesh. *Tropical Medicine & International Health: TM & IH* 23 (1): 10–25. https://doi.org/10.1111/tmi.12999.

Hart, O.E. and Halden, R.U. (2020). Computational analysis of SARS-CoV-2/COVID-19 surveillance by wastewater-based epidemiology locally and globally: feasibility, economy, opportunities and challenges. *Science of the Total Environment* 730 (August): 138875. https://doi.org/10.1016/j.scitotenv.2020.138875.

Havelaar, A.H. and Melse, J.M. (2003). "Quantifying Public Health Risk in the WHO Guidelines for Drinking-Water Quality: A Burden of Disease Approach." RIVM report 734301022/2003. World Health Organization.

Howard, G. and Bartram, J. (2003). "WHO | Domestic Water Quantity, Service Level and Health." WHO/SDE/WSH/03.02. http://www.who.int/water_sanitation_health/diseases/wsh0302/en/index.html.

Hrudey, S.E. and Hrudey, E.J. (2010). *Safe Drinking Water: Lessons from Recent Outbreaks in Affluent Nations*, 1e. London, UK: IWA Publishing.

Johnson, S. (2007). *The Ghost Map: The Story of London's Most Terrifying Epidemic--and how it Changed Science, Cities, and the Modern World*. Riverhead: Trade.

Knox, R.C., Canter, L.W., and Sabatini, D.A. (1993). *Subsurface Transport and Fate Processes*, 1e. CRC Press.

Manahan, S. (2009). *Environmental Chemistry*, 9e. Boca Raton: CRC Press.

Millennium Ecosystem Assessment (Program) (ed.) (2005). *Ecosystems and Human Well-Being: Wetlands and Water Synthesis: A Report of the Millennium Ecosystem Assessment*. Washington, DC: World Resources Institute.

Morris, R.D. (2008). *The Blue Death: The Intriguing Past and Present Danger of the Water you Drink* Reprint edition. New York: Harper Perennial.

Ravenscroft, P. (2006). *Predicting the Global Extent of Arsenic Pollution of Groundwater and its Potential Impact on Human Health.* UNICEF https://www.researchgate.net/publication/313628997_Predicting_the_global_extent_of_arsenic_pollution_of_groundwater_and_its_potential_impact_on_human_health.

Rosa, L., Giuseppina, M.I., Mancini, P. et al. (2020). First detection of SARS-CoV-2 in untreated wastewaters in Italy. *Science of the Total Environment* 736 (September): 139652. https://doi.org/10.1016/j.scitotenv.2020.139652.

Salzman, J. (2017). *Drinking Water: A History*. Revised and Updated edition. Overlook Duckworth.

US EPA (2010). "National Pollutant Discharge Elimination System (NPDES) Permit Writers' Manual." EPA-833-K-10-001.

US EPA, OW (2015a). "National Primary Drinking Water Regulations." Overviews and Factsheets. US EPA. 2015. https://www.epa.gov/ground-water-and-drinking-water/national-primary-drinking-water-regulations.

US EPA, OW (2015b). "Secondary Drinking Water Standards: Guidance for Nuisance Chemicals." Overviews and Factsheets. US EPA. September 2, 2015. https://www.epa.gov/sdwa/secondary-drinking-water-standards-guidance-nuisance-chemicals.

USGS (2021). "Water, the Universal Solvent." 2021. https://www.usgs.gov/special-topic/water-science-school/science/water-universal-solvent?qt-science_center_objects=0#qt-science_center_objects.

WHO (2011). "WHO | Guidelines for Drinking-Water Quality, Fourth Edition." WHO. 2011. https://apps.who.int/iris/bitstream/handle/10665/44584/9789241548151_eng.pdf;jsessionid=79ADD4AC28079D4AFF0D28584BD6EF30?sequence=1

WHO (2017). *Guidelines for Drinking-Water Quality*, 4e, Incorporating the 1st Addendum. Geneva, Switzerland: World Health Organization https://www.who.int/publications-detail-redirect/9789241549950.

WHO (2020). "Cholera: A Snapshot." 2020. https://www.who.int/westernpacific/health-topics/cholera.

WHO (2021a). "Drinking-Water." 2021. https://www.who.int/news-room/fact-sheets/detail/drinking-water.

WHO (2021b). "Population Using Improved Sanitation Facilities." 2021. https://www.who.int/data/gho/data/indicators/indicator-details/GHO/population-using-improved-sanitation-facilities-(-).

WHO (2021c). "Typhoid." 2021. https://www.who.int/news-room/fact-sheets/detail/typhoid.

Whorton, J.C. (2011). *The Arsenic Century: How Victorian Britain Was Poisoned at Home, Work, and Play*. USA: Oxford University Press.

4a The Practice of Water Security: Education for Sanitation, Water, and Health

Environmental health engineer Ben Fawcett has worked in many roles over more than three decades in the WaSH sector, such as development manager, program director, teacher, consultant, researcher, author, mentor, and humanitarian, just to name a few (Figure 4a.1). He has assisted many nongovernmental organizations (NGOs) (e.g. Oxfam, Red Cross, Water Aid, Save the Children) and governments in the participatory planning, implementation, and evaluation of development, humanitarian, and emergency relief programs related to water and sanitation in countless countries around the developing world.

Figure 4a.1 Ben Fawcett stands beside Fawcett Creek, New South Wales, Australia. Source: Courtesy of Ben Fawcett.

In 2008, Ben published a book with Maggie Black that provided a rationale for discussing a topic that has commonly been regarded as prohibited. Entitled *The Last Taboo: Opening the Door on the Global Sanitation Crisis*, the book first chronicled the history of human waste disposal. It then proceeded to discuss the effects of urbanization on the sanitation crisis, the development and design of toilets, and the economic implications of sanitation programs. The book is still an important contribution for academicians and practitioners, helping students and field workers overcome their reluctance to address this critical issue that impacts the lives of so many global citizens.

Beginning in 1996, Ben served as the Director of the Masters Program in Engineering for Development at the University of Southampton, United Kingdom. There he mentored approximately 150 students who later became engineers with NGOs and government agencies across the globe. Later Ben taught WaSH (water supply, sanitation, hygiene) to Masters students in Integrated Water Management (MIWM) in Brisbane, Australia. His goal was to give students a good understanding of both key principles and approaches to all aspects of WaSH, increasing their interest and skills in the connection of sanitation and hygiene (Figure 4a.2).

Figure 4a.2 Students build a sanitation slab for a rural community. Source: Courtesy of Ben Fawcett.

Ben speaks of two unique online courses that he helped to develop. One introduces the concepts of behavior change – so vital in bringing about sustainable improvements in the WaSH sector. The second one summarizes the basic, up-to-date principles and processes of WaSH in developing communities, aiming to interest students in furthering their studies. "Both courses continue to run at least once a year and attract students from many countries and backgrounds."

Ben has retired from full-time academia but continues to supervise individual student research projects in Cambodia, Costa Rica, El Salvador, Ghana, Kenya, Laos, Mozambique, Nepal, Niger, Solomon Islands, and Thailand focusing on a wide range of WaSH subjects (Figure 4a.3). He continues to lobby for more work in sustainable development of sanitation and hygiene, particularly in urban slums in developing countries, to find ways to better facilitate behavior change in sanitation and hygiene practices, and to support those changed behaviors through effective local government systems.

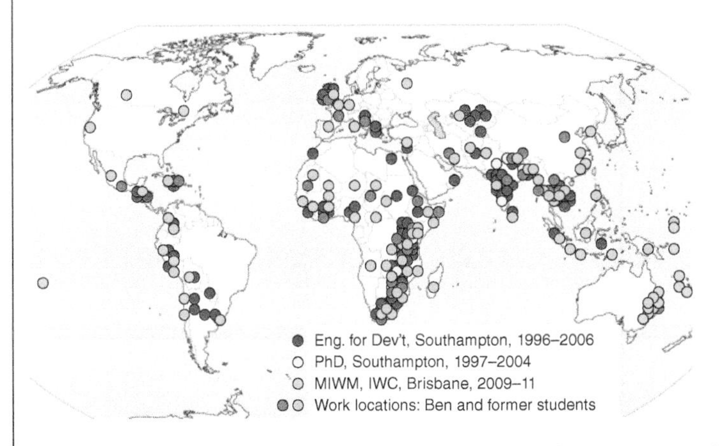

Figure 4a.3 Map shows work locations of Ben and his former students. Source: Courtesy of Ben Fawcett.

Ben Fawcett is the recipient of the 2011 OU International Water Prize

5

The Context of Water Security – Water Equity

The objective of this chapter is to introduce water equity as a critical component of the water security paradigm. When water access is inequitable, water security and, in fact local, national, and transnational peace, is in jeopardy. Water inequity often occurs when there is an imbalance in social, gender, political, economic, or natural forces. While one immediately thinks of developing countries when it comes to water inequity, affluent nations are also vulnerable to such forces. Water rights and pricing structures can either impede or promote water security. Finally, water-equity issues can also be transnational in the case of basins that transcend political boundaries. This chapter will introduce the reader to water-equity issues that are germane to every corner of the world.

Learning Objectives

Upon completion of this chapter, the student will be able to:

1. Understand the concept of water equity and its importance to water security.
2. Understand the disparities in developing country water access between rural and urban settings and as a function of income.
3. Understand gender issues related to water and its impact, positive and negative, in pursuit of gender equity
4. Recognize that the Flint, Michigan, water crisis represents an affluent nation example of water inequity, and understand historical forces that have helped shape this situation.
5. Understand broader examples of United States' lack of water access, understand how this situation came to be, and recognize steps forward to address this situation
6. Understand the global and US disparity of water pricing, both on a unit cost basis and as a percentage of household income, and recognize how tiered pricing structures can help address or exacerbate water inequities for the poor.
7. Understand the nature of US water rights (riparian, prior appropriation, and others) and recognize how these historical approaches can unintentionally promote water inequities.
8. Recognize that water basins that transcend geopolitical boundaries are subject to water inequities based on treaties negotiated to manage these basins.
9. *Understand some of the basic epidemiological concepts used to measure disease and effects within a population.

Fundamentals of Water Security: Quantity, Quality, and Equity in a Changing Climate, First Edition.
Jim F. Chamberlain and David A. Sabatini.

Companion website: www.wiley.com/go/chamberlain/fundamentalsofwatersecurity

5.1 Introduction

Having discussed water quantity and quality as key elements of water security, we now turn our attention to the topic of water equity. An obvious question is "What do we mean by water equity?" We can start by considering its opposite – what is water inequity? The simplest way to answer this question is to acknowledge the "haves and have nots," i.e. while certain people have adequate quantity and quality of water, other people lack one or both. Behind this question is yet a more fundamental issue – is water a human right or is it an economic commodity? Unlike certain commodities (e.g. food), where there are alternatives, there is no alternative to water. As such, one could argue that water is a basic human right, like the air we breathe. Nonetheless, a counterargument is that yes, the water for drinking is itself a human right and should be free. But if you want the water to be convenient (e.g. piped to your house, or your yard, or the village center), adequate (enough for your basic needs), reliable (available year-round and at all hours of the day), and safe (free of pathogens and chemicals of concern), then you should be willing to pay for these amenities or "add-ons." Some would equate this with paying the "true value" of water access. Thus, clean, convenient, reliable, adequate, safe water becomes an economic commodity that bears a cost and an associated price.

What are some drivers of water inequity? One driver would be *socioeconomic*, whether it be in a developing country (e.g. between rural versus urban, or within urban – formal versus informal or peri-urban regions in the city) or in an affluent nation (e.g. low-income families in the United States). This latter scenario was recently illustrated (year 2020) when utilities forgave delinquent utility payments due to the importance of water and hygiene in fighting the ongoing coronavirus pandemic. The inequity could be *gender-related*, as is the case when women bear the burden of fetching water in developing countries. Likewise, inequities can result from *other social categories* – such as race, ethnicity, caste, or class. Further, geographic inequities can result from *spatial variability* – e.g. locations within a state or country that are more prone to drought or flooding. Finally, water-equity issues can be *political*, either international (e.g. disputes to Nile River water between Ethiopia and Egypt) or intranational (e.g. water laws that favor first in time or allocation via proximity). All of these are water-equity issues that will be discussed in this chapter.

5.2 Socioeconomic and Rural/Urban Water Inequity in Developing Countries

Undoubtedly the starkest example of water inequity exists in developing countries. As discussed in Chapter 1, lack of water and sanitation is a global crisis with over 3/4 of a billion and over 2.5 billion impacted, respectively, leading to the inclusion of Sustainable Development Goal 6 – "Ensure availability and sustainable management of water and sanitation for all." Figure 5.1 shows water access at various service levels (piped, other improved, unimproved) for urban versus rural sub-Saharan Africa communities as a function of economic quintiles (poorest to richest in segments of 1/5 of the population). The service levels range from piped on premise, whether that be a tap stand in the yard or indoor plumbing, to other improved service, which could be a well, spring, rainwater catchment, etc., or unimproved

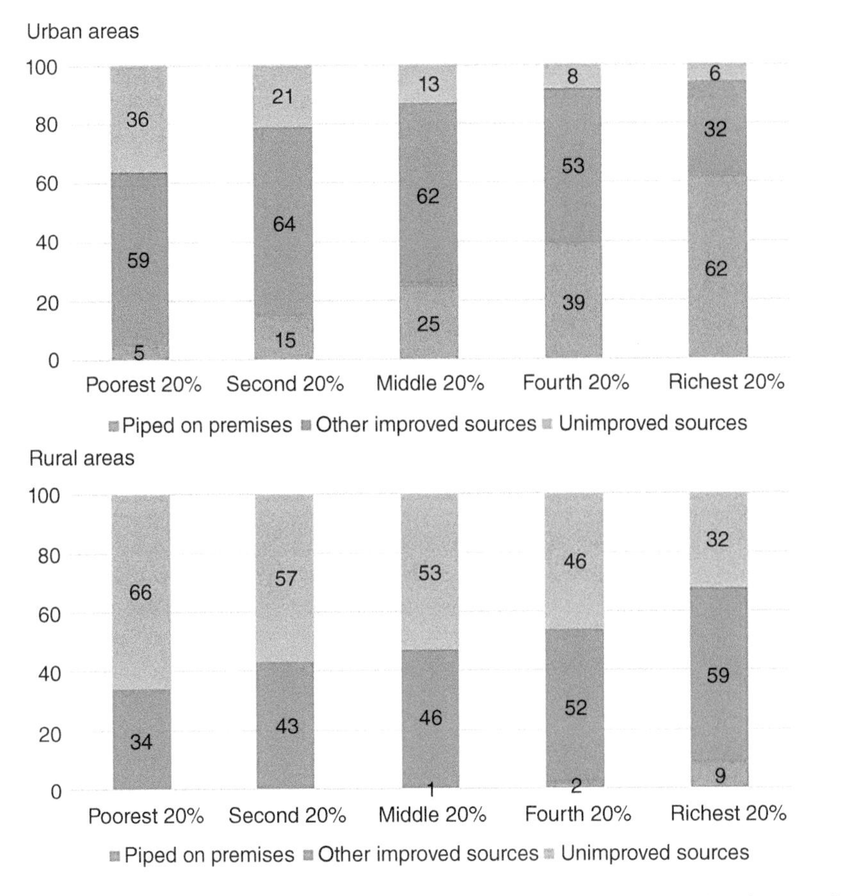

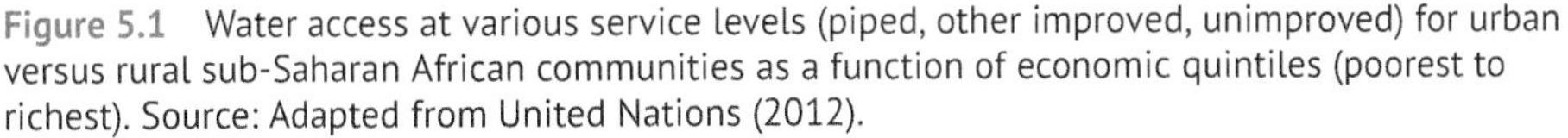

Figure 5.1 Water access at various service levels (piped, other improved, unimproved) for urban versus rural sub-Saharan African communities as a function of economic quintiles (poorest to richest). Source: Adapted from United Nations (2012).

source, such as a pond, river, or lake. It should be noted that, in this definition, an improved source may still have water-quality concerns (e.g. pathogens in the spring water or rainwater catchment).

The data in Figure 5.1 illustrate the impact of socioeconomic status on access to improved water sources. For example, in the urban context, as the target population increases from the bottom quintile (poorest 20% of the urban population) to the top quintile (wealthiest 20%), the reliance on unimproved water sources declines from 36 to 6%. At the same time, water piped on the premise increases from 5 to 62% for the bottom and top quintiles, respectively. Thus, water inequity is apparent. But even more stark is the contrast between urban and rural water access, also shown in Figure 5.1. We observe that the top quintile of rural dwellers has comparable reliance on unimproved water sources as the poorest quintile of the urban dwellers. And as we transition from the wealthiest to poorest quintile in the rural setting, we observe that 2/3 of the poorest people are relying on unimproved water sources, an astounding observation given the wealth of the global community.

Thus, from Figure 5.1 we observe that the rural poor are the most disadvantaged when it comes to access to improved water sources. The lack of improved water leads to increased burden of illness, decreased ability to attend school, and continued entrapment in the poorest income quintile. Conversely, providing access to improved and safe water sources improves health, education, and income-earning potential, which ultimately helps promote peace and stability.

Figure 5.2 expands our view from sub-Saharan Africa to a global perspective. Consistent with our discussion above, Figure 5.2 shows that numerous countries in Africa have less than 75% access to a basic drinking water service. Afghanistan, Papua New Guinea, and Yemen also report that perhaps as much as 50% of their populations do not have basic drinking water. In contrast, Australia, Europe, and most Asian countries have 90% or greater access to improved drinking water sources. The Middle East and North Africa (MENA) region, which will experience population growth along with a drier climate, is especially vulnerable. Per capita water availability is expected to drop by 23% in Jordan, 35% in Iraq, 38% in Morocco, and 40% in Yemen by the year 2025 (Devlaeminck et al. 2017). This additional water stress will exacerbate the political tensions already present in the region.

To illustrate the disparity within a given country, consider the case of piped water in Kazakhstan (Black 2016). While on average 61% of the people have access to piped water on premises, this number escalates to 90% for urban dwellers and reduces to 28% for rural people, demonstrating that piped water is not equally distributed between rural and urban settings. Further, significant disparity exists even within the urban and rural settings based on economic status. While overall 90% of urban dwellers have access to piped water on premises, 99% of the wealthiest and only 65% of the poorest urban dwellers have such access. And while overall 28% of rural people have access to piped water on premises, 50% of the wealthiest and only 15% of the poorest have such access. Thus, we see great disparity in access to piped water, ranging from a low of 15% for the rural poor to 99% for the urban wealthy (Black 2016). This water-access disparity can cause even further health and livelihood disparities within a given setting.

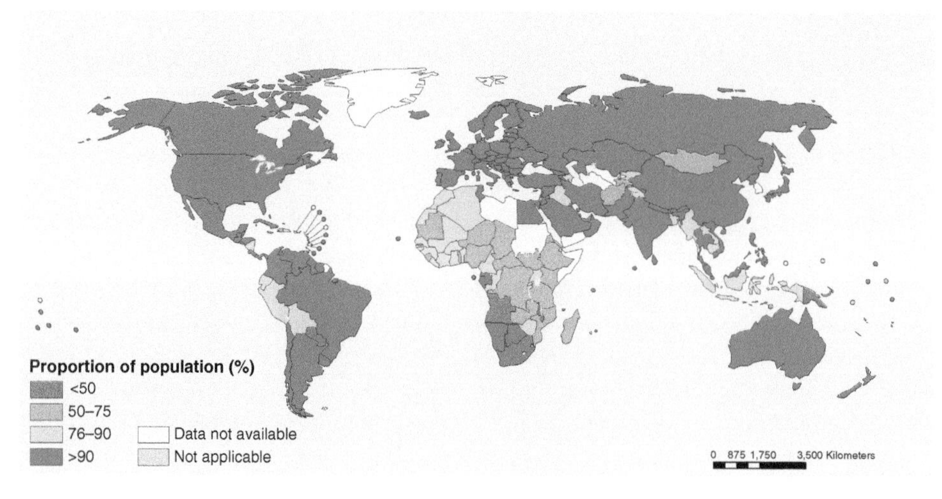

Figure 5.2 Percentage of people using at least basic drinking water services in 2015 (UNICEF/WHO 2019). Source: WHO.

5.3 Water and Gender Equity

Having discussed rural versus urban inequities, we will now turn our attention to water- and gender-equity issues. As many rural villagers lack piped water, the burden for fetching water falls primarily to women and girls, who together account for over 80% of water collection (WHO/UNICEF 2017) (Figure 5.3).

What are the consequences of this gender divide in fetching water? The carrying of full water containers, as heavy as 40 pounds, on the heads or backs of women and girls has a physical impact on them, contorting their spines and leading to problems in childbirth and later in life (Caruso 2017). Further, they are often fetching water for hours every day as the majority of unimproved water sources was found to be of at least 30 minutes roundtrip collection time (WHO/UNICEF 2017). This investment of time can make them late for school or unable to work, further exacerbating their comparative disadvantage to men and boys. Moreover, these practices often result in the occurrence of sexual and physical violence to women, further illustrating the gender divide in water and sanitation access (Truelove 2011).

Figure 5.4 shows pictures of women and children fetching water in Ethiopia and South Africa. One picture shows two young boys carrying a water jug that might weigh as much as they do, requiring them to use a stick through the handle of the jug to together carry the jug.

Table 5.1 provides insight into the effects of water supply and sanitation services on females across their life cycle: from prenatal and neonatal to preschool and primary age, secondary and higher education, working and childbearing to elderly. While water quality and adequacy are issues germane across the various stages of life cycle, other factors (e.g. menstrual hygiene, childbearing hygiene, menopause, toileting needs of elderly) are unique to various stages of the female lifecycle, as are the implications for well-being.

Relative to toileting needs of the elderly, one of the authors (DAS) had a first-hand view of this while visiting a public restroom in Singapore (Figure 5.5). When entering the restroom, the stall doors were initially open, displaying a squatting slab to the left (common in developing countries) and a toilet to the right. After initially assuming that the toilets were intended for Western visitors, he noticed the symbol indicating that the toilet was for use by elderly people. One common saying is that "those who can, squat." Only when no longer able to squat would someone resort to using a toilet. One implication is that in cases where only

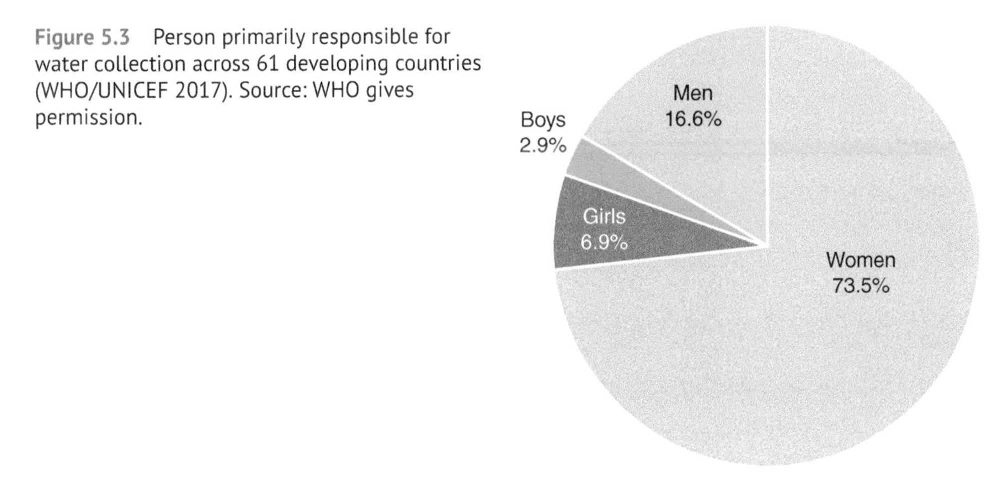

Figure 5.3 Person primarily responsible for water collection across 61 developing countries (WHO/UNICEF 2017). Source: WHO gives permission.

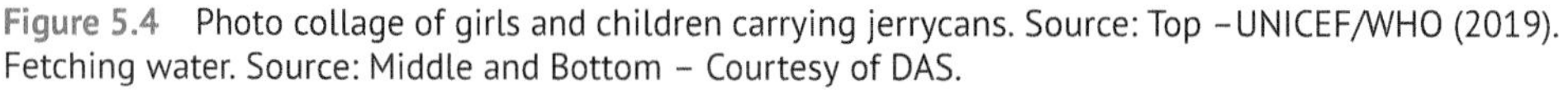

Figure 5.4 Photo collage of girls and children carrying jerrycans. Source: Top –UNICEF/WHO (2019). Fetching water. Source: Middle and Bottom – Courtesy of DAS.

squatting slabs are provided, the elderly might be severely disadvantaged, creating a scenario of age-based water inequity.

Thus, while water supply and sanitation are necessary irrespective of age and gender, there are aspects and implications of water supply and sanitation that do depend on age and gender. The failure to recognize, as well as address, these age- and gender-specific implications leads to water inequities and can threaten water security.

5.4 Not Just Developing Countries: The Case of Flint, Michigan

However, water-equity challenges are not limited to developing countries. Affluent nations, such as the United States, also experience water inequities. Myths about water access in the

Table 5.1 Water supply and sanitation services: illustrative effects on females across the lifecycle.

Place in life cycle	Issues related to water/sanitation	Implications for well-being
Prenatal and neonatal (<1000 days)	• Water quality and adequacy • Hygiene – mother and child	• Increased neonatal mortality • Increased disease burden
Preschool and primary age	• Water quality and adequacy • Hygiene – mother and child	• Females fetching water unable to attend school • Health and nutrition – child stunting, disease, drowning
Secondary and higher	• Water quality and adequacy • Menstrual hygiene • Privacy	• Health and nutrition • Safety – harassment, violence • Dignity, self-respect
Working and child-bearing age	• Water access/affordability • Hygiene of birthing services • Privacy • Voice in decision-making	• Maternal health • Toll of water collection • Safety – violence, dignity • Empowerment/status
Elderly	• Water quality and adequacy • Menopausal hygiene • Toileting needs of the elderly • Physical access to services	• Health • Dignity and self-respect • Stress on mainly female health providers

Source: Das and Hatzfeldt (2017).

Figure 5.5 Public restroom in Singapore: Interior view (left) shows squatting slab on the left and toilet on the right; Exterior view (right) shows that younger people use squatting slab and the toilet is intended for elderly. Source: Pictures courtesy of DAS.

global North abound and include notions that water access is universal, clean, affordable, trustworthy, and uniformly or equitably governed in high-income countries (Meehan et al. 2020). The recent water crisis in Flint, Michigan, is a prime counter-example. Located near Detroit, Flint shares a common history with its big-city neighbor as a bastion of growth in the evolution of the US automotive industry. General Motors (GM) and the United Auto workers were major entities in Flint, making it an automotive innovation hub. At one time, over a hundred different Flint manufacturing facilities served the auto industry. At its peak in the mid-twentieth century, Flint had one of the highest per capita incomes in the nation, leading to a high quality of life including an excellent school system along with the Flint Symphony Orchestra and the Flint Institute of the Arts (Clark 2019). While reaching a peak population of 200 000 people in the 1960s, Flint's population began to decline in the late twentieth century due to the closing of the GM plant and the ripple effect of automotive service companies doing likewise. The population dipped below 100 000 in 2013, the first time its population had dropped below this level since the 1920s. However, while the city population (and associated tax base) was declining precipitously, the cost of operating the city infrastructure remained largely the same, placing an increasing burden on the remaining citizens (an unfortunate reality experienced also by many rural communities). This set the stage for a major water crisis in 2014–2015 (Clark 2019).

In April 2014, seeking to decrease the water costs for Flint citizens, water managers shifted the city's water-supply source from Lake Huron to the Flint River. Even though water quality in the new water source would have justified it, corrosion inhibitors were not added in the water-treatment scheme (Hanna-Attisha et al. 2016; Pieper et al. 2018). Corrosion inhibitors react in the water to form a protective coating on pipe interiors. Without such a coating, elemental lead began leaching from the aging lead pipes. As a result, Flint citizens were exposed to elevated lead levels. The lead exposure continued for some time as the initial sampling method underestimated the lead levels in the water. The incidents of elevated lead levels in blood corresponded to higher lead levels in the water. In addition to lead exposure, the water supply change produced conditions ideal for Legionnaires' disease (a bacterial disease), further extending the health risks and death toll (Clark 2019).

While the city eventually returned to the original water supply, concerns remained due to the lead-leaching pipes still in the ground. The resources generated by the Governor's emergency orders enabled an aggressive program to locate, excavate, and replace lead-bearing pipes in the city.

The Flint story, tragic as it is, becomes even more heart-wrenching when considering water inequity. As Flint citizens migrated to the suburbs in the late-twentieth century, the lower income minorities remained in the inner city. Historical policies and practices limited city sections in which minorities could live and own property, and thus minority peoples living in these areas (predominately African–American) bore the brunt of the water tragedy (Clark 2019). The highest water lead levels, and correspondingly the greatest blood lead level increase, corresponded to the most disadvantaged neighborhoods (Hanna-Attisha et al. 2016). The Flint water could be seen as an extension of a lack of economic opportunity for African–Americans resulting from failed housing, education, and social-service policies as noted five decades earlier in the Kerner report (Kerner 1968). Two researchers point to the racial capitalism that "trapped" vulnerable populations in Flint, Michigan, resulting in environmental racism as the saga unfolded (Pulido 2016; Ranganathan 2016). The Flint, Michigan, water crisis thus shows the importance of water equity as part of a tapestry of social-equity issues that promotes a more unified, prosperous, and peaceful society.

5.5 Closing the Gap: US Water-Access Needs

Water-quality concerns in the United States are not isolated to Flint, Michigan. Patel and Schmidt (2017) indicate that throughout the United States, minority and low-income populations are more likely to live in rural or older housing areas prone to contamination so that, even when tap water is safe, the fear of contamination causes them to avoid drinking tap water. Brooks et al. (2017) also noted racial/ethnic and socioeconomic disparities in tap water consumption among US adults.

A 2019 report addresses the topic of "Closing the Water Access Gap in the United States: A National Action Plan." (US Water Alliance 2019). The report states that while over the past century water and sanitation have become widely available in the United States, they remain out of reach for some of the most vulnerable population – including communities of color, lower income people in rural areas, tribal communities, and immigrants. The report states that over two million Americans lack access to running water, indoor plumbing, and basic wastewater services, emphasizing that closing the water-access gap would enable these communities to thrive.

How could this be true in such a wealthy nation in modern times? The report states that leaders of these communities failed to sufficiently access water and wastewater infrastructure funding when the United States made historic investments in these systems in the mid- to late-twentieth century (US Water Alliance 2019). While these major investments helped provide safe and reliable water and wastewater services to nearly every American, serving to all but eradicate water- and sanitation-related deaths, current federal funding for water infrastructure pales in comparison to these former investments. Thus, communities that failed to benefit from these past investments are hard-pressed to close the gap, creating a virtually hidden water and sanitation crisis threatening the health and well-being of millions of US citizens.

The report states that race and poverty are the strongest predictors of water and sanitation access, or the lack thereof. While only 0.3% of white households lacked complete (indoor) plumbing, 0.5% of African–American and Latinx households and 5.8% of Native American households lacked the same. Thus, while African–American and Latinx homes are nearly twice as likely to lack indoor plumbing than white households, Native American homes are 19 times more likely to be without these basic necessities. Further, inequities exist within these communities; access was higher among families with higher income. These trends sound eerily akin to what we discussed above for developing countries.

Why were these communities left behind during times of water infrastructure investment? The report cites several examples. In the early 1900s, the federal government subsidized drinking water and irrigation for settlers in the West, often at the expense of Native American tribes. This led to water-access challenges in tribal areas such as the Navajo Nation. In the 1950s, Zanesville, Ohio, did not construct municipal water lines in African–American neighborhoods. In the 1960s, Roanoke, Virginia, did not extend water and sanitation lines to neighboring Hollins, a majority African–American town. In California's Central Valley, rural Latinx communities were discouraged from incorporating and therefore did not receive the same funding to build infrastructure that neighboring towns did. While these formal discriminatory practices may have ceased by now, these communities continue to suffer the consequences of lost opportunities, with little hope of catching up (US Water Alliance 2019).

The US Water Alliance report goes on to cite six case studies illustrating the need to close the gap (Figure 5.6). The *Navajo Nation (#1* on Figure 5.6*)* in the Southwest is a compelling

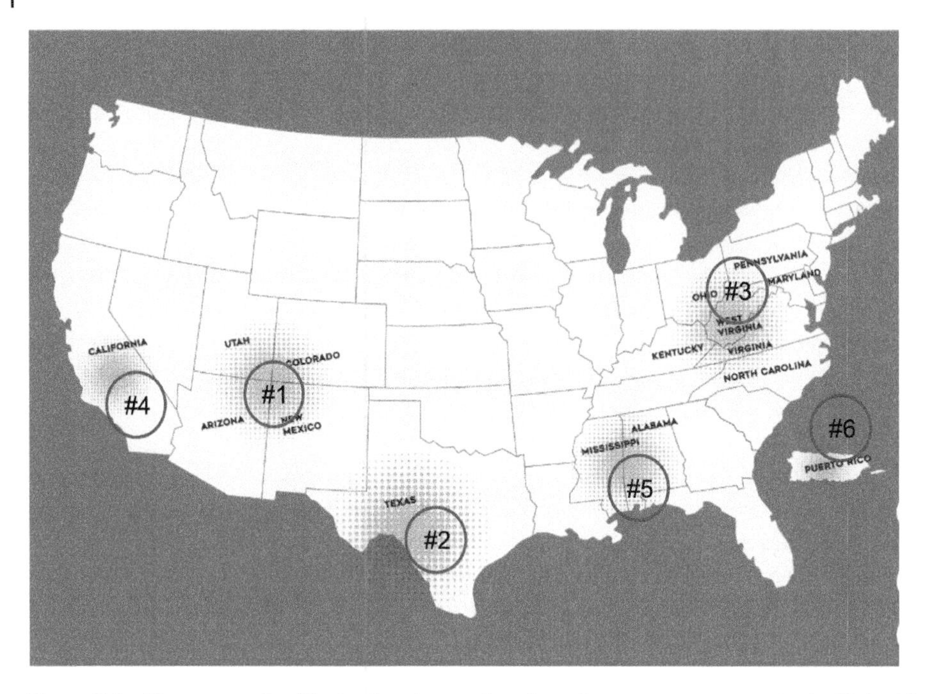

Figure 5.6 Six case studies illustrating the need to close the water-access gap in the United States. Source: Reproduced with permission from DigDeep.

case study. On this Nation in New Mexico, residents drive 40 miles several times each week to haul water for drinking, cooking, and bathing. In one area, residents report that groundwater depletion requires them to visit four or five locations to find the life-sustaining water they need. Female elders reported stockpiling water for emergencies and during winter months when hauling water is difficult. One resident reported fuel bills of $200 a month to haul water. Many of those interviewed reported having less than 10 gallons of water at home at any given time, making it difficult to balance hygiene and consumptive (basic hydration) needs. While the region itself has adequate water resources, the Navajo were historically excluded from compacts that allocated water distribution and an estimated 30% of people in the Navajo Nation lack access to running water and must haul water. Reliance on uranium-contaminated groundwater has resulted in a doubling of gastric cancer rates in mining areas, and high prevalence of diabetes can be linked to the easier access to sugary drinks than to clean drinking water. In fact, the rate of Type-2 diabetes is two to four times greater in the Navajo Nation as compared to the white US population.

The *Texas colonias (#2)* are residential areas located along the United States–Mexico border, occurring also in California, Arizona, and New Mexico. The colonias appeared in the mid-twentieth century as informal peri-urban or rural subdivisions, with many now surrounded by urban or suburban communities. Across the southern United States, colonias have a cumulative population of about half a million people (Texas alone has 2300 colonias) and are mostly Latinx. Nearly 2/3 of adults and over 90% of the children are US citizens. Residents are often low-income with many having informal employment. Colonias grew quickly because they provided affordable housing. Since these informal communities often lack infrastructure, families haul water by car or on foot. Purchase of trucked water cost upward of $250 per month for water used mainly for bathing and cleaning because of water-quality concerns.

For many years, an Appalachian town in *West Virginia (#3)* relied on a water system that obtained water from a stream receiving untreated toilet waste. Hauling water became so arduous that some residents reduced their water consumption to just five gallons per day while others showered using rain gutters to reduce their water hauling. In early 2018, local creeks turned black from time to time, which seemed to overlap with nearby mining and hydraulic fracturing ("fracking") operations. Rural Appalachian communities thus face three key water challenges: lack of household water access, poor water quality, and lack of wastewater services.

In the *Central Valley of California (#4)*, residents fill bottles at public taps because their household water is not safe to drink. In the state of *Alabama (#5)*, parents warn their children not to play outside because their yards are flooded with sewage. In *Puerto Rico (#6)*, wastewater regularly floods the streets of low-income neighborhoods. The plight of these regions is similar to those described above for developing countries and are hard to fathom in the twenty-first century United States.

The US Water Alliance report closes by proposing a plan of action to provide equitable water access, to "fill the gap," stating that to do so will require action by the water sector, government agencies, philanthropic organizations, nonprofits, and the public. The plan identifies four components: i) reimagine the solution, ii) deploy resources strategically, iii) build community power, and iv) foster creative collaboration. These components bear similarities with many of the concepts employed in addressing the global water crisis in developing countries.

Water access will often become strained or disrupted during or following a hazardous event, such as a natural disaster like a tornado or disease outbreak or a manmade disaster such as a chemical spill. One helpful preparation is to identify socially vulnerable communities that will need the most attention due to poverty, lack of transportation, crowded housing, or language barriers. The Centers for Disease Control and Prevention uses census data to determine the social vulnerability of every census tract in the United States (CDC 2020). This database along with color-coded maps can help public health officials be aware of locations that might be in greatest need for emergency shelters, clean water delivery, medicine and bedding, etc. Other nations can do this as well for their large urban areas.

Given that the gap analysis above identified Native American communities as experiencing great disparity in water access, it is instructive to further evaluate the plight of Native American communities in the United States. The Indian Health Service (IHS), which is responsible for water and sanitation services in Native American communities, provides an annual report on the status of these services. The deficiency in such services is ranked on a scale of 1–5, as depicted in Table 5.2. These Deficiency Levels (DLs) are used in the following discussion.

Figure 5.7 shows the fiscal year (FY) 2018 results for over 409 000 American–Indian/Alaska Native (AI/AN) homes included in the IHS Home Inventory Tracking System database (House outline A). Outline B represents 130 153 AI/AN homes identified as requiring some

Table 5.2 2018 deficiency service levels as defined by the IHS.

Level 1	An Indian tribe/community with a sanitation system, which complies with all applicable water supply and pollution-control laws; routine repair and replacement needs
Level 2	An Indian tribe/community with a sanitation system, which complies with all applicable water supply and pollution-control laws; deficiencies relate to capital improvements needed to meet the needs of tribe or community for domestic sanitation facilities
Level 3	An Indian tribe/community that has an inadequate or partial water supply and sewage-disposal facility that does not comply with applicable water supply and pollution-control laws or that has no solid waste disposal facility
Level 4	An Indian tribe or community with a sanitation system, which lacks either a safe water supply system or a sewage-disposal system
Level 5	An Indian tribe or community that lacks a safe water supply and a sewage-disposal system

Source: IHS (2018).

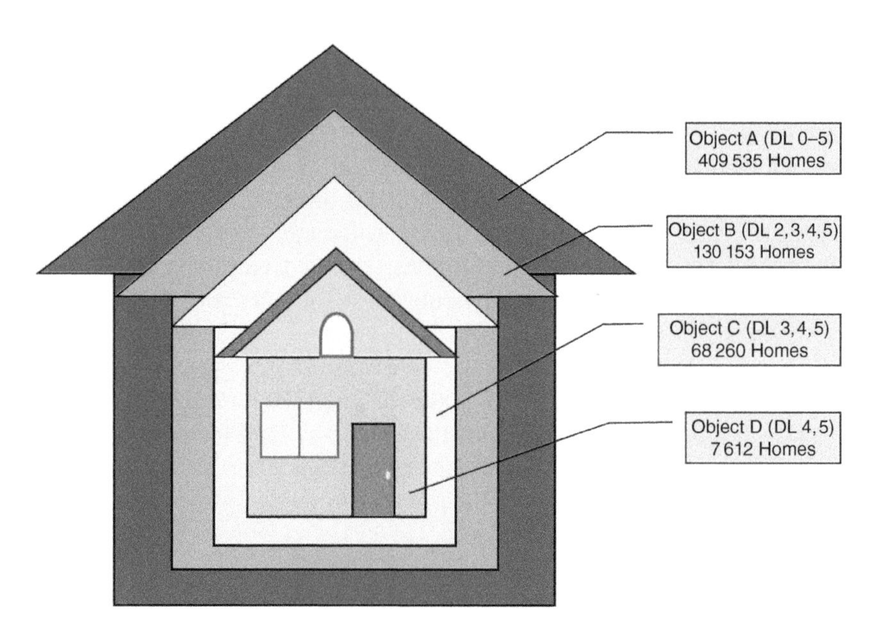

Figure 5.7 American Indian/Alaska Native (AI/AN) homes requiring sanitation facility improvements according to five deficiency levels (DL) defined above. Source: Original drawing adapted from IHS (2018).

form of sanitation facility improvement (classified as DL 2, 3, 4, or 5). The area of Outline C represents the 68 260 AI/AN homes that lack access to adequate sanitation facilities (DL 3, 4, or 5). Outline D represents the 7612 AI/AN homes that lack a safe water supply system and/or sewage disposal system (DL 4 and 5). From Outlines B, C, and D we observe that roughly half (206 025) of AI/AN homes inventoried require some level of sanitation improvement, with 19% (75 872) lacking adequate water supply or sanitation facilities and 2% (7612) lacking both a safe water supply and/or sewage system.

We thus see that affluent nations are not exempt from water-access variability and corresponding water inequities, not only across socioeconomic but also racial and ethnic groups. We will next look at economic disparities in water costs and associated water availability.

5.6 Equity Issues in Water Pricing

At the beginning of the chapter, we discussed whether water is a basic human right and thus should be free or whether water that is accessible, convenient, reliable, and safe is a commodity and thus warrants a fee. And, if water is a human right, should the state make sure that any fees be structured such that low-income citizens can afford this basic necessity? Black (2016) reported on the relative cost of water as a function of its source for five different major cities representing five different countries and three different continents (Africa, Asia, and South America) (Black 2016) (Table 5.3). The water sources vary from tanker truck to vendor to standpipe to household connection. Notably, within a given city the prices vary by as much as an order of magnitude while across the five cities (and countries) prices vary by up to two orders of magnitude. And bottled water is even more expensive (Walter et al. 2017). Given that the cheapest price shown is for a household connection ($0.23/1000 l), one might wonder why household connections are not more common in developing countries. The reason is that the poorest communities do not have the initial (capital) resources necessary to construct community-based systems and benefit from the resulting economies of scale. These communities must resort to more expensive water options, such as kiosks or vendors.

Another interesting parameter is the cost of household piped water as a percentage of disposable income for select countries. Table 5.4 shows this parameter for developed countries where this value ranges from 0.2 to 1.4%, with the US value being 0.3%. The average US household pays more for cell phone service and cable television (with convenience and leisure benefits) than for safe, reliable water and its disposal (with human health benefits). The amount that citizens in developed countries pay for water is a clear demonstration of the value of community infrastructure and economies of scale. This contrasts with remote villages in developing countries where lack of resources prevents community infrastructure. In a study we conducted in Cambodia, a literature search revealed rural water expenditures to be between 5 and 29% of household income. We used a target of 7% as the desirable upper limit for use in our study (Chamberlain and Sabatini 2014). Even so, this level (7%) is five

Table 5.3 Relative cost of water as a function of source and country.

City	Water source	Cost (US$/1000 l)
Luanda, Angola	Tanker, distant source	20.00
	Tanker, nearer source	4.00
Nairobi, Kenya	Vendor	6.50
	Standpipe	1.00
Kampala, Uganda	Bicycle vendor	5.40
	Kiosk	1.50
Bandung, Indonesia	Vendor	3.60
	Public tap	0.38
Lima, Peru	Household connection	0.28
	Truck	2.40

Source: Black (2016).

Table 5.4 Relative cost of piped water as a percentage of net disposable income.

Countries	% Disposable income
Poland	1.4
Germany	0.8
United Kingdom, France	0.7
Australia	0.6
Japan, United States, Canada, Norway, Sweden, South Korea	0.3
Mexico, Italy	0.2

Source: OECD (2009).

times higher than largest percentage listed in Table 5.4 and over 20 times higher than values reported for the United States and Canada. The sad reality is that the world's poorest people, in rural villages of developing countries, pay the most, on an absolute and relative basis, for life sustaining water – a clear and present water-equity issue.

Variability in pricing and economy of scale is evidenced in US water rates as well. For example, the water rate in Norman, Oklahoma (population 125 000), home of the University of Oklahoma, is currently $3.35 per 1000 gal ($0.90 per 1000 l). A smaller community of 6000 just 20 miles from Norman pays more than $5/1000 gal. According to Boyer et al. (2016), $4.90 per 1000 gal was the average water rate for communities across Oklahoma, like the smaller community near Norman (Boyer et al. 2016). Some communities have a rate two to three times higher than this average value (Vogel 2015). This diversity of rates can reflect economy of scale, availability, proximity and quality of water, infrastructure costs, etc. While the competitive rates for the three largest Oklahoma cities (Norman, Oklahoma City [OKC], Tulsa – all around $3/1000 gal) demonstrate the potential advantage of economy of scale, there are smaller communities in Oklahoma that have lower rates than Norman and OKC (Vogel 2015), as would be true, for example, for communities with nearby, high-quality ground water requiring limited treatment. Thus, economy of scale is not the only factor affecting water rates, as discussed further below.

As mentioned above, OKC, which is adjacent to and five times larger than Norman, has a water rate comparable to Norman ($3.06/1000 gal), although OKC imports the majority of its water from two lakes each about 100 miles away. Both Norman and OKC have increasing tiered or block rates (Griffin 2016), each having four tiers as shown in Table 5.5.

The base fee (sometimes referred to as meter fee) helps cover costs that are independent of water use; e.g. costs of meter reading (whether manual or software cost if automated), billing, customer service, and, in certain cases, capital/debt retirement cost (Griffin 2016), which could explain the higher base fee for OKC versus Norman. While it is logical that the base fee equally distributes fixed costs to all customers, it should be noted that elevated base fees may place undue burden on lower income families. For the increasing tier rate structure, with increasing monthly consumption the water rates increase, helping to promote conservation while charging larger water users incrementally more to help cover the higher resource and treatment costs necessary to meet these higher use rates (assuming that the least

Table 5.5 Monthly tiered water-rate structures in Norman and OKC, Oklahoma (as of June 2020).

Water usage	Monthly water fee – Norman, OK (pop ~123 000)	Monthly water fee – OKC, OK (pop ~644 000)
Base (meter) fee	\$6.00	\$17.22
<5 kgal	\$3.35/kgal	\$3.06/kgal
5–15 kgal	\$4.10/kgal	\$3.42/kgal
15–20 kgal	\$5.20/kgal	\$4.55/kgal
>20 kgal	\$6.80/kgal	\$5.91/kgal

Note: kgal = 1000 gal.

costly water sources are already being used and future expansions will require using more expensive options). The increasing tiered rate structure also helps keep down the rate for lower water users that may include lower income families. For all these reasons, increasing tiered rates have emerged as the dominant fee structure (Black, and Veatch Management Consulting, LLC 2019; Griffin 2016). (See also Chapter 15.)

A survey of water rates for the 50 largest US cities was conducted in 2019 (Black, and Veatch Management Consulting, LLC 2019). For most of the 50 largest US cities, the monthly water bill ranged from \$2.67 to \$8.00/kgal, with a few approaching or exceeding \$10.67/kgal. These variations in water fees may reflect the accessibility of the water source, the quality of the raw water, the costs of operation and maintenance, potential subsidies from or to other parts of the city budget, infrastructure integrity/deferred maintenance, progressiveness of management, etc. Figure 5.8 shows the monthly water bill for 11 of these cities, based on 7500 gal/month (average household use) and including the Environmental Protection Agency (EPA) affordability target of an upper limit of 2.5% of the city's median household income (MHI) for water bill alone or 4.5% of MHI for combined water and sewer bill (AWWA 2013). Monthly water bills in these cities are on average below the affordability target. However, the selection of MHI as a comparative indicator is not without controversy. Some believe that there is no clear linear relationship between MHI and the incidence of poverty, and it does not account for population levels that allow for greater economies of scale (AWWA 2013).

In addition to the variability in water rates, it is also interesting to note the variation in the city-specific values of affordability index, reflecting variations in median income. While most community rates in the study lie far below EPA's affordability target of 2.5% of median income, several communities (Detroit, Houston, Philadelphia, New Orleans) are precariously close to this target level, reflecting lower average income levels relative to water rates in these cities. The same survey also looked at the residential sewer bills for these same communities and found multiple communities exceeding the EPA recommendation of 2.0% of median income, illustrating the importance of sanitation services to community welfare. Thus, disparity in water costs is not limited to developing countries, but is also evidenced in affluent nations, and is important to water and sanitation equity and security.

The comparison of the retail price for bottled water versus tap water is also enlightening. A 20-oz bottle of water might cost \$1.50 or ~\$9.60 per gallon (\$9600/1000 gal). This price

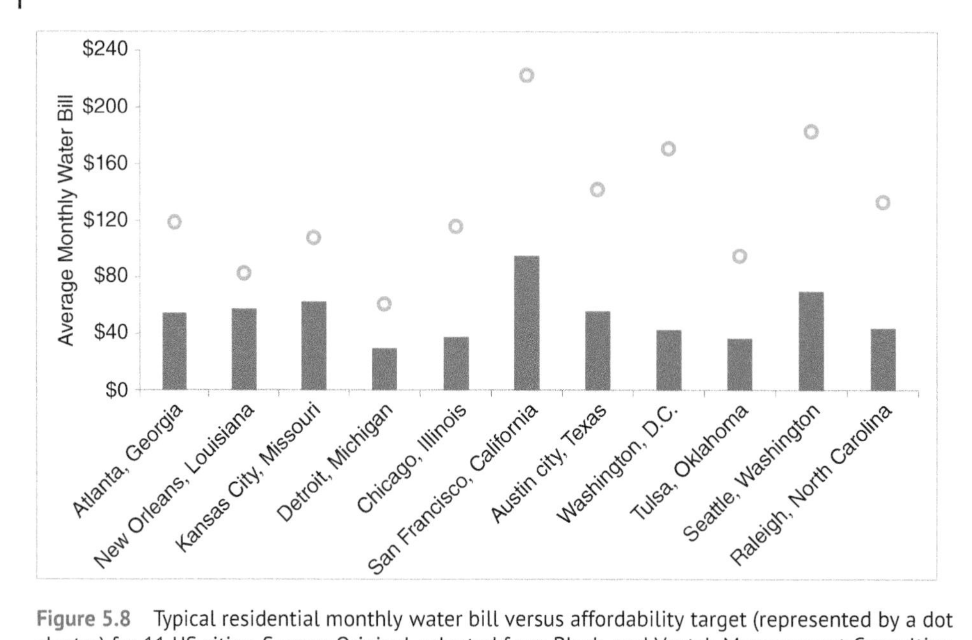

Figure 5.8 Typical residential monthly water bill versus affordability target (represented by a dot cluster) for 11 US cities. Source: Original, adapted from Black, and Veatch Management Consulting, LLC (2019).

contrasts with ~\$4/1000 gal of piped public water as discussed above. Thus, the bottled water is approximately 2400 times more expensive than public water provided to the home in the United States. Of course, we do not shower, wash our clothes, or flush our toilet with bottled water. Nonetheless, recognizing the disparity in cost of bottled versus tap water should impress us with how efficient our local utility is at providing safe, reliable water at such a reasonable price. In addition, it can be argued that delivering water through pipes is more environmentally benign than the manufacturing and shipping of plastic bottles, which eventually become solid waste, usually after a single use, and end up in a landfill, roadside ditch, or the ocean.

The equitable access to water can be limited by unit pricing, which can vary widely by location, source, and scale of delivery. A later chapter discusses the economics of water in more detail, but here we have highlighted the various factors that are involved.

5.7 US Water Rights – Equity Issues

US water rights can be categorized into two main approaches: riparian (common in eastern US states) or prior appropriations (common in western US states) (Getches et al. 2015). **Riparian water rights**, with its origins in English common law, allocates water to those with land adjacent to the path of the water. Historically, property with a riparian water right was strategic as it allowed the owner to operate water-driven mills, consume reasonable quantities of water, and use the water for fishing, boating, and hunting. The "reasonable use" aspect of riparian rights was introduced in early court rulings to consider reasonable needs of all riparian users. If the water is inadequate to meet the reasonable needs of all riparian users, then equitable reductions in use were anticipated. Over time, riparian rules have been adjusted by statutes and case law, and today riparian users are often required to obtain state permits. In some US states, even non-riparian users may acquire permits (Getches et al. 2015).

Prior appropriations with regard to water rights can be summarized by "first in time, first in right." Settlement in the western United States differed from the east in that the west was settled on federally owned land. Private ownership occurred later, after water uses had already been allocated. This is combined with the fact that rainfall and available water are scarce in the west (commonly delineated as west of the 100th meridian) (Getches et al. 2015). Furthermore, while industry developed along rivers in the east (e.g. water mills), mineral miners often needed water at locations (mines) far removed from surface water. But since miners could not own land, even if their operation was near a surface body, riparian rights did not apply since the land belonged to the government. Thus, these early entrepreneurs followed a similar approach to water as they did with minerals – first in time, first in right. Early court decisions reinforced this approach, which worked for farmers as well as miners (Getches et al. 2015). These rights remained valid as long as the beneficial use continued – leading to the adage "if you don't use it, you lose it." This contrasts with riparian rights, which would enable those few people fortunate enough to locate along the limited western waterways to monopolize the water, limiting development and expansion. For this reason, prior appropriation took root in this region and continues today, albeit with administrative agency oversight (Getches et al. 2015).

The prior appropriation doctrine presents certain threats to sustainable water security:

- Based on date of first diversion, those with seniority rights may take their full allocation even if it depletes downstream flow (e.g. during low flow or a drought). These appropriations need not consider downstream needs, e.g. hydropower production or aquatic ecosystems (Huffaker 2016).
- The "if you don't use it, you lose it" philosophy discourages adoption of water-efficient practices, which might also put the farmer at an economic disadvantage to his competitors. In this regard, a mandate to utilize such water-saving devices would level the playing field and make it easier for environmentally conscious farmers to make such changes and reduce stress on water resources.
- Those with sufficient resources can keep drilling deeper and deeper wells, causing even further lowering of the water table (groundwater "mining"). An interesting saga unfolds when neighbors continue this process, each lowering the water table below their neighbor's newly installed well, causing them to install an even deeper well, and the cycle continues – leading to "the deepest well wins" – until the aquifer is completely "mined" (depleted), that is. Sometimes called the "rule of capture," this scenario is a specific example of the tragedy of the commons in which, in the end, everyone is worse off (Hardin 1968).

Thus, while the prior appropriation doctrine may have been historically "appropriate" to enable rapid development of the west, this doctrine is ill-suited to respond to current priorities, which include the recognition of ecosystem services and alternate uses of the valuable resource. Unfortunately, a devastating drought may be required to motivate significant change in policy and enforcement (Huffaker 2016).

This brief discussion of water rights illustrates the negative impact that two opposing US policies have on the equitable allocation of water, thus hindering the pursuit of water equity and a water-secure future. For more details, see Getches et al. (2015) and Huffaker (2016).

5.8 Transnational Water Equity – Two Examples

Transnational water-equity issues emerge where water basins do not correspond to geopolitical boundaries and encompass multiple countries. Thus, water is frequently a topic of multinational treaties, as illustrated in Figure 5.9. One may presume that the original intent of these treaties was to formalize water equity among the signatories. But the equity of these transnational agreement is debatable and can lead to potential conflict among signatories. Two example cases, one in the United States (Colorado River basin) and one in Africa (Nile River basin), will be briefly introduced here.

In 1944, the USA–Mexico Treaty guaranteed Mexico 1.8 billion m^3 a year from the Colorado River. However, this treaty did not stipulate the quality of this water. In the 1950s, the allocated water became more saline due to the diversion of more Colorado River water to new areas. In response to this situation, a 1973 international agreement stipulated that the water allocated to Mexico should be of similar quality to that used in the United States, causing the United States to install a desalination plant. A new agreement in 2014 resulted in water traversing the final stretch of the delta for the first time since 1998. As can be seen, water-quantity and water-quality issues were important topics of discussion relative to the original treaty in 1944, with subsequent agreements promoting equity in the interpretation of the original treaty (Black 2016).

The Nile River is the longest river in the world, exceeding the length of the Amazon River by 132 miles (USGS 1990). With headwaters in Ethiopia (the Blue Nile) and Uganda (the White Nile), the Nile River drains water from 11 different countries before ending its journey in Egypt as it empties into the Mediterranean Sea. A 1929 agreement limited Nile water rights to Egypt and Sudan, neglecting water rights to the other countries through which the Nile traverses and leading to tenuous multinational relations. Later, agreements have attempted to rectify this inequity, with mixed results. Ethiopia's construction of its Grand Ethiopian

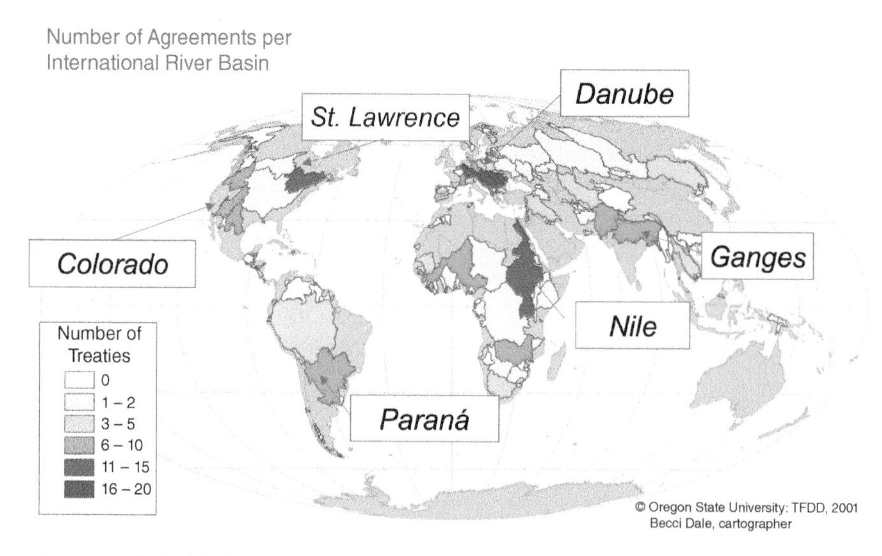

Figure 5.9 Multinational treaties still in effect in transnational basins across the world. Source: Adapted from OSU (2021).

Renaissance Dam (GERD) is a recent example of the continued pursuit of equitable sharing of the Nile River among countries in its basin. In the Khartoum Declaration of 2015, Egypt recognized Ethiopia's right to dam the Blue Nile, a positive step toward water equity, but challenges remain in the equitable distribution to all nations touched by the Nile River (Black 2016).

5.9 Conclusion

In this chapter, we have introduced the subject of water equity and presented several examples of water inequity to illustrate the importance of this topic. We started by describing water disparities in developing countries (e.g. between rural and urban, or within urban – from poorer informal or peri-urban regions versus wealthier regions in the city). We then went on to discuss water and gender equity, highlighting the case of women and girls who bear a disproportionate burden in fetching water in developing countries, inhibiting their ability to advance in society. Next, we discussed the fact that water-equity issues are not limited to developing countries, citing the Flint, Michigan, case and the US Water Alliance Report discussing the crisis in six regions of the United States, all severely lacking basic water and sanitation services. The plight of these regions is eye-opening and illustrates that, even today, water equity is a real and critical issue in affluent nations.

Water-pricing issues illustrate that the world's poorest, living in rural areas of developing countries, pay the most for water, on an absolute basis and as a percentage of their household income. But, just as above, disparities in water cost are not limited to developing countries. We presented variations in US water costs both within a state (Oklahoma) and across the nation (the 50 largest US cities), illustrating how fee structures can either favor or disfavor low-income citizens.

Finally, we closed this chapter by discussing that water-equity issues can relate to water rights, whether that be transnational or within a nation. The reader is thus now aware of the variety of water-equity issues, the pervasive nature of these issues, both in developing and affluent nations, and the importance of recognizing and appropriately responding to these water-equity issues in pursuit of a water-secure future.

Figure 5.10 is a humorous illustration by Kate Ely, a hydrologist with the Confederated Tribes of the Umatilla Indians in the northwest United States. She illustrates how water in the Columbia River basin flows toward capital, represented as dollar bills that intercept water at crucial junctures in the hydrologic cycle. Water becomes an economic commodity that may become out of reach to much of the world's citizens who live on less than $2 a day.

We may be tempted to be cynical or discouraged about the truth that is represented here. But the reality remains that the lives and health of millions of people are impacted by the equitable access to water across the globe.

*Foundations: Epidemiology, Randomized Control Trials (RCTs), and Measures of Association

Epidemiology is the study of the distribution and determinants of health outcomes in specified populations, and the application of this study to control for health problems:

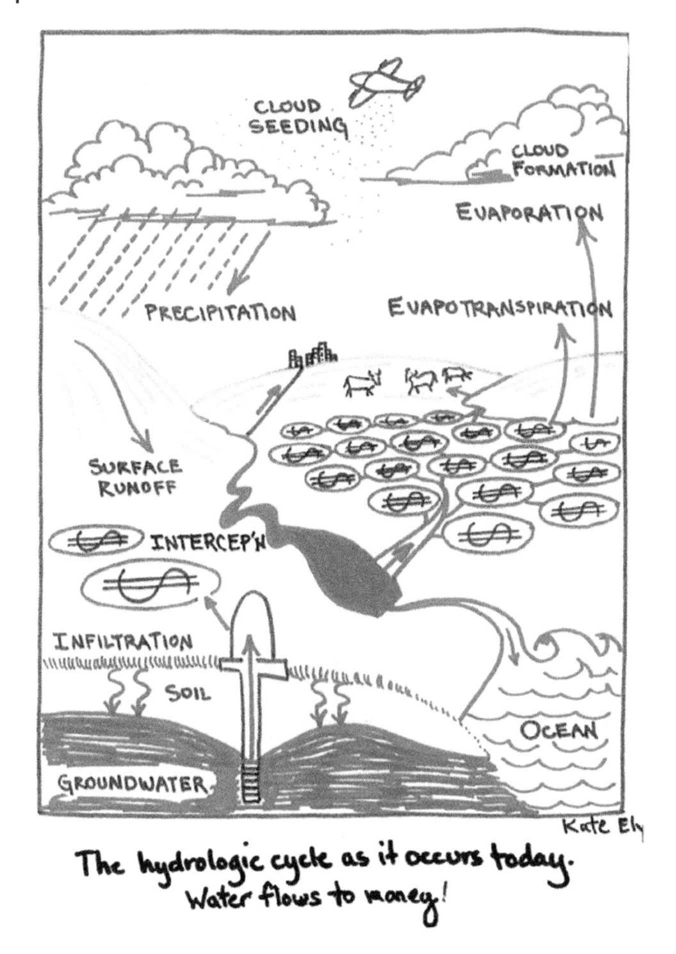

Figure 5.10 Water no longer flows by gravity alone. In reality, it flows toward money. Source: Used with permission of the artist, Kate Ely.

- *Distribution* is the description of place, time, patterns, and trends of disease.
- *Determinants* are the factors or exposure variables by which one can quantify disease or illness.
- *Health outcomes* can be either positive (studying a health intervention) or negative (studying cases of disease from exposure in a population).
- Specified *populations* are groups rather than individuals.
- *Application* is the purpose of the study: to detect causality, improve health outcomes, and to test effectiveness of an intervention.

For water managers and scientists, the science of epidemiology contains a suite of tools that can be used to measure the efficacy of interventions used to improve water security.

Interventions may be large water infrastructure projects or small improvements on the household scale. When a water-based intervention is introduced to a village or region, it is very important to quantitatively measure and evaluate the effectiveness of the intervention. Does the intervention do what we thought it would do? For example:

- Does a water-treatment system and piped water supply improve health factors as compared with villagers who do not have piped water?
- Does the provision of safe water at a school result in increased number of student hours in school?
- Is the ceramic water filter supplied to households actually being used and, if so, does it reduce incidences of diarrhea?

In this section, we examine some important epidemiological tools that can be used to evaluate water-security improvement measures.

The **randomized control trial (RCT)** is the gold standard in health interventions. Individuals or households are randomly assigned either to be part of an intervention or to be in a control group that is not part of the intervention. The two groups should be as identical as possible with the exception of one group receiving the intervention (ceramic filter, piped water, hand-washing station, etc.) and the other not receiving it. The intervention may be considered an "exposure" to a safer way of obtaining or using water.

For example, a government program in Addis Ababa sought to reduce incidents of gastrointestinal illnesses in school children by installing hand-washing stations in several of its schools, denoted as "study schools." One year after installation, its team of inspectors visited each study school and surveyed students and teachers while recording the incidences of diarrhea as reported. In order to accurately assess the intervention, the inspectors should be careful to consider:

- Rate of incidence of diarrhea at a similar school *without* a hand-washing station, or rate of incidence in the study school *before* the installation of a station as a way of comparison
- Manner of selection of the study schools. Was there a bias in selecting the schools, such as selecting those that are in higher income areas that may already have a lower incidence of diarrhea because of improved sanitation?
- Source of water used to supply water to the hand-washing station – is it clean or potentially contaminated at times?
- Usage of soap and hand-washing practice that is employed
- Other potential causes of diarrhea, such as contaminated food, by considering the F-diagram (given in Chapter 4).

If the trial is truly *random* and is able to compare results with a *control*, then the success of the intervention may be more adequately measured.

Measures of association are the quantifiable ways in which disease is described as being present in a population. *Prevalence* is the amount of disease currently within a population at a specific point (snapshot) in time. *Incidence* is the number of new cases within the population at a specific point in time, while *incidence rate* is the number of new cases over a period of time. An *odds ratio of incidence* can describe the ratio of cases to non-cases, i.e. the odds of contracting the illness as compared to not contracting it in the study population.

A table of outcomes can be used to record and calculate each of the measures described above (Table 5.6). In this table, the total population is given by the sum of all groups, $a + b + c + d$.

Using the variables of this table,

$$\textit{Prevalence} = \frac{(a + c)}{(a + c + b + d)} \tag{5.1}$$

$$\textit{Incidence rate} = \frac{(a + c)}{[(a + c + b + d) \times (\text{time at risk})]} \tag{5.2}$$

Table 5.6 A 2 × 2 table used to calculate various measures of association between an exposure or intervention and the presence of disease.

	Diseased	Nondiseased
Exposed (Case group)	a	b
Nonexposed (Control group)	c	d

Table 5.7 A table of outcomes following the introduction of a biosand filter in a rural region in Cambodia.

	Diseased	Nondiseased
With filter (Case group)	50	550
Without filter (Control group)	114	486

$$\textit{Overall odds ratio of incidence} = \frac{(a + c)}{(b + d)} \tag{5.3}$$

For example, consider that a rural region in Cambodia (population 1200) is studied prior to any intervention. The four villages in the region were very similar regarding demographics, history, and environmental setting. This group of communities is accustomed to using rainwater in the wet season and surface water sources in the dry season for most of their drinking and cooking needs. The prevalence of frequent diarrhea was reported by 230 people, or about 19% of the total population. A random group of households was selected to receive a household biosand filter, representing about 1/2 of the region's villagers. The remaining group of households did not receive biosand filters but was promised one at a later date. After two years, the outcome results were tabulated and presented in Table 5.7.

Prior to the intervention, the odds ratio of frequent diarrhea was 230/970 = 23.7% (diseased divided by nondiseased). After the intervention, the odds ratio in the case group was 9% (50/550), a reduction of 14.7% from the pre-intervention odds ratio. Using the above equation, the overall odds ratio post-intervention would be 15.8%.

Access to safe water for drinking and hygiene is essential for human health and is a basic human right. Interventions are sometimes needed to ensure that this access is being provided in an affordable and equitable manner. The tools of epidemiology can assist in objectively measuring and evaluating these critical interventions.

End-of-Chapter Questions/Problems

5.1 Based on Figure 5.1, discuss the following:

a. Is it better to be in the wealthiest quintile of a rural setting in a developing country or among the poorest quintile in the urban setting? In your opinion, why is this true?

b. What could be exceptions to this conclusion?

5.2 Do a search and find the most recent "Joint Monitoring Program" bulletin and discuss recent trends in access to water and sanitation in developing countries with a focus on:

a. Rural versus urban context,
b. Disparities within the rural and urban context, and
c. Gender-equity issues.

5.3 Referring to Table 5.1, discuss briefly the importance of water and sanitation in each stage of a woman's life cycle.

5.4 Considering the Flint, Michigan, case study, propose four to five steps that could have been taken along the way to prevent the situation from occurring in the first place. Consider not only technical but also social, government policy and oversight agency steps.

5.5 Considering the six case studies (see Figure 5.6) introduced in the "Closing the gap: US water-access needs" discussion, select one of these case studies and research the current situation and any recent, proposed or possible steps that could be taken to alleviate the situation.

5.6 Do an Internet search and find the most recent IHS Annual report on sanitation deficiencies. Using this report, discuss the current status of water and sanitation among AI/AN communities, including those regions with the greatest deficiency in sanitation access.

5.7 For your state of residence, or another state of interest to you, select two communities of small population (<8000), two of medium size (population will vary depending on the state), and two of the largest communities in this state. For each of the six communities,

a. Do a web search and provide a table of the base (meter) fee (independent of water usage) and the water-use fee (including tiered rates).
b. Using these values, calculate the monthly water fee for each of the following water uses: (i) 3750 gal/month, (ii) 7500 gal/month, and (iii) 15 000 gal/month.
c. Comment on your results in terms of any disparity in fees, and potential reasons, as well as the impact of these rate structures on low-income people who would likely have lower water uses.

5.8 Relative to US water rights, answer the following questions:

a. Discuss the similarities and differences in riparian rights and prior appropriation rights.
b. Historically, why were prior appropriations initially desirable and in what way is this approach less desirable today?

5.9 Referring to Figure 5.11, select one of the river basins of interest to you that has more than one treaty and provide a description of the basin including countries in the basin, water-quantity and -quality information about the river, and an early treaty as well as a recent treaty regarding the river basin.

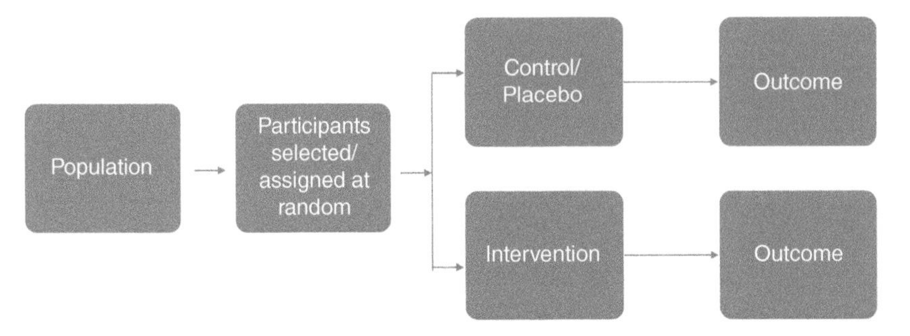

Figure 5.11 The RCT is the basis of evaluation for many studies in public health. Source: Original design, JFC author.

5.10 Provide a one-page description of the following basins and their related international treaties:
a. Colorado River basin
b. Nile River basin

5.11 A piped water system is brought to the edge of a large village in Nicaragua. Households are allowed to tap into the system and bring the water to their own house as long as they pay for the cost of materials and provide their own labor to tap into the main line. Thereafter, they are charged a fee per month based on the amount of water used. Discuss what is fair and what is not fair about this system.

5.12 A village in India (population 14 522) has water wells in two primary areas. Area A has wells that are suspected of containing arsenic and is used by 3300 residents. The remaining residents use wells in Area B that are presumed to be free of arsenic contamination. A household survey revealed that 489 residents in Area A had initial symptoms of arsenic poisoning. In Area B, there were 159 residents that displayed similar symptoms.
a. Develop a table of outcomes using the data given.
b. What are the odds of manifesting arsenic poisoning symptoms for residents in Area A and in Area B?

Further Reading

Allen, E., Maria Morazan, I., and Witt, E. (2018). Actively engaging women is helping solve the global water crisis. *Journal of Water, Sanitation and Hygiene for Development.* 08 (4): 632–639.

Clark, A. (2019). *The Poisoned City: Flint's Water and the American Urban Tragedy*. Picador: Reprint Edition.

Das, M.B. and Hatzfeldt, G. (2017). *The Rising Tide: A New Look at Water and Gender*. Washington, DC: World Bank.

Linton, J. and Budds, J. (2014). The Hydrosocial cycle: defining and mobilizing a relational-dialectical approach to water. *Geoforum* 57 (November): 170–180. https://doi.org/10.1016/j.geoforum.2013.10.008.

Devlaeminck, D., Adeel, Z., and Sandford, R. (ed.) (2017). *The Human Face of Water Security*, 2017e. New York, NY: Springer.
UNICEF (2017). *Thirsting for a Future: Water and Children in a Changing Climate*. New York, NY: UNICEF.

References

AWWA (2013). *Affordability Assessment Tool for Federal Water Mandates*. AWWA / WEF.
Black, M. (2016). *The Atlas of Water: Mapping the World's Most Critical Resource*, 3e. University of California Press.
Black & Veatch Management Consulting, LLC. (2019). "50 Largest Cities: Water & Wastewater Rate Survey, 2018–2019."
Boyer, T.A., Hopkins, M., and Moss, J.Q. (ed.) (2016). Willingness to pay for reclaimed water: a case study in Oklahoma. In: *Competition for Water Resources: Experiences and Management Approaches in the US and Europe*, 1e, 261–277. Waltham, MA: Elsevier.
Brooks, C.J., Gortmaker, S.L., Long, M.W. et al. (2017). Racial/ethnic and socioeconomic disparities in hydration status among US adults and the role of tap water and other beverage intake. *American Journal of Public Health* 107 (9): 1387–1394.
Caruso, B. (2017). "Women Still Carry Most of the World's Water." The Conversation. 2017. http://theconversation.com/women-still-carry-most-of-the-worlds-water-81054.
CDC (2020). "Social Vulnerability Index Fact Sheet." October 15, 2020. https://www.atsdr.cdc.gov/placeandhealth/svi/fact_sheet/fact_sheet.html.
Chamberlain, J.F. and Sabatini, D.A. (2014). Water-supply options in arsenic-affected regions in Cambodia: targeting the bottom income quintiles. *Science of the Total Environment* 488–489 (August): 521–531. https://doi.org/10.1016/j.scitotenv.2013.12.011.
Clark, A. (2019). *The Poisoned City: Flint's Water and the American Urban Tragedy*. Reprint Edition. Picador.
Das, M.B. and Hatzfeldt, G. (2017). *The Rising Tide: A New Look at Water and Gender*." 119074. The World Bank http://documents.worldbank.org/curated/en/901081503580065581/The-rising-tide-a-new-look-at-water-and-gender.
Devlaeminck, D., Adeel, Z., and Sandford, R. (ed.) (2017). *The Human Face of Water Security*. 2017th ed. New York, NY: Springer.
Getches, D., Zellmer, S., and Amos, A. (2015). *Water Law in a Nutshell*, 5e. St. Paul, MN: West Academic Publishing.
Griffin, R.C. (2016). *Water Resource Economics, Second Edition: The Analysis of Scarcity, Policies, and Projects*, 2e. MIT Press https://smile.amazon.com/Water-Resource-Economics-Analysis-Scarcity/dp/0262034042/ref=sr_1_1?dchild=1&keywords=Griffin+Water+Resource+Economics&qid=1613429558&sr=8-1.
Hanna-Attisha, M., LaChance, J., Sadler, R.C., and Schnepp, A.C. (2016). Elevated blood Lead levels in children associated with the Flint drinking water crisis: a spatial analysis of risk and public health response. *American Journal of Public Health* 106 (2): 283–290. https://doi.org/10.2105/AJPH.2015.303003.
Hardin, G. (1968). The tragedy of the commons. *Science* 162 (3859): 1243–1248. https://doi.org/10.1126/science.162.3859.1243.

Huffaker, R. (ed.) (2016). Institutional aspects and policy background of water scarcity problems in the United States. In: *Competition for Water Resources: Experiences and Management Approaches in the US and Europe*, 1e, 283–290. Waltham, MA: Elsevier.

IHS (2018). "Annual Report To the Congress of the United States On Sanitation Deficiency Levels for Indian Homes and Communities Fiscal Year 2018."

Kerner, O. (1968). *Report of the National Advisory Commission on Civil Disorders*. Washington, D.C: U.S. White House.

Meehan, K., Jepson, W., Harris, L.M. et al. (2020). Exposing the myths of household water insecurity in the global north: a critical review. *WIREs Water* 7 (6): e1486. https://doi.org/10.1002/wat2.1486.

OECD (2009). *Managing Water for all: An OECD Perspective on Pricing and Financing*. OECD.

OSU (2021). "Map and Image Galleries | Program in Water Conflict Management and Transformation | Oregon State University." 2021. https://transboundarywaters.science.oregonstate.edu/database-and-research/galleries.

Patel, A.I. and Schmidt, L.A. (2017). Water access in the United States: health disparities abound and solutions are urgently needed. *American Journal of Public Health* 107 (9): 1354–1356. https://doi.org/10.2105/AJPH.2017.303972.

Pieper, K.J., Martin, R., Tang, M. et al. (2018). Evaluating water lead levels during the Flint water crisis. *Environmental Science & Technology* 52 (15): 8124–8132. https://doi.org/10.1021/acs.est.8b00791.

Pulido, L. (2016). Flint, environmental racism, and racial capitalism. *Capitalism Nature Socialism* 27 (3): 1–16. https://doi.org/10.1080/10455752.2016.1213013.

Ranganathan, M. (2016). Thinking with Flint: racial liberalism and the roots of an American water tragedy. *Capitalism Nature Socialism* 27 (3): 17–33. https://doi.org/10.1080/10455752.2016.1206583.

Truelove, Y. (2011). (Re-)conceptualizing water inequality in Delhi, India through a feminist political ecology framework. *Geoforum* 42 (2): 143–152. https://doi.org/10.1016/j.geoforum.2011.01.004.

UNICEF/WHO (2019). "Progress on Household Drinking Water, Sanitation and Hygiene, 2000–2017: Special Focus on Inequalities." New York.

United Nations (2012). *The Millenium Development Goals Report 2012*. United Nations.

US Water Alliance (2019). *Closing the Water Access Gap in the United States: A National Action Plan*. US Water Alliance.

USGS (1990). "Rivers of the World: World's Longest Rivers." 1990. https://www.usgs.gov/special-topic/water-science-school/science/rivers-world-worlds-longest-rivers?qt-science_center_objects=0#qt-science_center_objects.

Vogel, J. (2015). Critical review of technical questions facing low impact development and green infrastructure: a perspective from the Great Plains. *Water Environment Research* 87 (9): 849–862.

Walter, C.T., Kooy, M., and Prabaharyaka, I. (2017). The role of bottled drinking water in achieving SDG 6.1: an analysis of affordability and equity from Jakarta, Indonesia. *Journal of Water, Sanitation and Hygiene for Development* 7 (4): 642–650. https://doi.org/10.2166/washdev.2017.046.

WHO/UNICEF (2017). *Safely Managed Drinking Water*. WHO https://www.who.int/water_sanitation_health/publications/2014/jmp-report/en.

6

Climate Change Impacts on Water Security

The objective of this chapter is to help the reader understand the impacts that a changing climate is currently having, and will likely have in the future, on water security. The rising air temperatures from increased greenhouse gas emissions have resulted in changes to the global hydrologic cycle. These changes affect water quantity, quality, and equity in unique ways, with some overlapping effects. Disadvantaged communities are especially prone to increased water insecurity, and socioeconomic inequities are heightened by the added stressor of climate change. Modern tools, such as remote sensing, can help refine predictions of both rainfall and runoff over long time periods. Climate change effects vary across the globe and thus must be considered with respect to latitude and region.

Learning Objectives

Upon completion of this chapter, the student will be able to:

1. Understand the observed and predicted changes to the climate and how these are important to water security.
2. Describe the differences between the paired descriptors of climate change/variability and stationarity/non-stationarity and why these are important in water-resource planning and water security.
3. Describe how warmer air temperatures can cause perturbations in the hydrologic cycle.
4. Describe the effects of climate change on both water quantity and quality.
5. Understand how climate change affects low-resource communities disproportionately, increasing the socioeconomic disparities that already exist among peoples.
6. Understand the modern tools of analysis and prediction for climate change effects on rainfall and runoff.
7. Appreciate the regional nature of climate change and understand that effects will vary throughout the globe.
8. *Perform simple calculations regarding dose and human health risk.

Fundamentals of Water Security: Quantity, Quality, and Equity in a Changing Climate, First Edition.
Jim F. Chamberlain and David A. Sabatini.

Companion website: www.wiley.com/go/chamberlain/fundamentalsofwatersecurity

6.1 Introduction

We have discussed water security in terms of water quantity, quality, and equity. An overarching factor that impacts all of these areas is the modern reality of a changing climate. Human beings and ecosystems have been adapting for many centuries to a climate that is now beginning to show patterns of more abrupt change in terms of air and ocean temperatures, precipitation, and extreme weather events. These patterns will have both near- and long-term effects on water for agriculture, production of energy, the well-being of ecosystems, human health and hygiene, and the geography of human settlements (Figure 6.1). In this chapter, we introduce the reader to the effects of climate change, while the effects on various sectors will be explored in more detail in subsequent chapters.

Fundamentally, water and climate are intimately connected throughout the hydrologic cycle – in the atmosphere, biosphere, hydrosphere (surface water and groundwater), cryosphere (solid forms of water, such as ice and glaciers), and soil moisture. Thus, changes in climate will inevitably result in changes to the water stocks and flows that make up this important cycle. Temperature and moisture patterns can shift more quickly than human and nonhuman species can adapt, causing water stress or insecurity at many levels.

There is indisputable evidence of a warming trend in recent decades as evidenced by the observed increase in global air and ocean water temperatures, widespread melting snow and ice, and the rise in sea levels. An increase in atmospheric greenhouse gas levels – carbon dioxide (CO_2), methane (CH_4), and nitrous oxide (N_2O) – has resulted in positive (warming) radiative forcing, as these gases/vapors absorb and radiate heat (Lindsey 2020). Of these gases, atmospheric carbon dioxide is the most important and is responsible

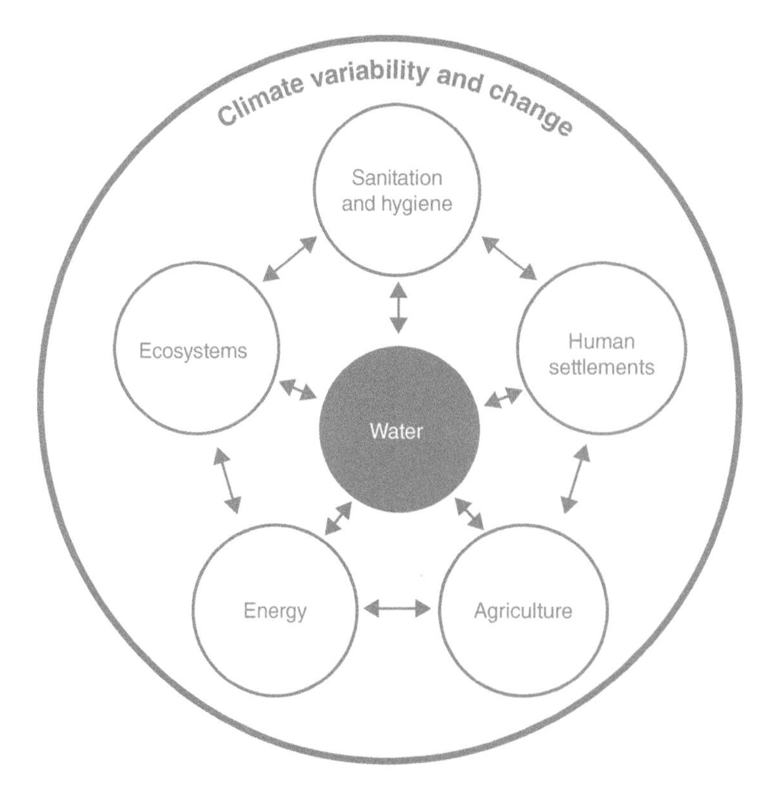

Figure 6.1 Climate variability affects the availability and quality of water for five important human sectors. Source: UNESCO (2020)/CC BY-SA 3.0 IGO/Public domain.

for about 2/3 of the total energy imbalance that is causing Earth's temperature to rise. The annual rate of increase in this constituent over the past 60 years is about 100 times faster than previous natural increases, such as those that occurred at the end of the last ice age 11 000–17 000 years ago (Figure 6.2) (Lindsey 2020). The timeline given in Figure 6.2 encompasses all of human history, including the beginnings of humanity in its present form (~315 000 years ago), the development of civilization (~50 000 years ago), and the dawn of agriculture (~10 000 years ago).

As seen in Figure 6.3, the average land surface temperature on Earth has increased by about 1.5 °C and sea surface water temperature has increased by 0.5 °C compared to the 1951–1980 average. Two-thirds of the warming has occurred since 1975, at a rate of roughly 0.15–0.20 °C

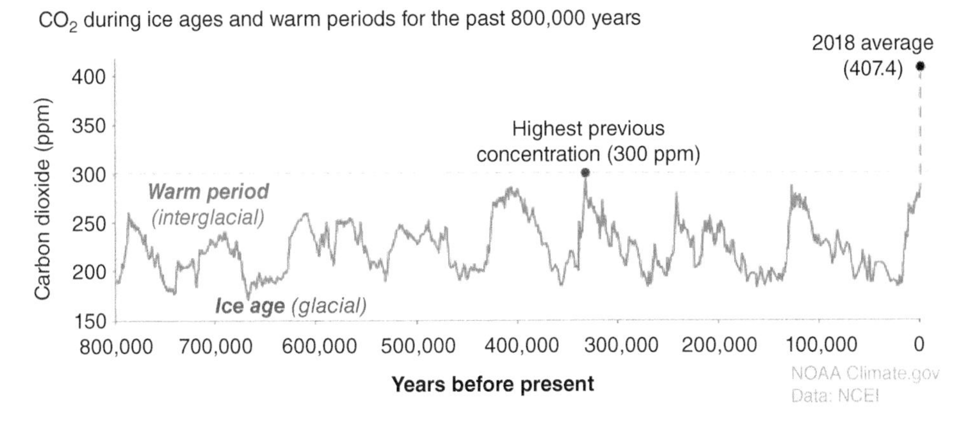

Figure 6.2 Atmospheric carbon dioxide concentrations have risen dramatically in the modern age, well above the highest previous concentration. Source: Adapted from Lindsey (2020).

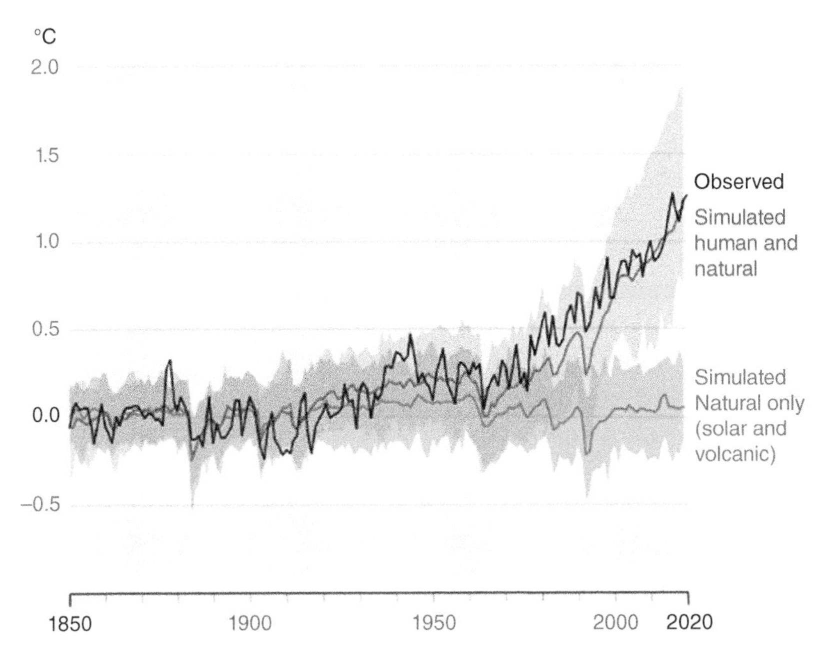

Figure 6.3 Observed land surface air and sea surface water temperatures, 1850–2020. Source: Based on NASA Goddard Institute (2021b).

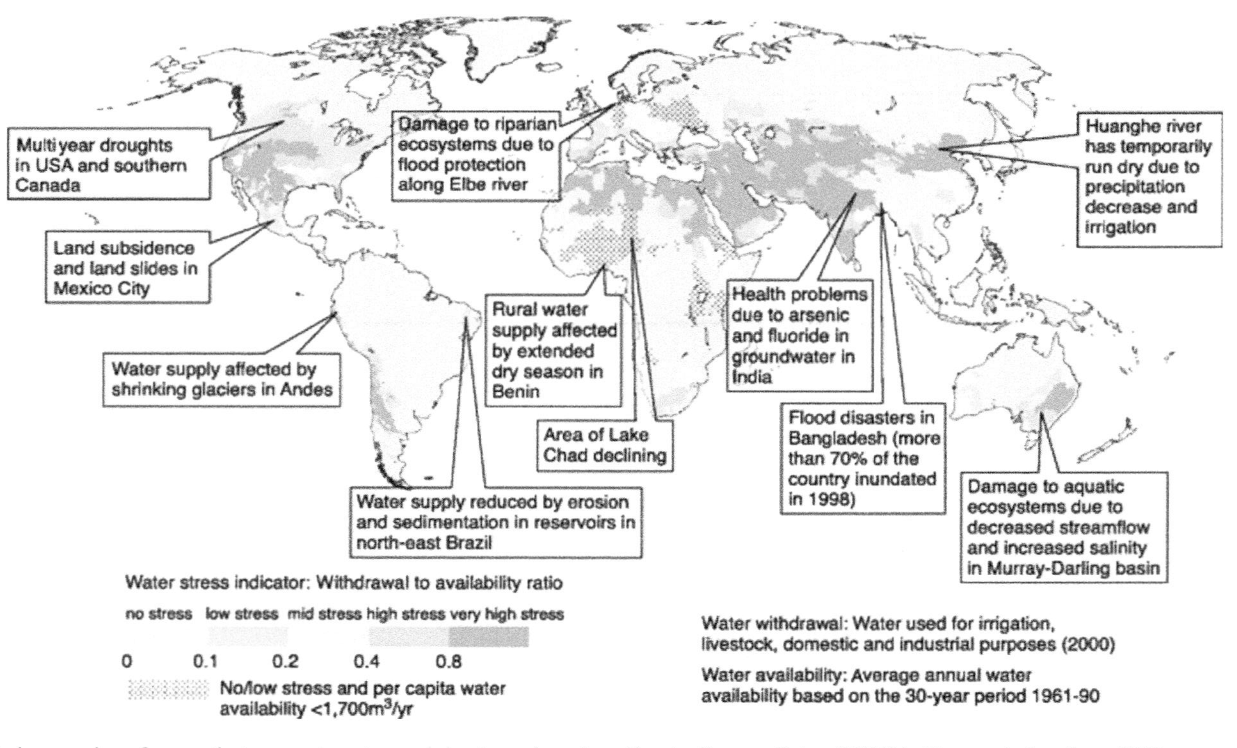

Figure 6.4 Global examples of areas that are water-stressed due to a changing climate. Source: Bates (2008)/with permission from IPCC.

(0.27–0.36 °F) per decade (NASA 2020). Recent analysis has predicted that the doubling of atmospheric CO_2 could result in temperature increases of 1.5–4.5 °C, with the high likelihood that the increase will be on the mid- to upper end of the range (Sherwood et al. 2020).

How do these anomalies manifest themselves in global effects to the human and ecological community? Several examples are shown in Figure 6.4. While droughts are more common in northern China and the western United States, flooding is a primary concern in Bangladesh. Shrinking glaciers affect the reliability and timing of water supply in the South American Andes and the Himalayas of India. Water quality is impacted by increased salinity of coastal groundwaters and erosion and sedimentation in reservoirs. This chapter describes more fully these and other types of impacts that climate change brings to water security.

Given the global impacts of climate change, the Intergovernmental Panel on Climate Change (IPCC) has come together as a body of 195 governments that are members of the United Nations or World Meteorological Organization (WMO) and are committed to studying the causes and impacts of climate change (IPCC 2013). Hundreds of scientists gather on a regular basis to analyze published reports and provide analysis and climate assessments to member states. These assessments are "policy-relevant" without being "policy-prescriptive" and serve to inform policymakers about the potential threats to water and food security (IPCC 2013).

Under smaller timeframes, such as by season or year, the climate is normally highly variable. The weather changes readily from hot to cold, wet to dry, windy to calm, and back again. Climate change, however, is the set of meteorological changes that happen over longer time periods, typically in decades. These changes are less easily discerned by the average person but are spoken of readily by people who have lived through the changes and who rely greatly on the weather, such as farmers, ranchers, and fishermen. They are also supported by data gathered by countless meteorological stations across the globe. The Mauna Loa Observatory in Hawaii, for example, has been continuously collecting weather data and monitoring CO_2 concentrations in the atmosphere since 1958. This current surface data, coupled with geologic data, such as ice core and sediment analysis, helps scientists extrapolate long-term trends and correlations of CO_2 gaseous concentrations and temperature oscillations over many centuries (Hansen 2009).

In this chapter we examine the impacts of a changing climate on water quantity, quality, and equitable access to disadvantaged peoples.

6.2 Observed and Projected Changes to the Hydrologic Cycle – Quantity of Water

The hydrologic cycle, described in Chapter 3, is the spatial and temporal transfer of freshwater between various stocks and pools – atmosphere to soils to surface and ground waters – and back again. The water is constantly moving between compartments, and the aboveground movement is stimulated primarily by air temperature differentials (in addition to gravity), which themselves result in wind currents and patterns. Changing climate impacts have been observed in the following hydrologic systems – precipitation, snow and land ice, ocean sea level and water quality, evapotranspiration (ET) and soil moisture, and stream/stormwater runoff.

Precipitation can be considered in terms of (i) overall quantity (an annual mean), (ii) the frequency of occurrence of extreme events, and (iii) the timing of the events. In the past 40 years, the quantity of precipitation over land has generally increased in the northern latitudes (between 30°N and 85°N) but has notably decreased in the 10°S to 30°N latitudes (Bates 2008). Concurrently, the *mean* rainfall has increased in the former but decreased in the latter latitudes. The variation can be accounted for by wind patterns that move moisture-laden air from regions near the equator to the northern latitudes. Further, these tropical and subtropical regions are home to over 60% of the world's population and are regions in which most of the water consumption is for the growing of food (World Economic Forum [WEF] 2011).

Stationarity in a time series occurs when statistical properties, such as mean, variance, and autocorrelation, are constant over a long period of time (UNESCO 2020). Stationarity, one might say, is a measure of the climate's natural variability. The design of water storage and conveyance structures, such as reservoirs, channels, pipes, and culverts, all assume stationarity in rainfall events. Thus, the magnitude of a 100-year rainfall event (with a 1% change of probability in any given year) can be predicted with reasonable certainty and used as basis for water resource design. Climate change has caused greater *non-stationarity* with shifting statistical parameters, which are not easily estimated. Structures that are based on historical data may now be over- or under-designed (Milly et al. 2008). If under-designed, the structure may result in collapse; if over-designed, the associated costs may be higher than is needed.

A second consideration is the frequency and magnitude of extreme events – especially floods and droughts. According to climate models, a greater increase is expected in the precipitation *extremes* rather than the *means*. This is because extreme precipitation is a function of water vapor availability, which is impacted by human activity (e.g. land-use change), while the mean depends primarily on the ability of the atmosphere to radiate long-wave energy (Bates 2008). Physical principles indicate that the atmosphere can hold 7% more water vapor for every 1 °C increase in temperature (Hornberger and Perrone 2019).

During the two decades from 1998 to 2018 floods and droughts caused more than 166 000 deaths, affected another three billion people, and caused total economic damage of almost US$700 billion (UNESCO 2020). Figure 6.5 shows the increase in weather-related catastrophes, with much of the chaos caused by storms and flooding. These hydrological events have increased by more than 50% in the past decade and are now occurring at an annual rate four times higher than in 1980 (UNESCO 2020).

Snow and land ice represent storage of water that is released during spring and dry summer months after a winter of frozen precipitation in the higher elevations. Nature's providence is evident in the way that solid water is slowly melted and released during the times most needed by humans, farm animals, and wildlife in the natural environment. A shift in the timing and volume of this melt can result in great hardships for farmers, ranchers, and villagers alike who have structured their livelihoods around this bounteous and dependable water source.

One fertile river valley is part of the Indus River basin of the Hindu Kush Himalayan region, a basin that spans into four nations – India, Pakistan, China and Afghanistan (Figure 6.6) (Adams 2019). It is home to 240 million people and derives nearly 80% of its water flow from melting glaciers (Ahmad 2019). Glacier volumes in this basin may decline by up to 90% in the current century as a result of decreased snowfall, longer melt seasons, and increased snowline elevations (Bolch et al. 2019). The resulting human and environmental impact will be staggering.

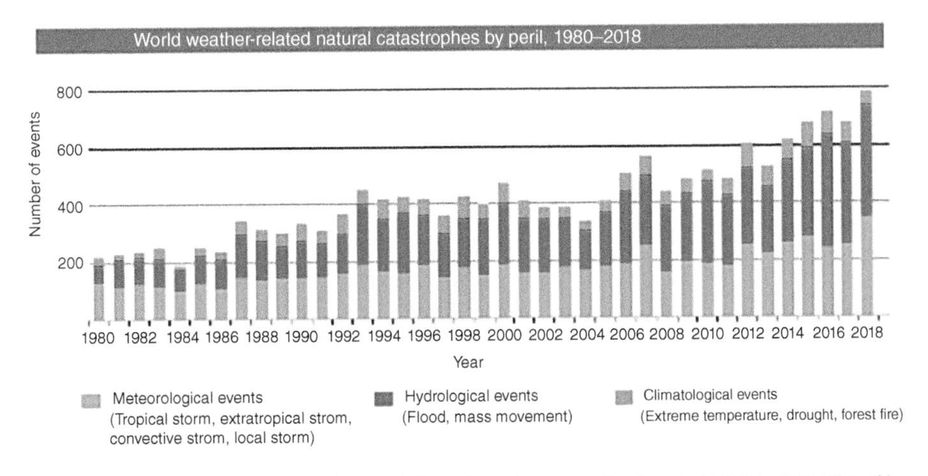

Figure 6.5 The number of global natural catastrophic events over the period, 1980–2018, has been steadily increasing. Source: UNESCO (2020)/CC BY-SA 3.0 IGO/Public domain.

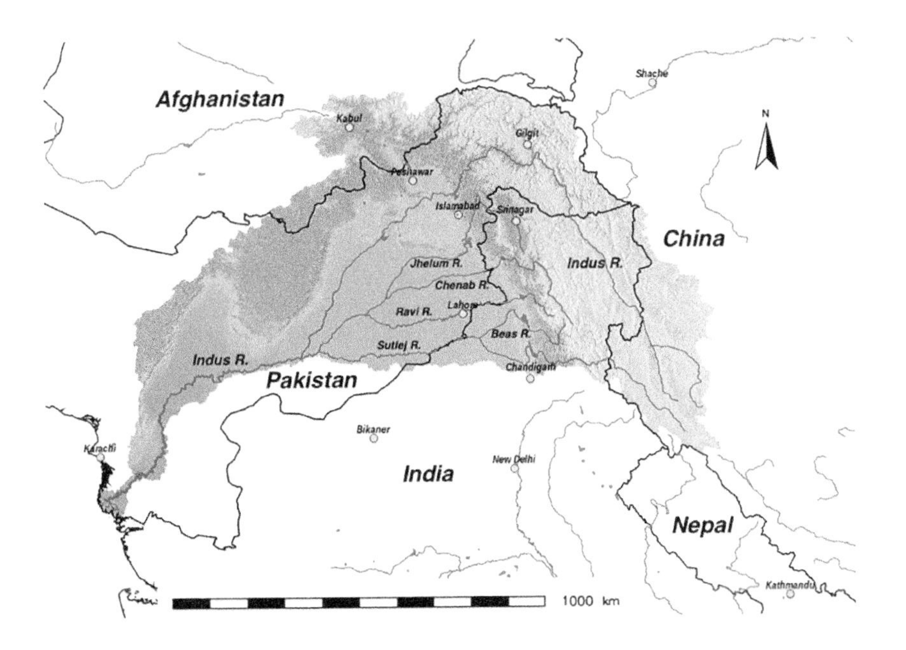

Figure 6.6 The Indus River basin of the Himalayan mountains in Asia. Source: Adams (2019)/with permission from ELSEVIER.

Figure 6.7 charts the evolution of the world's most important glaciers tracked by the World Glacier Monitoring Service. Data from all 11 glaciers in the US states of Alaska and Washington indicate a negative mass balance as well, with a mean loss of 870 mm/year (UNESCO 2020).

Permafrost is a layer of rock, soil, and ice that remains frozen throughout the year in many northern latitudes. The gradual melting of this layer allows connectivity between surface

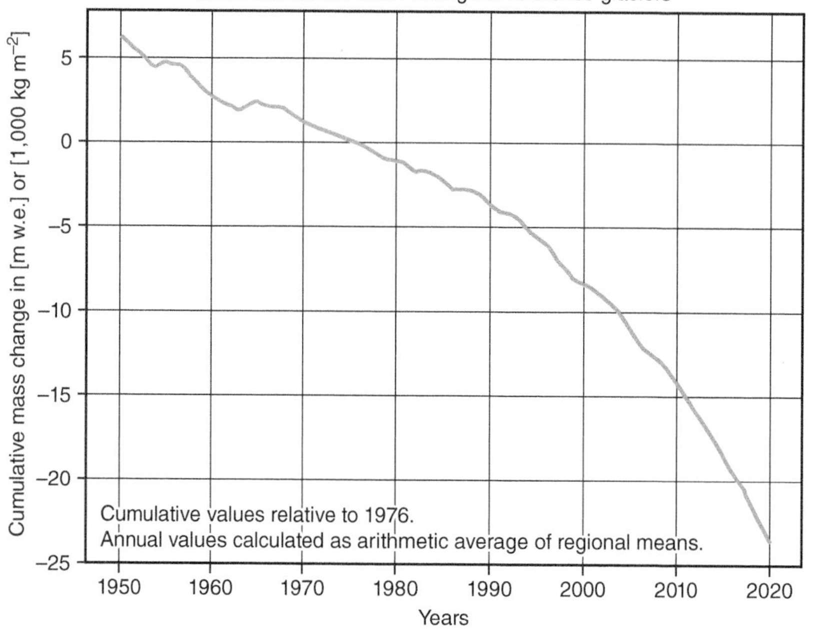

Figure 6.7 The change in cumulative mass balance (in water equivalents) for a set of global reference glaciers, showing yearly loss relative to 1976. Source: WGMS (2021)/with permission from University of Oklahoma.

water and deep soil. This can lead to drainage of lakes as well as subsidence and damage to built structures, including pipelines and water conveyance systems. The local hydrology is changed as soil moisture is released, affecting drainage patterns and ecological habitat.

Sea level is also impacted by a changing climate. Sea-level rise is mostly a result of two processes – (i) the melting of land glacial ice (~46%) and (ii) the expansion of water due to heating (34–75%, according to estimates) (Bates 2008; Church 2013). The IPCC's Special Report on the Ocean and Cryosphere in a Changing Climate (SROCC) predicts that the ocean's level will rise by 0.29–1.1 m (0.9–3.6 ft) by the end of this century (Portner 2019). However, other credible estimates are double this rate, and sea level could rise as much as 2.4 m (7.8 ft) if rapid loss of Antarctic ice has occurred (USGCRP 2017).

Both affluent nations and low-lying island nations are especially vulnerable to the rising sea level. Coastal cities are prime real estate in most countries, and in 2014, 127 million Americans, almost 40% of the population, lived in counties along the US coast (NOAA 2014). Rising sea levels mean greater likelihood of the extensive flooding of properties, especially when added to a storm surge from an extreme event. In addition, this rise may force salt water further into freshwater aquifers.

Evaporation, plant transpiration, and changes in soil moisture are dependent upon several factors – water and air temperatures, humidity, and wind speed. As global air temperatures rise, evaporation over water bodies and soil surfaces increases, adding more moisture to the atmosphere. As stated above, The warmer air can hold more moisture as well – about 4% more moisture for every rise of 1 °F (7% for every rise of 1 °C) (Darling and Snyder 2018). This moisture may fall as precipitation in downwind regions, shifting the locus of agricultural production to more northern latitudes. In such benefiting regions, overall ET could then increase due to a longer plant-growing season. This increase of ET could also lead to exacerbated periods of drought.

Evaporation rate can be measured using a shallow pan of standard size and construction and applying a pan coefficient to adjust for seasons and changing conditions. Recorded pan evaporation data is scarce, but extrapolation from such data has shown that actual evapotranspiration (AET) has increased during the second half of the twentieth century over most dry regions of the United States and Russia, a result of larger atmospheric moisture demand due to higher temperature.

Stormwater runoff and river discharge are the final output factors in a water budget, after water evaporation, uptake by plants, and soil moisture storage are accounted for (Chapter 3). Runoff has both positive benefits and negative impacts. Surface water flow fills rivers, ponds, and reservoirs with water that may be stored and used for irrigation, household use, or power generation. The negative impacts come from severe flooding, which happens when waterways cannot contain the extreme flows of water, causing damage and destruction to homes, farms, fields, and businesses. Changes in river flows result from the change of precipitation timing (seasonality) as well as whether the precipitation falls as snow or rain. Glacial melt will cause near-term rise in river flows but will gradually decline in future decades (Bates 2008). An analysis of five decades of European river flooding has shown clear patterns of change in flood timing. Northeastern Europe is seeing earlier spring snowmelt floods while Mediterranean areas of southern Europe see later spring runoff peaks. Researchers link this change of flood timing to warmer temperatures over the period of study (Blöschl et al. 2017).

Intense late summer rainfall, 70–100% above normal, caused flooding of over 14 000 square miles in Pakistan in the year 2010 (Figure 6.8). Unites States Agency for International Development (USAID) estimated that the floods affected more than 18 million people, destroyed 1.7 million homes, and caused almost 2000 deaths (Scott 2011). This intense rainy season was caused by several factors, including La Niña conditions and stagnant weather patterns over Russia. But the event highlights the severe vulnerability of communities who have planned their dwellings, food supply, economies, and water conveyance systems around knowledge of past rainfall patterns that are now so unpredictable. The consequences are more than inconvenient; they are devastating.

Figure 6.8 Even months after the Pakistani rains stopped in September 2010, homes and businesses remained inundated as there were few means of water dispersal to deal with the flooding. Source: Scott (2011).

Climate change is not the only major variable to affect the water quantity variable in water security. Population growth and associated increased water demand are major determinants in regionalized water scarcity (Vörösmarty et al. 2000). But adaptation to a changing climate will necessarily be part of the strategy in going forward.

6.3 The Effects of Climate Change on the Quality of Water

A warmer climate has both direct and indirect effects on water quality in the form of increased stormwater runoff, fresh/saline mixture due to sea-level rise, increased ocean acidity, heat pollution from warmer water temperatures, and the development of algal blooms.

Increased stormwater runoff from urban areas carries greater amounts of sediment, organic compounds, grease and oils, and lawn chemicals into receiving water bodies. In rural areas, increased runoff carries more nutrients, pesticides, and salts, in addition to sediments. This additional chemical loading is mostly left untreated in developing regions, and the deterioration of water quality directly affects downstream users. In affluent nations, more robust water treatment and monitoring may be necessary to adapt to these changing conditions.

Ocean water quality changes as the balance is altered between fresh/saltwater and atmospheric CO_2 exchange. Sea-level rise increases the possibility of saltwater intrusion along coastal zones, especially when adjacent freshwater aquifers are being drawn down due to pumping (Figure 6.9). A resultant brackish mixture (in the zone of transition) with increased levels of total dissolved solids (TDS) will be harder to treat to the level of drinking water standards. In addition, freshwater input to the ocean changes the salinity and, thus, the

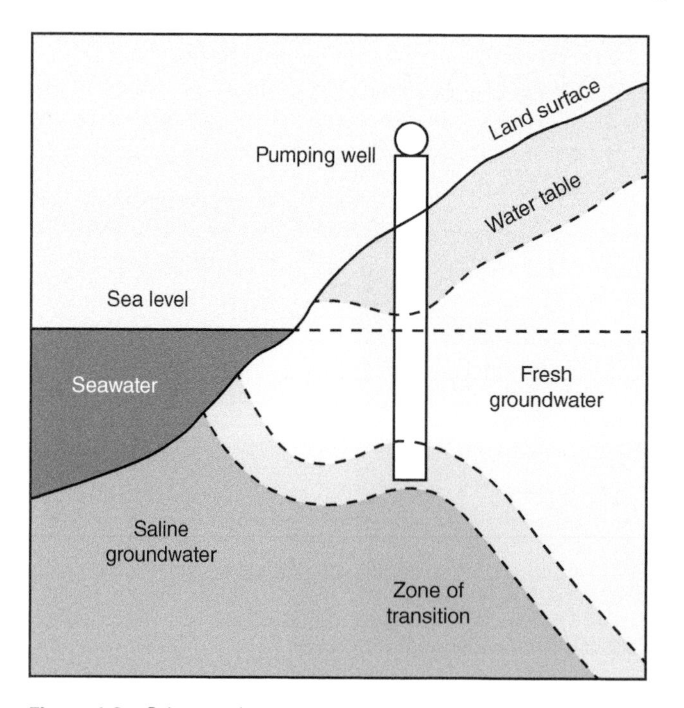

Figure 6.9 Saltwater intrusion can be exacerbated by sea-level rise and pumping of freshwater in coastal areas. Source: OP US EPA (2016)/with permission from EPA.

density of sea water (Bates 2008). The density-driven ("thermohaline") ocean circulation has a substantial impact on the surface temperature, precipitation, and sea level in the North Atlantic Ocean. Though unlikely in the twenty-first century, a disruption in this circulation would affect the climate in Northern European countries, including the United Kingdom (Caesar et al. 2018).

An increase in ocean *acidity* is correlated with an increase in atmospheric CO_2 that reacts with water molecules, forming carbonic acid and lowering the pH of the ocean. Since the start of the Industrial Revolution, the ocean's pH has dropped by 0.1 pH units. This drop may seem insignificant at first glance, but the pH scale is logarithmic, which means the acidity has increased by about 30%. Such a shift in environment makes it more difficult for marine life to extract calcium from the water in order to form shells and skeletons (Lindsey 2020).

Heat is another form of pollution that affects both freshwater and saltwater sources. In the past 50 years, the water temperatures of lakes in North America, Europe, and Asia have risen by as high as 2.0 °C (Bates 2008). These higher temperatures affect both the lake stratification and the timing and extent of the fall and/or spring turnovers, which recirculate biological organisms and nutrients (especially phosphorus) and may result in dissolved oxygen sag. This has the potential to result in fish stress and/or kills.

In addition, when water temperature is increased, it becomes less effective as a coolant for power plants. A heat wave in the Midwestern US (2006) forced nuclear power plants to reduce their power output because of warmer waters of the Mississippi River (UNESCO 2020). This loss of thermal efficiency can result in a 7–12% loss of thermoelectric power in most global regions and some losses as high as 81–86% in decades to come (van Vliet et al. 2016).

Finally, warmer temperatures result in *bleaching of coral* and an increase in *algal blooms*. The warmer reefs are bleached and weakened as they force out the microscopic algae, which give them their vibrant color. Coral reefs support some of the most diverse ecosystems on earth as they provide shelter, spawning grounds, and protection for many aquatic species. They also provide protection to coastal communities from damaging storm surge (WWF 2021). Algal blooms can occur in both freshwater lakes and coastal waters, which affect the taste, color, and odor of water and can result in toxicity to humans and wildlife. Harmful algae are impacted by climate change in two ways. First, toxic blue-green algae thrive in warmer water, which reduces mixing, allowing algae to grow thicker and faster. Second, algal blooms are a positive feedback mechanism as the blooms absorb sunlight and increase the temperature of the water even more (OW US EPA 2013b). At their worst, decaying algal blooms create toxic dead (anoxic) zones in the water, resulting in fish kills and doing much damage to coastal economies.

6.4 The Impacts of Climate Change on the Equitable Access to Water and Consequences to Disadvantaged Communities

The impacts of rising air temperatures have both direct and indirect effects that exacerbate the social and economic inequities that already exist between peoples. Vulnerable communities already live with narrow margins of safety, and climate change–induced shifts in water quantity and quality place an added burden on these communities.

Vulnerable communities lack the necessary resources to: (i) respond adequately to sea-level rise and extreme precipitation events, (ii) prevent food insecurity in agrarian households, (iii) reduce child and infant mortality, (iv) compensate for the reduced capacity to generate hydropower electricity, and (v) combat the greater susceptibility to waterborne and water-related diseases. Each of these is discussed in turn.

- Many resource-poor communities simply do not have the resources to adapt to the rising sea level and to an increase in the number of extreme weather events. Whereas affluent nations can build seawalls, elevated platform buildings, flood reservoirs, concrete stormwater channels, and other protective structures, resource-poor communities are essentially at the mercy of the sea. In addition, these communities need more time to recover from natural disasters, such as flooding.

  Small island developing states (SIDS) – e.g. Haiti, Grenada, Dominica, and many others – are characteristically more vulnerable to climate disasters because of socioeconomic constraints and environmental exposure. SIDS are indeed on the "front lines" of climate change. Climate change tortures these nations with escalating tides, cyclones/hurricanes, flooding, damaged crops, increased disease, and the loss of freshwater supplies (Sadat 2009). Sea-level rise threatens to inundate valuable coastal lands, including beaches and beachfront property (Figure 6.10). The Maldives, consisting of over 1000 islands to the southwest of India, is the world's lowest lying nation. On average the islands are only 1.3 m above sea level (Astaiza 2012). Many of the 325 000 residents (plus 100 000 expatriate workers) could lose their homes if the predictions come true. Although the SIDS nations are small in population, collectively they account for more than one-quarter of the world's nations (Sadat 2009).
- Food insecurity increases in low-resource communities who often have agriculture-based economies, with households growing food either for their own personal consumption or for external sale. Hydrologic changes due to climate change may result in soil erosion, overextraction of groundwater, increased salinization of irrigation and surface waters, and overgrazing of drylands (Bates 2008). Traditional crops and food sources may become suboptimal in the changing climate regime. In addition, labor-intensive agriculture often occurs in low-lying, flood-prone lands that are replenished with riverine silt deposition. While these changes may be met with resource-driven solutions and alternatives in more affluent nations, they can be devastating in marginal communities that are already living in the throes of water insecurity.

Figure 6.10 Sea-level rise threatens to inundate valuable coastal lands, including recreational beaches that bring in valuable tourism dollars to SIDS nations. Source: Moore (2019)/with permission from IPCC.

- Child and infant mortality can increase with lack of clean drinking water and the increased prevalence of diarrhea. As climate patterns shift the intensity and timing of rainfall, drinking water sources become susceptible to runoff pollution or may dry up altogether, forcing residents to turn to less-desirable alternatives. Lacking adequate health care and resources, these residents cannot mitigate the changes as easily as residents of affluent nations.
- Hydropower plays a significant role in the electrical generation portfolio of many developing nations, including Brazil, Turkey, China, and India (Yüksel 2009). Climate change–induced shifts in rainfall and runoff patterns can adversely impact hydropower production as a lack of storage and water elevation results in a decline in electricity production. A loss of hydropower may incentivize the use of power production using fossil fuels, thereby furthering the acceleration of greenhouse gas emissions.
- Climate change shifts the spatial distribution, intensity, and seasonality of vector-borne diseases by shifting the vector habitats and ability to thrive. Mosquito activity may decrease during drought, but it may be introduced into wetter areas where individuals have not developed immunity against diseases such as malaria and dengue fever. In these areas, epidemics may become more frequent. In addition, the distribution of schistosomiasis, using snails as intermediate hosts, has been altered in China by warmer global temperatures (Bates 2008).
- Because of these factors, climate change is already a greater threat to resource-poor communities and will continue to be so into the foreseeable future.

6.5 Perspectives from Three Global Regions

Climate change does not affect all nations equally, and its impacts can even vary for different regions within a given nation. Thus, the resultant challenges to water security are quite location specific. To illustrate this, we give summaries here of observed and projected aspects of climate change in three regions. [This section is adapted in large part from the IPCC report, *Climate Change and Water* (Bates 2008). The reader is encouraged to read the full report for more detail.]

The continent of Africa. The number of extremely wet seasons in east Africa is expected to increase due to climate change. Most models project a decrease in runoff in north African streams and rivers but an increase in east Africa. Much uncertainty exists in models for the rest of the continent. About 25% (200 million people) of Africa's population currently experience water stress. Water availability at less than 1000 m^3/person-year is expected by 2025 in nine eastern and southern African countries, including Kenya, Egypt, and South Africa. This prediction is based mostly on population growth but will be exacerbated by climate change. Egypt's agricultural sector, for example, consumes 85% of the total annual water resource. Thus, a loss of available water will impact Egypt's food and economy.

The Sahel (sub-Saharan) region of the African continent is the vast semiarid stretch between the Sahara Desert to the north and the tropical savannas to the south. Recent decades have witnessed loss of rainfall and periods of drought (Figure 6.11). Its 10 countries include Mali, Niger, Nigeria, Chad, and Sudan, and its 2014 population of 920 million is expected to double by the year 2050 (May 2015). The demographic boom, coupled with increasingly stressed water and food resources, may result in conflicts both within and among nations.

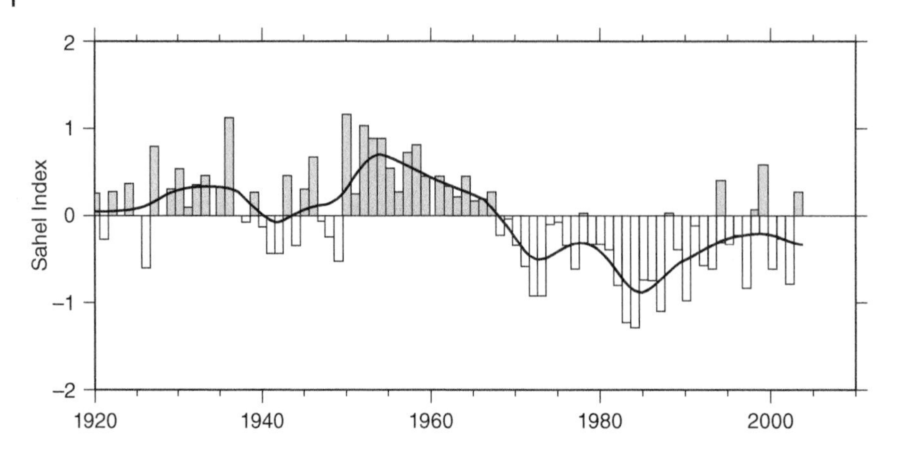

Figure 6.11 Sahel regional rainfall, 1920–2003 (vertical bars), with decadal variation shown by the smooth black line. Source: Bates (2008)/with permission from IPCC.

Mountain ecosystems atop Mt. Kilimanjaro (Tanzania) have been impacted by dry climactic conditions. The forest line has shifted downward by several hundred meters during the twentieth century, and the frequency and intensity of forest fires on the mountain slopes have reduced the fog-capturing capabilities and, thus, upset the water balance on the mountain (Bates 2008).

India and other Asian lands. Similar to Africa, many Asian nations have a high population with a fast growth rate, low levels of development, and a poor coping capacity. In general, the frequency of intense rainfall events has increased over much of the region, creating flooding, landslides, and mud flow. But the number of rainy days and the total amount of precipitation have decreased (Bates 2008). One can easily see the challenges to agriculture resulting from these conditions. In parts of China, lakes and rivers are drying up. In India, Pakistan, Bangladesh, and Nepal, a changing climate aggravates the already present water insecurity due to increasing demand and inefficiencies of water supply and treatment (Bates 2008).

India is home to nearly 18% of the world's population with over 1200 people per square mile (compared to 94 per square mile in the United States). Thus, it deserves special consideration regarding the impacts of climate change, which are heightened by increased population density. In one year alone (2018–2019), as many as 2400 Indians lost their lives to extreme weather events, such as floods and cyclones (hurricanes). But there is much regional disparity in this large country. In the summer of 2019, residents of Chennai (southeastern coast) were praying for rain, while those in Mumbai (western coast) were suffering the effects of far too much rain.

And India is hot. If climate change is left unchecked, average temperatures could climb from the current 25.1 °C (77.2 °F) to 29.1 °C (84.4 °F) by the end of this century (Padmanabhan et al. 2019). The number of extremely hot days (>35 °C, >95 °F) has already increased in Delhi to 1613 in this decade (2009–2018) from 1009 in 1959–1968. That is about 161 days each year, nearly half the year! Such extreme heat results in loss of work productivity in many regions of the world (Figure 6.12). The increase in temperatures also brings increased evaporation and loss of soil moisture. This factor coupled with a shifting time of the monsoon is especially hard on poorer farmers who are not able to irrigate.

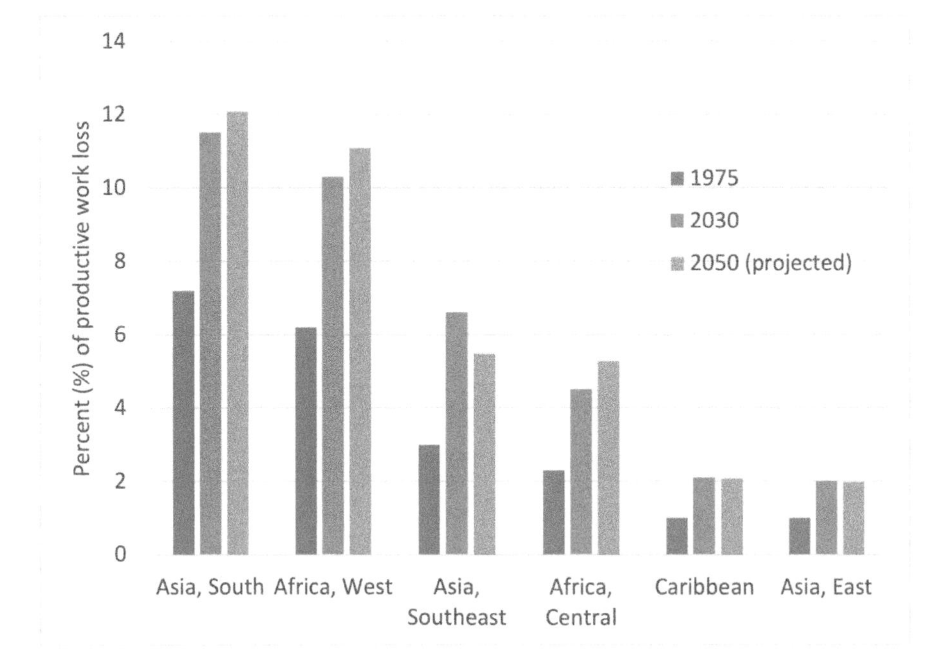

Figure 6.12 Additional extreme heat days result in productive work loss, up to 12% in South Asian countries. Source: Adapted from Kjellstrom (2014).

North American continent. The North American continent is a temperate zone with an abundance of freshwater resources and of species biodiversity. Projected changes in rainfall, soil moisture, and streamflow show that some areas will benefit and others will be hampered by climate change effects. In general, more precipitation will fall as rain rather than snow, and so spring and summer streamflow will be reduced. In addition, the peak streamflow from snowmelt will happen earlier in the spring due to warmer temperatures. There are already more extreme precipitation events, and widespread thawing of the permafrost in northern Canada and the Rockies. Increased periods of drought in the western United States are already stretching the water supply for populated cities such as Phoenix and Los Angeles.

The Colorado River watershed supplies water for seven US states, two Mexican states, and 34 Native American tribes. The upper 15% of the basin is in the most rain- and snow-rich area and supplies most (85%) of the flow. Business-as-usual scenarios estimate that the provided water will decrease by 30% of its current level, and the conditions of the Colorado River Compact, signed by all users, would be met only about 60–75% of the time (Bates 2008) (Figure 6.13). Thus, the projected water demand will exceed the water supply in coming decades. Through an intricate series of dams, reservoirs, water channels, and groundwater withdrawals, the population of this region (~38 million) is still growing with few alternative sources of water. As in many other regions of the globe, human and ecosystem survival will depend upon water reuse, conservation, efficient management, and human ingenuity in a diminishing water supply.

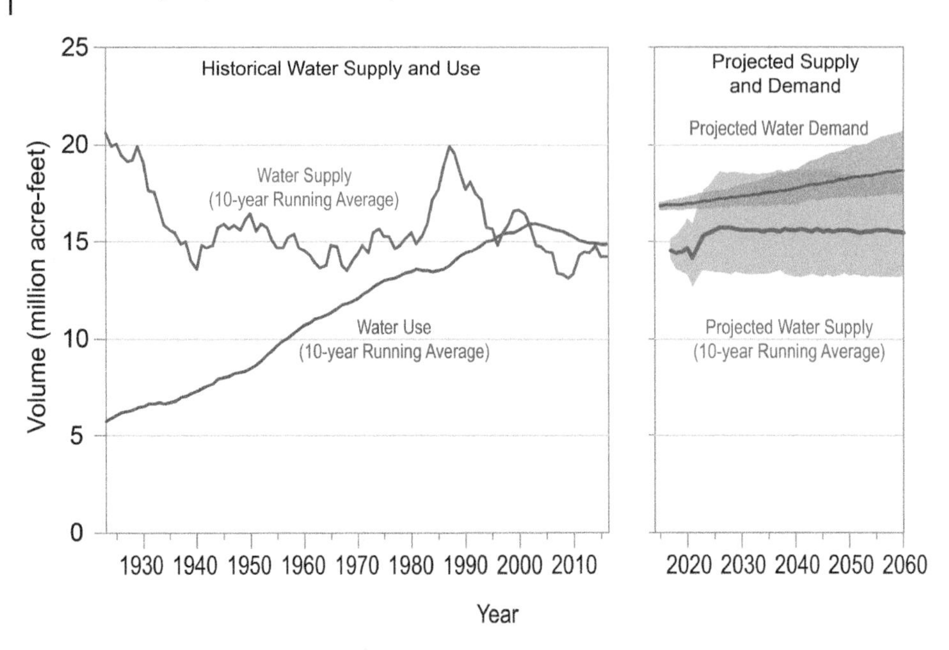

Figure 6.13 Historical supply and demand for the Colorado River, along with projected supply and demand. Source: USGCRP (2018).

6.6 Prediction of Natural Disasters Resulting from Climate Change

As a first hydrologic indicator of drought or flooding, runoff volume is the most important. A severe loss of runoff can result in depletion of both surface and groundwater resources needed for household, agricultural, and commercial uses. Likewise, severe excess runoff can result in flooding, deterioration of water quality, and landslides. The prediction of these disasters, then, has become a science of utmost importance. Especially in resource-constrained countries, very little actual rainfall data is collected and major streams and rivers are left ungauged for runoff estimates. This section describes some modern tools and techniques of analysis that can be combined with climate-change scenarios to predict runoff volume and thus anticipate these twin natural disasters.

- The Thornthwaite monthly water balance model is elegant in its clarity and requirement of few input parameters (Figure 6.14). Required parameters are mean monthly temperature and precipitation as well as the latitude of the location of interest in order to establish day length. From these parameters and some threshold temperatures (for transition between rain and snow), the model can estimate potential evapotranspiration (PET), a critical parameter for combining with other site characteristics and data for estimating AET, soil moisture, and runoff. The direct runoff factor is a function of the land cover, especially with regard to impervious surfaces, and can be adjusted in the model's input.
- Remote sensing technology can now be used to improve the accuracy, spatial coverage, and resolution of precipitation estimates (Hong et al. 2007). These estimates are in real time (sub-daily) and at scales of tens of kilometers. The data can be used in combination with the Natural Resources Conservation Service (NRCS) Curve Number method for estimating

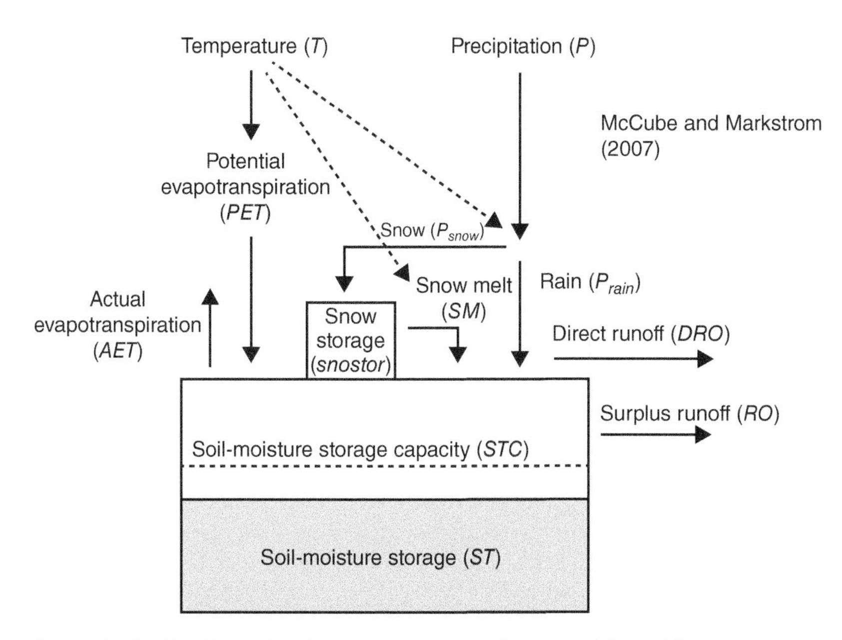

Figure 6.14 The Thornthwaite monthly water balance model provides a water accounting that can be used to predict runoff flows and soil-moisture storage. Source: McCabe and Markstrom (2007). USGS public domain.

runoff over a land area. The curve number (CN) is a dimensionless number between 0 and 100 that characterizes the surface permeability of the watershed topography. The CN is a function of soil type, antecedent soil moisture, and land use. Remote sensing can give an estimate of CN over large areas, which may then be refined by field surveys.

Using remote sensing to estimate CN values over latitudinal zones, runoff estimates were made based on the Thornthwaite model and compared to observed runoff data from the Global Runoff Data Center (GRDC) over a nine-year period, 1998–2006. (Hong et al. 2007). Three data sets were employed and modeled as shown in the three panels of Figure 6.15. The simulations (red) matched the general trends (blue), but the peaks were not as pronounced. As the model is more widely used and refined, the fluctuations may be more accurately estimated.

The US National Aeronautics and Space Administration (NASA) agency uses a satellite, GRACE-FO, to track and measure water movement and mass changes on the Earth's surface across the globe. By using the force of gravity in monitoring the changes in surface and underground water storage, ice sheet and glacier volumes, and changes in sea level, this satellite can give an accurate and integrated accounting of the evolution of the Earth's water budget (NASA 2021a).

- Global emission scenarios, developed by the IPCC Working Group III, are a suite of long-term emissions scenarios that incorporate inputs and impacts from the three main driving forces – technological change, demographics, and economic developments. When modeled, these scenarios provide policy-makers with a range of anticipated conditions with which to formulate short- and long-range policy. For example, clusters of scenarios are built around energy technologies as being (i) fossil fuel intensive, (ii) predominantly nonfossil fuel, and (iii) balanced between the two. These clusters are combined with socioeconomic storylines that vary according to predictions regarding population and

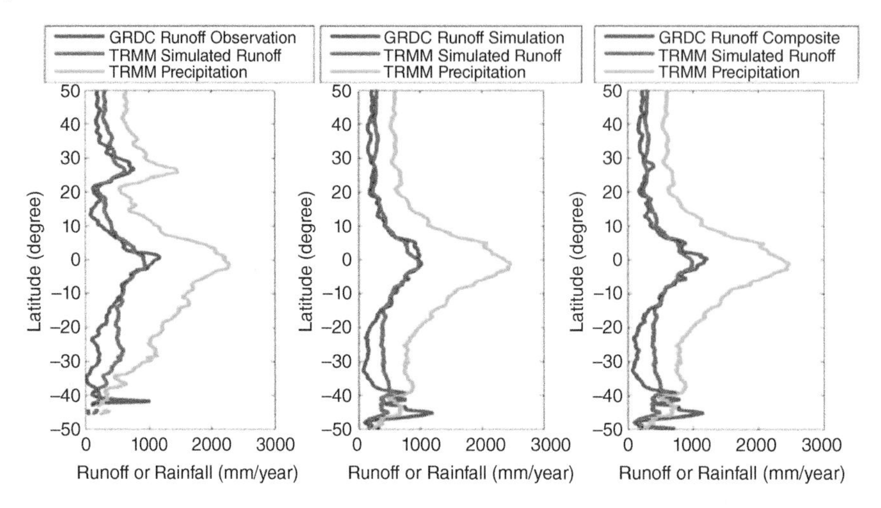

Figure 6.15 Comparison of simulated (TRMM) and observed (GRDC) global data to estimate mean runoff for zonal latitudes. Panels show data that are (left) observed, (middle) WBM (water-balance model), and (right) composite data from a third-party researcher. Source: Hong et al. (2007)/with permission from John Wiley & Sons.

per capita income predictions (IPCC 2000). The CO_2 emissions predicted for the year 2100 range from 10 times the levels in 1990 to close to 0 with some forms of intervention (Figure 6.16).

These three elements, when used together, can provide a scientific basis for the estimation of drought and flooding over long-term time intervals. The two case studies in which these elements have been utilized to predict drought and/or flooding are given in the following text.

The *Nzoia River basin in Kenya* is a major source of water for more than three million people who rely heavily on cereal and sugarcane-farming. The basin has a semiarid climate and is one of the major contributing sub-basins to Lake Victoria in East Africa.

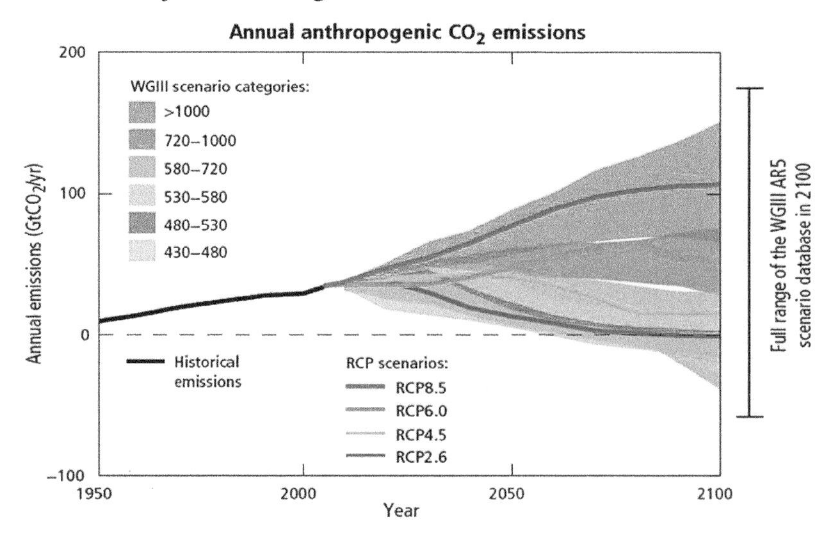

Figure 6.16 Global CO_2 emissions over a range of scenarios, 1950–2100. Source: IPCC (2021a)/with permission from IPCC.

Table 6.1 Model results reveal that runoff is more sensitive to temperature increase than to precipitation increase.

Water cycle component	Baseline	Change relative to 1990–1999 baseline		
	1990–1999	2020	2060	2090
Temperature (°C)	19.48	+0.81 (+4.2%)	+1.63 (+8.4%)	+2.10 (+10.8%)
Runoff (mm/month)	55.57	−7.77 (−14%)	−9.30 (−16.7%)	−10.03 (−18%)

Source: Adapted from Adhikari and Hong (2013).

Using the Thornthwaite monthly water balance model and actual precipitation and temperature data, values of runoff were determined for the baseline period, 1990–1999 (Adhikari and Hong 2013). The modeled results varied from the observed runoff by about 10%, capturing the seasonal fluctuations in general by overpredicting runoff somewhat during the rainy season. Using three IPCC emission scenarios, the model was extended into the decades of the 2020s, 2060s, and 2090s. The second half of the century is expected to be wetter, with precipitation increases of 5–15%. However, the results show that, in spite of increased precipitation due to additional moisture in the atmosphere, runoff volume in the basin was reduced because of higher air temperatures (Table 6.1). ET was predicted to increase, subsequently reducing runoff and potentially causing drought in the region over the long term. A sensitivity analysis confirmed that runoff volume is more sensitive to temperature change than to increased precipitation.

The *Asian water towers* are the high mountain ranges that supply ice and snow melt each spring, feeding five major river basins. The basins together provide water to 1.4 billion people, over 20% of the world's population. Researchers used a normalized melt index (NMI) to compare current discharges to predicted discharges over the period, 2046–2065 (Immerzeel et al. 2010). The NMI is a ratio of upstream glacier/snow melt discharges to the natural downstream discharges. This latter amount is water that is available to farmers for irrigation and households for cleaning, cooking, and drinking. This index provides a quantifiable measure of the correlation between available surface water and snow melt in these basins. The discharge generated by snow and glacial melt is highest in two of the western basins – the Indus (151%) and the Brahmaputra (27%).

Using temperature and precipitation estimates based on the A1B emissions scenario for 2046–2065, basin runoff was modeled for the five critical Asian water tower basins (watersheds). Whereas the Yellow River may receive more upstream discharge during the two decades of study, the Indus and Brahmaputra basins show a loss of discharge, which may result in increased water and food insecurity for millions of people in India and Pakistan.

6.7 Conclusion

The Sixth Assessment Report of the IPCC, released in 2021, presented sobering news on the state and future of the climate. Each of the four previous decades were warmer than any preceding decade (IPCC 2021b). No part of the world has been left untouched by the warming, which has resulted in intense heat waves, flooding, and drought while diminishing glacial ice and critical snow cover in the Northern Hemisphere (Meyer 2021). The report warned that while ocean acidification and the rise in land temperatures might be reversible, the rise of sea levels is not (IPCC 2021b). Such a scenario is likely to impact increase the disruptions associated with food insecurity and human migrations away from low-lying regions. Furthermore, if current practices are continued, the planet may see warming of up to 3 °C (5 °F) by the end of the current century, makin the threats even more frightening (Meyer 2021).

As we have seen, climate change is a global phenomenon, but its effects are inherently local. Even gradual rises in temperature and loss of precipitation are enough to create suffering and hardship for people who live close to the land. Models have shown that for each degree of rising temperature due to global warming, an additional 7% of the global population is expected to suffer a 20% decrease in renewable water resources (Cisneros 2014). Subsistence farmers, by definition, are growing crops and raising livestock to support the immediate family with very little surplus to sell. Thus, climate change is ultimately a crisis in social equity.

The practice of water security is greatly enhanced by keen observations coupled with advanced tools for modeling and measurement resulting in more accurate predictions of future water supply. Even relatively simple tools can provide important diagnostic information about water quantity and potential changes. As one example, modeling has estimated and compared the relative impacts on runoff between increased rainfall and an increase in temperature, with the latter being found more important (above, [Adhikari and Hong 2013]). Only with such tools and predictions can climate-change challenges be sufficiently described.

In subsequent chapters, water security is discussed in terms of water of sufficient quantity and quality that is available for agriculture, energy, industry, and ecosystem health. The availability of water in each of these sectors is affected by the impacts of climate change.

*Foundations: Human Health Risk Assessment – Calculations of Dose and Human Health Risk from Drinking Water

Much of the work of environmental scientists and engineers is directed toward reducing or eliminating human health risk from drinking water and other environmental factors. Although developed nations have the resources to treat their drinking water, many people throughout the world still resort to drinking the only water that is available, water that is often untreated or treated poorly. How does one determine the health effects of poor drinking water quality? What level of a water contaminant is considered safe? These questions are posed and answered by the practice of human health risk assessment, an important sub-discipline within the field of environmental health.

An astute observer might answer the above two questions by noting that the risk depends upon the harmful nature of the contaminant, the amount (quantity) of exposure of the human receptor to the contaminant, and the length of exposure – the time that the receptor has been drinking the water. These three factors can be measured by:

- Assessment of the toxicity of the chemical or biological hazard,
- Assessment of the human response that varies by dose, and
- Assessment of the extent of human exposure (concentration over time).

The resultant risk characterization is the numeric estimate of risk that can be compared to other risks, to background risk, and to the reduction of risk following an intervention.

The *hazard assessment* is a review and analysis of all known health effects that can be contributed to a particular chemical. The chemical effects may be of a *chronic* nature, such as long-term exposure to cancer-causing (carcinogenic) substances, or of an *acute* nature, such as chemicals that are considered poisons at high doses, such as arsenic at high concentrations. (Arsenic also produces chronic health effects from long-term exposure to low levels.)

The *dose–response assessment* more specifically analyzes the correlation between health effects and level of dose. Laboratory animals, usually mice or rats, are exposed to chemicals at various dosages and the effects are observed. Often enough, the true threshold – when health effects begin – cannot be observed experimentally. Thus, there exists a lowest observed adverse effect level (LOAEL) and a no-observable adverse effect level (NOAEL), the latter that represents the highest dose at which there are no observed adverse health effects (Figure 6.17). The true threshold is assumed to lie somewhere in between.

In this assessment, a dose is the amount of chemical that is able to cross a protective boundary (such as skin or stomach) and can interact with an organ or metabolic function in the body of the receptor (Mihelcic and Zimmerman 2009). The dose that results in an observable effect may be determined first in laboratory animals and then extrapolated to the human receptor. A safety factor is often applied to account for uncertainty, lack of adequate data, and the unique responses of living human beings. The dose has units of mg chemical per kg body weight per day, or mg/kg-day.

In the case of carcinogens, a dose–response assessment results in a slope factor expressed in terms of probability of occurrence per dose rate with units of inverse mg/kg-day or $(\text{mg/kg-day})^{-1}$. Thus, to obtain the overall risk, we would multiply the calculated dose by the slope factor, which results in an overall risk expressed as a probability of contracting cancer. This is discussed below as risk characterization.

In the case of noncarcinogens, the adverse endpoint may be a health condition, such as chronic fatigue, weight loss, a learning disorder, or liver disease (Mihelcic and Zimmerman 2009). The analysis results in a reference dose (RfD), which is the maximum allowable exposure. This dose can be expressed mathematically as:

$$RfD = \frac{\text{NOAEL}}{\text{UF}} \tag{6.1}$$

where

NOAEL = No observable adverse effects level
UF = Uncertainty (safety) factor, usually 10–1000

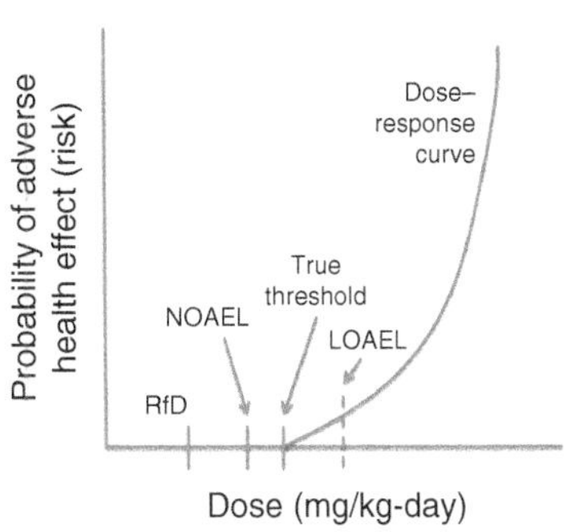

Figure 6.17 The probability of an adverse health effect begins at the true threshold dose and rises as the dose increases. Source: JFC.

The Integrated Risk Information System (IRIS) database of US EPA provides both slope factors and reference doses for chemicals to be used in the hazard analysis just described (ORD US EPA 2013a).

The *exposure assessment* is used to determine the frequency and level of contact between the human receptor and the chemical in the environment. In the case of water, the primary exposure is ingestion of drinking water, although recreational swimming may also result in inadvertent exposure.

If the contact rate, contaminant concentration, and body weight are constant over time, then the total dose (D_T – mg_c/kg) and average daily dose (D_{avg} – mg_c/kg-da) can be found by:

$$D_T = \frac{CR \times C \times t_e}{BW} \tag{6.2}$$

$$\dot{D}_{avg} = \frac{D_T}{t_{avg}} = \frac{CR \times C \times t_e}{BW \times t_{avg}} \tag{6.3}$$

where

CR = Contact rate (e.g. liters/day)
C = Concentration of contaminant at the receptor (e.g. mg_c/l)
BW = Body weight of the receptor (kg)
t_e = Time of exposure
t_{avg} = Time to be averaged, or lifetime (e.g. 70 years)

For example, consider the following pieces of information about a groundwater well with arsenic:

- Concentration of arsenic in water = 20 μg/l = 0.02 mg/l
- CR = 2 l/day consumption rate of water
- BW = 70 kg for an adult male
- t_e = Duration = 40 years, until he gets clean water to drink
- t_{avg} = Lifespan of 70 years

$$\dot{D}_{avg} = \frac{2\ ^{l}/_{d} \times 0.02\ ^{mg}/_{l} \times 40 \text{ years}}{70 \text{ kg} \times 70 \text{ years}} = 3.2 \times 10^{-4}\ ^{mg}/_{kg-day}$$

The *risk characterization* builds upon the results of the first three assessments. For carcinogenic substances, the maximum acceptable environmental risk of contracting cancer is often set at 1 in a million (1×10^{-6}) of an adverse effect. An unacceptable risk is 1×10^{-4} of an adverse effect (i.e. 1 person out of 10 000 would develop an adverse effect). So, the range of acceptable risk = 10^{-4} to 10^{-6} over background risk levels (Mihelcic and Zimmerman 2009).

Expressed mathematically, risk (R) = Cancer slope factor (ρ) × dose (D):

$$R = \rho D \tag{6.4}$$

For example, consider the presence of benzene in a drinking water source. The slope factor for benzene is equal to 0.055 $(mg/kg\text{-}da)^{-1}$ (ORD US EPA 2013a). If the calculated dose over a 70-year lifetime = 0.29 mg/kg-da, then the risk = $0.055 \times 0.29 = 1.59 \times 10^{-2}$. According to this characterization, 1.59 out of 100 individuals are at risk to develop cancer. Or, over a population of 10 000 people, 159 will develop cancer. This is an unacceptable risk and thus should be mitigated through intervention.

Similarly, for noncarcinogens a hazard quotient (HQ) is defined as the ratio of average daily dose to the reference dose:

$$HQ_{\text{water}} = \frac{\text{Average daily dose}}{RfD} = \frac{\dot{D}_{\text{avg}}}{RfD} \tag{6.5}$$

If the HQ is greater than 1, then an intervention is warranted. If there are multiple hazards, a hazard index (HI) is the sum of all hazard quotients for a site:

$$HI = \sum_{\text{i}} HQ_{\text{i}} \tag{6.6}$$

These analyses are tools that can help the water manager determine when intervention is needed as well as the efficacy of a reduction in chemical exposure. Risk characterization can give a quantifiable measure of the potential threats to human health in drinking water, thus serving to improve one of the primary motivations for enacting water security.

End-of-Chapter Questions/Problems

6.1 Describe three ways in which higher air temperatures affects the hydrologic cycle.

6.2 The MENA region is composed of countries in the Middle East and North Africa. These countries have abundant oil and natural gas resources but are also arid and semiarid in climate. Use the Internet to describe the effects of climate change on the countries in this important region.

6.3 The historical 100-year flood level is often used for design of bridges, culverts, and other water conveyance infrastructure.
   a. How is the 100-year flood level expected to change in the northeastern United States?
   b. What are the implications of this change for water managers and engineers?

6.4 Use the Internet to describe a specific case in which climate-change effects have been linked to major changes in water quality for a local or regional population.

6.5 The Maldives nation is extremely vulnerable to sea-level rise.
   a. What steps are being taken to adapt to sea-level rise in the Maldives?
   b. What plans are being considered to secure the safety and well-being of its citizens?

6.6 The IPCC is composed of three working groups of scientists who examine climate change from various points of view. Use the IPCC website to:
   a. Name and describe each of the three subgroups.
   b. What is the nature of each of their reports and with what frequency do they publish reports?

6.7 Sea-level rise is partly attributed to the melting of *land* ice.
   a. Why is there no rise associated with the melting of *sea* glaciers, such as those found in winter in the Arctic Sea water? Use concepts of physics to explain your answer.

b. If not contributing to sea-level rise, what is another positive contribution to global warming that comes from melting sea ice?

6.8 The chapter describes the use of modeling to predict runoff in the Nzoia River basin of Kenya. Examine and describe the predictions for the Arkansas-Red River Basin (US) using work done by the Arkansas-Red Basin Forecast Center (ARBFC).

6.9 Compare the NMI for three large river basins that depend at least in part on glacial melt. What significance do the values of this index have for the people who live in the basins? Be as specific as possible.

6.10 Describe three ways in which climate change can affect the short-term and long-term health of disadvantaged populations.

6.11 Using the US EPA's IRIS database:

a. Calculate the HQ for dioxin (2,3,7,8-Tetrachlorodibenzo-p-dioxin) at a dose rate of 5.8×10^{-9} mg/kg-day.

b. What are the potential health effects of your calculated HQ?

6.12 Arsenic has been found in a groundwater well that has been used over many decades. Calculate the cancer risk that corresponds to a calculated dose rate of 3.2×10^{-4} mg/kg-day. Use the US EPA's IRIS website for toxicity information on arsenic.

Further Readings

Hansen, J. (2009). *Storms of My Grandchildren: The Truth About the Coming Climate Catastrophe and Our Last Chance to.* Save Humanity: Bloomsbury Press.

IPCC (2021). Summary for Policymakers. In: Climate Change 2021: The Physical Science Basis. Contribution of Working Group I to the Sixth Assessment Report of the Intergovernmental Panel on Climate Change. Cambridge University Press. In Press.

UNESCO, UN-Water (2020). *United Nations World Water Development Report 2020: Water and Climate Change.* Paris: UNESCO.

References

Adams, T.E. (2019). Chapter 12 - water resources forecasting within the Indus River Basin: a call for comprehensive modeling. In: *Indus River Basin* (ed. S.I. Khan and T.E. Adams), 267–308. Elsevier https://doi.org/10.1016/B978-0-12-812782-7.00013-8.

Adhikari, P. and Hong, Y. (2013). Will Nzoia Basin in Kenya see water deficiency in coming decades as a result of climate change. *British Journal of Environment & Climate Change* 3 (1): 67–85.

Ahmad, O. (2019). The indus – a river of growing disasters. *The Third Pole* (blog). https://www.thethirdpole.net/2019/02/04/the-indus-a-river-of-growing-disasters.

Astaiza, R. (2012). 11 Islands That Will Vanish When Sea Levels Rise. Business Insider. 2012. https://www.businessinsider.com/islands-threatened-by-climate-change-2012-10.

Bates, B. (2008). *Climate Change and Water — IPCC*. IPCC https://www.ipcc.ch/publication/climate-change-and-water-2.

Blöschl, G., Hall, J., Parajka, J. et al. (2017). Changing climate shifts timing of European floods. *Science* 357 (6351): 588–590. https://doi.org/10.1126/science.aan2506.

Bolch, T., Shea, J.M., Liu, S. et al. (2019). Status and change of the cryosphere in the extended Hindu Kush Himalaya region. In: *The Hindu Kush Himalaya Assessment: Mountains, Climate Change, Sustainability and People* (ed. P. Wester, A. Mishra, A. Mukherji and A.B. Shrestha), 209–255. Cham: Springer International Publishing https://doi.org/10.1007/978-3-319-92288-1_7.

Caesar, L., Rahmstorf, S., Robinson, A. et al. (2018). Observed fingerprint of a weakening Atlantic Ocean overturning circulation. *Nature* 556 (7700): 191–196. https://doi.org/10.1038/s41586-018-0006-5.

Church, J.A. (2013). *Climate Change 2013: The Physical Science Basis. Contribution of Working Group I to the Fifth Assessment Report of the Intergovernmental Panel on Climate Change*. IPCC.

Cisneros, B.E. (2014). Freshwater resources. In: *Climate Change 2014: Impacts, Adaptation, and Vulnerability* (ed. JFC), 229–269. IPCC.

Darling, S.B. and Snyder, S.W. (2018). *Water Is … The Indispensability of Water in Society and Life*. World Scientific Publishing Co.

Hansen, J. (2009). *Storms of My Grandchildren: The Truth About the Coming Climate Catastrophe and Our Last Chance to Save Humanity:* Bloomsbury Press. https://www.amazon.com/Storms-My-Grandchildren-Catastrophe-Humanity/dp/1608195023/ref=sr_1_2?crid=17T9R969464GQ&dchild=1&keywords=storms+of+my+grandchildren+by+james+hansen&qid=1597769511&sprefix=Hansen+James+Stor%2Caps%2C188&sr=8-2.

Hong, Y., Adler, R.F., Hossain, F. et al. (2007). A first approach to global runoff simulation using satellite rainfall estimation. *Water Resources Research* 43 (8): https://doi.org/10.1029/2006WR005739.

Hornberger, G.M. and Perrone, D. (2019). *Water Resources: Science and Society*. Baltimore: Johns Hopkins University Press.

Immerzeel, W.W., van Beek, L.P.H., and Bierkens, M.F.P. (2010). Climate change will affect the Asian water towers. *Science* 328 (5984): 1382–1385. https://doi.org/10.1126/science.1183188.

IPCC (2000). *Emissions Scenarios: Summary for Policymakers : A Special Report of IPCC Working Group III*. Geneva: WMO (World Meteorological Organization) : UNEP (United Nations Environment Programme).

IPCC (2013). "IPCC Factsheet: What Is the IPCC?" 2013. https://www.ipcc.ch/about.

IPCC (2021a). "AR5 Synthesis Report — IPCC." 2021. https://www.ipcc.ch/report/ar5/syr/synthesis-report.

IPCC (2021b). "Summary for Policymakers. Climate Change 2021: The Physical Science Basis. Contribution of Working Group I to the Sixth Assessment Report of the Intergovernmental Panel on Climate Change." IPCC.

Kjellstrom, T. (2014). "Productivity Losses Ignored in Economic Analysis of Climate Change - United Nations University." 2014. https://unu.edu/publications/articles/productivity-losses-ignored-in-economic-analysis-of-climate-change.html.

Lindsey, R. (2020). "Climate Change: Atmospheric Carbon Dioxide." Climate.Gov. 2020.

May, J. (2015). "Demographic Challenges of the Sahel – Population Reference Bureau." 2015. https://www.prb.org/sahel-demographics.

McCabe, G.J. and Markstrom, S.L. (2007). "A Monthly Water-Balance Model Driven By a Graphical User Interface." Open-File Report 2007–1088. Open-File Report. USGS.

Meyer, R. (2021). "It's Grim." The Atlantic. August 9, 2021. https://www.theatlantic.com/science/archive/2021/08/latest-ipcc-report-catastrophe/619698.

Mihelcic, J.R. and Zimmerman, J.B. (2009). *Environmental Engineering: Fundamentals, Sustainability, Design*, 1e. Wiley.

Milly, P.C.D., Betancourt, J., Falkenmark, M. et al. (2008). Stationarity is dead: whither water management? *Science* 319 (5863): 573–574. https://doi.org/10.1126/science.1151915.

Moore, R. (2019). "IPCC Report: Sea Level Rise Is a Present and Future Danger." NRDC. 2019. https://www.nrdc.org/experts/rob-moore/new-ipcc-report-sea-level-rise-challenges-are-growing.

NASA (2020). "World of Change: Global Temperatures." Text.Article. NASA Earth Observatory. January 29, 2020. https://earthobservatory.nasa.gov/world-of-change/global-temperatures.

NASA (2021a). "GRACE-FO Mission Brochure." GRACE-FO. 2021. https://gracefo.jpl.nasa.gov/resources/71/grace-fo-mission-brochure.

NASA Goddard Institute (2021b). "Data.GISS: GISS Surface Temperature Analysis (v4): Analysis Graphs and Plots." 2021. https://data.giss.nasa.gov/gistemp/graphs_v4.

NOAA, National Oceanic and Atmospheric Administration (2014). "What Percentage of the American Population Lives near the Coast?" 2014. https://oceanservice.noaa.gov/facts/population.html.

Padmanabhan, V., Alexander, S., Srivastava, P. (2019). "The Growing Threat of Climate Change in India." Livemint. July 21, 2019. https://www.livemint.com/news/india/the-growing-threat-of-climate-change-in-india-1563716968468.html.

Portner, H.-O. (2019). "IPCC Special Report on the Ocean and Cryosphere in a Changing Climate - Summary for Policymakers." IPCC.

Sadat, N. (2009). "Small Islands, Rising Seas." *United Nations* (blog). United Nations. 2009. https://www.un.org/en/chronicle/article/small-islands-rising-seas.

Scott, M. (2011). "Heavy Rains and Dry Lands Don't Mix: Reflections on the 2010 Pakistan Flood." Text.Article. NASA Earth Observatory. April 6, 2011. https://earthobservatory.nasa.gov/features/PakistanFloods.

Sherwood, S., Webb, M.J., Annan, J.D. et al. (2020). An assessment of Earth's climate sensitivity using multiple lines of evidence. *Reviews of Geophysics* e2019RG000678. https://doi.org/10.1029/2019RG000678.

UNESCO (2020). "Water and Climate Change." https://www.unwater.org/publications/world-water-development-report-2020.

US EPA, OP (2016). "Climate Adaptation and Saltwater Intrusion." Overviews and Factsheets. US EPA. 2016. https://www.epa.gov/arc-x/climate-adaptation-and-saltwater-intrusion.

US EPA, ORD (2013a). "Integrated Risk Information System." Reports and Assessments. US EPA. 2013. https://www.epa.gov/iris.

US EPA, OW (2013b). "Climate Change and Harmful Algal Blooms." Overviews and Factsheets. US EPA. September 5, 2013. https://www.epa.gov/nutrientpollution/climate-change-and-harmful-algal-blooms.

USGCRP (2017). *Climate Science Special Report*. Washington, DC: U.S. Global Change Research Program https://science2017.globalchange.gov/chapter/12.

USGCRP (2018). "Fourth National Climate Assessment - Chapter 9: Agriculture." 2018. https://nca2018.globalchange.gov.

van Vliet, M.T.H., Wiberg, D., Leduc, S., and Riahi, K. (2016). Power-generation system vulnerability and adaptation to changes in climate and water resources. *Nature Climate Change* 6 (4): 375–380. https://doi.org/10.1038/nclimate2903.

Vörösmarty, C.J., Green, P., Salisbury, J., and Lammers, R.B. (2000). Global water resources: vulnerability from climate change and population growth. *Science* 289 (5477): 284–288. https://doi.org/10.1126/science.289.5477.284.

WGMS (2021). "The State of Global Glaciers – World Glacier Monitoring Service." https://wgms.ch/global-glacier-state.

World Economic Forum (WEF) (2011). *Water Security: The Water-Food-Energy-Climate Nexus.* Island Press.

WWF (2021). "Everything You Need to Know about Coral Bleaching—and How We Can Stop It." World Wildlife Fund. https://www.worldwildlife.org/pages/everything-you-need-to-know-about-coral-bleaching-and-how-we-can-stop-it.

Yüksel, I. (2009). Hydroelectric power in developing countries. *Energy Sources, Part B: Economics, Planning, and Policy* 4 (4): 377–386. https://doi.org/10.1080/15567240701756897.

6a The Practice of Water Security: On the Front Lines of Sanitation in Rural Africa

Some of the greatest challenges in children's health depend on a most basic need – a clean and safe place to use the toilet (Figure 6a.1). There is no question that the lack of sanitation and hygiene leads to the transmission of pathogens to the human body, and this leads to gastrointestinal illness, dehydration, malnourishment, and even death. For this reason, Goal 6 of the Sustainable Development Goals (SDGs) is to "ensure availability and sustainable management of water and sanitation for all" (United Nations [UN] 2021).

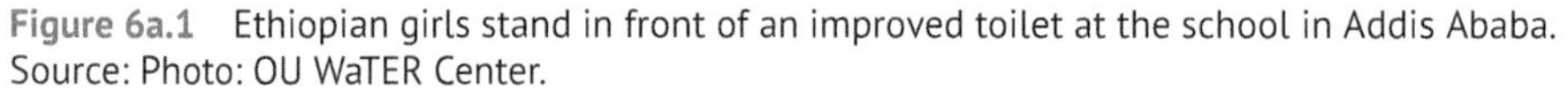

Figure 6a.1 Ethiopian girls stand in front of an improved toilet at the school in Addis Ababa. Source: Photo: OU WaTER Center.

Ada Oko-Williams is a senior WaSH manager for WaterAid in London (Figure 6a.2). She has seen first-hand the burden that is placed on children, especially girls, when they have to attend school without a toilet. She writes that "one in four people do not have access to even a basic toilet and 620 million of the world's schoolchildren—almost twice the population of the United States—do not have decent toilets at school" (Oko-Williams 2019). This situation has a major impact on their chance to receive an education and thus break the cycle of poverty. Without a toilet at school, children often will run home or go outside to relieve themselves in potentially dangerous situations. This is especially true of girls who may be in their period yet have to share limited toilets with other children, both male and female. These toilets may lack privacy, may not lock, may not be cleaned regularly, and may not have adequate water for cleansing.

Diarrhea annually kills over 300 000 children under the age of 5 years, and those children who do survive the extreme undernutrition will likely see their growth stunted. This results in emotional, cognitive, and social impairment. Ada notes that "a quarter of all stunting is attributed to five or more episodes of diarrhea in the first two years of life" (Oko-Williams 2019). In Ethiopia, 60% of primary schools do not have toilets, while in Nigeria, despite having Africa's largest economy, 52% of people do not have decent toilets.

Figure 6a.2 Ada Oko-Williams teaches a class (top) and awards a student prize for Art Competition at the 2013 OU International Water Conference (bottom). Source: Photos (top): Sanitation Learning Hub (Mercer 2018), (bottom): OU WaTER Center.

Ada and WaterAid work with local communities and partners to build toilets in schools to keep children in lessons and to provide girls with a safe space to manage their feminine hygiene. They also work to persuade governments and decision makers that building and maintaining toilets is a good investment that can drastically improve the health and lives of their people (Oko-Williams 2019). Some African governments, such as those in Burkina Faso and Zambia, have acknowledged the problem and are making decent toilets in schools a priority. Only in this way can children receive the health and proper education that they deserve.

Ada admits that "it may not be the most glamourous of inventions, but the humble toilet saves lives and creates futures" (Oko-Williams 2019). A safe and effective toilet also demands a sufficient quantity of water that is of suitable quality for washing hands and cleaning the human body. Thus, it is an important component of water security.

Ada Oko-Williams is the recipient of the 2013 OU International Water Prize

References

Mercer, E. (2018). "Ways Forward for Rural Sanitation in Africa." Sanitation Learning Hub. October 22, 2018. www.ids.ac.uk/opinions/ways-forward-for-rural-sanitation-in-africa.

Oko-Williams, A. (2019). "Children's Potential Is Wasted without Toilets at School." *The BMJ* (blog). December 12, 2019. https://blogs.bmj.com/bmj/2019/12/12/ada-oko-williams-childrens-potential-is-wasted-without-toilets-at-school.

United Nations (UN) (2021). "THE 17 GOALS | Sustainable Development." 2021. https://sdgs.un.org/goals.

Part III

Competing Uses of Water and Threats to Security

7

Water for Food

The objective of this chapter is to describe the demand upon the world's water supply for food production. Food production, in the form of either plants or livestock, is not possible without water. As people are lifted out of poverty, their diets become more diverse, calorie-dense, and water-intense. The agricultural water cycle is an interplay of green water (rain and soil moisture) and blue water (surface and groundwater stocks). While water management can utilize various tools for greater water conservation and irrigation efficiency, crop selection is important as well as management of soil moisture and field drainage. In some parts of the world, food production faces competition with nonfood crops. Finally, the reader will consider some case studies in which these concepts are illustrated.

Learning Objectives

Upon completion of this chapter, the student will be able to:

1. Understand the global population and dietary trends that already affect the security of water for food.
2. Understand the linkages between water security and food security.
3. Understand the distinction between and the importance of blue and green water in the water cycle.
4. Understand the concepts of virtual water and the water footprint as it applies to foods and food crops.
5. Identify and describe several adaptation strategies for conserving water for food.
6. Understand the concepts of water productivity and physical and economic water scarcity.
7. Identify points of food wastage along the journey for field to table.
8. Be familiar with responses to agricultural water stress in the two nations of India and Israel.
9. *Calculate contaminant concentrations in a well-mixed lake.

7.1 Introduction

Food for human consumption as well as forage and grain for animal nourishment require water for production. There is no substitute. The global world population is expected to

Fundamentals of Water Security: Quantity, Quality, and Equity in a Changing Climate, First Edition.
Jim F. Chamberlain and David A. Sabatini.

Companion website: www.wiley.com/go/chamberlain/fundamentalsofwatersecurity

rise to eight billion over the next 15 years (to 2035), and the majority of these people will live in an urban setting. A notable exception is sub-Saharan Africa, which will still be mostly rural and populated with villagers who depend overwhelmingly on local agriculture (Molden 2007). Populations are outgrowing the ability of local suppliers to produce enough food, thereby increasing food imports. As seen in Figure 7.1, there is great disparity between nations regarding their dependency on imported food. Over the last 20 years, regions as different as Central America, China, and sub-Saharan Africa have increased their food importation, while Brazil, Canada, Australia, and Russia have increased their exportation of food. Food imports create an external water demand from the supplying country. For example, Europe imports large amounts of sugar and chocolate (cacao), two water-intensive crops, potentially resulting in water stress in remote regions of production (Hoekstra 2008).

As dietary habits become more urban and "westernized," they tend to include more water-intensive components, such as beef, pork, and dairy products. To further exacerbate the matter, land that was once used to grow grain for food and feed is now growing first-generation biofuel feedstock (corn, soybean), which is more water-intensive than petroleum (see Chapter 8). Thus, water used for nonfood production will increasingly become a driver of water stress across the globe. Due to both the increased populations and water required for their diet, the total water demand for agriculture is estimated to double by 2050 (Water Initiative 2011).

Recall that the majority (~70%) of the world's freshwater is withdrawn and mostly consumed in the production of food. But not all food is created equal. The water that is consumed in food production either is evapotranspired by the plant or becomes part of the food product itself, the embedded or "virtual water" contained in the food. This water requirement, called the "water footprint of food," varies according to the product that is grown and can range from 214 l/kg of tomato to over 15 000 l/kg of beef (Figure 7.2).

The Green Revolution (1950s–1960s) marked a rapid increase in food productivity due to improvements in fertilizers, pesticides, and high-yield crop varieties. This revolution began

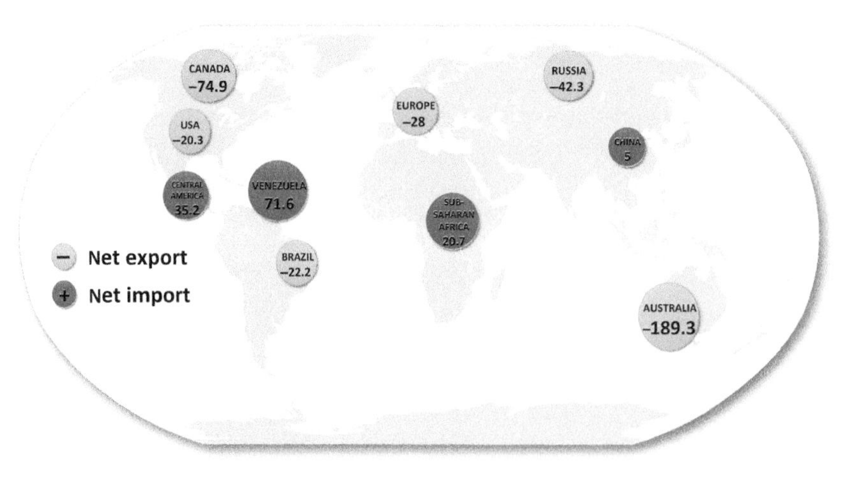

Figure 7.1 Change in food import dependency rate (in percent change) for selected countries and regions in the period 1997–2017. Source: Adapted from The Economist (2020).

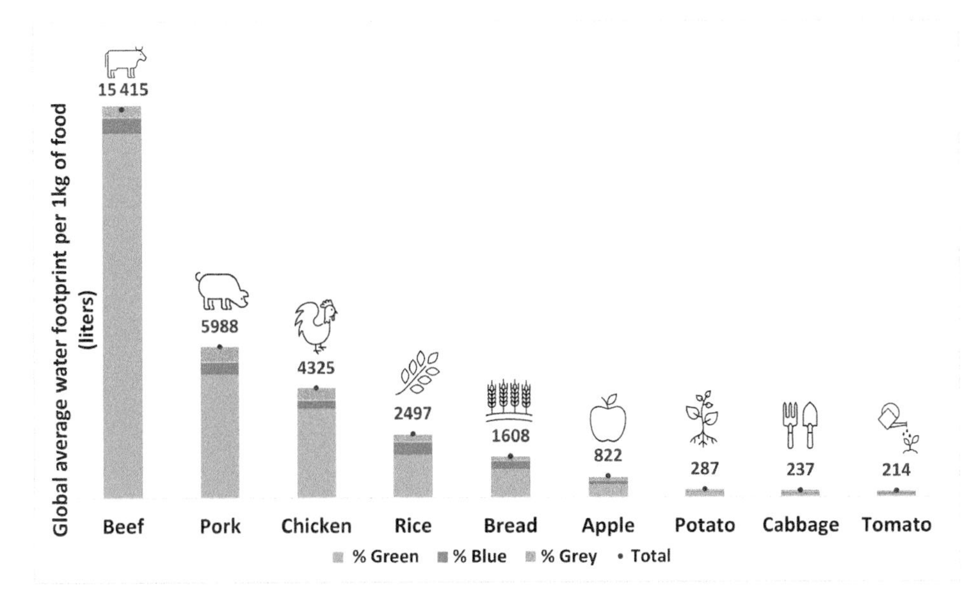

Figure 7.2 The amount of water (liters) needed to produce a kilogram of food varies greatly. Source: Original figure, adapted from WFN (2020).

in 1943 with cooperative efforts of the Mexican government and the Rockefeller Foundation that resulted in the International Maize and Wheat Improvement Center, known by its Spanish acronym CIMMYT (Centro Internacional de Mejoramiento de Maíz y Trigo). To cite a few examples, the productivity of maize (corn) on a land area basis (hectograms per hectare) has increased in China but decreased or held steady in sub-Saharan Africa and India, reflecting the disparity in achievement (Figure 7.3).

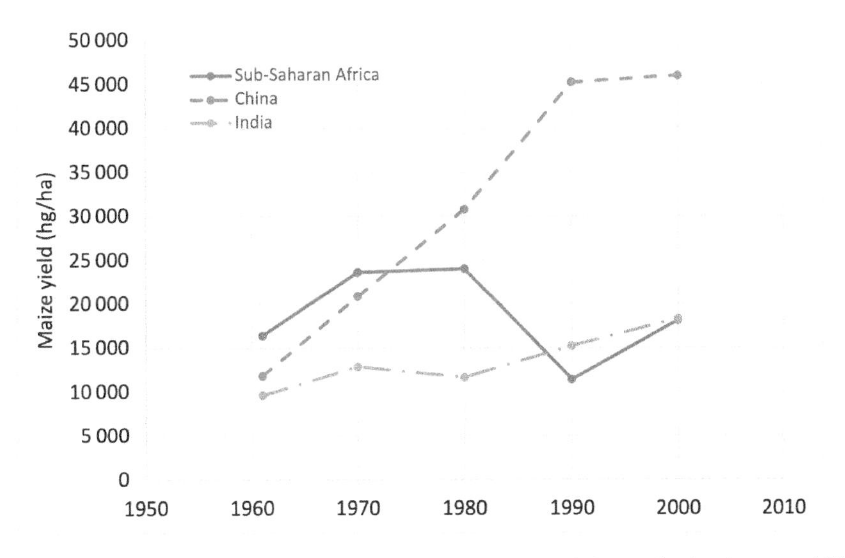

Figure 7.3 Variation in maize yield – sub-Saharan Africa, China, India, for the years 1961–2000. Source: FAOStat (2020). Adapted based on reference.

Overall, the Revolution has allowed for unimagined global population expansion. The *Limits to Growth* report from the Club of Rome in 1972 used computer simulations to model economic and population growth in a world of constrained resources. The gradual depletion of nonrenewable resources was predicted to result in a peaking of industrial and agricultural production, followed by subsequent decline of both production and population (Bardi 2011). In response, the report advocated a lowering of the environmental impact per unit of gross domestic product (GDP), increasing the efficiency of each unit of natural resource that was utilized. The Former Secretary-General of the United Nations, Kofi Annan, recently called for a "Blue Revolution" in agriculture in which every drop of water now produced a higher yield, or "more crop per drop" (Lankford et al. 2013).

This chapter will examine each of these themes more closely and help the reader understand both the changing demand and the fluctuating supply of water for food.

7.2 Evolution of the Global Diet

The Westernization of diet is characterized by an increase in energy-dense foods, such as animal products, plant oils, and sugars, with a concomitant decrease in cereals, grains, and vegetables. A study of the diets of 152 countries (representing 98% of the global population) from 1960 to 2009 was conducted by the International Center for Tropical Agriculture (CIAT). The study found that national diets, on average, increased in caloric, fat, and protein content and in food weight (CIAT 2019). From 1970 to 2000, the average caloric content grew from 2400 to 2800 kcal/person/day, an increase of over 16% (Molden 2007). National diets have become more homogeneous and are relying on a subset of truly global foods, such as wheat, rice, maize, and barley (cereals). Plants grown for oil, such as soybean, palm oil, and sunflower, increased in abundance while rye, sorghum, cassava, and sweet potatoes were in decreasing demand. Maize and soybean have become major food crops for livestock.

Researchers have compared actual diets across global regions, which reveal both the disparity in types of food consumed and water used (Table 7.1). Residents in the Organization for Economic Co-operation and Development (OECD) countries average 3300 kcal/day and consume more meat, milk, and dairy products, accounting for 3/4 of their food water footprint. Latin Americans consume 2900 kcal/day with a higher proportion coming from cereals and pulses. ("Pulses" are legumes, such as beans, peas, chickpeas, lentils, and peanuts.) In contrast, over half of the food water footprint from sub-Saharan Africans is from cereals and pulses, and their average water footprint is only 30% that of OECD residents. The total water in the final column represents the water footprint of the daily diet (further defined below).

The evidence is clear that water security is intricately linked to diet composition and food demand and production. As countries increase in standard of living, the general trend is toward foods that are more water intensive, such as meat, milk, and dairy.

Table 7.1 Comparison of daily diets and water usage for three global regions. Water use is measured as water consumed.

Typical diet	Serving size (g)	Water footprint (l/kg food)	Total water (l)
Latin America			
Meat (carcass weight)	61	15 500	945.5
Milk and dairy, excl. butter (fresh milk eq.)	111	5000	555.0
Pulses, dry	11.3	4055	45.8
Cereals, food	138	3400	469.2
Vegetable oils, oilseeds, and products (oil eq.)	13.6	2364	32.2
Sugar and sugar crops (raw sugar eq.)	42	1500	63.0
Roots and tubers	63	250	15.8
Other food (kcal/person/day)	264		
Total food (kcal/person/day)	2898		**2126.4**
OECD countries			
Meat (carcass weight)	80	15 500	1240.0
Milk and dairy, excl. butter (fresh milk eq.)	202	5000	1010.0
Pulses, dry	2.9	4055	11.8
Cereals, food	167	3400	567.8
Vegetable oils, oilseeds, and products (oil eq.)	19	2364	44.9
Sugar and sugar crops (raw sugar eq.)	34	1500	51.0
Roots and tubers	77	250	19.3
Other food (kcal/person/day)	458		
Total food (kcal/person/day)	3360		**2944.7**
Sub-Saharan Africa			
Meat (carcass weight)	10.1	15 500	156.6
Milk and dairy, excl. butter (fresh milk eq.)	31	5000	155.0
Pulses, dry	10.5	4055	42.6
Cereals, food	125	3400	425.0
Vegetable oils, oilseeds, and products (oil eq.)	9.4	2364	22.2
Sugar and sugar crops (raw sugar eq.)	10.7	1500	16.1
Roots and tubers	187	250	46.8
Other food (kcal/person/day)	126		
Total food (kcal/person/day)	2238		**864.1**

Sources: Alexandratos and Bruinsma (2012), Mekonnen and Hoekstra (2010), Hoekstra (2008), Chapagain and Hoekstra (2010).

7.3 Agricultural Water Cycle

Water that falls as precipitation is the primary input of water to a watershed basin or to a farmer's field. This water can be intercepted by the plant canopy as it falls from the sky. Or it may run off into a river or lake or infiltrate the ground and enter the aquifer as groundwater. These latter stocks of water are often called "**blue water**" and is the water that is referenced when discussing potential stocks for irrigation.

The water may also be retained as water in the upper layer of the unsaturated soil zone, often called a "**green water**" resource. This water can seep into a surface water or groundwater body or be taken up through the plant structure resulting in plant productivity. The amount of water used depends upon the vegetation that dominates the land surface – trees, crops, shrubs, and/or grasses. Green water contributes significantly more than blue water to the global water consumption of human beings, especially for agricultural products (up to 87%) (Hoekstra and Mekonnen 2012).

For this reason, some researchers believe that an overlooked component of agricultural sustainability is the conservation of productive green-water flow (Falkenmark et al. 2009). The productive flow is the water that transpires through the plant, carrying vital nutrients needed for plant production, while the evaporative flow is lost to the system. These two flows are usually measured together as evapotranspiration (ET), but measures can be taken to limit the nonproductive losses. When green water is managed well, the prospects for water sustainability are improved. Later in this chapter, we will consider ways in which green water can be better conserved.

7.4 Virtual Water and the Water Footprint of Food

We have seen that the bulk of freshwater requirement is not for household drinking or cooking but for the growing of food in agriculture, generating energy, and producing industrial products. This is water that must be available locally, as the transport of water across great distances for agricultural purposes is usually not economically feasible or desirable. So, a trade of *actual* water between water-rich and water-poor nations is not commonly practiced. This has led many to a kind of "Malthusian pessimism" regarding the future potential of water-scarce regions of the world (e.g. Gleick 1993).

Bulk foods, however, can be transported more easily, as can goods and products that consume varying amounts of water in production. This "**virtual water**" is the sum total of the freshwater consumed in the growing (in food products) and production of the product on a per mass basis. Life cycle analysis is used to track water usage in various stages of production. In this case, actual water is not exchanged, but embedded (or virtual) water is traded and tracked, giving water-poor countries a more sustainable outlook. The accounting has some uncertainties and shortcomings, however, as illustrated in Figure 7.4.

As we saw in Figure 7.2, there is a wide disparity of water required to grow various foods on a "per kilogram" basis, i.e. the water footprint. Rice is a staple in many countries and only requires a fifth as much water per kilogram as growing the same amount of beef. The Western diet is most egregious from a water consumption standpoint because of the high percentage of beef consumed. Some scientists, however, think the beef water usage (15 415 l/kg) could be revised downward. This number, according to the Water Footprint Network (WFN), is mostly (94%) rainwater (green water) falling from sky (WFN 2020). When cattle are grazed – versus being raised in a concentrated animal feeding operation (CAFO) – this water is consumed as

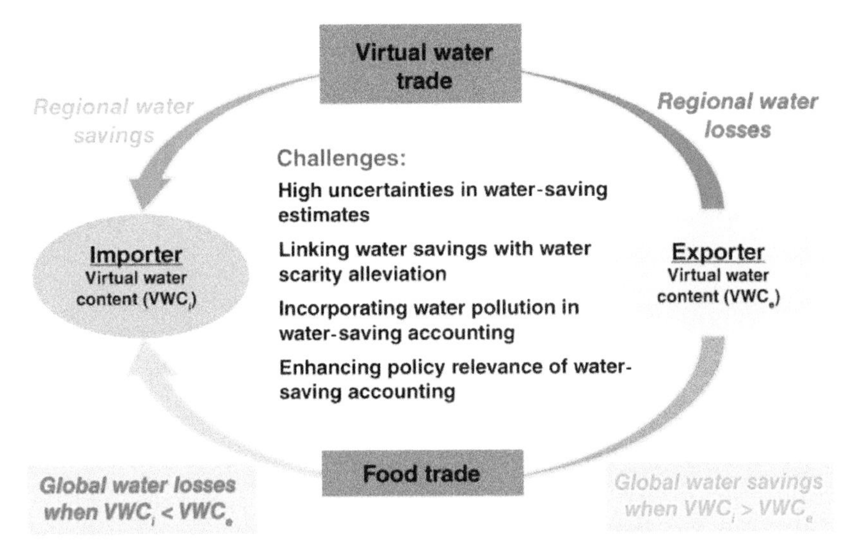

Figure 7.4 Water savings and losses related to virtual food trade highlight the differences in water productivities between food importers and exporters. Source: Liu et al. (2019)/with permission from John Wiley & Sons.

pasture and rangeland grasses. But, as the environmental scientist Sandra Postel points out, this rain would be bound up in the grasses whether the animals are eating them or not (Postel 2017). She feels that a more accurate water estimate would include the actual volume of water that the cattle consumed, water used to irrigate pastureland, and the water used to process the animals for final marketing. This new estimate would result in a water consumption of 3680 l/kg or about 1/4 of the original estimate. This amount is only about as much as it takes to produce two eggs! Thus, we can see that the elements to be included in the life cycle assessment are of great importance in determining the overall assessment of water usage.

From a global perspective, water availability for local food production is a function of both blue- and green-water resources. Figure 7.5 highlights the countries that currently have ample freshwater resources (shades of green) as well as those that are in deficit (red/orange) and that are close to deficit (yellow). A broad swath of countries from north Africa to India

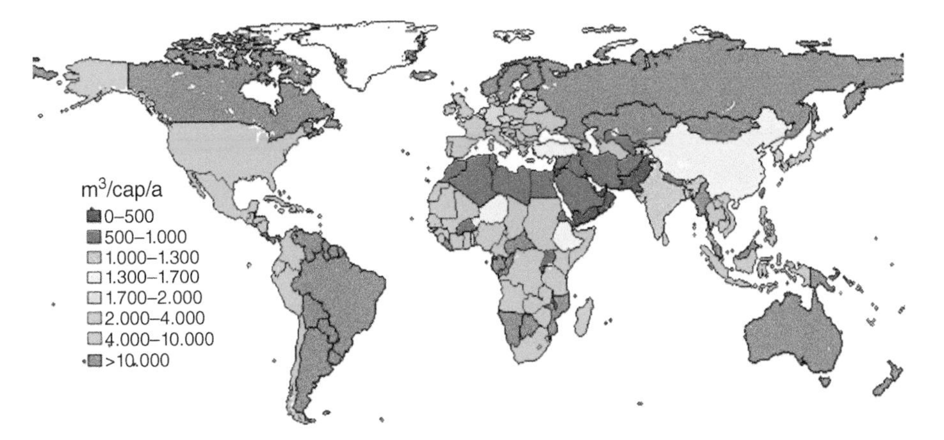

Figure 7.5 Estimates of future (Y2050) freshwater availability for food self-sufficiency, including both "blue" and "green" sources. Those countries with <1300 m^3/cap-year are in deficit. Source: Falkenmark et al. (2009)/with permission from Springer Nature.

are facing freshwater deficits, while China, Turkey, and a few sub-Saharan countries are approaching this critical scenario (Falkenmark et al. 2009). The two greatest factors that could reduce deficit are increases in water productivity and expanded irrigation in countries that are economically poor but water-rich, such as many sub-Saharan African nations (Falkenmark et al. 2009). Methods to do this will improve water equity and are thus discussed in more detail in the following section.

7.5 Managing Water for Food in a Changing Climate

As stated, current global food consumption is estimated to be at around 2800 kcal/person/day, but the Food and Agriculture Organization (FAO) estimates that the average consumption will increase to 3000 kcal/person/day by the year 2030, with 20% of this going toward animal products (Falkenmark and Rockström 2011).

Agricultural productivity is highly dependent upon locally available moisture and air temperature, two parameters that are most affected by global climate change. As detailed in Chapter 6, the global mean surface temperature is predicted to rise by 1.4–5.8 °C (2.5–10.4 °F) by the year 2100 due to increased CO_2 in the atmosphere. This increase may result in mixed blessings regarding food production. High temperatures and increased drought can result in a loss of **net primary production** (NPP). Increased basin runoff, raising the potential for flooding, would also result from reduced plant transpiration. On the other hand, increased atmospheric CO_2 would suggest higher growth (photosynthetic) potential. A recent study using open-air field trials cautions that the increased yield from CO_2 will be offset by the losses due to other climate factors (Long et al. 2006). The balance of these contributions will likely vary from one locale to the other.

Regardless, the combination of climate change and predicted population growth will result in agricultural water stress in many critical regions. An analysis of five countries was made using the thresholds of water shortage (1000 m^3/cap-year) and 100% green-water availability to meet the agricultural demand. India and Ethiopia will cross both thresholds with increased populations and drier climates. Kenya will be able to manage its population growth with a climate that is expected to get wetter under most climate models (Niang et al. 2014). Similarly, Nigeria has abundant rainfall to match its rapid population growth but will need to manage its green water efficiently in order to feed the growing populace. Iran is an arid country that is already water-stressed in its blue-water regime and its situation is expected to worsen (Rockström et al. 2009).

Water security will demand sustainable water-management strategies for agriculture in a changing climate. We continue this section with some strategies for achieving this goal.

a. Adaptation strategies for increased water productivity. Researchers with the International Water Management Institute (IWMI) estimate that the water needs for food can be met by adopting the following strategies (Molden 2007):

- Crop selection and rotation
- Management of soil moisture and field drainage
- Improved irrigation techniques, such as drip and deficit irrigation

Crop selection and crop rotation. The water needs of crops vary considerably on a per growing season basis. Thus, crop selection itself can conserve freshwater consumption. For example, sugarcane requires 1500–2500 mm (mm) of rain per growing season, whereas soybean, maize, and peanuts require only about 450–800 mm of water per season (Spuhler and Carle 2020).

Crop selection is especially important in arid and semiarid regions and can make a critical difference in the flow of a river that is used for irrigation. The Verde River in central Arizona is an important water supply, along with the Salt River, for the city of Phoenix as the two come together to form the Gila River. Normally flowing year-round, the Verde now has stretches that run dry due to seven major irrigation canals that divert water to farms, ranches, and even residential lawns. One proposed solution was conversion of local crop production from corn and alfalfa to barley. Barley, used for animal feed and brewing beer, requires about 1 acre-foot (ac-ft) of water per year per acre, a fraction of the water needed to grow corn (2 ac-ft) and alfalfa (4 ac-ft) (Rogers 2020). A malt house was built so that there would be market for the new crop. Thanks to a local retiree and The Nature Conservancy, the switch to growing barley in the watershed has reduced the water demand on the Verde River by 78.5 million gallons (240 ac-ft) in one year (Rogers 2020). Since the price of barley on the global market is very close to corn and wheat, there was no substantial loss of economic income (Tricase et al. 2018).

An alternative way of quantifying crop water needs is by using **water productivity** as a reciprocal indicator. Water productivity is defined as the yield of a crop – in physical yield (kg), economic yield ($US), or nutritional yield (calories) – per unit of water (m^3). It is mostly a function of crop product, climate (in degree-days), and soil type. Agricultural value of water varies between crops and locations. Crops such as flowers and strawberries can pay the grower more than $500/ac-ft, cotton can pay $40–100/ ac-ft, and pasture can pay $30/ac-ft and less.

Table 7.2 illustrates the range of water productivity for some more common agricultural products. One can see the highest physical yield per cubic meter of water in the three vegetables listed – potatoes, tomatoes, and onions (shaded in yellow). The highest economic

Table 7.2 Water-productivity metrics for selected crops and agricultural products.

	Water productivity		
Product	**Physical yield (kg/m^3)**	**Economic yield ($US/ac-ft)**	**Nutritional yield (cal/m^3)**
Wheat	0.2–1.2	49–370	660–4000
Rice	0.15–1.6	62–222	500–2000
Maize (corn)	0.30–2.00	37–271	1000–7000
Lentils	0.3–1.0	111–370	1060–3500
Potatoes	3–7	370–863	3000–7000
Tomatoes	5–20	925–3700	1000–4000
Onions	3–10	370–1230	1200–4000
Apples	1–5	986–4930	520–2600
Olives	1–3	1233–3700	1150–3450
Beef	0.03–0.1	111–370	60–210
Fish (aquaculture)	0.05–1.0	86–1665	85–1750

Source: Adapted from Molden (2007).

yield is in the fruits, such as apples and olives, with tomatoes also being a lucrative crop (in green). The nutritional yield is highest in corn and potatoes (in blue), perhaps the reason that these crops are staples in many countries.

Crop rotation may happen from year to year or from season to season within a calendar year. In addition to planting a more water-efficient crop, rotation also helps to reduce the buildup of pests and disease, which may occur in the case of monoculture plantings. In addition, legumes can fix nitrogen in the soil, while crops with higher crop residue can reduce surface crusting and water runoff (Mosali 2013).

Management of soil moisture and field drainage. The greatest potential for increased crop yield and water productivity is in the areas of rainfed agriculture, where most of the world's rural poor are already living (Molden 2007). Thus, proper management of water and soil moisture can yield significant benefits. Soil moisture (green water) can be conserved using various techniques, such as the following:

- Placing (or leaving) a *mulching material* on the soil to conserve moisture by reducing evaporation. The mulching material can be the crop residue after harvest, or it can be locally available organics (such as straw, wood chips, leaves), or it can be inorganic (volcanic rock, plastic sheeting). The mulch also reduces weed growth and prevents against soil erosion during heavy rainfalls.
- Using *low-till or no-till cultivation* to replace conventional plowing. Direct planting techniques also help to conserve soil moisture and improve soil structure.
- Adding a *soil amendment* to increase the organic fraction of the soil. This addition helps the soil retain both water and vital plant nutrients, thereby improving the soil tilth (ability to grow crops).
- Enhancement of *soil drainage* can prevent waterlogging and provide replenishment for aquifers and downstream ecosystems.

Improved irrigation techniques – drip and deficit irrigation. There is evidence that the expansion of irrigation has resulted in falling food prices and greater nutritional gains for much of the world. Even with conventional irrigation methods, yield gains can result in a multiplier effect on the economy of 2.5–4.0 (Molden 2007). These gains are not just from an increase in food but also from an increased demand for other goods and services from greater farm income. One study found that irrigation and farmer literacy were the two main factors in alleviating rural poverty in India (Bhattarai and Narayanamoorthy 2003). In addition, irrigation increases land value, enables production of vegetables and fruits, and slows the expansion of deforestation by increasing land productivity (Weckler 2021).

Proper irrigation depends on knowledge in four main areas:

- Crop – daily water-use requirements
- Water – flow of water; length of time to irrigate
- Soil – soil-moisture level at time of irrigation; water-holding capacity of the soil
- Field – slope variations across the planted acreage (NRCS 2012)

Conventional methods of irrigation are not very water efficient as large volumes of water are delivered through earthen canals or sprayed from hose or pivot systems. In many surface-irrigated (or flood-irrigated) schemes, water is delivered to the farm at a fixed interval for a set period, and charges are assessed per delivery regardless of the actual amount used. Under these conditions, farmers tend to take as much water as they can while they can. This often results in overirrigation, which not only wastes water but also causes problems connected with the disposal of return flows, waterlogging of soils, leaching of nutrients, and excessive elevation of the water table, requiring expensive drainage works (Weckler 2021). In spray systems, typically only about 30–50% of the withdrawn water is actually transpired through the crops, resulting in a loss of 50–70% through evaporation and/or drainage (Molden 2007).

Drip irrigation is the slow delivery of water directly to the plant roots, either from the surface or below the surface. This type of irrigation, while more expensive, is much more efficient in its usage of water, with one field study in the United States showing its efficiency at 60% over furrow (gravity) irrigation and 10–20% over sprinkler-type irrigation (Amosson et al. 2011). A comparison of the most common types of irrigation methods is shown in Table 7.3.

A second strategy for saving water without affecting yield is by optimizing the timing of irrigation. This approach is called **deficit irrigation** and requires intimate knowledge of the crop's behavior during its growth cycle. Water is applied only during critical growth stages of the plant when it is most needed. Some crops are more amenable to this practice, including cotton, maize, wheat, and sugar beet. Other crops, such as potatoes, are more sensitive and less adaptable (FAO 2000).

Both drip and deficit irrigation are examples of greater irrigation efficiency. Drip irrigation maximizes the spatial efficiency of water delivery, while deficit irrigation maximizes the temporal efficiency.

b. Competition with bioenergy crops. To reduce their greenhouse gas emissions, nations around the world have been incentivizing the production of feedstock as first-generation biofuels to replace fossil-based fuels. These feedstocks include corn (maize), sugarcane, soybean,

Table 7.3 Comparison of various types of irrigation techniques.

	Surface irrigation	High-pressure sprinkler irrigation	Microirrigation
Examples	Flood, furrow	Center pivot, spray	Drip, subsurface drip, low-pressure sprinkler
Energy requirements	Gravity flow across fields	Uses motors to pump water through high-pressurized sprinklers	Uses motors to force water through pipe holes directly to root systems
Water-supply application	Large applications, intermittent	Small applications, continuous	Small applications, continuous
Efficiency	Limited amount of precision; least efficient	Can be linked to precision applicators to increase efficiency; midrange in efficiency	Often linked to precision applicators to increase efficiency; considered the most efficient
Potential water losses	Evaporation, deep percolation, runoff	Evaporation, wind drift, deep percolation	Little to none if managed correctly
Cost/net investment per acre (Amosson et al. 2011)	Low $162/acre	Moderate $250–410/acre	High $890/acre

Source: Modified from Hornberger and Perrone (2019).

cassava, and wheat, crops that can also be used in food supply. Thus, the consumption of a crop for fuel will inevitably compete with its consumption for food. With a limited amount of arable land, this scenario will naturally result in rising food prices as the demand exceeds the available supply. One modeling study in 2008 used two biofuel expansion scenarios to predict the impact on caloric consumption and malnourishment due to reduced accessibility to caloric food (Rosegrant et al. 2008). The increased malnourishment will hit hardest in sub-Saharan nations, but it is also significant in South Asian and East Asian and Pacific nations. This is not a direct water-security issue per se but is certainly a challenge for food security.

The direct impact of bioenergy crops on water consumption will depend on both crop and region. For example, the crop switch in the United States is mostly from soybean to corn. Corn requires more water than soybean in the Northern and Southern Plains, where much of the production occurs. The aquifers in these regions – e.g. the Ogallala Aquifer – are already stressed with a water level drop of 100 ft in some areas. The switch from food to fuel would exacerbate this stress (National Research Council 2007).

c. *Water losses in food wastage.* We have looked at the embedded (virtual) water that is represented by a kilogram of food by type. Export of food then is export of water; wastage of food is wastage of water. There are several places in our food's journey "from field to fork" where losses are encountered (Figure 7.6). The accumulation of these losses amounts to over 50% of the food that is harvested but never makes it to the dinner table. Some of this lost matter is transferred to other places within the system – for animal feed, bioenergy, or for soil amendment between seasons. But the inefficiencies of harvesting, processing, packaging, and distribution all result in a loss of both food and water for human consumption. These estimates show that there are many opportunities for improvement.

The wastage intensity varies across regions. In developing countries, most of the wastage is early in the food chain as harvesting technologies, storage, and distribution practices are not as efficient. In industrialized countries, the wastage is heavier on the downstream end in urban centers with high populations. Distance to market and a more complicated food chain in the West are also drivers of increased food wastage (Lundqvist et al. 2008).

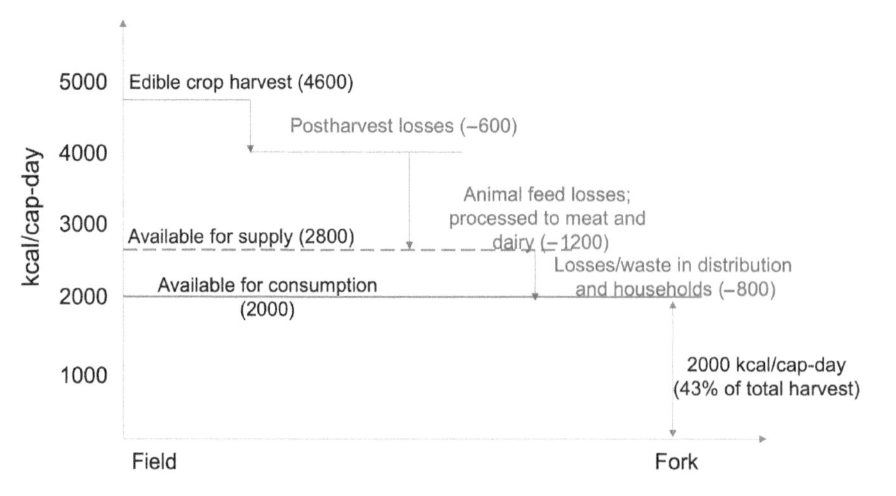

Figure 7.6 Gross estimates of typical losses on a global scale due to food wastage. Less than half of the kilocalories available for consumption remain by the time the food reaches the table. Source: Original drawing, adapted from Lundqvist et al. (2008).

7.6 Brief Case Studies – India and Israel

Two brief case studies will help to illustrate how two countries are coping the increased strain on water resources for food production.

India has 17% of the world's population and only 4% of the Earth's readily available freshwater. But its grain production, together with that of China and the United States, has an impact on international food markets akin to Organization of the Petroleum Exporting Countries' (OPEC's) impact on oil markets (Solomon 2010). Unfortunately, India uses twice as much water per bushel to grow their wheat as does China or the United States. Food and water security in this populous country are thus closely linked with that of the world.

Following the British irrigation projects of the colonial era, India embarked upon a post-1948 frenzy of dam building and large irrigation schemes. Along with high-yielding seeds and fertilizer improvements of the Green Revolution, India's living standards rose steadily until the bust of the 1970s and 1980s (de Fraiture et al. 2009). At this point, the ballooning population along with ineffective government bureaucracy resulted in sagging farm productivity. Irrigation water was not delivered when promised, and most large dams ended up costing much more than expected. The general discontent came to a head in the Narmada River basin in 1989 when 60 000 peasants protested yet another large Indian government water scheme, which included 30 major dams and a massive irrigation superstructure but would displace many Indians while reaping few local economic benefits (Solomon 2010).

India's hydrologic cycle depends largely on a consistent monsoon season, a few months in which 80% of the rainfall and runoff occurs. An early large monsoon can cause flooding and destruction; a late small monsoon can bring famine. Traditionally, Indians have built their own small- and medium-sized local storage units that can hold water through the dry months between the monsoons. These may be as shallow and simple as a terraced catchment on a sloped hillside or as deep and complex as a stone stepwell at a depth of several stories (Figure 7.7).

Groundwater depletion is an especially acute problem in India. In 1975, India had 800 000 wells; by the year 2000, it had about 22 million wells (Solomon 2010). More importantly, the wells in the late 1970s were mostly hand-dug, shallow wells. But with the proliferation of mechanized drilling and government-subsidized electricity, the newer wells were much deeper and higher producing resulting in falling water tables. In the breadbasket regions of India – Punjab and Haryana – water tables are falling over 3 ft/year. Indeed, nationwide India relies on groundwater mining for over half of its irrigation water with withdrawal rates estimated to be twice the rate of natural recharge (Solomon 2010). Such a scenario is unsustainable and frightening, especially given the continued population growth that is expected to reach 1.6 billion by 2050, surpassing China's population (Hub 2019).

Because of disenchantment with government dams, traditional water catchment practices are returning and being improved, including the low-tech water *johads*. These johads can be built with local materials and village labor. A pit is dug into the ground that will naturally collect water as it runs down a slope and into a stone wall or earthen embankment. The johad intercepts monsoon rains, replenishes the local groundwater table, and provides water during the dry season for crops and domestic needs. By the year 2015, 8600 johads had been built, replenishing aquifers and bringing water to 1000 villages (Vansintjan 2015).

Figure 7.7 A stepwell in rural India used to store rainwater during the dry season. Source: Photo credit: By Abhishek Pandey / CC BY-SA 4.0, Wikimedia Commons.

India has thus returned to reliance on very ancient techniques of small, decentralized rainwater collection of water for agricultural use.

The nation of Israel has ~6.5 million people and an annual rainfall of ~550 mm, which is concentrated in the winter months. From May to October, the growing season receives very little rainfall. Water efficiency is, thus, not a luxury but a necessity for domestic agriculture.

Israel has made great strides in agricultural water efficiency, developing a semblance of sectoral water security. It was not always so. For much of the twentieth century, the Jordan River, only about 4% the size of the Nile, was tapped for the water needs of four political entities – Israel, Palestine, Syria, and Jordan. The river once carried 1.3 billion cubic meters of water a year but now carries only 20–30 million cubic meters, ~2% of the original flow (Moser 2010). Much of the water supply shortfall is met by the incessant pumping of regional groundwater aquifers. The lowering of the water levels in the Gaza Aquifer led to infiltration of sewage and seawater to such an extent that Palestinian residents of the Gaza Strip were prone to health problems associated with this contaminated water (Solomon 2010).

Irrigation water is metered and prices are among the highest in the world at $260–290/ac-ft, which provides an incentive for wise water usage (Arlosoroff 2002). Israeli farmers currently practice drip irrigation on 75% of its irrigated fields (Figure 7.8), and the expanded practice of wastewater treatment and reuse (up to 85%) provides an additional nutrient-rich water source for farmers (Siegel 2017). Both solutions utilize modern technologies, but such are now available and becoming less expensive as they are more widely adopted.

Figure 7.8 An example of drip irrigation piping in Shefa Farm, Israel. Source: Juandev / CC BY-SA 3.0 / Wikimedia Commons.

Israel has thus turned to a portfolio of approaches that encourage agricultural water conservation and maximization of efficiency.

7.7 Conclusion

We close this chapter by taking a global look at the distinction between physical and economic water scarcity (Figure 7.9). **Physical water scarcity** (dark blue) denotes regions in

Figure 7.9 Areas of physical and economic water scarcity across the globe. Source: WWAP (2012). United Nations – free to use to support the mission.

which more than 75% of freshwater is withdrawn for agricultural, industrial, or domestic purposes, implying a water demand that is approaching unsustainability. Areas approaching physical scarcity (light blue) are tending toward unsustainable water use. **Economic water scarcity** denotes areas in which there is malnutrition and limited access to water even though water resources are sufficient to meet local demand. This limited access is due to socioeconomic constraints and can be remedied by changes in institutional practice. The areas highlighted in dark blue, then, can benefit greatly from practices discussed in this chapter, leading to improved water equity, better nutrition, and more water for food and agriculture.

With the exception of fisheries, all agriculture is land-based, and decisions made regarding land use are also decisions that impact both food and water. Water security, then, encompasses not only direct usage and conservation of water but also usage and conservation of land that is needed to grow our food and sustain our world's populations. Agriculture is the world's largest user of freshwater. By improving this one sector alone, we will be greatly enhancing global water security.

*Foundations: Water-Quality Aspects and Stratification of Lakes and Reservoirs

Although groundwater is often of superior quality, many people still obtain their water for domestic usage – water for drinking, cooking, and cleaning – from surface water sources. And many of these sources are lakes or reservoirs, either natural or man-made. In the United States, for example, surface water sources (lakes and rivers) provided water for 22% of the community water systems (CWS) and 68% of the population, or about 196 million people (US EPA 2008). Globally, approximately 144 million people still collect their water directly from unimproved surface water sources (WHO 2019). Thus, it is important for water-security practitioners to be familiar with the dynamics of a lake/reservoir system so that water-quality issues may be properly addressed.

Precipitation that is not absorbed by plants or soil often flows along the ground, eventually making its way into rivers, streams, and lakes. Along the way, it is collecting both natural and anthropogenic contaminants that include pathogenic bacteria, pesticides, and nutrients (nitrogen and phosphorus) used for fertilizers. This section considers both the fluid dynamics and the most important water-quality parameters in a lake system.

Water-quality aspects. A lake will naturally accumulate silt and organic matter as it ages. A young lake has a low nutrient content and, therefore, low plant productivity. Such a lake is called *oligotrophic*, a term meaning "few foods." As the lake slowly accumulates nutrients from rainfall and basin runoff, it begins to develop a murkiness from the proliferation of phytoplankton, made up of single-celled algae and cyanobacteria. In addition, organic matter gradually decays and depletes the store of dissolved oxygen (DO). The lake may then become *eutrophic* ("many foods"), shallower and warmer as the accumulating silt and organic debris fill in the margins where more and more plants take root (Figure 7.10). Eventually the lake will become a marsh or bog, but the process may take thousands of years. The input of anthropogenic sources of nutrients from municipal and industrial wastewater, septic systems, and agricultural runoff, however, accelerates the process considerably. Such a process is termed *cultural eutrophication* and can quickly degrade water quality as algal blooms die and decay and fish die off from lack of oxygen. The most recent EPA report on water quality

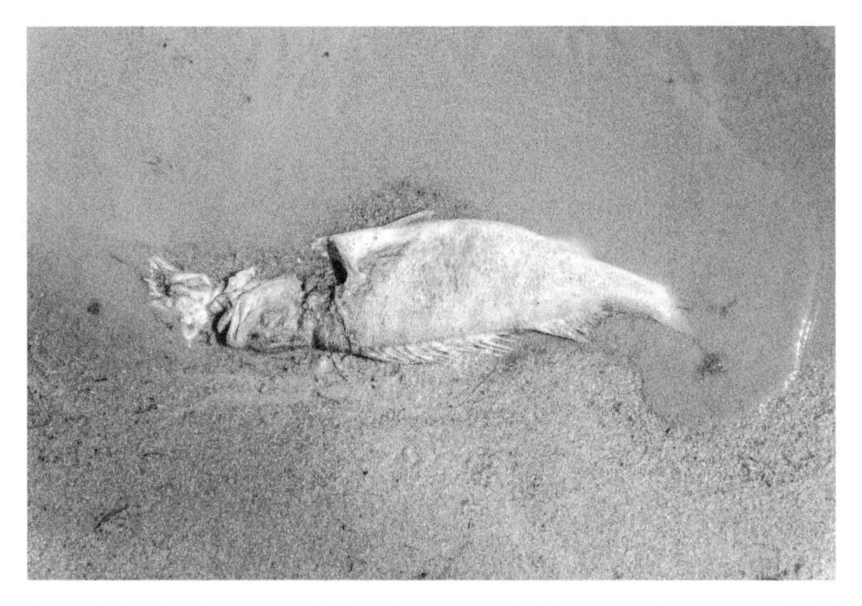

Figure 7.10 A highly eutrophic lake, characterized by blue-green algae and fish kills, in the Yunnan Province of China. Source: Dmytro/Adobe Stock.

revealed that 55% of US freshwater lakes are either eutrophic or hypereutrophic based on concentrations of *chlorophyll-a*, a measure of the amount of algae in a water column (US EPA 2017). Most of this impairment is from the addition of nutrients (nitrogen and phosphorus) from agricultural activities and urban runoff, two nonpoint sources that are very hard to contain.

How do we protect a lake's water quality? Plants need both sunlight and nutrients for growth. It is neither feasible nor desirable to shield a lake from sunlight. So, the only option is to limit the amount of nutrients that a lake receives. In 1840 Justus Liebig, a German chemist and a principal founder of organic chemistry, determined that the growth of a plant is dependent upon the amount of foodstuff that is available in limited quantity. This idea, known as *Liebig's law of the minimum*, resulted in the concept of a limited nutrient without which plants could not grow in a specific environment (Masters and Ela 2007). In coastal waters, this limited nutrient tends to be nitrogen; in freshwaters, it is phosphorus. About 40% of US lakes are currently in a disturbed condition from phosphorus, the limiting nutrient in freshwater conditions (US EPA 2017). Agricultural practices – such as terracing and more precise application – that limit the loss of field nutrients are needed to minimize the threat of cultural eutrophication.

In contrast to a stream or river, a lake is a closed water body that does not easily carry away unwanted contaminants. Reservoir degradation can occur due to toxic organic chemical pollution from point sources that are located in close proximity to the shoreline. Some of these contaminants have a very long residence time until they are degraded or settled and absorbed in the sediment layer. About 86% of shoreline miles of the US Great Lakes are impaired for designated lake uses, mostly from polychlorinated biphenyls (PCBs), dioxins, and pesticides. In addition to agriculture, these magnificent lakes are impaired by legacy sediment pollution as well as atmospheric deposition. Fish tissue can be highly contaminated with PCBs and mercury through the process of biomagnification.

What level of nutrients are needed for algal photosynthesis? Consider the following representative chemical equation for algal growth:

$$106\,CO_2 + 16\,NO_3^- + HPO_4^{2-} + 122\,H_2O + 18\,H^+ \rightarrow C_{106}H_{263}O_{110}N_{16}P + 138\,O_2$$

Using stoichiometry with the knowledge of the gram molecular weights of N (14) and P (31), we can see that the ratio of N to P is:

$$\frac{N}{P} = \frac{(16\text{ moles})(14\text{ g/mole})}{(1\text{ mole})(31\text{ g/mole})} = 7.2 \tag{7.1}$$

Thus, about seven times more nitrogen is needed than phosphorus in order to grow algae. So, if the concentration ratio of N/P is much greater than 7.2, then the body of water is phosphorus limited. If the ratio is much lower than this, the water is nitrogen limited. On an absolute basis, empirical studies have shown that phosphorus concentrations in excess of 0.015–0.020 mg/l and nitrogen concentrations in excess of 0.3 mg/l are likely to cause algal blooms, if other conditions are met (Masters and Ela 2007).

In order to estimate the concentration of a nutrient in a lake's water body, we can use a mass-balance approach with a few simplifying assumptions (Figure 7.11). In Figure 7.11, Q is flowrate, C is the concentration of a nutrient (N or P), S is any additional input from an external source, and v_s is the settling rate of the nutrient. We can first assume that a lake will be relatively well-mixed on an annual or seasonal basis, that the flow into the lake will equal the flow out of the lake ($Q_{in} = Q_{out} = Q$). Flow into the lake may be from an incoming stream and outflow may be over a spillway. The concentration of a contaminant out of the lake will then be equal to the concentration in the lake itself ($C_{out} = C$). The settling rate, v_s, is difficult to quantify, but empirical estimates are ~10–16 m/year (Masters and Ela 2007).

Focusing on the water phase in the lake, and assuming conservation of mass, we observe the following:

$$\begin{aligned}&\text{Rate of addition of N/P (inflow from nonpoint and point sources)}\\&\qquad = \text{Rate of removal of N/P (outflow, loss to sediment):}\end{aligned}$$

$$QC_{in} + S = QC + v_s AC \tag{7.2}$$

Solving for C, concentration in the lake, we have:

$$C = \frac{QC_{in} + S}{Q + v_s A} \tag{7.3}$$

This value can then be compared with the nutrient levels given above for algal growth.

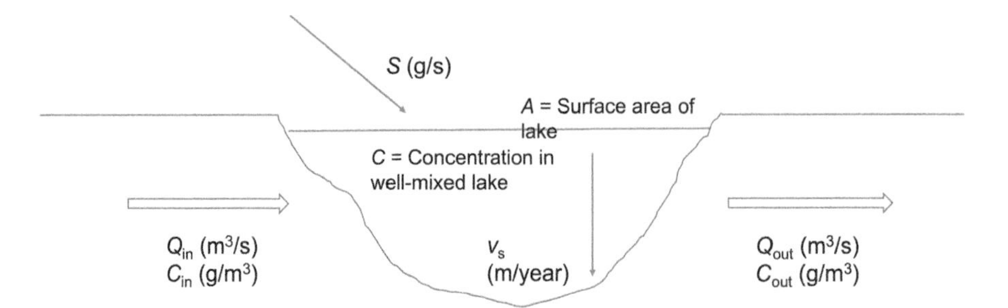

Figure 7.11 Schematic of a well-mixed lake, showing flowrate variables, concentrations, inputs, and settling velocity. Source: JFC original.

Lake stratification. The maximum density of water is reached at 4 °C. Thus, ice at 0 °C is slightly less dense and floats atop the water around it that is slightly warmer and denser. Thus, during winter months, fish and other organisms are able to survive below the icy surface. In the warmer months, the water density decreases with increasing temperature. As a result, a deep lake will stratify into a cold layer near the bottom (*hypolimnion*) and a warmer layer at the top (*epilimnion*). In between is a transition layer known as the *thermocline (or metalimnion)* where the temperature gradient is more pronounced. Swimmers may notice this layer as they dive deeply and experience the sudden drop in water temperature.

We have seen how oxygen depletion can be caused by the excessive growth and decay of algal plants. But a second phenomenon, known as "lake overturn," can contribute to the problem. In the summer, thermal stratification results in essentially two lake layers as described above, with warmer water on top of colder water. In temperate climates, the Fall season brings cooler air temperatures. Slowly, the epilimnion layer cools as well until the water temperature stratification disappears. With autumn winds and/or a passing storm, the lake bottom gets churned up and complete mixing of the lake is approached. In cold climates, the same thing will happen again in the spring when the ice melts, the spring winds blow, and water mixing results. Residents in these climates who use these lakes as drinking water sources will notice variations in taste and smell during these periods of overturn.

For purposes of water quality, DO is also affected by stratification and lake overturn. Because of reaeration and photosynthesis, DO in the epilimnion is greater than in the hypolimnion. If the lake is clear and oligotrophic, then there is the possibility of some DO in the hypolimnion. Otherwise, if the lake is murky and eutrophic, then the DO in lower layers is oxygen-starved and will result in fish die-offs. During the Fall or spring overturn, both nutrients and oxygen are circulated and mixed throughout the water column, giving the lake a "breath of fresh air."

The above is true only of lakes in temperate zones. Tropical lakes, with no cold weather season, remain stratified with cooler water staying in the bottom layer, becoming more and more anaerobic as the lake tends to eutrophication. Decomposition of organic matter in the hypolimnion can create a dangerous concoction of gases such as methane (CH_4), hydrogen sulfide (H_2S), and CO_2. These gases tend to stay dissolved in the bottom waters until a land-shaking event, such as an earthquake or landslide, triggers a release. Such an event happened twice in Cameroon, central Africa, in a period of two years, 1984–1986. Lake Monoun exploded in a limnic eruption of carbon dioxide, killing 37 people in August 1984. Two years later, again in August, the eruption of Lake Nyos asphyxiated over 1700 people and 3500 livestock. A cloud of carbon dioxide traveled as far as 16 miles from lake and, being heavier than air, descended upon several villages, displacing the air and suffocating the people and animals (Rensberger 1986). Prevention of such catastrophes and protection of lake water quality depend upon good watershed management, as discussed elsewhere in this text.

End-of-Chapter Questions/Problems

7.1 Consider data given in Figure 7.1 and similar data from the Internet. Give a reason why the following countries have a relatively high/low food import dependency:

a. Japan – high rate
b. Brazil – low rate
c. Saudi Arabia – high rate
d. Nigeria – low rate

7.2 Use the Internet and data in Table 7.1 to compare the water footprints of the following meals:
a. A typical meal in El Salvador
b. A typical meal in rural India
c. The last meal you ate at a nice restaurant

7.3 Find an LCA for the production of beef, pork, or chicken. Describe the various phases of production and consumption of water in each phase.

7.4 Find a good summary article on *The Limits to Growth* report from the Club of Rome in 1972. List five of its major findings or recommendations.

7.5 In your own words, what is the difference between blue and green water? Why does it matter to make this distinction as it affects water security?

7.6 Consider that an average diet is about 2800 kcal per capita per day. Using data in Table 7.1, calculate the approximate water usage cost per 2800 calories for diets consisting of the following four foods:
a. Diet One: beef, potatoes, onions, and corn
b. Diet Two: fish, rice, tomatoes, and onions
c. Compare the two diets and comment on the difference.

7.7 The Murray–Darling Basin is seen as Australia's breadbasket and is the source of 40% of the nation's agricultural income. What are producers and water managers doing in this region to increase their water-use efficiency? Describe these measures and provide your references.

7.8 Almost 1/5 (18%) of the world peoples live in China. What measures are being taken in China to increase the efficiency of its farming practices?

7.9 According to the map of physical and economic water scarcity, explain why the following is true:
a. Australia has one region of high physical water scarcity.
b. Much of northern India is of high economic water scarcity.

7.10 Compare the two concepts of physical and economic water scarcity.
a. Which is more important and why?
b. Which type is easier to address and what measures should be taken?

7.11 Total nitrogen was recorded in a rural pond as 5 μg/l (insert Greek letter). What concentration of total phosphorus input would enable the growth of algae in this lake?

7.12 What is the approximate concentration of nitrogen (mg/l) in a well-mixed 5-acre lake at the end of a rainy year? The flow into the lake is 12.5 m^3/hours with an incoming average nitrogen concentration of 0.2 mg/l. Use a settling rate of 14 m/year. The input from precipitation is 18 mg/month of N. Set up your calculations on a monthly basis.

Further Reading

Garrido, A. and Ingram, H. (ed.) (2014). *Water for Food in a Changing World*, 1e. Routledge.
Hoekstra, A.Y. and Mekonnen, M.M. (2012). The water footprint of humanity. *Proceedings of the National Academy of Sciences* 109 (9): 3232–3237. https://doi.org/10.1073/pnas.1109936109.
Molden, D. (ed.) (2007). *Water for Food, Water for Life: A Comprehensive Assessment of Water Management in Agriculture. London*. Sterling, VA: Earthscan.
Smil, V. (2000). *Feeding the World: A Challenge for the Twenty-First Century*. MIT Press.
Water Initiative, The World Economic Forum Water (2011). *Water Security: The Water-Food-Energy-Climate Nexus*, 2e. Washington, D.C: Island Press.

References

Alexandratos, N. and Bruinsma, J. (2012). "World Agriculture towards 2030/2050: The 2012 Revision." ESA Working Paper No. 12-03.
Amosson, S.H., Almas, L., Girase, J.R. et al. (2011). *Economics of Irrigation Systems*." B-6113. Texas: A&M AgriLIFE Extension.
Arlosoroff, S. (2002). "Integrated Approach for Efficient Water Use Case Study: Israel." Presented at the World Food Prize Conference, Des Moines, IA, USA.
Bardi, U. (2011). *The Limits to Growth Revisited* 2011th Edition. Springer.
Bhattarai, M. and Narayanamoorthy, A. (2003). Impact of irrigation on rural poverty in India: an aggregate panel-data analysis. *Water Policy* 5 (5–6): 443–458. https://doi.org/10.2166/wp.2003.0028.
Chapagain, A.K. and Hoekstra, A.Y. (2010). "The Green, Blue and Grey Water Footprint of Rice." Value of Water, Report No. 40. UNESCO-IHE.
CIAT (2019). "The Changing Global Diet." CIAT. 2019. https://ciat.cgiar.org/the-changing-global-diet.
Falkenmark, M. and Rockström, J. (ed.) (2011). Chapte 6: Back to basics on water as constraint for global food production. In: *Water for Food in a Changing World*, vol. 333. New York: Routledge.
Falkenmark, M., Rockström, J., and Karlberg, L. (2009). Present and future water requirements for feeding humanity. *Food Security* 1 (1): 59–69. https://doi.org/10.1007/s12571-008-0003-x.
FAO (2000). "Deficit Irrigation Practices." 2000. http://www.fao.org/3/y3655e02.htm#b.
FAOStat (2020). "FAOSTAT." 2020. http://www.fao.org/faostat/en/#data/QC.
de Fraiture, C., Fuleki, B., Giordano, M., and Suhardiman, D. (2009). "Trends and Transitions in Asian Irrigation: What Are the Prospects for the Future?" IWMI - FAO workshop on Asian irrigation (Bangkok). https://www.researchgate.net/publication/254394397_Trends_and_Transitions_in_Asian_Irrigation_What_Are_the_Prospects_for_the_Future.
Gleick, P. (1993). *Water in Crisis: A Guide to the World's Fresh Water Resources*, 1e. New York: Oxford University Press.
Hoekstra, A.Y. (2008). The water footprint of food. In: *Water for Food*, J. Förare (Ed.). The Swedish Research Council for Environment, Agricultural Sciences and Spatial Planning (Formas). http://www.waterfootprint.org/Reports/Hoekstra-2008-WaterfootprintFood.pdf.
Hoekstra, A.Y. and Mekonnen, M.M. (2012). The water footprint of humanity. *Proceedings of the National Academy of Sciences* 109 (9): 3232–3237. https://doi.org/10.1073/pnas.1109936109.

Hornberger, G.M. and Perrone, D. (2019). *Water Resources: Science and Society*. Baltimore: Johns Hopkins University Press.

Hub, IISD's SDG Knowledge (2019). "India's Population Expected to Surpass China's by 2050: World Population Data Sheet | News | SDG Knowledge Hub | IISD." October 2019. http://sdg.iisd.org/news/indias-population-expected-to-surpass-chinas-by-2050-world-population-data-sheet.

Lankford, B., Bakker, K., Zeitoun, M., and Conway, D. (2013). *Water Security: Principle, Perspectives and Practices*. Earthscan.

Liu, W., Antonelli, M., Kummu, M. et al. (2019). Savings and losses of global water resources in food-related virtual water trade. *WIREs Water* 6 (1): e1320. https://doi.org/10.1002/wat2.1320.

Long, S.P., Ainsworth, E.A., Leakey, A.D.B. et al. (2006). Food for thought: lower-than-expected crop yield stimulation with rising CO2 concentrations. *Science* 312 (5782): 1918–1921. https://doi.org/10.1126/science.1114722.

Lundqvist, J., de Fraiture, C., and Molden, D. (2008). *Saving Water: From Field to Fork - Curbing Losses and Wastage in the Food Chain*. Stockholm: SIWI.

Masters, G. and Ela, W. (2007). *Introduction to Environmental Engineering and Science*, 3e. Upper Saddle River, N.J: Pearson.

Mekonnen, M.M. and Hoekstra, A.Y. (2010). *The Green, Blue and Grey Water Footprints of Farm Animals and Animal Products*. UNESCO-IHE.

Molden, D. (ed.) (2007). *Water for Food, Water for Life: A Comprehensive Assessment of Water Management in Agriculture. London*. Sterling, VA: Earthscan.

Mosali, J. (2013). "Crop Rotation Yields Many Benefits." Noble Research Institute. 2013. http://www.noble.org/news/publications/ag-news-and-views/2013/november/crop-rotation-yields-many-benefits.

Moser, P. (2010). "Jordan River Could Die by 2011: Report." 2010. https://phys.org/news/2010-05-jordan-river-die.html.

National Research Council (2007). *Water Implications of Biofuels Production in the United States*. https://doi.org/10.17226/12039.

Niang, I., Ruppel, O.C., Abdrabo, M.A. et al. (2014). "Climate Change 2014: Impacts, Adaptation, and Vulnerability. Part B: Regional Aspects. Contribution of Working Group II to the Fifth Assessment Report of the Intergovernmental Panel on Climate Change." IPCC.

NRCS (2012). "Improving Irrigation Water Use Efficiency | NRCS Arizona." 2012. https://www.nrcs.usda.gov/wps/portal/nrcs/az/newsroom/releases/nrcs144p2_065174.

Postel, S. (2017). *Replenish: The Virtuous Cycle of Water and Prosperity*. https://smile.amazon.com/Replenish-Virtuous-Cycle-Water-Prosperity/dp/1642830100/ref=sr_1_1?dchild=1&keywords=Replenish+Postel&qid=1593794678&s=books&sr=1-1.

Rensberger, B. (1986). "Cameroon Lake Victims Died of Asphyxiation." *Washington Post*, September 13, 1986. https://www.washingtonpost.com/archive/politics/1986/09/13/cameroon-lake-victims-died-of-asphyxiation/7d9fda57-b5d3-4092-bc41-86b6ebada234.

Rockström, J., Falkenmark, M., Karlberg, L. et al. (2009). Future water availability for global food production: the potential of green water for increasing resilience to global change. *Water Resources Research* 45 (7): https://doi.org/10.1029/2007WR006767.

Rogers, J. (2020). "Barley Dreams: An Arizona Town Gambles on Beer to Save Water in the Verde River." *The Nature Conservancy*, 2020.

Rosegrant, M.W., Zhu, T., Msangi, S., and Sulser, T. (2008). Global scenarios for biofuels: impacts and implications. *Review of Agricultural Economics* 30 (3): 495–505.

Siegel, S. (2017). *Let There Be Water*. Griffin. https://smile.amazon.com/Let-There-Be-Water-Water-Starved/dp/1250115566/ref=sr_1_1?crid=2SIV12GFQR65K&dchild=1&keywords=let+there+be+water+by+seth+siegel&qid=1598400233&s=books&sprefix=Siegel+let+there+%2Caps%2C177&sr=1-1.

Solomon, S. (2010). *Water: The Epic Struggle for Wealth, Power, and Civilization*, 1e. New York: Harper.

Spuhler, D. and Carle, N. (2020). "Crop Selection | SSWM - Find Tools for Sustainable Sanitation and Water Management!" 2020. https://sswm.info/sswm-solutions-bop-markets/improving-water-and-sanitation-services-provided-public-institutions-0/crop-selection.

The Economist (2020). "The World's Food System Has so Far Weathered the Challenge of Covid-19." *The Economist*, May 9, 2020. https://www.economist.com/briefing/2020/05/09/the-worlds-food-system-has-so-far-weathered-the-challenge-of-covid-19.

Tricase, C., Amicarelli, V., Lamonaca, E., and Rana, R.L. (2018). "Economic Analysis of the Barley Market and Related Uses." In *Grasses as Food and Feed*. https://doi.org/10.5772/intechopen.78967. IntechOpen. https://www.intechopen.com/books/grasses-as-food-and-feed/economic-analysis-of-the-barley-market-and-related-uses.

US EPA (2008). "Factoids: Drinking Water and Ground Water Statistics for 2007." EPA 816-K-07-004. Washington, D.C. http://www.epa.gov/safewater/data.

US EPA, OW (2017). "2017 National Water Quality Inventory Report to Congress." Reports and Assessments. US EPA. December 11, 2017. https://www.epa.gov/waterdata/2017-national-water-quality-inventory-report-congress.

Vansintjan, A. (2015). "Water Johads: A Low-Tech Alternative to Mega-Dams in India." *No Tech Magazine*, 2015. https://www.notechmagazine.com/2015/06/water-johads-a-low-tech-alternative-to-mega-dams-in-india.html.

Water Initiative, The World Economic Forum Water (2011). *Water Security: The Water-Food-Energy-Climate Nexus*, 2e. Washington, D.C: Island Press.

Weckler, P. (2021). "Water Usage in Agriculture - Comments," April 27, 2021.

WFN (2020). "Water Footprint Network." Water Footprint Network. 2020. https://waterfootprint.org/en.

WHO (2019). *Progress on Household Drinking Water, Sanitation and Hygiene 2000–2017: Special Focus on Inequalities*. WHO http://www.who.int/water_sanitation_health/publications/jmp-report-2019/en.

WWAP (ed.) (2012). *Managing Water under Uncertainty and Risk. Vol. 3: Facing the Challenges*. United Nations World Water Development Report, 4.2012,3. Paris: UNESCO [u.a.].

8

Water and Energy

In this chapter, we introduce the water–energy nexus and show that it is a crucial component of the water-security paradigm. Just as with water, energy is critical to the quality of life we enjoy, both in our personal lives and as a global society. Energy production requires water in each of its multiple phases – from raw materials extraction to production to transmission. Understanding the energy cycle enables us to better understand the water demands of this sector, whether it be for production of electricity (e.g. thermoelectric, hydroelectric, renewables) or for transportation fuels (e.g. gasoline, biofuels). Any steps we take to conserve energy will reduce the associated water (and carbon) footprints. Shifting to renewable energy resources, such as wind and solar, can have the same benefit. In addition to the water footprint of energy, there is also an energy footprint of water. Given that obtaining, treating, and transporting water require energy, it is imperative that we improve the energy efficiency of this process. Given the critical importance of the water–energy nexus, identifying a path forward is vital to the future of water, and societal, security.

Learning Objectives

Upon completion of this chapter, the student will be able to:

1. Understand why the water–energy nexus is important to individual and societal well-being.
2. Understand the role water plays in the energy-production process.
3. Understand the economics of energy production and transmission and how this impacts the preferred energy source with its associated water impacts.
4. Recognize the role of water in thermoelectric, hydroelectric, and renewable energy production.
5. Understand the role of water in the production of transportation fuels.
6. Recognize the tradeoffs between cost and water/environmental impacts of selecting an energy source as well as a path forward to a more secure water–energy nexus.
7. Understand the energy demands of various drinking water sources, treatment systems, and transport.
8. *Perform simple calculations regarding pumps, pressure, and power generation.

Fundamentals of Water Security: Quantity, Quality, and Equity in a Changing Climate, First Edition.
Jim F. Chamberlain and David A. Sabatini.

Companion website: www.wiley.com/go/chamberlain/fundamentalsofwatersecurity

8.1 Introduction

While in the previous chapter we discussed the water–food nexus, in this chapter we will focus on the water–energy nexus. Energy is critical to modern society as we know it, from an individual, city, national, and global perspective. As an individual, we rely on a steady flow of energy to power, cool, and heat our homes; charge our cell phones; and fuel our transportation. At a city level, energy enables a community to support businesses, schools, libraries, entertainment, and houses of worship. At a national level, energy is a major driver of the GDP and economic vitality of a nation. From a global perspective, energy sources can either lead to or interfere with international cooperation and trade. Thus, energy, much like water, is pervasive throughout every level of society.

Yet energy is not produced in a vacuum without inputs and consequences. In certain settings, water is the driving force to produce energy – from the waterwheel to the modern hydroelectric plant. When an energy source is utilized to produce thermoelectric power (as in natural gas or coal-powered plants), water is an essential ingredient – from the boiler-based generators to the water-based cooling tower. Even when using renewable energy from solar or wind, water is still critical to the production of windmill blades and solar panels. Further, water is a major ingredient for producing and refining petroleum for transportation fuels. When considering biorenewable energy (e.g. ethanol from corn), water is necessary for growing and processing the feedstock. Thus, water is intricately linked to the energy-production process. Likewise, energy is a vital part of the water process. For example, drinking water relies on energy to transport water from the source to the treatment plant, to operate the treatment processes, and to deliver the treated water to the consumer. Thus, the water–energy nexus is pervasive and critically important.

8.2 Energy and Water for Individual and Societal Health and Well-being

In his book *Thirst for Power: Energy, Water and Human Survival*, Michael Webber proposes that water and energy are the two most critical inputs to modern civilization (Webber 2016). Webber states that upon arriving at this position, he realized that it deviated from Maslow's well-known hierarchy of food, water, and shelter. Nonetheless, he concluded that, while Maslow's hierarchy is fundamentally important for personal health and well-being, at a *societal* level water and energy are paramount because sufficient water and energy are vital inputs to growing the necessary food, producing adequate shelter, etc. Conversely, in the absence of energy and water, individual and societal needs cannot be met. Webber's position was reinforced when Nobel Laureate Rick Smalley of Rice University announced his "Top Ten Problems of Humanity for the Next 50 Years," listing energy and water as the first and second most important, followed by food, environment, poverty, terrorism and war, disease, education, democracy, and population (Smalley 2003). Smalley and Webber both agree that an abundance of cheap and safe energy is critical to addressing global challenges while promoting prosperous societies and that water is pivotal to providing this energy.

As further confirmation of the water–energy nexus, both water and energy are highlighted in the National Academy of Engineering list of Grand Challenges for Engineering in the twenty-first century, accounting for 3 of the 14 challenges (NAE 2008). At the global scale, the UN Sustainable Development Goals include "Clean Water and Sanitation" as SDG 6 and

"Affordable and Clean Energy" as SDG 7 (see Figure 1.1). A well-established energy system supports multiple sectors: from the individual to businesses, medicine, and education to agriculture, infrastructure, communication, and high-technology sectors. Nonetheless, globally one in seven people still lack electricity, mainly located in rural areas of the developing world. In sub-Saharan Africa, an estimated 573 million people still lack access to electricity (UNDP 2020). Without electricity, women and girls spend hours fetching water, clinics are unable to store vaccines and other medications, schoolchildren are unable to do homework at night, and people cannot run competitive businesses. The health and well-being of some three billion people are adversely impacted by the lack of clean cooking fuels, relying instead on wood, charcoal, dung, and coal, which cause indoor air pollution and associated respiratory illness. Thus, expanding infrastructure and incorporating advanced technologies to provide clean and more efficient energy will globally encourage growth, improve health, and protect the environment. Further, energy is the main contributor to climate change. The production of energy releases around 60% of all global greenhouse gases, reinforcing the need for not only cheap but also clean energy sources (UNDP 2020).

8.3 Energy Production and Trends: A Primer

In former times, energy was harvested in the form of horses and oxen, windmills, and waterwheels. With the advent of the steam engine, energy production expanded and became more mobile as compared to windmills and waterwheels. Energy production was now less dependent on location (next to a river) or environmental conditions (sufficient wind or water levels). The discovery of petroleum and natural gas reserves increased the energy supply and convenience of transportation and heating fuel as compared to wood and coal. More recently, wind and solar energy are emerging as viable clean energy sources. Biofuels, produced from renewable sources such as corn and soybean, are also of growing interest. Thus, as we think of energy, we must consider its multiple sources and consumption from multiple sectors, including the individual (home, transportation), community (streets, parks, infrastructure), industrial (production, distribution), and societal perspective.

Today, electricity generation comes largely from power plants using turbine-driven generators. A turbine captures the energy of a moving fluid, be it liquid (water) or gas (steam, combustion gases, air) and converts it to mechanical energy - rotating the blades of a shaft attached to a generator. The generator then converts the mechanical (rotational) energy to electrical energy. Most of the largest power plants in the United States use steam turbines, in which steam is produced by burning a fuel in a boiler with the steam driving a turbine generator. Heat exchangers are then used to capture "waste" heat, recycling as much of the heat as possible, thereby improving the system efficiency. This process results in **thermoelectric power** since it is heat being transformed into power.

The three major categories of energy for electricity generation are fossil fuels (coal, natural gas, petroleum), nuclear energy, and renewable energy sources. Table 8.1 provides a breakdown of electricity generation information for the United States in 2019 and globally in 2018, showing that natural gas and coal together accounted for 62% of US fuel source. Air-quality concerns along with low costs have helped make natural gas the dominant US fuel source. Among the renewables, wind and hydropower lead the way, followed by solar, biomass, and geothermal. Globally, coal was the number one fuel source while natural gas and renewables (primarily hydropower) account for 24 and 25%, respectively.

Table 8.1 US (2019) and global (2018) electricity generation by energy source.

Fuel source	United States (%)	Global (%)
Coal	24	38
Natural gas	38	24
Nuclear	20	10
Renewables	17	25
Oil	1	3

Source: Based on IEA (2020).

Expanding on this single-year snapshot, Figure 8.1 shows a seven-decade trend of US electricity generation from 1950 to 2019, demonstrating that coal was the dominant energy source for electricity generation in the 1950s and remained so into the 2000s. As mentioned above, air-quality concerns with coal coupled with lower natural gas prices led to the reversal, with natural gas now being the dominant source. Nuclear energy began to appear in the United States in the 1970s, growing in the late 1980s and 1990s to its current level. Renewables have been a contributor throughout this period, with hydropower being dominant initially and other renewables emerging in the 1990s and since, as discussed below.

Figure 8.2 shows the same time period (1950–2019) with a focus on renewables, demonstrating the growing interest in biofuels starting in the 1990s, wind in the 2000s, and solar in the 2010s. Projections to the future suggest that natural gas and renewable energy will continue to increase as a fuel source, while coal will continue to decline, and the other fuels will remain relatively constant (EIA 2019). The growing emphasis on sustainable energy sources (renewables) will reduce the carbon footprint and greenhouse gas emissions of the energy sector and, as discussed below, will at the same time significantly decrease the water footprint of the energy sector.

In addition to electricity generation, which is a major US energy consumer (39%) (NAS 2020b), transportation fuels are the next largest US energy consumer. In the United

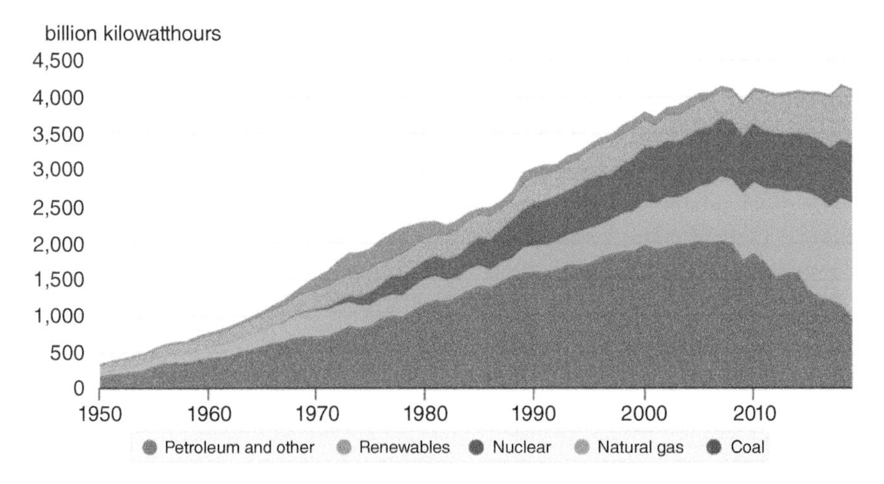

Figure 8.1 US electricity generation (billions of kWh) by energy source: 1950–2019. Source: US EIA (2020a).

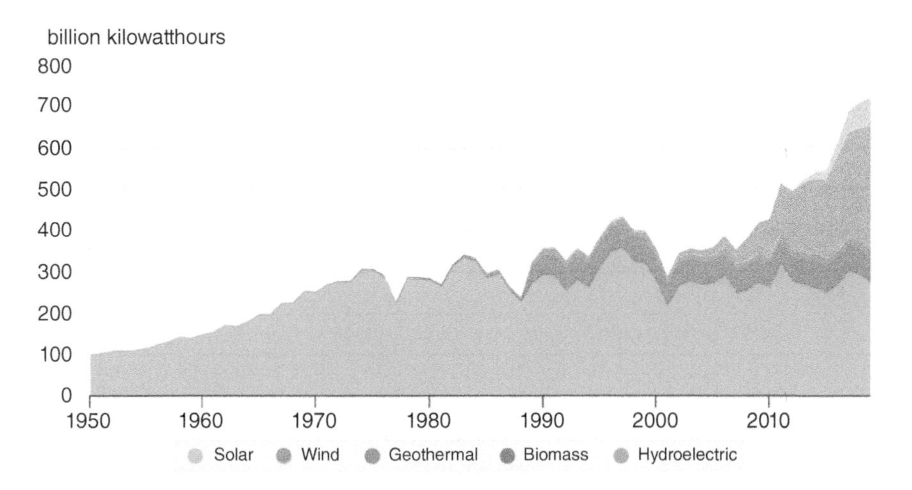

Figure 8.2 US electricity generation (billions of kWh) by renewable energy sources: 1950–2019. Source: EIA (2020).

States, 28% of the yearly energy consumption is to transport people and goods. Within the transportation sector, cars, light trucks, and motorcycles are the largest consumer (58%), followed by other trucks (23%), aircraft (8%), boats and ships (4%), and trains and buses (3%) (NAS 2020a). The United States, with less than 5% of the world's population, accounts for more than 1/5 of the world's automobiles. In 2014, cars, motorcycles, trucks, and buses drove over three trillion miles in the United States – farther than driving to the sun and back 16 000 times (NAS 2020b). Over the next quarter century, US driven miles are projected to increase by about 23%; however, vehicle efficiency has the potential to offset these additional miles in terms of fuel consumption.

With this introduction to energy production and use, and having discussed trends in energy sources, we can now discuss the water footprint of the different energy sources. Anything that we do to conserve energy will also reduce the associated water demand. Also, by shifting to energy sources and production processes with lower water footprint, we can reduce water stress. But first, we look at energy pricing in recognition of the fact that energy costs are a major factor in selecting an energy source with its associated water impacts.

8.4 Energy Prices: Supply and Demand

There are similarities regarding supply costs and demand patterns for electricity and water. We begin by presenting these for generation and distribution of electricity.

Electricity prices reflect fuel costs along with power plant, transmission, local distribution, and logistical (e.g. meter reading, billing) costs. These rates may vary from country to country, within a country, seasonally, and, during peak season, based on time of use (TOU). Pricing based on season and TOU reflects the fact that electrical power systems must be designed to satisfy peak demand and are thus "overdesigned" (underutilized) during normal operation.

Among states in the United States, higher electricity rates reflect local conditions including fuel costs, power plant capacities and efficiencies, regulatory variations, and increased reliance on cleaner fuels, including renewables. The state of Hawaii falls into the highest rate category due to its dependence on petroleum fuels which are largely imported to the island

(US EIA 2020d). Transmission lines send electricity over long distances and thus operate at higher voltage, while distribution lines are for local connections and deliver electricity on demand at lower voltage.

For most OECD countries, the peak season for electricity consumption is the hot summer season, especially where air conditioning is prevalent. During peak season, more expensive generation facilities (such as an older, less-efficient facility) or additional generation capacity at existing plants are brought on line to meet peak demand. The peak demand period thus likely results in higher production costs. Power providers in the United States generally offer a TOU rate option in which the power used during the peak period (e.g. 2 : 00 to 7 : 00 p.m.) is more expensive than the flat and nonpeak rates. This option is intended to motivate customers to mitigate peak demands on the system and minimize their need to use less-efficient generation systems and/or build an additional generation facility. In addition, measures that encourage conservation during peak demand can mitigate the high incremental cost of providing the next unit of electricity during peak demand time as well as the associated water stress.

Economics often dictates the energy of choice, i.e. "the cheapest fuel wins." Thus, reducing the cost of low water-intensity energy options will do much to improve water security. In addition, efforts to conserve energy will have the ripple effect of reducing the associated water footprint.

8.5 Water for Thermoelectric Power Generation

The largest water user (by withdrawal) in the United States is thermoelectric cooling, accounting for over half of the fresh surface water withdrawals and the majority of the saline surface water withdrawals (Delgado et al. 2015). These two surface water sources contribute the vast majority of thermoelectric cooling water with groundwater contributing a minor portion. It is important to make a distinction between consumptive and nonconsumptive use. In thermoelectric systems, *withdrawal* is the amount of water a power plant takes in from a source such as a river, lake, or ocean for cooling purposes. *Consumption* is the amount of water lost through evaporation during the cooling process. Agricultural irrigation is the largest consumptive user of freshwater. Most of the irrigation water use is either incorporated into the plant itself or evaporated into the atmosphere, while the majority of power plant cooling water is withdrawn and then discharged to the surface water body from which it was taken. Nonetheless, even though cooling water is mainly nonconsumptive, its withdrawal from the water source impacts that entity via deterioration of water quality (Pan et al. 2018). The cooling water becomes heated; chemicals may exist due to corrosion, anticorrosion agents, or biofouling; and oxygen levels are greatly reduced. Recall that increased water temperature reduces oxygen solubility, thereby impacting the receiving water's ecosystems.

According to a United States Geological Survey (USGS) study, estimates for thermoelectric water withdrawals in the year 2015 were 18% less than 2010 levels. This decline in withdrawals, even in times of increasing energy demand, can be attributed to decommissioning of lower efficiency systems, a shift from coal to natural gas as a fuel source, and new power plants that are using more water-efficient power generation and cooling-system technologies. On average, 15 gallons (gal) of water were used to produce 1 kW-hour (kWh) of electricity in 2015, compared to almost 19 gal/kW-hour in 2010 (Dieter et al. 2018).

Thermoelectric plant water withdrawals in the United States vary widely according to both geographical region and methods of cooling. Figure 8.3 shows that thermoelectric withdrawals are greatest in the eastern states, accounting for 84% of total thermoelectric-power withdrawals in the United States and 70% of the related net power generation (Dieter et al. 2018). The density of thermoelectric plants in the eastern states is reflective of both US population density and the prevalence of alternative fuel sources (hydroelectric, wind, solar) in the western states (as discussed later).

However, not all thermoelectric power plants are created equal, especially as it relates to water use and cooling systems (Figure 8.4):

- Power plants equipped with *once-through cooling systems* accounted for 96% of total water withdrawals and 37% of net power generated in the United States in 2015 (Dieter et al. 2018). At the same time, once-through systems had consumptive use of 1% (99% of water withdrawals were returned to their source).
- Newer thermoelectric plants are using *recirculating (closed-loop) cooling systems,* which withdrew much less water (4% of total thermoelectric-power withdrawals) and produced most (63%) of the power in the year 2015
- *Dry air cooling* reduces the overall water requirement by 90%, making the water footprint very attractive. But since the heat capacity of air is much less than water, the dry cooling towers must be larger, making these systems less efficient and two to four times more expensive (Delgado et al. 2015).

Table 8.2 summarizes the pros and cons of the three methods of cooling systems. As the table shows, there are tradeoffs between the cooling options, implying that with increasing water stress future systems will likely trend toward more water-efficient processes and alternate water sources.

In light of the significant water withdrawals and ecological impacts of thermoelectric power generation, the use of alternate water sources is of growing interest. Not surprisingly,

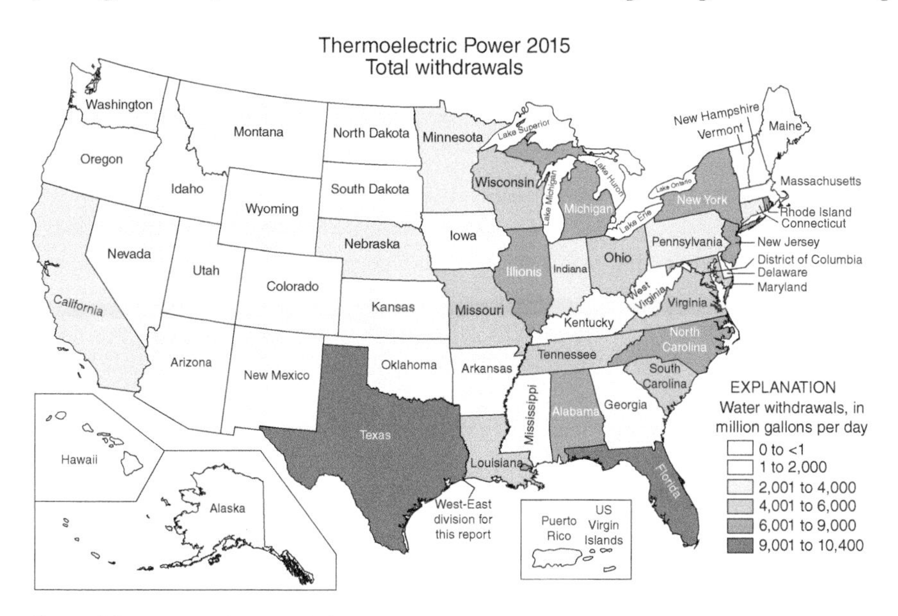

Figure 8.3 Water withdrawals for thermoelectric power generation in the United States by state – 2015. Source: Dieter et al. (2018). USGS (Public domain).

Figure 8.4 Coal power plant in China. Steam is generated from water used to cool this coal-fired power plant in China. Source: chungking/Adobe Stock.

Table 8.2 Comparison of once-through wet cooling, recirculating/wet cooling towers and dry (air) cooling towers.

Cooling type	Water withdrawal	Water consumption	Capital cost	Plant efficiency	Ecological impact
Once-through (wet) cooling	Intense	Moderate	Low	Most	Intense
Recirculating, closed-loop (wet) cooling	Moderate	Intense	Moderate	Moderate	Moderate
Dry (air) cooling	None	None	High	Least	Low

Source: Rodriquez et al. (2013).

saline withdrawals are more common in coastal states. Figure 8.5 depicts existing and proposed water sources and types (US DOE 2014). While currently fresh surface water is the dominant source of cooling water, there is an increasing move toward using less freshwater. This would be replaced with alternate water sources, including ground water, plant discharge, and reclaimed water. Diversifying the future portfolio of cooling water sources will free up fresh surface water for other uses, thereby increasing water security.

While existing freshwater withdrawals may require some level of treatment (such as anticorrosion, antiscaling, or antifouling agents), use of alternative waters requires increased scrutiny of water-quality issues. Further, different water sources (e.g. reclaimed water, industrial process water, agricultural runoff water) will require additional and varied treatment steps prior to use. Some of these water-quality issues are germane to nearly all freshwater withdrawals (e.g. removal of suspended and dissolved solids) while others are unique to specific source waters (e.g. removal of organics, pesticides, heavy metals). Thus, while cooling water seems like a very simple thing, a multitude of water-quality considerations must be considered (Pan et al. 2018).

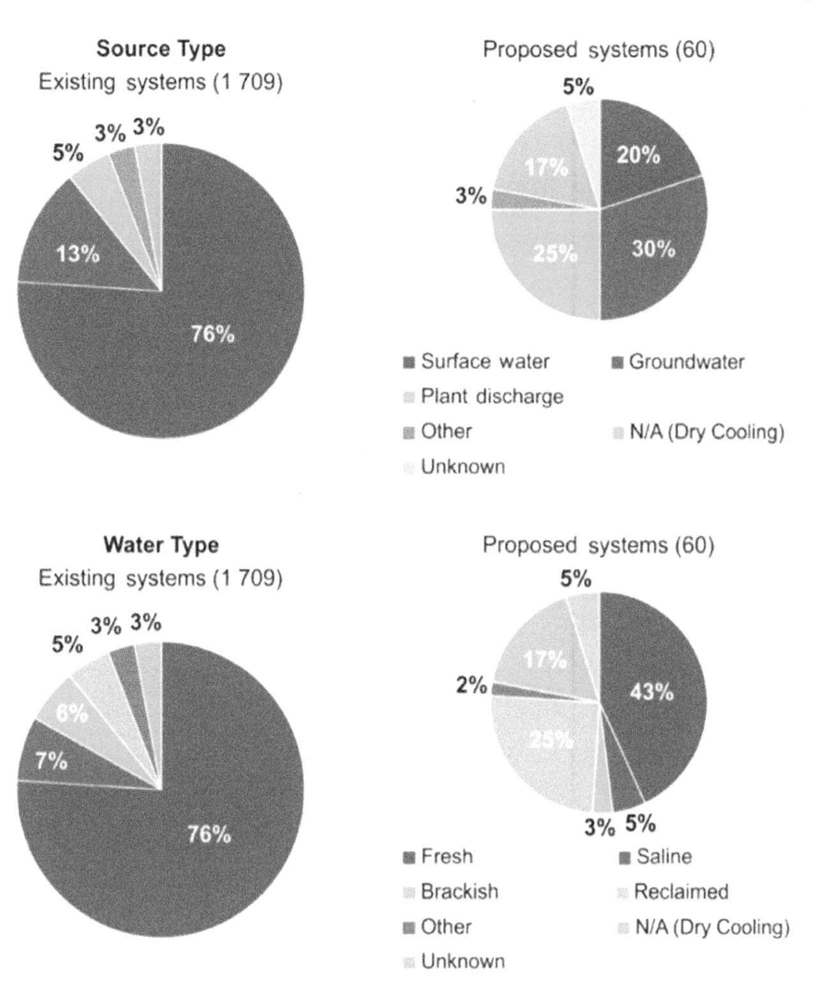

Figure 8.5 Cooling water withdrawals for thermoelectric power plants by source and type – existing and proposed systems. Source: US DOE (2014). Public domain.

8.6 Water and Renewable Electricity: Hydroelectric Power

Having evaluated the water withdrawal of thermoelectric power generation, let us pivot to hydroelectric and alternative renewable power-generation facilities. **Hydroelectric power** generation globally provided 16% of electricity and accounted for the highest level of **renewable energy** in 2018 (IEA 2020). Hydroelectric power production tripled in magnitude from 1973 to 2013. Although over half of the world's hydroelectric power is produced by just four countries (Brazil, Canada, China, and the United States), a number of countries rely heavily on hydropower (Figure 8.6) being greater than 60% in some cases. Hydroelectric-generation capacity expanded in North America and Europe between 1920 and 1970 when thousands of dams were built, a portion of them specifically for hydroelectric generation. Developed nations largely ceased dam construction in the 1970s since the best sites were already being developed and amid concerns about the growing environmental and social costs. Today, more dams are being removed in North America and Europe than are being built (Moran et al. 2018).

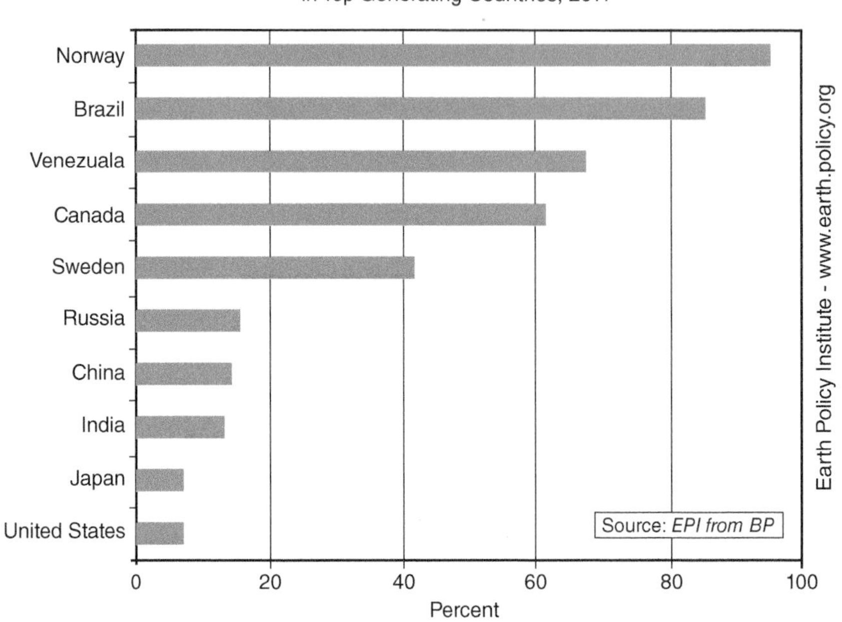

Figure 8.6 Share of electricity from hydropower in the top 10 generating countries, 2011. Source: Alley (2021).

As a consequence, since the 1970s new hydropower dam construction has shifted to the developing world and China. For example, since 1973 China's hydropower production has increased from 3% of global hydropower generation to 24% in 2013, with capacity doubling from 2010 to 2020. This generation capacity is maximum in the wet season, when flows and water levels are at their highest. When fully operational the Grand Ethiopian Renaissance Dam (GERD), the largest in Africa, will produce 6000 MW of electricity, doubling Ethiopia's current generation capacity. This will enable millions of Ethiopians to be grid-connected for the first time. Viewed as more than an engineering project, Ethiopians have come to see the GERD as a source of national pride. But Egyptians are concerned over how filling the massive reservoir, twice the size of London, will affect downstream Nile River flow, a source of international debate and even strife (The Economist 2020).

Most of all the US dams were built for flood control, municipal water supply, and irrigation water. Only a small number of US dams were built specifically for hydropower generation. About half of total US utility-scale hydropower generation is concentrated in the western US states of Washington, California, and Oregon. As a slight variation, pumped-storage hydropower takes advantage of the low-value electricity generated in off-peak hours by using this to raise water to an elevated storage reservoir, which can then flow back through the hydroelectric plant during peak demand. Even though the process is a net loss in terms of electricity generation, it is a net gain economically by generating high-value electricity during peak demand. In 2019, there was nearly 23 000 MW of total pumped-storage hydropower production capacity in 18 US states, with five states accounting for 61% of the national total (US EIA 2020c).

Hydroelectricity is often considered climate friendly as it does not emit greenhouse gases and provides an alternative to burning fossil fuels for power. However, dams, reservoirs, and hydroelectric generators can have substantial environmental and social impacts. Most clearly, hydroelectric reservoirs may obstruct fish migration. In addition, a dam and its reservoir can change natural water temperatures, water chemistry, river flow characteristics, and silt loads. These environmental changes alter the course and ecology of the river, thereby harming the native plants and animals dependent on the river ecosystem. Further, reservoirs formed by dam construction can inundate important natural areas, agricultural land, or archeological sites as well as cause the displacement of peoples (US EIA 2020b). For example, China's Three Gorges Dam, the world's largest operating hydropower facility, sets records for the number of people displaced (more than 1.2 million), number of cities and towns flooded (13 cities, 140 towns, 1350 villages), and length of reservoir (more than 600 km) (International Rivers 2020).

While the actual production of hydropower electricity is generally nonconsumptive, evaporative and seepage losses from the reservoirs do occur. The degree of losses varies depending on site location (humid versus arid) and design. In an arid environment, reservoirs with large footprints (surface area) can experience significant evaporative losses. Lake Hefner, a 2500-acre water supply lake for Oklahoma City, loses 20% of its capacity each year to evaporation alone. Run-of-the-river hydropower minimizes evaporative losses but do not reap the benefit of water storage. These units cannot optimize operation for efficient generation during peak demand (Delgado et al. 2015).

Further, during the dam-construction phase, the production of concrete and steel infrastructure can result in significant greenhouse gas emissions. However, given the 50–100 year lifetime of a dam and hydropower plant, these emissions can be offset by the long-term emissions-free hydroelectricity (US EIA 2020b).

A well-known environmental impact of hydropower in the northwest United States is the disruption of the salmon spawning cycle. These valuable fish swim back upriver from the sea to reproduce in their spawning grounds in the beds of rivers and streams. Both the salmon and its annual migration play a critical role in the religious identity of the Columbia River native peoples (Columbia River Inter-Tribal Fish Commission [CRITFC] 2021). Since dams can interfere with this migration, different approaches have been adopted to mitigate this problem, including the construction of fish ladders and "elevators" that help fish move around or over dams to the spawning grounds upstream. Further, turbines have been developed that reduce fish mortality to less than 2%, a significant decrease from previous levels of 5–10% (US EIA 2020b).

8.7 Water and Renewable Electricity: Wind and Solar Power

While renewable energy options are of increasing interest due to their reduced carbon footprint and their renewable nature, they also have the added benefit of a lower water footprint. Wind energy is of special interest in Oklahoma, where, according to the theme song from the popular musical play *Oklahoma*, "... the wind comes sweeping down the plain ..." The windy plain states in the central United States (along with California) account for the majority of wind production, mirroring a map of prevailing wind patterns. In 2019 wind accounted for 7.3% of US electricity generation, putting it nearly on par with hydroelectric power (6.6%), with solar and biomass producing 1.8 and 1.4% of electric power, respectively (see Figure 8.2).

The water intensity of electricity generation incorporates the water demand at each of three stages: raw material, transformation, and delivery. The renewables (solar, wind, hydroelectric) have minimal water demand in terms of raw material, much less significant than for coal and nuclear energy sources. Wind and photovoltaics have minimal water use during transformation (energy production), with hydroelectric suffering from severe evaporative losses, as discussed above. Delivery of the electricity is seen to have minimal water demand for all sources. Thus, as expected, wind and solar are quite attractive from a water-intensity perspective.

8.8 Life Cycle Water Intensities – Electricity and Transportation Fuels

We summarize the above discussion by providing a graphical comparison of **water intensities** for both electrical production and transportation fuels. Figure 8.7 compares the water intensity and life cycle unit costs for the most common sources of electricity. The sources are arranged according to cost and dispatchability. Four conventional dispatchable (available on demand) sources are followed by five renewable and mostly non-dispatchable sources that are dependent upon wind, sun, or rainfall. Offshore wind, natural gas combustion, biomass, and coal are all more expensive on a life cycle basis than onshore wind and solar. Water usage follows a slightly different pattern with biomass and hydroelectric being the largest overall water users and solar and wind having very low water-usage intensities. The water usage for hydroelectric is over twice that of conventional sources due to evaporation from hydropower reservoirs. The water usage of biomass for electrical production is a higher order of magnitude than conventional sources, with its water evapotranspiration rate of 209 m^3/MWh.

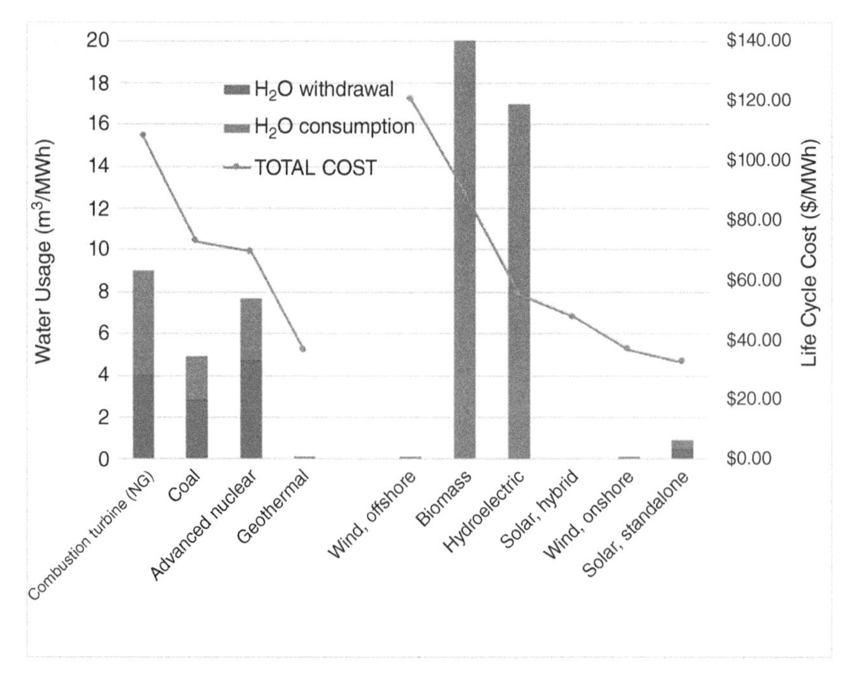

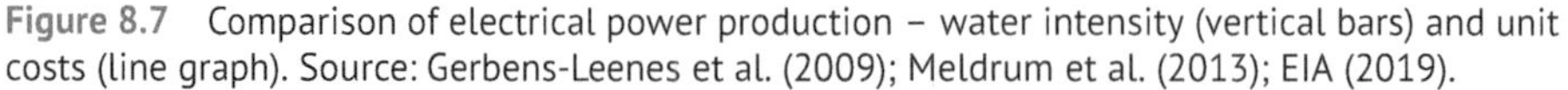

Figure 8.7 Comparison of electrical power production – water intensity (vertical bars) and unit costs (line graph). Source: Gerbens-Leenes et al. (2009); Meldrum et al. (2013); EIA (2019).

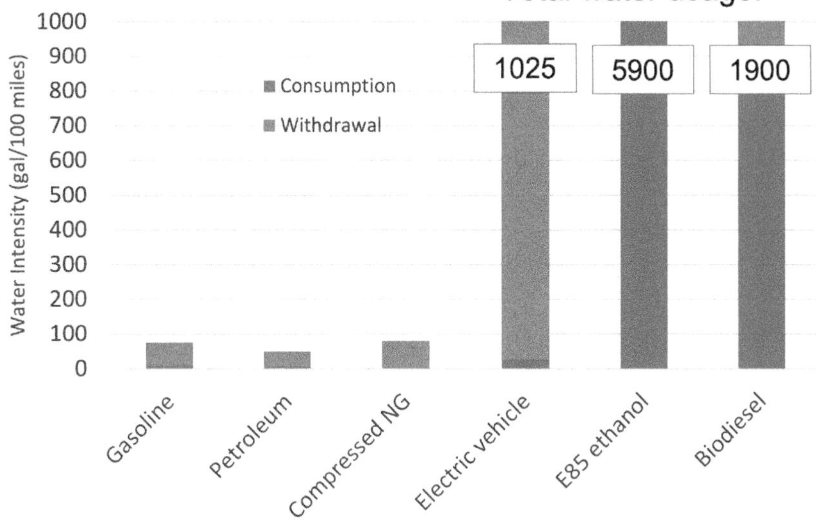

Figure 8.8 Water intensity (gal/100 miles) of common transportation fuels. Source: Adapted from King and Webber (2008).

Transportation fuels have varying degrees of water demand during both the raw materials stage (obtaining or growing the energy source) and the transformation (refining) stage. Petroleum-based fuels have the smallest overall water footprint, while biofuels have the largest water footprint during the raw material stage (growing the feedstock). On a consumption basis alone, E85 ethanol is the largest water user by orders of magnitude (Figure 8.8). These comparisons suggest that while climate change concerns warrant a consideration of biofuels for their low carbon footprint (in relative terms), water usage should also be taken into account.

8.9 Energy for Water – Access, Treatment, and Transfer

While thus far we have discussed the water footprint of energy, in this section we will focus on the energy footprint of water. Obtaining water as well as processing and delivering it to the consumer require energy inputs (raw materials, transformation, and delivery in Figure 8.9). Assuming that the readily accessible water – local, high quality, and requiring minimal treatment – has often already been tapped, new water sources frequently require water transfer by piping raw water a long distance and/or providing advanced treatment. For example, brackish water, seawater desalination, and water recycling and reuse all require energy-intensive treatment. As indicated by the range of values in Figure 8.9, the energy "cost" of water transfer and treatment is highly dependent on the proximity, elevation difference, and quality of the source water. Converting seawater into drinking water is technically viable and highly attractive except for the cost associated with distillation and reverse osmosis processes (to be discussed in more detail in Chapter 12).

Water transfer is a concept that dates back to the Roman aqueducts and before. While aqueducts were able to use gravity flow to transport the water, in most instances pumping is required to overcome friction and/or lift the water from the source to the consumer.

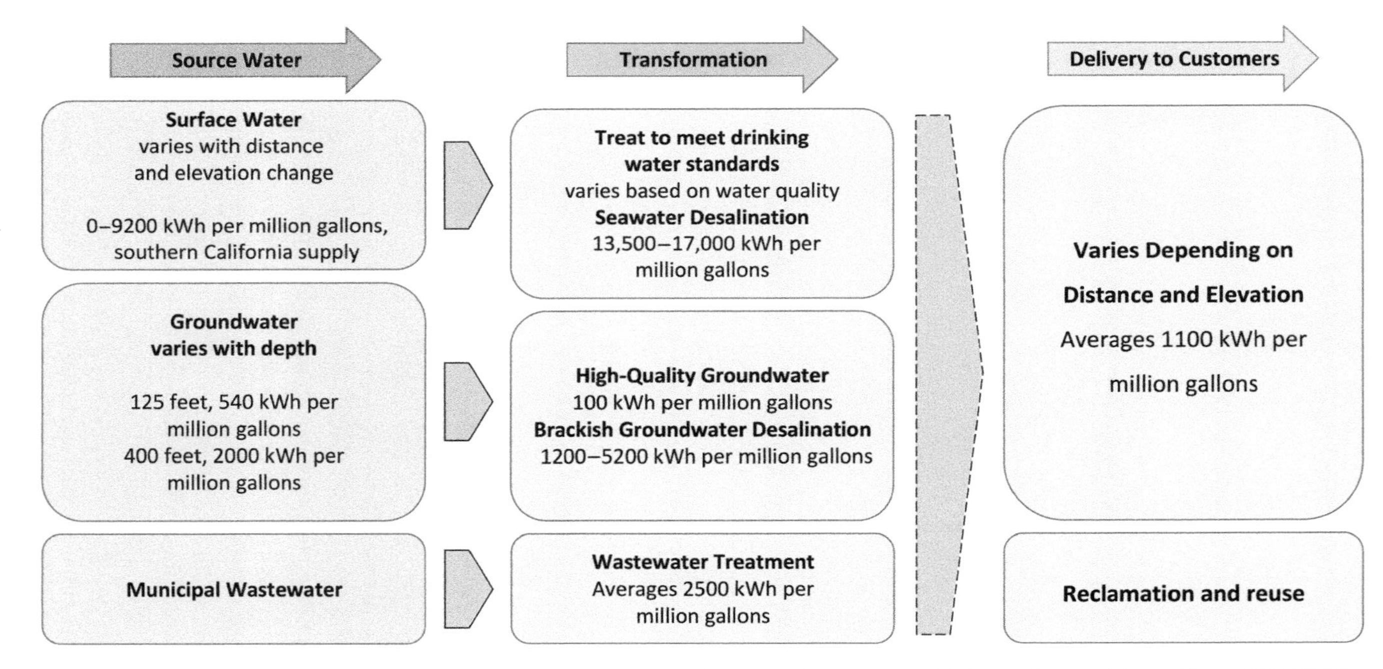

Figure 8.9 The energy footprint of drinking water and wastewater – raw materials, transformation, and delivery. Source: Adapted from WEF (2011).

Table 8.3 Break-even distances for two levels of drinking water reuse treatment versus water transfer for cities of three sizes given elevation differences common in Oklahoma.

City	Population	Water supply (MGD)	Break-even distance (miles) – level I (BAC)	Break-even distance (miles) – level II (MF/RO/UV-AOP)
Small	~25 000	4.9	8–10	18
Medium	~125 000	18.2	12	26
Large	~625 000	100	15	42

"MGD" – million gallons per day; "BAC" – biologically activated carbon; "MF" – microfiltration; "RO" – reverse osmosis; "UV-AOP" – ultraviolet light-based advanced oxidation process.
Source: Chamberlain et al. (2020)/with permission from John Wiley & Sons.

As transfer distances become longer, the cost of reusing locally available water (e.g. recycled, reclaimed, reuse water) is receiving additional attention. For a given set of conditions, Table 8.3 shows breakeven distances for two levels of advanced treatment of reclaimed water, showing that if a small community has to go further than 10–18 miles, a medium-sized town more than 12–26 miles, and a large city more than 15–42 miles to obtain a new water source, it is cheaper to install advanced treatment and recycle or reuse local water than to transfer water beyond these distances (Chamberlain et al. 2020). By implementing water reuse, energy is conserved, money is saved, and greenhouse gas emissions are reduced. While these exact values will vary depending on local conditions, the analysis approach can be adapted accordingly.

8.10 Conclusion – A Path Forward

The US Department of Energy (DOE) has identified six strategic pillars for addressing the water–energy nexus (Table 8.4). These six pillars include improving the water efficiency of energy production, optimizing the energy efficiency of water management, increased use of nontraditional water sources in the energy sector, and improving water quality, ecosystem, and seismicity relative to energy production. A number of these initiatives have been discussed above. Increased seismicity (Pillar 5) is a topic of interest in some areas as deep well injection of produced waters has greatly increased seismicity leading to limitations on the amount of produced water that can be safely injected.

Table 8.4 Six strategic pillars to address the water–energy nexus as identified by the US DOE.

1	Optimize freshwater efficiency of energy production, electricity generation, and end use systems
2	Optimize the energy efficiency of water management, treatment distribution, and end use systems
3	Enhance the reliability and resilience of energy and water systems
4	Increase safe and productive use of nontraditional water sources
5	Promote responsible energy operations with respect to water quality, ecosystem, and seismic impacts
6	Exploit productive synergies among water and energy systems

Source: US DOE (2014).

Building on this earlier study, DOE then identified Grand Water Challenges to be addressed by 2030. These five challenges include cost-effective desalination, transforming produced water to a resource, near-zero water impacts for thermoelectric plants, resource recovery from domestic wastewater, and development of small, modular energy-water systems for unique settings. Produced water is a topic of growing interest due to recent advances and increased utilization of fracking operations in gas and oil production. Fracking serves to open up channels in shale formations to release oil and natural gas trapped in these formations. Fracking fluids include propellants, which serve to keep these channels open once the high-pressure injection process ceases. The water injected to produce the fractures combines with the formation water that flows back to the well once the high-pressure injection process ends. This fracking and flowback water is often of poor quality (e.g. brackish to brine along with chemicals added to enhanced the fracking process). To reduce the amount of fresh water required, recycling and reuse of produced waters along with potential use of marginal waters (e.g. lower quality ground water) for fracking are of growing interest.

Recognizing and understanding the water–energy nexus is an important step toward promoting water security. By looking for ways to minimize the water requirements for energy production, we can increase water security and create a more sustainable future for our global society.

*Foundations: Pressure, Pumps, and Power Generation

In this Foundations section we discuss pressure, pumps, and power generation, all critical elements in the water–energy nexus.

Pressure is force per unit area of a fluid. Let us consider a pipe with a cross-sectional area of 1 ft^2 that is 1 ft tall. If we place this pipe vertically on a flat surface and fill it with water, what will be the pressure at the bottom of the water column? The volume of water (V) in the pipe is 1 ft^3 and the cross-sectional area at the bottom of the pipe (A) is 1 ft^2. Recalling that the density of water (γ) is 62.4 lb_f/ft^3, we find that the pressure in psi (pounds per square inch) is as follows:

$$P = \frac{\gamma V}{A} = \frac{62.4\frac{lb_f}{ft^3} * 1\ ft^3}{1\ ft^2}\left(\frac{1\ ft}{12\ in}\right)^2 = 0.433\ \frac{lb_f}{in^2}\ (psi) \tag{8.1}$$

Thus, 1 ft of water produces 0.433 psi of pressure. Likewise, 10 ft of water produces 4.33 psi of pressure and a water tower 150 ft tall would produce 63 psi of pressure. This pressure (63 psi) is typical of an average pressure we experience in our homes. To achieve this pressure in relatively flat areas, water towers are typically elevated to the height of a 15-floor building (each floor being roughly 10 ft). During peak water use (typically around 6 : 00 p.m.), the actual pressure we experience may be less due to the cumulative friction losses of many households using water at the same time. This is why water from a lawn sprinkler may not shoot out as far during peak demand time – a combination of the lower water level in the water tower and all the friction losses associated with peak flow in the system.

Now let us consider a dam and reservoir – e.g. the Hoover Dam in Nevada, built to generate electricity and provide irrigation water for farmers in California and Arizona. The target water hydraulic head (level) in the reservoir is 590 ft, so the pressure at the bottom of the dam would be 255 psi. That is a lot of pressure! The dam itself along with pipes transferring this water needs to be designed to handle this pressure, which is three to four times higher than what we experience in our pipes at home.

Pumps and power. But how do we size a pump to lift the water into an elevated water tank. This leads us to a power equation. Fundamentally, power (P) is equal to work (W) per unit time (t), and work is a force (F) applied over a given distance (d). Thus, we have the power equation below, where Q is flow (volume per time), r is force per volume, H is elevation or distance, and $Q\gamma H$ is thus force × distance/time or power:

$$P = Q\gamma H \tag{8.2}$$

Let us suppose that we want to fill our water tank at a flow rate of 100 gallons per minute (gpm) and our water tower is 150 ft tall. Thus,

$$P = 100\text{ gal/minute} \times 62.4\text{ lb/ft}^3 \times 1\text{ ft}^3/7.48\text{ gal} \times 150\text{ ft} \times 1\text{ minute}/60\text{ seconds}$$
$$= 2085\text{ ft-lb/second}$$

Now that is a strange set of units! But recognizing that 1 horsepower (HP) = 550 ft-lb/second, we calculate:

$$2{,}085\text{ ft-lb/second} \times 1\text{ HP}/550\text{ft-lb/second} = 3.8\text{ HP}$$

This is the energy that must be imparted to the water to lift it 150 ft at this rate. Given that pumps often have efficiencies in the range of 75–90%, we would propose a 5.0 HP motor (3.8/0.75 = 5.0). Looking at the power equation, if we want to fill the tank twice as fast, or if the tank is twice as tall, we would need to double the size of the pump. Now granted, those of you who have designed pumps recognize that this is the simple version of pump design and recognize that one must consider pump curves to more accurately design a pump system. But as an introduction to pump design, the power equation is a "powerful" concept.

Let us take the power equation one step further. Rather than filling an elevated storage tank, let us assume that we have a reservoir of water that is generating electricity. Referring to the power equation, we see that the power that can be generated is related to the height of water in the reservoir (the elevation difference between the water level in the reservoir and the discharge from the turbines) and the flow (Q) of water, not considering friction losses and efficiencies. Thus, the taller the reservoir, the greater the head (H) and thus potential power. Further, the flow (Q) can be strategically released from a reservoir to correspond with the power demand; limited flow can be released from the reservoir at night when the power demand is minimal while maximum flow can be released during peak power demand. As discussed earlier, this is one advantage of a dam and reservoir for power generation versus a "run of the river" scheme where it is the flowing water that accounts for the majority of electricity generated (Q is high and H is relatively small). However, the river flow cannot be modulated to correspond with peak electricity demand as can be done by strategic water release from a dam-reservoir system. For the Hoover Dam the power-generation capacity is on the order of 2000 MW (megawatts), which is a combination of the hydraulic head (590 ft) and the water flow (Q) through the hydroelectric system. In drought years, the hydraulic head in the reservoir will be lower and the power generation decreases accordingly.

The concept of pumping water to higher elevations and releasing elevated water to generate electricity come together in the pumped-storage scheme. In this scheme, water is pumped to an elevated storage reservoir at night when electricity is cheap and then flows back through the turbines to generate electricity during the day when it is more valuable. While this system has a net loss in energy (friction and inefficiencies mean that less electricity is generated than is used in this scheme), from an economic perspective the increased value of daytime electricity more than offsets the energy losses. Given that the cost ratio of peak period electricity to nonpeak is often 6 : 1, this economic benefit more than compensates for energy losses in this TOU scheme.

End-of-Chapter Questions/Problems

8.1 Pick a country in North America, Europe, Africa, and Asia and identify the major sources of electricity generation (e.g. indicate percentage of coal, natural gas, nuclear, hydroelectric, wind, solar). Comment on your results.

8.2 Select four major US cities and three international cities and summarize their electricity pricing schemes (e.g. base fee, summer versus winter rates, tiered rates, TOU rates if applicable). Comment on your results.

8.3 Based on the most recent USGS study on Estimate Use of Water in the United States, pick a state in the northeast, southeast, central, and western regions of the United States and summarize their (a) total water withdrawals, (b) total water withdrawals per capita, and (c) breakout of water withdrawals for major water withdrawal categories (e.g. public supply, irrigation, thermoelectric generation, industrial, livestock, domestic) for each state. Comment on your results.

8.4 What are the benefits and challenges of using alternate sources and types of power plant cooling water?

8.5 Considering three major dams – the US Hoover Dam, the Chinese Three Gorges Dam, and the Ethiopian Grand Renaissance Dam – summarize (a) basic information about each dam (when built, duration of construction, height, size, impounded water, power generation) as well as (b) societal and environmental impacts of each dam.

8.6 What are the advantages and challenges of wind and solar photovoltaics as renewable energy sources from a cost, environmental, and water footprint perspective?

8.7 From an energy-use perspective, when is water reuse advantageous versus water transfer and why?

8.8 Referring to Table 8.4, pick three of the six pillars to address the water–energy nexus and discuss the advantages and challenges associated with each of them.

8.9 (a) For a water tower 180 ft above ground surface, what would the water pressure (psi) be in a ground-level pipe at the base of the tower? (b) For a water reservoir 300 ft deep at the dam, what would the water pressure (psi) be at the base of the dam?

8.10 Redo questions 10 (a) and (b) assuming seawater with density = 64 lbs/ft^3 versus freshwater.

8.11 What pump size (horsepower) would you need to fill a water tank with 180 of elevation at a flow rate of 80 gpm using a pump efficiency of 90%?

Further Reading

Delgado, A., Rodriguez, D.J., and Sohns, A.A. (2015). *Thirsty Energy: Understanding the Linkages between Energy and Water*. Washington, D.C: World Bank.

NAS (2020a). "How We Use Energy, Transportation — The National Academies." 2020. http://needtoknow.nas.edu/energy/energy-use/transportation.

NAS (2020b). "Our Energy Sources — The National Academies." 2020. http://needtoknow.nas.edu/energy/energy-sources.

US DOE (2014). *The Water-Energy Nexus: Challenges and Opportunities*. U.S Department of Energy.

US DOE (2020). "Water Security Grand Challenge." Energy.Gov. 2020. https://www.energy.gov/water-security-grand-challenge/water-security-grand-challenge.

Webber, M.E. (2016). *Thirst for Power: Energy, Water, and Human Survival*. Yale University Press.

References

Alley (2021). "Global Use of Hydroelectricity | EARTH 104: Earth and the Environment (Development)." 2021. https://www.e-education.psu.edu/earth104/node/1060.

Chamberlain, J.F., Tromble, E., Graves, M., and Sabatini, D. (2020). Water reuse versus water conveyance for supply augmentation: cost and carbon footprint. *AWWA Water Science* 2 (1): e1170. https://doi.org/10.1002/aws2.1170.

Columbia River Inter-Tribal Fish Commission (CRITFC) (2021). "Salmon Culture | Pacific Northwest Tribes, Columbia River Salmon." CRITFC 2021. https://www.critfc.org/salmon-culture/tribal-salmon-culture.

Delgado, A., Rodriguez, D.J., and Sohns, A.A. (2015). *Thirsty Energy: Understanding the Linkages between Energy and Water*. Washington, D.C: World Bank.

Dieter, C.A., Maupin, M.A., Caldwell, R.R. et al. (2018). "Estimated Use of Water in the United States in 2015." USGS Numbered Series 1441. Circular. Reston, VA: U.S. Geological Survey. http://pubs.er.usgs.gov/publication/cir1441.

EIA (2019). "Annual Energy Outlook 2019 with Projections to 2050." US EIA.

EIA (2020). "U.S. Energy Facts Explained - Consumption and Production - U.S. Energy Information Administration (EIA)." 2020. https://www.eia.gov/energyexplained/us-energy-facts.

Gerbens-Leenes, P.W., Hoekstra, A.Y., and van der Meer, T. (2009). The water footprint of energy from biomass: a quantitative assessment and consequences of an increasing share of bio-energy in energy supply. *Ecological Economics* 68 (4): 1052–1060.

IEA (2020). "World Gross Electricity Production, by Source, 2018 – Charts – Data & Statistics." IEA (International Energy Agency). 2020. https://www.iea.org/data-and-statistics/charts/world-gross-electricity-production-by-source-2018.

International Rivers (2020). "Asia." International Rivers. 2020. https://www.internationalrivers.org/where-we-work/asia.

King, C.W. and Webber, M.E. (2008). Water intensity of transportation. *Environmental Science & Technology* 42 (21): 7866–7872. https://doi.org/10.1021/es800367m.

Meldrum, J., Nettles-Anderson, S., Heath, G., and Macknick, J. (2013). Life cycle water use for electricity generation: a review and harmonization of literature estimates. *Environmental Research Letters* 8 (1): 015031. https://doi.org/10.1088/1748-9326/8/1/015031.

Moran, E., Lopez, M.C., Moore, N. et al. (2018). Sustainable hydropower in the 21st century. *Proceedings of the National Academy of Sciences* https://doi.org/10.1073/pnas.1809426115.

NAE (2008). "Grand Challenges - 14 Grand Challenges for Engineering." 2008. http://www.engineeringchallenges.org/challenges.aspx.

NAS (2020a). "How We Use Energy, Transportation — The National Academies." 2020. http://needtoknow.nas.edu/energy/energy-use/transportation.

NAS (2020b). "Our Energy Sources — The National Academies." 2020. http://needtoknow.nas.edu/energy/energy-sources.

Pan, S.-Y. (ORCID:0000000320824077), Snyder, S.W., Packman, A.I. et al. (2018). "Cooling water use in thermoelectric power generation and its associated challenges for addressing water-energy nexus." *Water-Energy Nexus* 1 (1). http://doi.org/10.1016/j.wen.2018.04.002.

Rodriquez, D.J., Delgado, A., DeLaquil, P., and Sohns, A. (2013). *Thirsty Energy*. Washington, DC: Water Papers – World Bank.

Smalley, R. (2003). "Top Ten Problems of Humanity for Next 50 Years." Presented at the Energy & NanoTechnology Conference, Rice University.

The Economist (2020). "The Grand Renaissance Dam: Showdown on the Nile." *The Economist*, July 2, 2020. https://www.economist.com/middle-east-and-africa/2020/07/02/the-bitter-dispute-over-africas-largest-dam.

UNDP (2020). "SDG Goal 7: Affordable and Clean Energy." UNDP. 2020. https://www.undp.org/content/undp/en/home/sustainable-development-goals/goal-7-affordable-and-clean-energy.html.

US DOE (2014). *The Water-Energy Nexus: Challenges and Opportunities*. U.S Department of Energy.

US EIA (2020a). "Electricity in the U.S. - U.S. Energy Information Administration (EIA)." 2020. https://www.eia.gov/energyexplained/electricity/electricity-in-the-us.php.

US EIA (2020b). "Hydropower and the Environment - U.S. Energy Information Administration (EIA)." 2020. https://www.eia.gov/energyexplained/hydropower/hydropower-and-the-environment.php.

US EIA (2020c). "Where Hydropower Is Generated - U.S. Energy Information Administration (EIA)." 2020. https://www.eia.gov/energyexplained/hydropower/where-hydropower-is-generated.php.

US EIA (2020d). "US Electricity by State." https://www.eia.gov/electricity/state/, accessed July 10, 2020.

Webber, M.E. (2016). *Thirst for Power: Energy, Water, and Human Survival*. Yale University Press.

WEF (2011). *Water Security: The Water-Food-Energy-Climate Nexus*, 2e. Washington, D.C: Island Press.

8a The Practice of Water Security: Women, Water, and Food Security

CARE USA was formed after the Second World War to send food parcels to Europe, parcels that affectionately became known as "care packages." The distribution of food (charity) eventually transformed into longer term programs that focused on food security (development). The organization, now simply CARE (Cooperative for Assistance and Relief Everywhere), has expanded to include refugee relief in places such as Syria and Jordan, sexual reproduction and maternal health, and water for agriculture. CARE also promotes local microfinancing, as in Access Africa, which supports a village-level savings and loan association. Peter Lochery served as CARE's Director of Water for two decades in which he expanded CARE's water programs to include research and advocacy (Figure 8a.1). He also focused on water for food, co-founded the Millennium Water Alliance (MWA), and addressed the issues that girls and women face due to poor water and sanitation.

Figure 8a.1 Peter Lochery gives the plenary address at the third biennial OU International Water Conference, Norman, Oklahoma, USA. Source: Courtesy of OU WaTER Center.

Regarding water, CARE and its allies are aware of the disproportionate burden of water collection and storage that falls upon women and young children. As Peter says, "As much as four hours a day in managing water uses up time which could be devoted to other work around the household, a business venture, education, and other cultural or leisurely activities. It also opens women up to sexual harassment and rape as they travel some distance by foot collecting water. In addition, in schools with inadequate sanitation, women and girls have a much harder time during puberty. There's nowhere for them to manage their menses. So, we have put a lot of effort in our work with schools into changing or improving facilities so that there are washrooms and school health clubs that can advise teenage girls, provide sanitary pads on emergency basis, etc." CARE's approach has been to work with the host government as a team to do the analysis. The team identifies the issues, agrees on what the changes in policy are that need to be put in place, and recommends adjustment on the government budget to meet these requirements. Peter says

"Our work, then, is not so much around service delivery. It's around analysis, research and advocacy done, of course, with our colleagues in the field, but also typically working with local and national government...We try to get them involved in addressing some of the problems and the barriers that are identified by girls at school, their parents, ministries of education, and local education authorities."

There is often a sort of tension between the delivery of service (such as clean water and sanitation facilities) and research and advocacy. Governments often have social compacts with their populations to provide services. A problem arises when NGOs (non-governmental organizations) go into a region and start providing those services directly. As Peter says: "We may be providing the wrong service and/or the service may have unintended consequences, while at the same time usurping the government's role or letting it off easy."

Food and water security come together in the concept of soil moisture (Figure 8a.2). The extent to which soil can capture and retain moisture is dependent on the condition of the soil. Thus, water security becomes soil-quality management, using landscape management approaches. CARE began using the term "water smart agriculture" as a way of trying to advocate for the appropriate balance between green water (the water in the soil) and blue water (the water from rainfall stored in lakes and rivers). The bulk of the food in the developing world is being produced by smallholders, many of whom are women and/or are not likely to have access to irrigation. And so, they are dependent upon the efficiency of rain-fed agriculture and the maintenance of soil moisture to increase capture and filtration rates and retention. Water security thus has a great impact on the lives of women regarding cleanliness, safety, livelihood, and food security.

Figure 8a.2 Soil moisture is the connecting link between food and water security. Source: sawitreelyaon/Adobe Stock.

Peter Lochery is the recipient of the 2015 OU International Water Prize

Source: "World Views" segment with Rebecca Cruise, KGOU Radio. Norman, Oklahoma. 2015.

9

Water for Industry

The objective of this chapter is to introduce the beneficial uses of water for the production of industrial goods and services. Each of these products has a "water footprint," which can be compared one to another, and the trade of these products represents a trade of "virtual water." These beneficial uses may put a strain on local water resources, especially in arid areas that are already experiencing water stress. Select industries – such as steel, sugar, and paper – are presented as examples illustrating how water is used and what kinds of contaminants are produced. Some companies are actively striving to reduce their local water demand, however, through various initiatives and production modifications. Several case studies will be explored that demonstrate both the contributions and cautions to water security that are represented by commercial industry. Industrial water use can be made more sustainable by an increase in water conservation, reuse, and recycling. All of these will lead to greater water security.

Learning Objectives

Upon completion of this chapter, the student will be able to:

1. Understand the role of private and commercial enterprises in water security.
2. Explain and quantify the water footprint of common industrial products.
3. Appreciate the benefits and limitations of "virtual water" as a political and economic construct.
4. Understand how water is used in typical industries, such as steel, sugar, and paper.
5. Appreciate better the water aspect of sustainability.
6. Have developed some familiarity with the major corporate social initiatives that are ongoing and exhibiting success.
7. Have increased knowledge of the applications of private–public cooperation for water security in various parts of the world.
8. *Understand the importance of dissolved oxygen in a receiving stream and be able to perform simple calculations.

Fundamentals of Water Security: Quantity, Quality, and Equity in a Changing Climate, First Edition.
Jim F. Chamberlain and David A. Sabatini.

Companion website: www.wiley.com/go/chamberlain/fundamentalsofwatersecurity

9.1 Introduction

Since the industrial revolution of the eighteenth to nineteenth century, citizens of the world have come to enjoy and expect many of life's amenities mass-produced by the world's industrial sector. Mechanized systems and chemical processes rely either directly or indirectly on the ready availability of water for cleaning, cooling, and creating!

Global industrial water demand accounts for about 19% of freshwater withdrawals (Black 2016). However, in the developed nations of Europe and North America, the percentage of demand is much higher, over 50% in the United States if electricity generation is included (Blackhurst et al. 2010) (Figure 9.1). This water is used to develop products that are often exported to international markets, representing a virtual water trade that scaffolds the world's economy. These percentages may not fully reflect the impact of industry, however, as a commercial enterprise often discharges water of degraded quality, even if partially treated, into a water commons, such as lakes and rivers. The impacts of industrial water consumption and degradation are most severe in low-income and water-stressed communities.

Water in industry has many uses, many of them non-substitutable. Water can be used as a resource (constituent) incorporated into the product, as in production of a beverage. It is needed for heating, cooling, and cleaning process equipment; for packaging and shipping; for transporting dissolved constituents away from the production; and as a solvent (UNESCO 2017). Most often the **indirect water usage**, that is, water consumed in the supply chain, far exceeds the **direct water usage** for anything other than agriculture and power generation (Blackhurst et al. 2010). Industries that have a high water usage and need for treatment include brewery and carbonated beverages, the dairy industry, sugar mills and refineries, textile manufacturing, pulp and paper mills, oil and gas, and the automotive and aircraft industries (United Nations 2009).

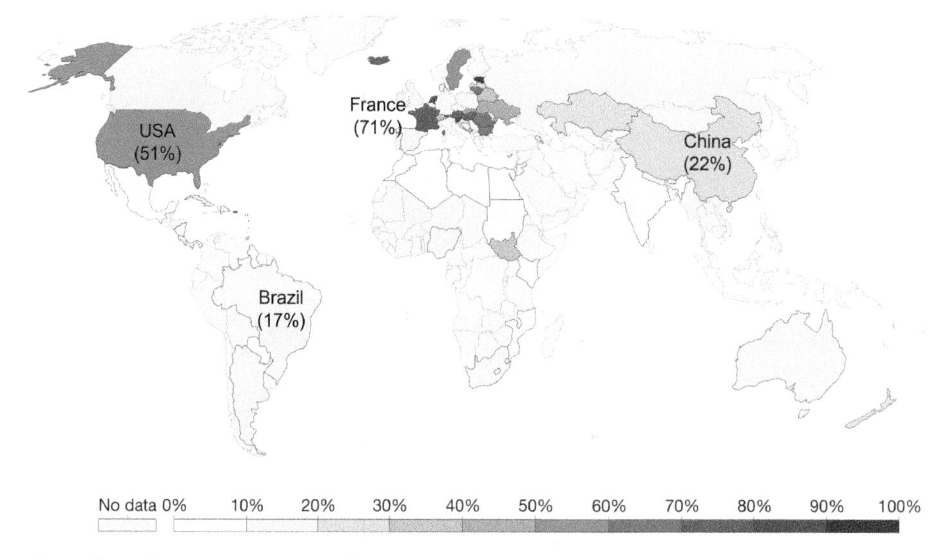

Figure 9.1 Annual percentage of freshwater withdrawn for industry varies between countries and regions, including electrical generation (year 2015). Source: Adapted from Ritchie and Roser (2017).

Unsurprisingly, not all industrial products are the same regarding the total amount of water consumed. Over 1000 l of water are consumed in the making of 1 kg of paper, whereas the production of 1 kg of sugar and steel require only 200 and 175 l, respectively. These quantities of water represent the "virtual water" (discussed below) of the product. In addition, the economic productivity of water (e.g. GDP per water consumed) varies from one industrialized country to another, with Germany reaping $40.2 USD per cubic meter of water consumed, while in the United States the amount is $18 USD (Postel and Vickers 2004).

Even within a given industry or sector, the water use varies from process to process. The production of 1 kg of cheese requires 5000–5500 kg of water, but to produce 1 kg of beef requires three times as much water, 16 000 kg of water on average (Hoekstra 2003). Biofuels (using corn or soybean) and petroleum are both produced to power-motorized transportation, but the water footprint (WF) varies considerably. Here, water intensities range from 28 to 36 gal H_2O/mile for corn ethanol (E85) and 8 to 10 gal H_2O/mile for soy-derived biodiesel, while the water intensity of petroleum is about 1.0–1.5 gal H_2O/mile, if a full life cycle analysis is considered (King and Webber 2008; National Research Council 2008).

Thus far, we have only addressed the *quantity* of water used in industry. However, the *quality* of water that is released after industrial use may be equally if not more important as about 70% of industrial-use water in developing countries is released directly and untreated into its rivers (Black 2016). This industrially impacted water resource can be damaged from further beneficial use. To avoid this occurrence, the US EPA issues effluent guidelines for industrial wastewater, basing these regulatory standards on existing treatment technologies. These guidelines are incorporated into National Pollutant Discharge Elimination System (NPDES) permits, which are the permits obtained before an industry can discharge to a public waterway or a municipal wastewater-treatment plant (US EPA 2014).

The private sector has a primary obligation to its owners and shareholders. Businesses do realize, however, that they rely on shared resources that are part of the public commons, and the preservation of these resources will be to the benefit of the business as well as the public. Thus, there is a growing awareness of water risk to industry, and leaders have begun shifting their attention to more sustainable production and management strategies within local water regimes. This chapter will examine each of these themes in greater detail.

9.2 Global Trends and the Need for Water Stewardship

Water has become increasingly scarce due to urbanization, population growth, and climate change. In addition, as global populations begin to enjoy more comfortable lifestyles, they are seeking nonagricultural products that are traded widely on the global market, such as computers, blue jeans, and automobiles. Businesses and factories will choose a location based on many factors, including availability of labor, government incentives, cheap and reliable energy, and transportation options. Adequate access to water and raw resources is important but may be only one consideration in a matrix of relevant requirements. The allure of a big city with a suite of potential investors and large pool of labor (made cheaper by the law of supply and demand) may over-ride long-term vision regarding water sustainability. A dense population brings competing water demands and climate change adds another dimension of uncertainty into long-range planning.

Corporate water stewardship (CWS) is the collection of internal and external actions taken voluntarily by the private sector that ensure that water will be used in ways that are:

- Socially equitable,
- Environmentally sustainable, and
- Economically beneficial (Hepworth and Orr 2013).

This notion of stewardship differs from Integrated Water Resources Management (IWRM) in that stewardship implies taking care of something you do not own. An industry both taps into and releases into the public water bodies in its proximity. The site may be using water from a service provider who is getting water from a reservoir. The water quality of this reservoir depends upon upstream point and nonpoint sources and upon adequate rainfall. Upstream erosion control is dependent upon land use/land cover in the upstream catchment. The site may release water into a receiving water body, perhaps after being treated by a wetlands system, becoming source water for downstream users. This water may have a reduced water quality relative to the receiving water body. But this same receiving source may have other inputs from an adjacent sub-watershed (AWS 2014). Thus, water is a highly complex public good, and its desirability as an economic resource provides incentive for both expropriation and good stewardship.

Later in the chapter, we discuss some examples of corporations who take these potential impacts seriously in their planning and mission statements.

9.3 Virtual Water and the "Water Footprint" Revisited

We have seen that the majority of freshwater consumption is not for household drinking or cooking but for the growing of food in agriculture and for producing industrial products. This is water that must be available locally, as the transport of water across great distances is not economically feasible or desirable. So, a trade of *actual* water between water-rich and water-poor nations is not commonly practiced.

Food, however, can be transported more easily, as can goods and products that consume varying amounts of water in production. This "embedded" water in food and products is given the label of "virtual water" and is the sum total of the freshwater consumed in the growing (in food products) and production of the product on a per mass basis. Life cycle analysis is used in tracking water usage in various stages of production. In this case, actual water is not exchanged, but embedded (or virtual) water is traded and tracked, giving water-poor countries a more sustainable outlook. Table 9.1 lists the virtual water content of six selected products.

The net import of embedded water can relieve some of the strain on water-stressed countries (Hoekstra 2003). Instead of using scarce resources to create a product locally, a country can import that product and use locally available water for more urgent needs. In this manner, *water scarcity* (*WS*) can be somewhat mitigated. On a global scale, it is more reasonable to produce water-intensive products in places where water is more naturally abundant and likely less expensive. Of course, the virtual water trade is not a function only of *water availability* (*WA*) but also of consumer demand. This explains how two neighboring nations with similar climates and water abundance may be different in levels of importing/exporting virtual water. Even a single nation may exhibit a temporal variability; for example, China

Table 9.1 Water use per tonne of product produced for selected industries (usage shown is the upper end of range based on process of production).

Product	Water use (cubic meters/tonne)
Paper	20 000
Sugar	4000
Steel	350
Petrol	40
Soap	35
Beer	25

Source: United Nations (2009).

was a major importer of virtual water in 2000 but by 2002 was a net exporter (Warner and Johnson 2007).

As discussed in Chapter 7, virtual water leads to the quantification known as the "**water footprint**," or the "cumulative virtual water content of all goods and services consumed by one individual or by the individuals of one country" (Warner and Johnson 2007). This tool can be used to assess the environmental impact of one's individual (or one's country) product choices. It is a metric that represents the overall water impact of a country and its citizens. An individual's WF is calculated by adding up the direct water usage along with the virtual water usage on a temporal basis. For example, a pair of jeans consumes water in the growing of cotton for the product, its manufacturing, and its periodic washings over time.

In similar fashion, a nation's WF can be quantified and compared with other nations. Several calculations and terms are employed in comparing water security in nations and regions (Table 9.2). *Water withdrawal* (*WU*) is the domestic usage within the country, although this is typically estimated as blue freshwater usage only. (Green water usage, transpired by plants out from the soil, is harder to estimate.) *Net virtual water import* (*NVWI*) is the amount of

Table 9.2 Virtual water and water footprint metrics for selected countries, 1995–1999.

Country	Population (millions)	Water domestic withdrawal (WU)	Water availability (WA)	Net virtual water import (NVWI)	Water footprint (WF)	Water scarcity (WS)	Water self-sufficiency (WSS)	Water dependency (WD)
		Mm^3/ year	Mm^3/ year	Mm^3/ year	m^3/ year/ capita	%	%	%
Afghanistan	27.8	35 704	50 000	−229	1377	71.4	100	0
Brazil	169	46 856	6 950 000	−9000	225	0.7	100	0
Costa Rica	3.7	1464	95 000	1257	729	1.5	53.8	46.2
Congo	49.6	51	832 000	86.6	3	Negligible	37.1	62.9
Israel	6.1	2277	2200	4598	1127	103.5	33.1	66.9

Source: Data from Hoekstra (2003).

water each year that is traded virtually through goods and products (virtual water import minus virtual water export). This amount will be negative if the nation is a net exporter of water. The per capita *WF* of a nation can then be calculated by:

$$\text{Water footprint } (WF) = (WU + NVWI)/\text{Population} \quad (9.1)$$

WA (*water availability*) is the amount of precipitation falling within a country's borders. *WS* (%) can then be defined as the ratio of water usage to WA:

$$WS\,(\%) = WU/WA \times 100 \quad (9.2)$$

Water dependency (WD) is the ratio of NVWI into the country to the total national water usage, including imports. This value represents the percentage of total water need that is being met by virtual imports:

$$WD\,(\%) = \frac{NVWI}{WU + NVWI} \times 100 \quad (9.3)$$

And finally, *water self-sufficiency (WSS)* is the national capacity to supply its citizens with water needed to meet the total domestic demand for water-dependent goods and services:

$$WSS\,(\%) = 100 - WD \quad (9.4)$$

Table 9.2 illustrates the diversity that exists among national dependencies. Brazil and Afghanistan are both net exporters of water, but Afghans have a much greater per capita footprint than Brazilians. Costa Rica is a water-rich country with rainfall of about 100 in./year with its mountainous regions receiving four times that amount on an annual basis. Yet, it still imports water due to its higher standard of living and reliance on other countries for high-value products. Finally, Israel and the Congo are both very water-dependent, in spite of very different climates and standards of living. Congo does not yet have the infrastructure to capture, transmit, and incorporate its water into durable goods for export. Israel has done much to reduce its domestic demand, but because of its extremely arid climate it still relies greatly upon imported water. For this reason, it is as water dependent as the Congo.

Figure 9.2 shows the virtual water trade between global regions. This map gives a visual look at the WD that exists based on virtual water trade.

Although resource scarcity and political instability seem to be a recipe for "water wars," there has been no outright military conflict over water in the past 4500 years, according to studies conducted by Oregon State University (Wolf et al. 2003). The concepts of "virtual water" and WF might help us understand the dynamics of political cooperation and water

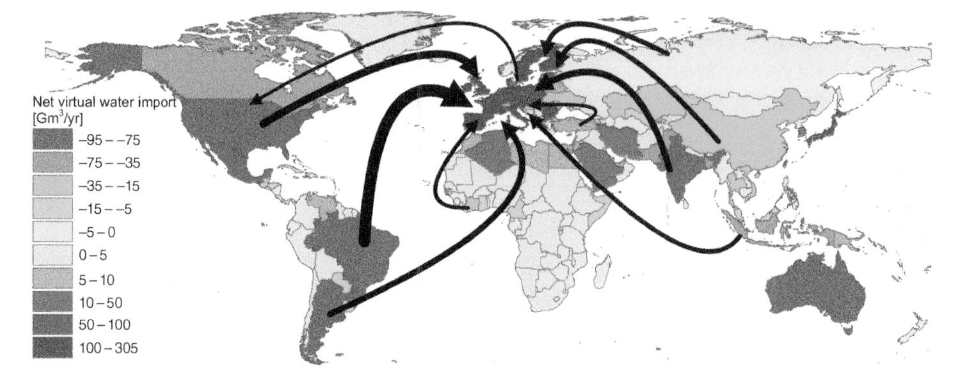

Figure 9.2 Arrows showing virtual water import into Europe. Nations in red are net importers; nations in green are net water exporters. Source: Water Footprint Network (2021) / Used with permission from Water Footprint Network.

appeasement. Trading in water-dependent goods is advantageous to those water-stressed regions of the world and is perhaps a basis for a more optimistic outlook.

9.4 Chemistry of Industrial Wastewater

Because of the variety of industries, from poultry to pharmaceuticals, industrial wastewater can carry a wide and diverse range of pollutants. The US EPA divides these toxic pollutants into three main categories. The **priority pollutants** are a comprehensive list of about a dozen heavy metals and over 120 organic pollutants, mostly pesticides and solvents, which can cause adverse health and/or ecological effects when released into the environment.

The Clean Water Act (with amendment) designated a listing of five **conventional pollutants**, i.e. pollutants that are defined as such by the Clean Water Act and are generally amenable to municipal water treatment (US EPA 2010). All other pollutants that are not priority or conventional fall into a third category of nonconventional. The conventional and nonconventional pollutants, along with typical treatment schemas, are listed in Table 9.3.

Table 9.3 Listing of pollutants in EPA's three categories, along with typical treatment schemas to reduce or eliminate the pollutants.

	Typical treatment schemas
Priority Pollutants	
Metals, e.g. arsenic, chromium, copper, lead.	Precipitation, ion exchange, adsorption
Organic compounds, e.g. pesticides (aldrin, dieldrin), ethylbenzene, methylene chloride.	granular activated carbon (GAC), biological removal
Conventional Pollutants	
Biochemical oxygen demand (BOD_5)	Oxidation ponds, activated sludge
Total suspended solids (TSS)	Sedimentation, filtration
Oil and grease	Preliminary treatment
pH	Addition of acidic/basic chemicals
Fecal coliform	Sedimentation, filtration, disinfection
Nonconventional Pollutants	
Ammonia (as N)	Nitrification, denitrification
Chromium VI (hexavalent)	Ion exchange
Chemical oxygen demand (COD)	Sedimentation, biological removal
COD/BOD_7	Sedimentation, biological removal
Fluoride	Ion exchange w/activated alumina
Manganese	Filtration
Nitrate (as N)	Denitrification, ion exchange
Organic nitrogen (as N)	Nitrification, denitrification
Pesticide active ingredients (PAI)	GAC
Phenols, total	Advanced oxidation processes
Phosphorus, total (as P)	BPR (biological phosphorus removal)
Total organic carbon (TOC)	Same as removal of BOD

Source: US EPA (2014) and Davis (2010).

Table 9.4 Effluent limitations derived from secondary treatment standards.

Discharge characteristic	Average monthly concentration	Average weekly concentration
BOD_5 (mg/l)	30 (or 25 mg/l $CBOD_5$)	45 (or 40 mg/l $CBOD_5$)
Total suspended solids (mg/l)	30	45
pH (units)	6.0–9.0	6.0–9.0

Source: Davis (2010) and US EPA (2010).

The US Congress requires secondary treatment before discharging wastewater into natural water bodies. The US EPA defines secondary treatment as that which produces the characteristics of the three pollutants given in Table 9.4.

An industrial wastewater may contain some combination of the priority, conventional, or nonconventional pollutants, affecting the water quality of downstream users. The permissible limits (NPDES effluent standards) are listed in the Code of Federal Regulations (CFR), which may be updated periodically. These are usually based on the best practicable control technology currently available (BPT) at the time of promulgation (US EPA 2010).

9.5 Examples from Five Industrial Sectors – Iron and Steel, Pulp and Paper, Sugar, Textiles, and Oil and Gas Production

Found in the United States and other developed nations, some of the most water-intensive industries are pulp and paper, food processing, production of chemicals, petroleum refining, and metals manufacturing. For this reason, three US states in which these industries are prevalent – Indiana, Texas, and Louisiana – account for 40% of the entire US industrial water use (Darling and Snyder 2018). These industries use water for source extraction, fabrication, cooling, cleaning, and transportation as well as incorporation into the product itself. Here we take a closer look at some of the more significant industries whose effluent has a potentially large impact on both water quantity and quality.

Iron and steel production begins with the smelting of pig iron in blast furnaces from which is released a considerable amount of oil, dust, acid, iron, and other metals. Water is sprayed to remove dust containing cyanides, sulfur compounds, phenol, dust, metal ions, ashes, slags, and ore particles. The wet cleaning of blast-furnace gas may release both total suspended and dissolved solids (TSS, TDS). Rolling mills produce wastewater during the cooling of bearings and shafts.

The water consumption is about 10 m^3 (2650 gal) per ton, but by recirculation it is possible to reduce this amount of water by a factor of two to seven. Wastewater from the iron and steel industry can contain up to 5% of grease and oil (by volume) when recirculation of the wastewater takes place. Sodium chloride is used for breaking up the emulsion, while additional chemicals are added as a flocculation agent (aluminum sulfate) or for saponification of the grease and oil (calcium hydroxide) (Kemmer 1988).

The *pulp and paper industry* produces not only paper but also heavier products such as cardboard, corrugated liner board, and shipping cartons and containers. Some mills also produce plywood and particle board. Pulp (raw fiber–containing sheets) is produced by using either mechanical or chemical methods to separate out the cellulosic fibers from the raw wood. Chemical pulp is used for materials that need to be stronger or combined with mechanical pulps to give different characteristics to a product.

A typical mill in the southeastern United States produces about 1000 tons of product per day. Water usage will range from 38 to 246 m^3/ton for pulp manufacturing and 19 to 57 m^3/ton for paper manufacturing (Kemmer 1988). Since a good portion of the raw material is organic, the *biochemical oxygen demand* (BOD) of wastewaters can be quite high. For a final paper product, bleaching with chlorine compounds is the traditional process.

Sugar is an essential substrate for the human diet and is in high demand. Sugar industries generate about 1000 l of wastewater per ton of sugarcane crushed. Wastewater from sugar industry, if discharged without treatment, poses pollution problems in both aquatic and terrestrial ecosystems. This effluent contains suspended solids, organic matter (BOD), and chemicals used for coagulation of impurities and refining of end products. Basic (calcium hydroxide) and acidic substances (phosphoric acid) are used to aid in clarification at different stages of production (Poddar and Sahu 2015).

The *textile industry* is truly a global industry. China is the largest exporter of textiles, with $109 billion of exports in 2017, slightly less than the next four exporters combined – the European Union (EU), India, USA, and Turkey (Lu 2018). All finished textile products, such as clothing, carpeting, or tire cord, begin with the fibers of wool or cotton, synthetic fibers, or some combination of these. Textile mills are engaged principally in receiving and preparing fibers, transforming the materials into yarn, thread or webbing, and then converting and finishing these latter into fabric or related products. Water processes are used to clean the fibers of dirt, grit, or grease; remove process impurities (such as metals); and finish the products to a desired appearance, feel, and durability (Kemmer 1988).

Wastewater is generated when textile mills use any of the following processes:

- Scouring and cleaning of raw wool
- Carbonizing, fulling, dyeing, bleaching, rinsing, fireproofing, and other similar processes
- Fabric and carpet finishing and coatings, including desizing, bleaching, mercerizing, dyeing, printing, resin treatment, waterproofing, flameproofing, and application of soil repellency and other special finishes

Today's *oil and gas production* can proceed using either conventional (vertical drilling) or a more recent unconventional method. This latter method is able to liberate natural gas and oil reserves from very tight (low-permeability) formations. This is done by adding a lateral leg of horizontal drilling in addition to hydraulic fracturing ("fracking"), a high-pressure opening of the tight shale formations. Both methods use large quantities of water and can be a threat to local water quality. As a vertical well is drilled down to depths of 1–3 km, care must be taken not to spoil a shallow aquifer that may be used for drinking water. Freshwater is used for this drilling through the shallow depths and is followed by deeper drilling using a "diesel mud" mixture of diesel fuel, bentonite, and water, especially when an expansive layer is encountered.

The case of unconventional drilling results in additional environmental impact. Once the driller is in the desired formation, he cases off the vertical well, turns in a curved direction toward horizontal, and begins using a saltwater brine for drilling the lateral, sometimes as long as 1–3 miles in length (Figure 9.3). Then the driller begins fracturing the rock with a high-pressure solution in order to break open the shale formation and release the natural gas. Fracturing fluids are mostly water and sand (99%), the latter of which is used to hold open

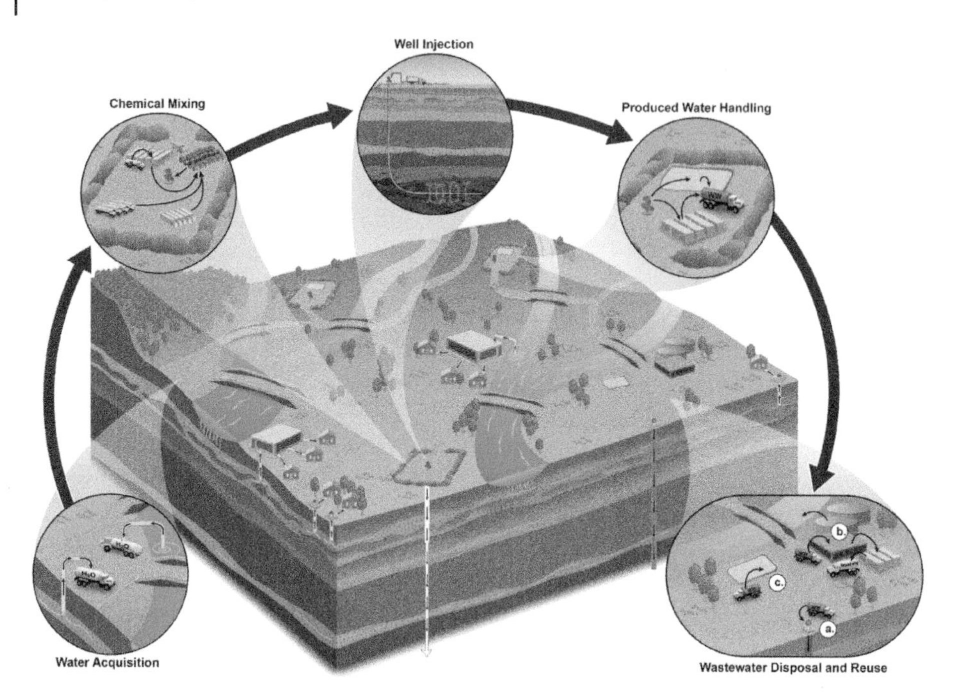

Figure 9.3 The water cycle that is used in hydraulic fracturing method of oil and gas production. Source: US EPA (2013).

the fractures that have formed. The remaining 1% of the fluid has acid and other chemicals that will keep the sand in suspension and prevent corrosion and unwanted microbial growth (Zoback and Arent 2014).

The potential water-related impacts of these processes include:

- Surface contamination from spills and leaks. Spilled liquids could be combination of crude oil and brackish water. Leaks (mostly from pipelines) could include dissolved gases and heavy metals.
- Subsurface contamination of freshwater aquifers. Nearby residents are justly concerned about natural gas leakage into drinking water aquifers.
- Excessive use of water resources in arid regions. This could result in excessive aquifer draw-down and depletion.
- Contamination from the disposal of "produced water" (injection water combined with oil, gas, salts, i.e. "brine") and wastewater. Residue from evaporation is high in chlorides and heavy metals.

Some studies show that poor well construction rather than the actual fracking process is the cause for methane leakage into adjacent aquifers (Zoback and Arent 2014). One study in the Barnett shale (Texas) showed regulatory exceedance levels of arsenic, selenium, strontium and TDS in private water wells located within 3 km of active natural gas wells (Fontenot et al. 2013).

Produced water that is typically injected into deep aquifers can be reused in the fracturing process by (i) reduction of TDS and some critical ions, such as Na, Cl, and Fe, and (ii) blending with fresh or brackish water (Silva et al. 2014). In 2012, about 90% of produced water from the Marcellus Shale play (Pennsylvania) was reused. However, supply often exceeds the demand

Table 9.5 Selected wastewater constituents and representative data for five industrial sectors.

	Total suspended solids (TSS)	Heavy metals	BOD/COD	Odor	Organics (general)
Iron and steel	M	M	V	I	V
Pulp and paper	M	V	M	C	M
Sugar	M	I	M	C	M
Textile	C	I	M	C	M
Oil and gas	M	M	M	I	M

M, major factor; C, contributing factor; V, variable; I, insignificant.
Source: Adapted from Kemmer (1988) and Vengosh et al. (2014).

for this blendstock, and the practice is not as common in other plays, such as the Barnett in north Texas.

Table 9.5 presents selected wastewater constituents for the five industries we have just discussed. BOD/chemical oxygen demand (COD), suspended solids, and organics are significant constituents for four of the industrial waste streams. If these constituents are not sufficiently removed, the wastewater effluent can severely degrade the quality of the receiving water body.

9.6 Life Cycle Assessment for Water Impacts – The Case of Corn-to-Ethanol

The quest for sustainability will often include a substitution of one product or process for another in order to reduce greenhouse gas emissions. In the past few decades, an alternative to fossil-based transportation fuels has evolved using traditional farm-grown crops. For both concern over global warming and a desire for greater fuel independency, the United States and other nations have begun promoting bioethanol, made from corn and sugarcane, and biodiesel, made mostly from soybean. A life cycle assessment (LCA) is an investigative tool that identifies and quantifies the environmental impacts of a product or process. All stages are evaluated, from manufacturing, usage, and consumption to disposal (if relevant). The results of LCAs can be used to compare one or more products on the basis of resource consumption and generation of pollution.

In the case of corn-to-ethanol, both water usage and water quality are impacted in the crop production and biorefinery stages. Crop production uses both surface and groundwater irrigation. Some of that water is lost through evapotranspiration (ET) and/or infiltration into the soil. Runoff from fertilizers and pesticides makes its way into streams, affecting human and ecological uses downstream.

An acre of corn loses 8000 l of water or more a day to ET, leaching and runoff (National Research Council 2007; Mowitz 2012). At a conversion rate of 10.2 l of ethanol per bushel (EIA 2015), this amounts to ~780 l of water per liter of ethanol. In addition, land use change to energy crops may release additional nitrates and pesticides, depending on the conversion. Corn requires more chemical fertilizer inputs to maintain expected yields as compared with other traditional crops. This results in more runoff laden with excess nitrates and pesticides, both of which are threats to human and ecosystem health. Nitrate can be converted into nitrite in the gastrointestinal (GI) tract, causing methemoglobinemia ("blue-baby syndrome"), which prevents the delivery of sufficient oxygen in the bloodstream. In addition, excessive nutrient loading may create dead zones in receiving waters, such as in the Gulf of Mexico near the mouth of the Mississippi River. These zones can destroy fisheries, the food and livelihood of many households. Pesticides in drinking water can cause long-term diseases such as cancer, organ damage, chemical hypersensitivity, asthma, and birth defects. Thus, the environmental impacts of increased corn acreage are not trivial and should be considered in evaluation.

The ethanol biorefinery uses water to convert corn into ethanol via the following pathway:

- The ground corn (or other grain) is mashed to the consistency of coarse flour and then cooked at high temperatures with a mixture of water and enzymes needed to break the starch polymers into glucose (sugar) molecules.
- This sugar is then fermented into a mash (called "beer") and carbon dioxide.
- The fermented mash is distilled into ethanol and solid stillage. Water is consumed in steam through the cooling towers as well as evaporation from drying the stillage.

As a result of these processes, the bioethanol refinery consumes about 4 gal of water per gallon of ethanol produced. This contrasts with 1.5 gal of water needed to produce 1 gal of petroleum (National Research Council 2007). When this is added to the water needed to grow the feedstock, the water consumption is over 500 times more water than to produce petroleum fuel for transportation on a per gallon basis. These numbers also reveal that ethanol refinement process improvement will do little to conserve water in the overall production schema since the bulk of the water is consumed in the growing of the plant. Figure 9.4 shows how the locations of ethanol-production facilities, while close to the supply of feedstock, may also be an endangerment to local and important aquifers.

This brief case study highlights the need to consider water security simultaneously with other metrics of sustainability. A solution that is favorable regarding climate-change considerations may not be amenable to water security.

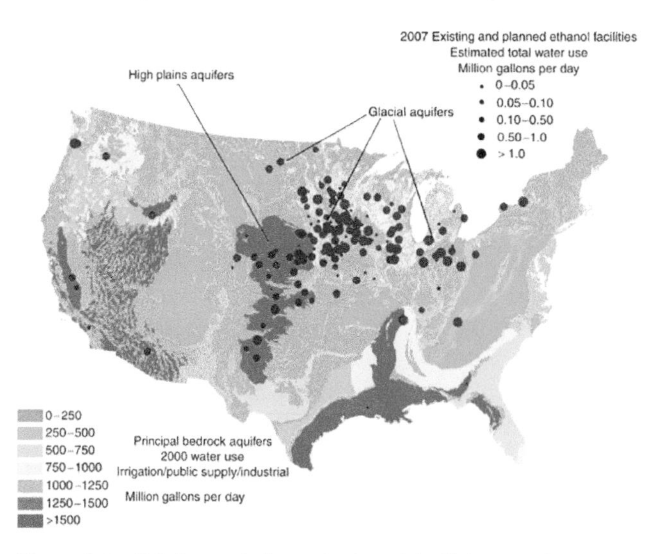

Figure 9.4 Existing and planned ethanol facilities are often located over aquifer that provide critical water supply for irrigation, industry, and public water supply. Source: Janice Ward, U.S. Geological Survey, personal commun., July 12, 2007.

9.7 Corporate Social Initiatives

Corporations across the globe are beginning to see the necessity in securing their water sources and becoming more intentional about sustainability across their water spectrum. We highlight here some of the major initiatives in this regard:

- The UN Global Compact formed the CEO Water Mandate in the year 2008, following a meeting of the World Economic Forum. This Mandate is a group of corporate leaders that has formally signed an agreement that its businesses directly and indirectly – through its supply chains – have a significant impact on water resources. To date (2020), there are over 170 endorsing companies who have signed on with over 980 projects around the world (Water Initiative 2011).
- The Water Resources Group (WRG) claims over 800 partners (both private and public), highlighting projects that save 367 million cubic meters of freshwater extraction avoided each year. Their work includes building the business case for greater water efficiency (Mexico, Peru) and working to ensure that the price of water reflects the true cost of providing the water (Bangladesh, Mongolia) (WRG 2019). The WRG does this through first establishing multi-stakeholder partnerships (MSPs) that can together design and implement the sustainable solutions to shared water challenges.
- JPMorgan Chase & Co. is an investment and financial services firm that has also taken an interest in water security via the World Resources Institute (WRI). Its report "Watching Water: A Guide to Evaluating Corporate Risks in a Thirsty World" (2008) identifies four major themes:
  - Exposure to water scarcity and pollution may actually be greater in companies' supply chains than in their own operations.
  - The power-generation, mining, semiconductor manufacturing, and food and beverage sectors are particularly exposed to water-related risks.
  - Corporate disclosure of water-related risks is seriously inadequate and should appear in the regulatory filings on which most investors rely.
  - Investors should assess the reliance of their portfolios on water resources and their vulnerability to problems of WA and pollution (Levinson 2008).
- Finally, the CDP Water Disclosure project seeks to inform investors about water-related risks and opportunities relevant to the business sector. They encourage full water transparency on the part of businesses as well as the setting of water targets. These targets are contextual in the sense of acknowledging that appropriate goals will vary according to the level of WS or abundance in a local basin. For example, Levi Strauss & Co. has set a target of reduced water use in manufacturing by 50% in areas identified as "high water stress" by 2025. The goal of General Motors Corporation is to reduce its water intensity in 2020 by 15% from a 2010 baseline (CDP 2020a).

The following are some additional industrial players that have made public their concerted efforts to reduce the consumption of freshwater and/or to provide mitigation efforts to offset necessary consumption (Water Initiative 2011):

- **Cisco Systems, Inc.**, is a multinational tech company that develops, manufactures, and sells networking hardware, software, telecommunications equipment, and other high-technology services and products. The company has revamped its San Jose (CA) campus to include the use of recycled water, native and drought-resistant plants, and irrigation controls. For its data centers in water-scarce areas, it incorporates resource constraints into its development plans (Water Initiative 2011).

- **Coca-Cola Company** has partnered with the World Wildlife Fund (WWF) to produce a set of freshwater-conservation goals while conserving seven of the world's most precious watersheds that include the Yangtze River basin in China and the Mesoamerican Reef catchments in Mexico and Central America. In addition, the company plans to reduce the amount of water used in manufacturing its products from 2.7 l/1 l product (in 2004) to 1.7 l/1 l by 2020. Some bottling plants currently use only 1.4 l of freshwater to make a product (Coca-Cola 2018). An analysis across the supply chain revealed the largest portion of the WF is in the field production of sugar beet (Coca-Cola 2018).
- **Patagonia** co-founded The Conservation Alliance back in 1989 and gives $100 000 annually to this fund. By 2019, the Alliance had delivered nine important conservation victories, protecting 420 755 acres and 82 river miles, removing one dam, and acquiring one popular climbing area (North America). Patagonia also recently bought water-restoration certificates (WRCs) to restore eight million gallons of water for the Middle Deschutes River (Oregon) to offset its own domestic water consumption.
- **Samsung Electronics** has instituted the 3R (Reduce, Reuse, Recycle) Strategy for water resources. They were able to reuse 56 154 tons of water in 2017, which is 16% more than the year before. Their goal is to reduce the water intensity-based water consumption from 60 to 50 tons/KRW* 100 million by 2020 (*South Korean currency). They have saved water by replacing old valves to stop leaks and by developing sophisticated control units so that only the necessary amount is used. In addition, they have standardized water profile in their production train so that the same water can be reused in multiple processes (Samsung 2020).
- **British Petroleum (BP)** estimated that around half of their major operations withdraw fresh water in areas considered water-stressed or water-scarce (BP 2018). These operations account for 23% of their total freshwater withdrawals. In Oman, an extremely water-scarce country, the company desalinates brackish water to be used for drilling and hydraulic fracturing. In selected refineries and petrochemical plants, they have studied freshwater withdrawal, consumption, and wastewater volume and saw a slight increase in efficiency for the fiscal year 2018. The results of the study in the selected refineries are being implemented into a long-term water-management program to reduce water demand. BP saw a small decrease in freshwater withdrawal, consumption, and wastewater volumes in 2018 primarily due to operational changes such as maintenance at selected refineries and petrochemical plants. Overall, there was a slight increase in consumption efficiency (BP 2018).
- **AstraZeneca** is a pharmaceutical company based in the United Kingdom and was one of only three companies to receive a "double A" rating by the CDP for both climate change and water security (CDP 2020b) (Figure 9.5). Their target goal is to maintain absolute water use at the 2015 level through the year 2025. They currently show an 18% WF reduction between 2015 and 2019. The company has prioritized water efficiencies in water-stressed areas, including sites in Puerto Rico, China, and the United Kingdom. In these and other locations, the company conducts water audits and establishes rainwater harvesting. The company's on-site and off-site wastewater-treatment methods are designed to remove most of the residual COD, which is the amount of oxygen required to chemically oxidize organic matter in wastewater. The 2019 COD discharge was 405 tons, a decrease of 26% from 2018. This reduction came mostly from a divestment of one of its biological manufacturing sites. The AstraZeneca company acknowledges that it has 11 manufacturing operations in areas

Figure 9.5 Clean water is needed for both the consumption and manufacturing of pharmaceutical medications. Source: motortion/Adobe Stock.

of high-water stress. In one of these sites, Chennai, India, a dry season often preceded a period of deadly flooding in the monsoon season, wreaking havoc on the city's water planning and supply. Here, the company teamed with Environmentalist Foundation of India (EFI) to restore two sites for water impoundment and storage, one of them a former dumping site in a low-lying area.

Without transparency, measuring and monitoring, and inclusion of suppliers in water accounting, water can easily remain an "invisible actor" in the business model of most industries. But CWS companies are beginning to show that the upfront costs of risk assessment and water usage reporting are paying good environmental and potential economic dividends.

9.8 Conclusion – The Future of Sustainable Industrial Water Use

Integrated water management methodology can be applied not only to a watershed but also to an industrial process or service, including its chain of suppliers. The lowest and easiest step on the triangle of resource management is conservation, the reduction of water usage through the improved operation of existing processes and equipment. The next step is to reuse water as reclaimed water that was formerly wasted. This is water that can be reused in its current state of water quality without additional treatment. The final step is to consider recycling of water that can be reclaimed by improving its water quality through treatment (Figure 9.6).

When examining the entire process, from feedstock to finished product, an industry can creatively address the need to reduce overall water consumption:

- At one newsprint site in North America, a switch to a higher yield fiber resulted in a reduction of water usage by 33%.
- A UK petroleum refinery saved 1000 m^3 of water per hour by identifying leaks and returning condensate to its boiler (a form of water re-use).

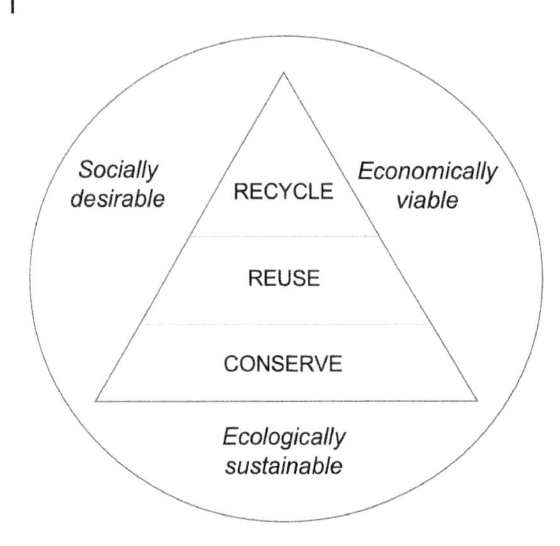

Figure 9.6 A pyramid of water resource management represents a set of priorities inside a circle of sustainability. Source: Adapted from NAS (2004).

- Chromate was once used as a biocide in cooling towers but was later found to produce harmful health effects, as highlighted in the work done by environmental activist, Erin Brockovich. The usage of a fluorometer (signaling a change in fluorescence) has helped identify harmful biological growth in cooling towers, and green biocides can now be used on an as-needed basis as identified by this more-precise monitoring.
- The use of ultrafine membranes and membrane bioreactors (combination membrane and activated sludge) have made more feasible the reduction of BOD and the subsequent recycling of industrial wastewater (NAS 2004).

These examples show the solutions that can emerge when water sustainability has a place in the management strategy.

The industrial uses of water are both important and highly varied. Water security for each sector has to be examined uniquely – by individual process and location – regarding both water quantity and quality. Analysis of WF and trading of virtual water may allow useful comparison among countries and sectors. As nations develop and become more urbanized, this demand for water will continue to grow and will remain a significant factor in the future of water sustainability.

*Foundations: The Dynamics of Oxygen in Receiving Water Bodies

Conservation of mass in mixing waters. The principle of conservation of mass says that mass can neither be created nor destroyed. A common way for a constituent's mass in water to be presented is in its concentration, C, which is mass divided by volume. For example, the concentration of ammonium might be written as $C_{NH4} = 22$ mg/l, that is, 22 mg of NH_4^+ per liter of water. The flux of mass into or out of a water body can be found by the product of the concentration, C (mass/volume), and its flowrate, Q (volume/time). Thus, mass flowrate, $\dot{m}_{in}$, is in units of mass per time.

Considering a stationary water body, such as a lake, we can write an equation for the change in mass of a water constituent over time using the following equation:

$$\frac{\Delta m}{\Delta t} = \dot{m}_{\text{in}} - \dot{m}_{\text{out}} \pm \dot{m}_{\text{rxn}} \quad (9.5)$$

where

$\dot{m}_{\text{in}}$ and $\dot{m}_{\text{out}}$ are mass fluxes into and out of the lake, respectively
$\dot{m}_{\text{rxn}}$ is mass that is gained or lost due to a chemical reaction

If the overall change and chemical reactions over time are negligible, then the equation can be reduced to:

$$\dot{m}_{\text{in}} = \dot{m}_{\text{out}} \quad (9.6)$$

If the receiving water body is a river or stream, then the flowrate of the river is considered as well as the flowrate of the constituent entering the river. For example, if an industrial waste pipeline is discharging chemical A into a river, and the river may already contain some background level of A, then the equation above can be used to determine the resulting concentration of A downstream below the pipe's outfall:

$$Q_{\text{pipe}} C_{\text{A,pipe}} + Q_{\text{river}} C_{\text{A,river}} = Q_{\text{final}} C_{\text{final}} \quad (9.7)$$

Of course, the constituent being analyzed may be a contaminant, such as an ammonium salt, or it may be a valuable component, such as oxygen.

Consumption of oxygen in receiving water bodies. Oxygen is extremely important for aquatic life. Dissolved oxygen in the water is consumed rapidly when organic matter, introduced as waste, is oxidized by bacteria. If the rate of oxygen consumption (*degradation*) exceeds the natural rate of oxygen recovery in the flowing stream (*reaeration*), then the flow will become anerobic (or anoxic). The degradation rate (k_L/day) can be estimated from laboratory testing, while the reaeration rate (k_R/day) is a function of stream turbulence and temperature.

The BOD is the actual amount of oxygen that is consumed by microorganisms in the wastewater during a specified period, usually five or seven days (BOD_5 or BOD_7). The demand may come from carbonaceous (carbon-containing) organic matter or nitrogenous (nitrogen-containing) organic matter. Excessive BOD can result in severe degradation of water quality in the receiving water body, affecting both aquatic life and downstream users.

One estimate of the BOD is obtained by calculating the *theoretical oxygen demand (ThOD)*, which assumes that all of the organic matter is broken down by microorganisms. The equations are written and balanced, and then the stoichiometry is used to determine the amount of oxygen consumed for a given concentration of organic matter.

The general equations are as follows:

$$\{CH_2O\} + O_2 \rightarrow CO_2 + H_2O \quad \text{(carbonaceous oxygen demand)} \quad (9.8)$$

$$NH_4^+ + 2O_2 \rightarrow 2H^+ + NO_3^- + H_2O \text{ (nitrogenous oxygen demand)} \quad (9.9)$$

So, for example, the oxidation of a waste stream that contains 100 mg/l of octane (a component of gasoline, C_8H_{18}) can be represented by the following:

$$C_8H_{18} + 12.5O_2 \rightarrow 8CO_2 + 9H_2O \quad (9.10)$$

The stoichiometry given can be used to determine the oxygen demand:

$$\frac{100 \text{ mg octane}}{\text{liter}} \times \frac{1 \text{ mmole octane}}{114 \text{ mg octane}} \times \frac{12.5 \text{ mmole } O_2}{1 \text{ mmole octane}} \times \frac{32 \text{ mg } O_2}{\text{mmole } O_2} = \frac{351 \text{ mg } O_2}{\text{liter}}$$

Thus, we can say that 351 mg/l of oxygen is consumed in this reaction. Because this demand is due to carbonaceous organic matter, it is called carbonaceous oxygen demand. A similar approach can be taken for nitrogenous organic matter. The sum of the two together is the total ThOD for a waste stream.

Understanding the oxygen sag curve. The oxygen sag curve is a graphical illustration of the gradual loss and recovery of dissolved oxygen in a stream as organic waste is consumed by bacteria (Figure 9.7). Natural waters can contain a maximum amount (saturation) of dissolved oxygen, the amount of which is dependent upon water temperature. The oxygen deficit, D, is the difference between the saturation dissolved oxygen at that water temperature (DO_s) and the actual amount of DO in the stream. When a pollution source is introduced, the stream's DO gets depleted from an initial deficit (D_0) to a critical point downstream at which point it begins to recover. The critical point occurs when the DO reaches a minimum (DO_{min}) at a downstream distance and point in time, x_c and t_c. Knowledge of this point will help in determining where and when greatest potential damage may occur to aquatic species.

Depletion of oxygen in the water is the most important cause of fish stress, which begins to occur at around 5 mg O_2/l for warm water fish and 8 mg O_2/l for cold water fish. Fish kills can begin at 2–3 mg O_2/l for many warm water species (Texas A&M 2020).

Prediction of dissolved oxygen in a stream. The Streeter–Phelps model is most commonly used to predict the effects of oxygen depletion in a flowing stream. In this model, the change in oxygen deficit, D, can be represented by:

$$\frac{dD}{dt} = k_L \times L_t - k_R \times D_t \quad (9.11)$$

where

k_L is the deoxygenation coefficient
k_R is the reaeration coefficient

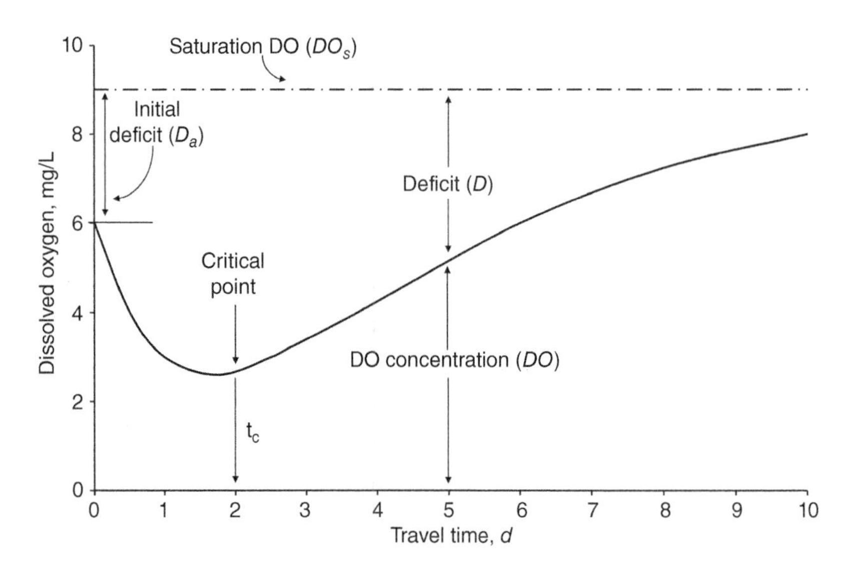

Figure 9.7 The Streeter–Phelps oxygen sag curve.

L_t represents the amount of BOD remaining at any time, i.e. the potential to consume oxygen, in mg O_2/l, at any time, t
D_t is the oxygen deficit (mg O_2/l) at any time, t

This equation can be solved for D at any point in time, t:

$$D_t = \frac{k_L\, x\, L_o}{(k_R - k_L)} \times \left(\epsilon^{-k_L t} - \epsilon^{-k_R t}\right) + D_o \times \epsilon^{-k_R t} \tag{9.12}$$

where

D_o is the initial oxygen deficit, $DO_s - DO_i$ (saturation level – initial level)
L_o is the initial oxygen demand, $L_t = L_o e^{-kLt}$, so that at time, $t = 0$, $L_t = L_o$ = the total potential oxygen demand

The water-quality conditions will be the worst at the critical time. Therefore, it is advantageous to solve the above equation for t_c:

$$t_c = \frac{1}{k_R - k_L} \times \ln\left[\frac{k_R}{k_L} \times \left(1 - \frac{D_o \times (k_R - k_L)}{k_L \times L_o}\right)\right] \tag{9.13}$$

Typical ranges and formulas for k_L and k_R are given in Table 9.6.
An example problem using these equations is given below.

Example Problem

After treatment, the ThOD of a waste is determined to be 28 mg O_2/l. The waste is discharged into a river, 3-m deep, flowing at a velocity of 2 m/second, water temperature of 22 °C, and background oxygen deficit of 4 mg O_2/l. What is the dissolved oxygen in the stream after one day (24 hours)?

1. First, calculate/determine the deoxygenation and reaeration coefficients. For k_L, we will use 0.30/day.
   Using the equation given in Table 9.6

$$k_R = \frac{3.9(2)^{0.5}([1.025]^{(22-20)})^{0.5}}{(3)^{1.5}} = 1.13/\text{day}$$

Table 9.6 Ranges and formulas for deoxygenation (k_L) and reaeration (k_R) coefficients.

Deoxygenation coefficient, k_L (day^{-1})	
Unpolluted river water	<0.05
Treated municipal wastewater	0.10–0.35
Untreated municipal wastewater	0.35–0.70
Reaeration coefficient, k_R (day^{-1})	
Can be approximated by: $k_R = \frac{3.9 v^{0.5}([1.025]^{(T-20)})^{0.5}}{H^{1.5}}$ v = *Mean stream velocity* $\left(\frac{m}{s}\right)$, T = *Temperature* (°C), H = *Average depth of flow* (m)	0.1–1.2, typical

Source: Ago (2018) and Mihelcic and Zimmerman (2009).

2. Using any standard table for oxygen saturation, we can say that the saturation dissolved oxygen at 22 °C is 9.07 mg/l. So the initial $DO = DO_s - D = 9.07 - 4 = 5.07$ mg/l.
3. We can use the equation given for D_t, with $t = 1$ day and $L_o = 28$ mg O_2/l:

$$D_t = \frac{k_L \times L_o}{(k_R - k_L)} \times \left(\epsilon^{-k_L t} - \epsilon^{-k_R t}\right) + D_o \times \epsilon^{-k_R t}$$
$$= \frac{0.3 \times 28}{(1.13 - 0.3)} \times (\epsilon^{-0.3} - \epsilon^{-1.13}) + 5.07 \times \epsilon^{-1.13}$$
$$= (8.4/0.83 \times 0.418) + (5.07 \times 0.323) = 5.87 \text{ mg } O_2/l$$

So, the dissolved oxygen in the stream is $9.07 - 5.87 = 3.2$ mgO_2/l.

This is a very low level of oxygen due mostly to the waste stream and high water temperature. Fish stress and fish kills would be likely.

End-of-Chapter Questions/Problems

9.1 Describe several (three to four) ways that water is used in industry.

9.2 What were some of the sectors in which China was exporting "virtual water" during the 2000s? (Use the Internet.)

9.3 Research the "water footprint" of the United States and report the amount of water used in agriculture, industry, and energy, and an average citizen's daily footprint.

9.4 List five companies from the CEO Water Mandate that you are familiar with then list then companies not on this list. Speculate on why the former are on the list and why these latter might not yet be on the list.

9.5 SABMiller Brewing Company is trying to reduce their water footprint. Use an Internet search to summarize and comment upon the steps that this major company is taking.

9.6 Pick a company that makes a product that you use daily or on a regular basis. Research that company's water-conservation efforts and respond with a paragraph detailing your results.

9.7 Pick two of the following commodities – bottled water (in a plastic container), pair of blue jeans, or laptop computer. Find an LCA/analysis on the Internet. Summarize the usage of water throughout the process of production.

a. Generate a table that gives the water consumed for each phase of production.
b. How much virtual water is represented on a per unit basis for the products selected?
c. How much virtual water is represented on a per kilogram basis (liters/kg) for the product you selected?
d. Comment on the differences and variation.

9.8 Calculate your individual "water footprint." Water calculators can be found using an Internet search.

a. How much of your water use was "virtual water" and how does this compare with the US average?

b. Recalculate your "water footprint" using values that you are able to cut out of your lifestyle. (e.g. instead of taking 11–15-minute showers, take 5–10-minute showers.) How much water could you be saving?

9.9 The textile industry uses billions of cubic meters of fresh water annually to produce clothes. Additionally, pollution from the clothing industry alone constitutes 1/5 of the industrial freshwater pollution globally.

a. Determine the approximate composition of your wardrobe – pairs of pants, cotton shirts, shoes, underwear, and socks.
b. Calculate the amount of water that it took to produce your wardrobe.

9.10 A sugar plant (crystalline cane sugar) releases organic waste at a flowrate/BOD_5 of 0.5 m^3/second/20 mg/l, respectively, into a river flowing at a rate of 9 m^3/second. The river has a background oxygen demand level of 10.5 mg/l.

a. Calculate the resultant oxygen demand of BOD_5 in the stream just below the point of release.
b. What would be the effects on fish (if any) due to this resultant oxygen demand?

9.11 An Indian textile finishing plant for processing and desizing a woven fabric has a COD in its effluent of 1240 mg/l. In India, the permissible limit to discharge is COD of less than 250 mg/l and BOD of less than 30 mg/l. Recommend a treatment to reduce COD to an acceptable level. Cite your sources used.

9.12 A brewery produces a wastewater that contains 15 000 mg/l of organics [$C(H_2O)$] and 50 mg/l of NH_3-N, ammonia as nitrogen (Mihelcic and Zimmerman 2009). Calculate the ThOD for this waste stream.

9.13 Consider Table 9.7, comparing oil and gas production in two active formations for one year (Silva et al. 2014).

a. Using data from the table, why do you suppose that more water is reused in the Marcellus shale than in the Barnett shale?

Table 9.7 Comparison of water management in Marcellus and Barnett shale plays, 2012.

	Marcellus (Pennsylvania)	Barnett (Texas)
Drilling water, MM gal	0.085	0.25
Hydraulic fracturing water, MM gal	5.5	3.8
New horizontal wells drilled	1365	660
Wells completed	540	500 (est.)
Active horizontal wells	3680	>10 000
Wastewater produced, MGD	3.1	2 (est.)
Source water availability	Abundant	Scarce
In-state salt-water disposal wells	8	12 000
Fraction of wastewater reused	0.87	0.05
Fraction of wastewater deep-well injected	0.12	0.94

MM = million gallons.
Source: Based on Silva et al. (2014).

Further Reading

Black, M. (2016). *The Atlas of Water: Mapping the World's Most Critical Resource*, 3e. University of California Press.

Darling, S.B. and Snyder, S.W. (2018). *Water Is . . . The Indispensability of Water in Society and Life*. World Scientific Publishing Co.

Hepworth, N. and Orr, S. (2013). Chapter 14: Corporate Water Stewardship. In: *Water Security: Principles, Perspectives, and Practices* (ed. B. Lankford, K. Bakker, M. Zeitoun and D. Conway). New York, NY: Routledge.

Kemmer, F. (1988). *The Nalco Water Handbook*, 2e. New York: McGraw-Hill Professional.

References

Ago, Florencemaejin #science • 2 Years (2018). "Streeter Phelps Equation." Steemit. 2018. https://steemit.com/science/@florencemaej/streeter-phelps-equation.

AWS (2014). *The AWS International Water Stewardship Standard*. Alliance for Water Stewardship.

Black, M. (2016). *The Atlas of Water: Mapping the World's Most Critical Resource*, 3e. University of California Press.

Blackhurst, B.M., Hendrickson, C., and i Vidal, J.S. (2010). Direct and indirect water withdrawals for U.S. industrial sectors. *Environmental Science & Technology* 44 (6): 2126–2130. https://doi.org/10.1021/es903147k.

BP (2018). "Responding to the Dual Challenge: BP Sustainability Report 2018."

CDP (2020a). "Companies Leading on Environmental Performance." 2020. https://www.cdp.net/en/companies.

CDP (2020b). "Companies Scores: The 'A' List 2019." 2020. https://www.cdp.net/en/companies/companies-scores.

Coca-Cola (2018). "Improving Our Water Efficiency - News & Articles." 2018. http://www.coca-colacompany.com/news/improving-our-water-efficiency.

Darling, S.B. and Snyder, S.W. (2018). *Water Is . . . The Indispensability of Water in Society and Life*. World Scientific Publishing Co.

Davis, M.K. (2010). *Water and Wastewater Engineering*. MacGraw-Hill https://www.amazon.com/Water-Wastewater-Engineering-Mackenzie-Davis/dp/0071713840/ref=sr_1_2?dchild=1&keywords=MacKenzie+Davis+water+and+wastewater&qid=1588104742&sr=8-2.

EIA (2015). "Corn Ethanol Yields Continue to Improve - Today in Energy - U.S. Energy Information Administration (EIA)." 2015. https://www.eia.gov/todayinenergy/detail.php?id=21212.

Fontenot, B.E., Hunt, L.R., Hildenbrand, Z.L. et al. (2013). An evaluation of water quality in private drinking water Wells near natural gas extraction sites in the Barnett shale formation. *Environmental Science & Technology* 47 (17): 10032–10040. https://doi.org/10.1021/es4011724.

Gleick, P. (1993). *Water in Crisis: A Guide to the World's Fresh Water Resources*, 1e. New York: Oxford University Press.

Hepworth, N. and Orr, S. (2013). Chapter 14: Corporate water stewardship. In: *Water Security: Principles, Perspectives, and Practices*. New York, NY: Routledge.

Hoekstra, A.Y. (2003). *Virtual Water Trade: The Proceedings of the International Experts Meeting on Virtual Water Trade*. The Netherlands: IHE Delft.

Kemmer, F. (1988). *The Nalco Water Handbook*, 2e. New York: McGraw-Hill Professional.

King, C.W. and Webber, M.E. (2008). Water intensity of transportation. *Environmental Science & Technology* 42 (21): 7866–7872. https://doi.org/10.1021/es800367m.

Levinson, M. (2008). Watching water: a guide to evaluating corporate risks in a thirsty world. *Global Equity Research* 60 pp.

Lu, S. (2018). "WTO Reports World Textile and Apparel Trade in 2017." *FASH455 Global Apparel & Textile Trade and Sourcing* (blog). 2018. https://shenglufashion.com/2018/08/16/wto-reports-world-textile-and-apparel-trade-in-2017.

Mihelcic, J.R. and Zimmerman, J.B. (2009). *Environmental Engineering: Fundamentals, Sustainability, Design*, 1e. Wiley.

Mowitz, D. (2012). "2,500 Gallons per Bushel." Successful Farming. 2012. https://www.agriculture.com/machinery/irrigation-equipment/drip-irrigation/2500-gallons-per-bushel_272-ar22678.

NAS (2004). *Water and Sustainable Development: Opportunities for the Chemical Sciences - A Workshop Report to the Chemical Sciences Roundtable. Water and Sustainable Development: Opportunities for the Chemical Sciences: A Workshop Report to the Chemical Sciences Roundtable*. National Academies Press (US). https://www.ncbi.nlm.nih.gov/books/NBK83737.

National Research Council (2007). *Water Implications of Biofuels Production in the United States*. https://doi.org/10.17226/12039.

National Research Council (2008). *Water Implications of Biofuels Production in the United States*. https://doi.org/10.17226/12039.

Poddar, P. and Sahu, O. (2015). "Quality and Management of Wastewater in Sugar Industry." *Applied Water Science*. https://www.researchgate.net/publication/272401609_Quality_and_management_of_wastewater_in_sugar_industry.

Postel, S. and Vickers, A. (2004). "Boosting Water Productivity." Presented at the Worldwatch Institute.

Ritchie, H. and Roser, M. (2017). "Water Use and Stress." *Our World in Data*, November. https://ourworldindata.org/water-use-stress.

Samsung (2020). "Resource Efficiency | Environment | Sustainability | Samsung US." Samsung Electronics America. 2020. https://www.samsung.com/us/aboutsamsung/sustainability/environment/resource-efficiency.

Silva, J., Gettings, R., Kostedt, W., and Watkins, V. (2014). Produced water from hydrofracturing: challenges and opportunities for reuse and recovery. *The Bridge (NAE)* 44 (2): 34–40.

Texas A&M, Agrilife Extension (2020). "Dissolved Oxygen | AquaPlant." 2020. https://aquaplant.tamu.edu/faq/dissolved-oxygen.

UNESCO (2017). "WWAP - World Water Assessment Program. Fact 1: Demographics & Consumption | United Nations Educational, Scientific and Cultural Organization." 2017. http://www.unesco.org/new/en/natural-sciences/environment/water/wwap/facts-and-figures/all-facts-wwdr3/fact1-demographics-consumption.

United Nations (2009). "WWDR3: 'Water in a Changing World." UNESCO.

US EPA (2010). "National Pollutant Discharge Elimination System (NPDES) Permit Writers' Manual." EPA-833-K-10-001.

US EPA (2013). "The Hydraulic Fracturing Water Cycle." Data and Tools. US EPA. 2013. https://www.epa.gov/hfstudy/hydraulic-fracturing-water-cycle.

US EPA, OW (2014). “Effluent Guidelines.” Collections and Lists. US EPA. United States. February 7, 2014. https://www.epa.gov/eg.

Vengosh, A., Jackson, R.B., Warner, N. et al. (2014). A critical review of the risks to water resources from unconventional shale gas development and hydraulic fracturing in the United States. *Environmental Science & Technology* 48 (15): 8334–8348. https://doi.org/10.1021/es405118y.

Warner, J.F. and Johnson, C.L. (2007). ‘Virtual water’-real people: useful concept or prescriptive tool? *Water International* 32 (1): 63–77. https://doi.org/10.1080/02508060708691965.

Water Footprint Network (2021). “Virtual Water Trade.” 2021. https://waterfootprint.org/en/water-footprint/national-water-footprint/virtual-water-trade.

Water Initiative, The World Economic Forum Water (2011). *Water Security: The Water-Food-Energy-Climate Nexus*, 2e. Washington, D.C: Island Press.

Water Resources Group (WRG) (2019). *2019 Annual Report: Building Trust, Growing Resilience – 2030 Water Resources Group*. World Bank Group https://www.2030wrg.org/2019-annual-report-building-trust-growing-resilience.

Wolf, A.T., Yoffe, S.B., and Giordano, M. (2003). International waters: identifying basins at risk. *Water Policy* 5 (1): 29–60.

Zoback, M.D. and Arent, D.J. (2014). Shale gas development: opportunities and challenges. *The Bridge (NAE)* 44 (1): 16–23.

10

Water for Ecosystems and Environment

In this chapter, we explore the water needs of organisms and plants that comprise aquatic ecosystems. These needs are often assigned a secondary importance, but the goods and services provided by ecosystems are both diverse and critical, especially to the most vulnerable human populations. Water quantity is considered by determining appropriate environmental flows, while water quality includes the various parameters that are significant for ecosystem functioning. A few case studies are presented to give the reader examples of how water for ecosystems is both managed and measured.

Learning Objectives

Upon completion of this chapter, the student will be able to:

1. Understand the many services and functions provided by ecosystems.
2. Understand the different types of aquatic ecosystems that exist primarily in and around water.
3. Understand how ecosystem services most directly affect the livelihoods and well-being of disadvantaged communities.
4. Recognize the human impacts on natural ecosystem functions.
5. Understand which water-quality parameters are important for the survival of ecosystem species.
6. Describe common methodologies that are used to determine environmental flows needed to sustain ecosystems.
7. Understand some case studies in which Integrated Water Resources Management has been used for appropriate water allocation and the protection of ecosystem services.
8. *Understand the calculations of ecological risk and the formulation of total maximum daily loads.

10.1 Introduction

Thus far, we have taken a primarily anthropocentric view of water security. The need for water has been defined as the *human* need for water – for food, health, energy, and industry. In modern times, this need has begun to disrupt the natural hydrologic cycle in ways that

Fundamentals of Water Security: Quantity, Quality, and Equity in a Changing Climate, First Edition.
Jim F. Chamberlain and David A. Sabatini.

Companion website: www.wiley.com/go/chamberlain/fundamentalsofwatersecurity

compromise water security for living, nonhuman species who also require water for life and sustenance. As populations expand, the immediate demand for human water consumption tends to overshadow ecosystem demand for a proper allocation. An ecosystem includes both the physical environment and the diverse populations of plants, animals, and bacteria that comprise the biological community within the system (Figure 10.1). In this chapter, we explore the connectivity of water and ecosystems, considering both the intrinsic worth of the environment and its connection with human health and well-being.

Freshwater environments are home to complex and species-rich ecologies that reflect the immense beauty and diversity of nature. Ecological diversity can be compared among three major biomes – terrestrial, marine, and freshwater – as a ratio of species diversity to the extent of habitat, called the "**relative species richness**." Coastal ocean environments are more dense and diverse because the shallower, warmer waters support a food-supply chain that has a base of photosynthetic organisms such as algae and plankton (Darling and Snyder 2018). But because of the vast ocean expanse, the richness ratio of marine life (0.2) is lower than that of freshwater (3.0) and terrestrial (2.7) biomes (Black 2016).

Ecological and associated species diversity can be seen spanning a single administrative unit as well, such as across a nation like Brazil. The tropical rain forest in the northwest region transitions to a savannah in the central region and then to grasslands and thorny scrubs in the southeastern region (Figure 10.2). Such a diversity means that environmental water policy promulgated by a central government should be adaptable to the needs of local ecosystems, which may vary widely one from another.

This chapter begins with describing the good and services provided by ecosystems. We present the important water-quality parameters that are used in assessing ecosystem health. Then we look at the impact that human activities are having on these systems. We continue by providing methods for estimating appropriate environmental flows, a needed component for overall integrated watershed management. The chapter closes with some case studies to show how some of this management happens in real world settings.

Figure 10.1 Ecosystem services include those given by wetlands and water-drenched landscapes. Source: IzzetNoyan/Adobe Stock.

Figure 10.2 Landscape along the Rio Cipo in the Serra do Cipo National Park in Minas Gerais, Brazil. Source: Christian Dietz/Adobe Stock.

10.2 Ecosystems Functions and Services

Aquatic ecosystems are dependent upon water present or supplied as rivers, springs, wetlands, coastal bodies of water (estuaries, lagoons), and groundwater. In broad categories, these aquatic ecosystems can be surface water or groundwater-dependent ecosystems. **Surface water ecosystems** are those that occupy lakes, ponds, or free-flowing streams. The water may be clear and swift, likely teeming with fish, aquatic invertebrates, and diatomaceous (single-celled) algae. Ecosystem interactions are complex, but their growth and sustenance depend upon water and the transport of vital nutrients and bacteria through the water. Indicator species are species used to indicate the relative health of an aquatic environment. These include benthic (bottom-dwelling) macroinvertebrates, such as caddisflies, mayflies, water pennies, and freshwater clams.

Groundwater-dependent ecosystems (GDEs) include springs, seeps, wetlands, aquifer and cave ecosystems, some terrestrial ecosystems, and those gaining reaches of streams that are sustained by groundwater. Because these systems lack the visibility of surface water sources, they have traditionally been given less attention in scientific study, public policy, and management considerations (Kreamer et al. 2015). The lack of information and location has resulted in an under-valuation of these critical systems, and so we focus our attention on these here.

Springs have long been considered sacred sites, places of divine encounters, from Native American holy places to the grotto of Lourdes in France. They are also convenient gathering places as well as points of military conflict because of their extreme sustenance value. Saladin's armies captured or killed the majority of an army of twelfth-century Christian crusaders as they desperately sought to quench their thirst in the springs of Hattin (Kreamer et al. 2015). But springs are also critical water sources for rural peoples across the globe, from Mississippi to Malawi, and the collapse of a perennial spring source may mean total loss of easily accessible, uncontaminated water. The ravages of a civil war in the

Democratic Republic of the Congo, for example, may lead villagers with no choice but to drink contaminated water (Shore 1999). Though not common knowledge, individual springs may support unique species (such as the pupfish in Ash Meadows, Nevada, USA) or provide an important habitat for a critical phase of life, such as the wintering in warm water by Florida manatees (Kreamer et al. 2015).

Wetlands are rich ecological biomes where water covers the soil permanently or is present at or near the surface of the soil for extended periods of time throughout the year, including during the growing season (US EPA 2015a). Wetlands serve vital functions as they filter surface water, sequester carbon, regulate flooding, and provide habitat for inland fisheries. These fisheries are often the primary food source for island and coastal communities. In addition, wetland-dependent bird species are numerous and include grebes, pelicans, cormorants, herons, flamingos, geese, ducks, and many others. An economic assessment of 63 million hectares of wetlands on a global basis estimated the value at $3.4 billion per year with the bulk of these benefits coming to Asia with an economic value of $1.8 billion per year (Brander and Schuyt 2010).

All of these water-based ecosystems contribute goods and services listed in Table 10.1. These include provision of basic needs, regulation of water-quality threats, habitat support of diverse species, and enhancement of culture.

A closer look at the table will reveal the deep dependence of low-resource communities on these goods and services and the potential inequities that arise from their loss. Rural populations that live "close to the land" and rely directly on subsistence farming and fishing are the

Table 10.1 Water-dependent (aquatic) ecosystem goods and services.

Type of service	Description/Examples
Provision of needs	• Household water supply – drinking, cooking, washing, and other needs • Supply of water for agriculture, industry, energy production, etc. • Aquaculture • Transportation • Fiber and fuel – fuelwood, peat, fodder • Biochemical – extraction of medicines from biota
Regulation of threats	• Climate regulation – source of/sink for greenhouse gases; influence local precipitation and temperature • Groundwater recharge and discharge • Water purification and waste treatment • Flood and storm protection • Erosion and sedimentation control
Support of habitat	• Maintenance of biodiversity – wildlife, waterfowl, clams and mussels, etc. • Fisheries habitat • Protection of endangered species • Nutrient cycling
Enhancement of culture	• Recreation – swimming, boating, etc. • Aesthetic appreciation • Spiritual and religious associations • Enhanced property values • Educational opportunities

Source: Compiled from Daily (1997), Hornberger and Perrone (2019), Millennium Ecosystem Assessment (2005).

ones who most directly benefit from these goods and services. Surface water is collected for household drinking, cooking, and cleaning. Peat (a wetlands moss) is used for heating and as a fuelwood in some regions. Mangrove seedlings are being replanted in coastal regions where these trees with their magnificent root systems preserve the shoreline, protect against flooding, and provide habitat for coastal fisheries. Structural measures, such as fish ladders, have attempted to preserve spawning species, but the decline of these fisheries has occurred nonetheless (Wescoat and White 2003). Finally, many indigenous populations attach a high religious or spiritual association to many historical waterbodies, such as sacred mountain lakes or streams. The resilience and survival of these communities may depend directly on access to these goods and services. In low-resource communities, periods of prevalent water stress will also likely result in aquatic ecosystem stress and the loss of these natural amenities.

10.3 Human Impacts on Natural Infrastructure

Humans have historically sought to magnify some ecosystem services at the unintended expense of other services. Over geologic time, both plant and animal species have adapted to natural infrastructure and environmental conditions, including occasional droughts and flooding. Abrupt changes in both structure and hydrology can either degrade or eliminate valuable services altogether. For example, the construction of a dam for hydroelectric power can produce desirable electricity and flood control, while diminishing the provisioning services for riparian communities and the quality of natural habitat downstream of the structure (Table 10.2).

Both marine (saltwater) and freshwater ecosystem services are affected by human activities. Marine services are directly affected first by overfishing, which affects both a valuable food source as well as related employment. The depletion of the stock of northern cod off the eastern coast of Canada in the early 1990s led to 30% unemployment and the loss of valuable fisheries production (Daily 1997). Freshwater ecosystems are most profoundly affected by water infrastructure projects, such as dams for flood control, hydropower, storage, and recreation. Dikes and levees are often built to provide local flood control but effectively shift the flood burden downstream while disrupting the soil fertility benefits of periodic flooding. Due mostly to urban sprawl, the United States is losing wetlands at a rate of about 50 000 acres per year, with a loss not only of the services provided by this crucial biome but also of the countless waterfowl and wildlife species that dwell in this water–land habitat (Earthtalk 2008). The increased emission of greenhouse gases exacerbates changes in climate, which affect runoff patterns in abrupt temporal time scales, too short for the adaptation of aquatic species.

Proper water-management decisions must take a “big picture” approach in order to assess the impact of major projects that affect both the water quality and quantity upon which delicate ecosystems depend. In many cases, the affected downstream human communities are disenfranchised with regard to the management decisions being made. These decisions result in social inequities and economic hardship for those who rely most on valuable ecosystems. The drive to control nature not only destroys worldly beauty but also the natural communities upon which much of human life depends.

Table 10.2 Threats to freshwater aquatic ecosystem services from human activities.

Human activity	Impact on aquatic ecosystems	Values/Services at risk
Dam construction	Alters timing and quality of river flows, water temperature, nutrient and sediment transport, delta replenishment; blocks fish migration	Habitat, recreation, and commercial fisheries; maintenance of deltas and their economies; diminishment of downstream domestic water supply
Dike and levee construction	Destroys hydrologic connection between river and floodplain habitat	Habitat, recreation, and commercial fisheries; natural floodplain fertility; natural flood control
River diversions	Depletes streamflows to damaging low levels	Habitat, recreation, and commercial fisheries; dilution of pollution; hydropower; transportation
Draining and loss of wetlands through surface diversion or aquifer depletion	Eliminates a key component of the aquatic environment	Natural flood control; habitat for fisheries and waterfowl; recreation; natural water-quality filtration
Deforestation/poor land-use management (e.g. increase of impervious surfaces)	Alters runoff patterns; inhibits natural recharge; fills waterbodies with silt	Water supply quantity and quality; flood control; fish and wildlife habitat; transportation; water for electricity generation
Uncontrolled pollution/release of metals and acid-forming pollutants to air and water	Diminishes water quality by altering chemistry of rivers and lakes	Water supply, habitat, commercial fisheries, recreation
Emission of greenhouse gases	Has potential to make dramatic changes in runoff patterns from increases in temperature and changes in rainfall	Water supply, hydropower, transportation, fish and wildlife habitat, dilution of pollution; recreation, fisheries, flood control
Population and overall consumption growth	Increases pressure to dam and divert more water, drain more wetlands, pump more aquifers, etc.; increases water pollution, acid rain, and potential for climate change; depletes fisheries	Virtually ALL ecosystem services are affected

Source: Modified from Daily (1997).

10.4 Water-Quality Needs for Ecosystem Species

Natural ecosystems have evolved over time into an equilibrated balance between aquatic organisms, plants, and water quality. Some water-quality parameters are distinctive in their importance for maintaining this balance. These are described in this section.

Because of their ability to breathe with gills, fish species require dissolved oxygen (DO) for survival. The desirable DO range is 2–5 mg/l, with coarse fish (e.g. pike, perch) being able to survive on the low end and salmonid species needing oxygen at the higher end

(Davie and Quinn 2019). This is why one would tend to find salmon in cooler waters only, with associated higher DO levels, while perch, pike, and carp can live in warmer water temperatures. Related to DO is the biochemical oxygen demand (BOD). This parameter measures pollution indirectly as the rate of oxygen depletion by bacteria and other microorganisms feeding on the organic matter in the water and consuming oxygen in the process. Untreated domestic sewage may have a BOD as high as 220–500 mg/l, while an unpolluted stream should have a value of less than 5 mg/l BOD (Davie and Quinn 2019).

The acidity of water is quantified by the pH, a function of the concentration of hydrogen ions; $pH = -\log [H+]$. The pH of natural waters may vary considerably, and rainwater will generally be slightly acidic, with pH usually 5–6, but may drop as low as 4 (Davie and Quinn 2019). At the other end of the acidity spectrum, rivers that drain rocks that are rich in carbonates may have a higher pH from the presence of bicarbonate ions. As the waters become more acidic, aquatic life becomes more narrowly constrained as shown in Table 10.3. A low pH is characteristic of waters that derive from mine works and tailings.

Nitrates and phosphates are the nutrients of greatest concern as these can cause an over-proliferation of aquatic plants, including algae. Phosphates are not as soluble as nitrates and so may cling to sediments and stay in river bottoms for a longer period of time as they are slowly released. The additional plant growth provides a benefit initially as the plants remove the nitrogen and phosphorus from the water column. However, at higher levels algae become a predominant species. As these species die off, bacteria that feed on the plant matter will deplete the oxygen in the waterbody to undesirable levels. Trophic classifications are used by limnologists to categorize various levels of nutrient loading. The range is from oligotrophic ("few nutrients") to mesotrophic ("moderate nutrients") to eutrophic ("good nutrients") to hypertrophic ("excess nutrients"). Table 10.4 gives the OECD (Organization for Economic Cooperation and Development) trophic levels for various water parameters. Chlorophyll is an indicator of algal growth and is measured at various depths in the water with the maximum being recorded.

Temperature, turbidity, and salinity also have effects on aquatic species. Increased water temperature will decrease the DO level potentially below the optimum level of 4–5 mg/l for most aquatic organisms. Especially in urban streams that drain impervious areas, a sudden thunderstorm can increase the temperature dramatically, resulting in organism mortality. Other factors that can increase water temperature are decreased base flow, reduced stream

Table 10.3 The effect of acidity as measured by pH on fish and aquatic species.

Effect on organisms or process	pH
Neutral, prevalence of indicator species	7.0
Mayflies disappear	6.5
Phytoplankton species decline	6.0
Waterfowl breeding species decline	5.5
Salmonid reproduction fails; most amphibia disappear	5.0
Caddis flies, stone flies, dragonflies, damselflies disappear; most adult fish harmed	4.5

Source: Davie and Quinn (2019)/with permission from Taylor & Francis.

Table 10.4 Selection of indicators used by the OECD for lakes and reservoirs' classifications.

Trophic level	Average total *P* (μg/l)	Dissolved oxygen (% saturation)	Max. chlorophyll (at depth) (μg/l)
Oligotrophic	10	> 80	8.0
Mesotrophic	10–35	40–89	8–25
Eutrophic	35–100	0–40	0.25–75
Hypertrophic	>100	0–10	> 75

Source: OECD (1982).

bank vegetation, and industrial discharges (DeBarry 2004). Higher temperatures can shift the composition of the algal community from diatoms to the less desirable blue-green algae (DeBarry 2004). Suspended sediment load can result in increased turbidity of the stream with a decrease in light penetration to the detriment of desirable submerged aquatic plants. This sediment also fills in gravel bottoms where fish and benthic organisms lay their eggs. Agricultural return flows can drastically increase salinity to dangerous levels. The Colorado River in west central Colorado has a salinity of about 300 mg/l. But irrigation return flows in the Grand Valley have a salinity of 1,500–10 000 mg/l as they pick up salts from highly saline soils and underground rock formations (Gillilan and Brown 1997). Such a composition threatens the existence of indigenous freshwater species who have adapted to lower levels of salinity.

In order to preserve species richness in aquatic systems, water managers will need to monitor not only environmental flows (discussed below) but also critical water-quality parameters that threaten to reduce the natural biodiversity of these systems.

10.5 Environmental Flows for Ecosystems Health

We have noted that the term "aquatic ecosystems" includes organisms and plants dependent upon rivers, lakes, and coastal bodies of water (estuaries, lagoons) as well as GDEs (Arthington et al. 2018). Accordingly the Brisbane Declaration and Global Action Agenda on Environmental Flows of 2018 (a revision of the Brisbane Declaration of 2007) defines **environmental flows** as "the quantity, timing, and quality of freshwater flows and levels necessary to sustain aquatic ecosystems which, in turn, support human cultures, economies, sustainable livelihoods, and well-being." (Arthington et al. 2018) The determination of desired instream flow will allow managers to calculate surplus water available for withdrawal, such as for irrigation and wastewater reuse. The assessment of environmental flows is both a social and a technical process. It is "social" in that the desired flows are dependent upon the society's intended uses – to support livelihood, cultural or spiritual practices, biodiversity, or all of the above. It is "technical" in that a team of specialists may be required to do a proper assessment, including biologists, hydrologists, and wildlife specialists.

How much flow water is required to sustain stream and riverine ecosystems? Potential answers come from a knowledge of both river hydraulics and stream ecology. Generally, the appropriate methodology falls into one of three types – historical flow, hydraulic rating, and habitat methods.

The *historic flow and extrapolation methods* set withdrawal limits based on a historic flow range. These assume that there is a linear relationship between flow and biological health. A widely used procedure is the Tennant/Montana method, named after Donald Tennant, a fisheries biologist who developed the technique through personal observations of western streams over two decades, 1960s–1970s (Gillilan and Brown 1997) (Table 10.5). When more detailed analyses are not possible due to time or funding constraints, these percentages can be used as "rules-of-thumb" for survey or planning purposes. Thus, for a river with typical western North American weather regimes, the winter flow should be at least 20% of the mean annual flow and the summer flow should be at least 40%. These indices can be adjusted based on various stream morphologies or can be indexed on a monthly basis to mean monthly flow (Jowett 1997). The indices can also be extrapolated from a large number of detailed studies for the particular stream (O'Keeffe and Le Quesne 2009).

The *hydraulic rating method* assumes a relationship between several hydraulic parameters, such as wetted perimeter, stream depth, and velocity on a single cross-section of the river. These parameters are linked to the ecological health of the stream and desirable flows are obtained accordingly.

Given the slope and channel characteristics, an average velocity of a stretch of stream can be found using the Manning's equation:

$$v = \frac{k \times R^{2/3} \times \sqrt{s}}{n} \tag{10.1}$$

where

v is velocity (m/second)
k is a constant depending on units (1 for SI units, 1.49 for Imperial units)
R is the hydraulic radius of the channel (m) = Ratio of flow area to wetted channel perimeter
s is the channel slope (m/m)
n is the Manning roughness coefficient (–)

Table 10.5 Percentage of mean annual flow needed to establish various levels of habitat quality using the Tennant/Montana method in western US rivers.

	Percentage of mean annual flow	
Habitat quality/health	**October–March (winter)**	**April–September (summer)**
Annual flushing/maximum flow	200	200
Optimum	60–100	60–100
Outstanding	40	60
Excellent	30	50
Good	20	40
Fair-to-severe degradation	≤10	≤30

Source: Gillilan and Brown (1997).

Flow, Q, is the product of flow area and velocity ($Q = AV$). Thus, as flow increases, both the velocity and wetted perimeter change, though at different rates. For very wide rivers, the hydraulic radius can be approximated as the average depth of water (Davie and Quinn 2019). The most common hydraulic method observes that wetted perimeter increases with flow up to an inflection point. For example, Tennant noted an inflection point at 10% of the mean flow in the rivers that he studied (Tennant 1976). The rating can be used alone or as a tool in the habitat method described next.

The suite of *holistic/habitat methods* utilizes the expertise of a range of hydrologic and ecological specialists who rely upon empirical observations and measurements of stream health in response to various flow regimes. The methods assume an optimum of productivity or benefit in a fish life cycle at various flow levels, and high flows may lead to a decreased biological response (Davie and Quinn 2019). The more sophisticated of these methods involve computer modeling, such as Physical Habitat Simulation System (PHABSIM) and Instream Flow Incremental Methodology (IFIM). Hydraulic simulation is used to describe the area of a stream having various combinations of depth, velocity, and channel index as a function of flow.

The method results in a weighted usable area (WUA), a surrogate for extent of habitat, which corresponds to stream discharge. For example, PHABSIM has been used to measure the usable habitat effects of streamflow for two rivers in the United Kingdom – the Piddle and the Wye. The data shows an optimum level of flow at 0.5–0.6 m^3/second, estimated by modeling, which can be used in water allocation decisions for a given waterway (Acreman 2001).

The relationship between flow and biological response using the above three methods may be compared for a hypothetical river. It should be noted that providing maximum flow may not also result in the optimization of biological response. The optimum biological response will occur at some percentage of the historic mean.

Table 10.6 summarizes these three commonly used methods for estimating instream flow, from the simplest (Historic flow) to the most complex (Holistic/Habitat) category.

The building block method (BBM) allows for an optimum base flow for the dry season and for the wet season, while providing for periodic floods on which many species depend for certain phases of their life cycles (Figure 10.3). This method relies upon both good historical flow and meteorological knowledge of the river basin.

It must be noted that with all of these methods, other important parameters are not yet included, such as temperature, DO, and water-quality parameters, including sediment load and salinity, as discussed above.

10.6 Application of Integrated Water Resources Management (IWRM) for Ecosystems

As we have said above, IWRM is the process of managing human activities and natural resources on a watershed basis. It takes into consideration the needs of the local environment and ecological species as well as the societal needs and economic benefits, as long as these are sustainable and not damaging to the other two sectors. We see that a larger, more diverse watershed is sure to result in the need for a more complex IWRM scheme, as it will include many more stakeholders, users, and types of water demands.

The US Forest Service has divided the United States into *domains* or ecoregions based on climate, *divisions* based on similar precipitation levels and patterns, and *provinces*

Table 10.6 Summary of three commonly used methods for determining environmental instream flow for ecosystem maintenance.

	Method		
Characteristic	**Historic flow**	**Hydraulic**	**Holistic/Habitat**
Examples	Tennant/Montana method	Wetted perimeter versus discharge	Building block method (BBM); PHABSIM, ISIM
Complexity	Simplest	Moderately complex	Most complex
Popularity in the United States	Second most popular in the United States (16 states)	Third most popular in the United States (6 states)	First most popular in the United States (24 states)
Data requirement	Flow record	Cross-section survey	Cross-section survey Habitat suitability criteria
Method of assessing flow requirement	Percent (%) of average annual or monthly flow Percent (%) exceedance sometimes given	Percent (%) habitat retention Inflection point	Percent (%) habitat retention Inflection point Optimum/minimum habitat (exceedance or percentage)
Stream hydraulics	Effect on width, depth, and velocity; dependent on morphology Maintenance of stream "character"	Effect on depth and velocity; dependent on morphology Maintenance of stream "character" in terms of wetted perimeter	Prescribed depth and velocity Potential loss of stream "character"
Ecological assumption	Close relationship between natural flows and indigenous ecology	Biological productivity related to wetted area of stream	Close relationship between stream habitat and ecology
Comments on comparative advantages and disadvantages	"Rule-of-thumb" flow assessment Flow always greater than zero and less than but related to natural flow; precludes enhancement	More interpretation required Flow dependent on channel shape Levels of protection difficult to relate to ecological goals	Application and interpretation are critical Species-dependent Allows trade-offs Flow assessment independent of natural flow; enhancement potential is recognized

Source: Modified from Jowett (1997) and O'Keeffe and Le Quesne (2009).

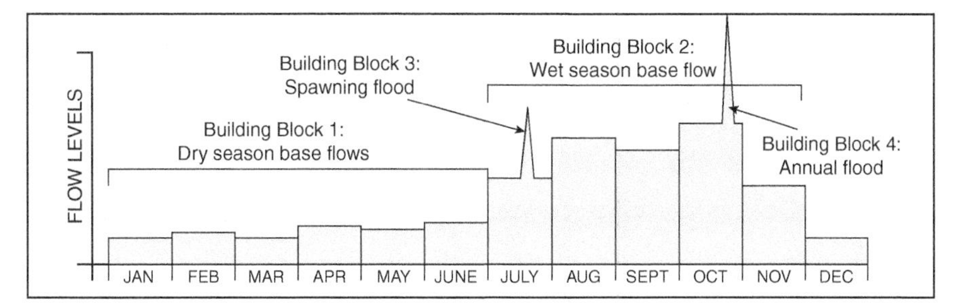

Figure 10.3 The Building Block Method of environmental flow assessment gives a monthly flow regime that includes flood events necessary for fish life cycle needs. Source: O'Keeffe and Le Quesne (2009).

based on similar vegetation and natural land covers (U.S. Forest Service 2021) (Figure 10.4). These divisions help planners and scientists to evaluate linkages between management and land/aquatic ecosystems of a similar nature. Each ecoregion (province and below) will contain ecosystems with similar organisms, habitat, and hydrology (DeBarry 2004).

The following two case studies give examples of collaboration in both water resources management and in determination of suitable environmental flows. Both cases demonstrate the critical need to include ecosystems and water quality in care for valuable waterbodies.

England and Wales. River systems in England and Wales, two water-rich regions of the United Kingdom, have traditionally been used as locally controlled sources of water supply and dilution mechanisms for waste disposal (Figure 10.5). Not much thought or attention was given to water quality or the ecosystems that depended on such. The Water Act of 1973 shifted water resource management from the local to the national level and created regional water authorities (RWAs) to manage water on a watershed basis. The intent was to provide a more holistic resource management with the RWA in charge of both waste disposal and water-pollution control within the basin boundaries. This created an uncomfortable "conflict of interest" scenario, which resulted in privatization of drinking water supply and wastewater treatment in the late 1980s (Davie and Quinn 2019).

In 2000, the European Union (of which the United Kingdom was a part) adopted the Water Framework Directive, which required each nation to set environmental objectives for

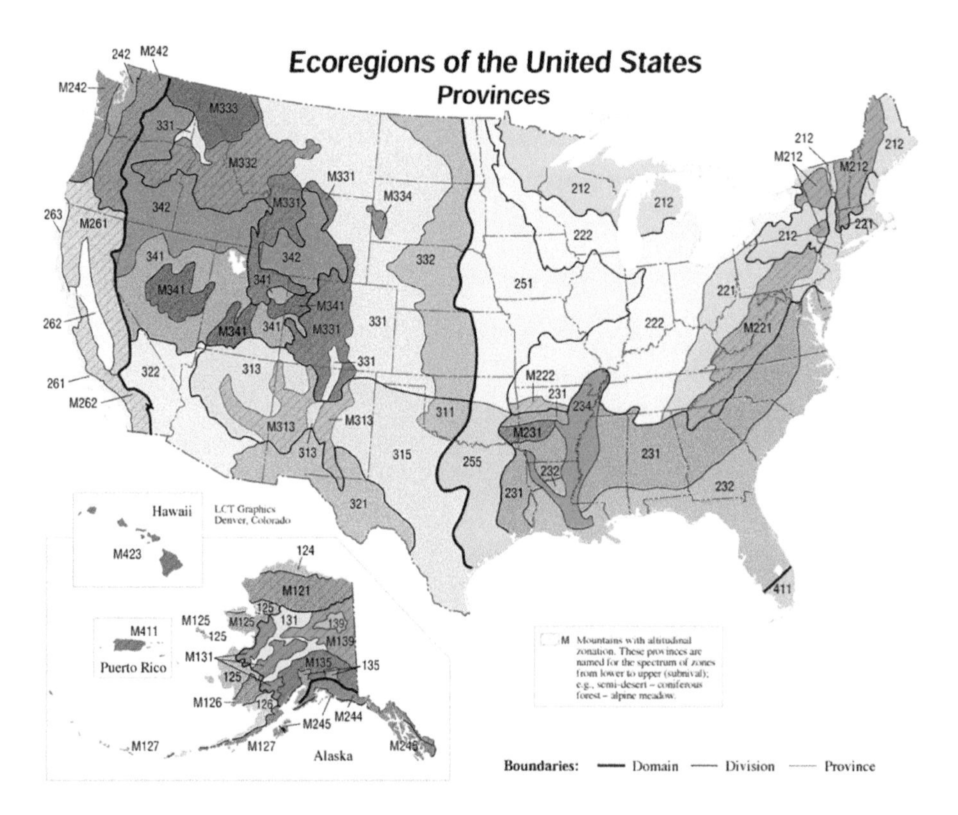

Figure 10.4 A map of ecoregions of the United States, showing three domains and various divisions and provinces (US Forest Service 2021). Source: Public domain.

Figure 10.5 The River Dee forms part of the border between England and Wales. Llangollen, view from the railway station. Source: Akke Monasso/CC BY-SA 2.0/Wikimedia Commons.

each of their waterbodies. These objectives addressed pollutant concentration limits, health of fish populations, and groundwater quantity. From 1961 to 2000, the amount of water abstracted for both England and Wales leveled off and then declined about 11% from its peak in 1988 in spite of a population increase of four million in that same period (Davie and Quinn 2019). Much of the decrease came from a decrease in industrial abstractions as well as reduction in water-supply leakage, a reduction that costs a considerable sum of capital infrastructure investment. The decline in water abstractions was good for the aquatic environment, allowing a more natural river regime. The combination of privatization of wastewater treatment, public basin-wide management, and clear environmental goals has resulted in a rise in percentage of rivers with "good" chemical water quality in both England (55–80%) and Wales (80–87%) over the period 1990–2009 (Davie and Quinn 2019).

Upper Ganga Basin of India. "There are only two things in this world – the sun and Mother Ganga. And of these, Mother Ganga is the only one that we can touch." This statement from a religious pilgrim bespeaks the cultural and religious importance of the Ganga River basin, the largest basin in India. In 2008, the World Wildlife Federation office in India (WWF-India) began training and coordinating local teams to conduct an environmental flow analysis (EFA) for the Upper Ganga River, a river that is critical to irrigation agriculture, ecosystems health, and religious culture (O'Keeffe et al. 2012). The Ganga is the "Holy Mother River of India" and the environmental flow requirements would thus need to meet the religious needs of people as well as biota of the river (Figure 10.6). The assessment group decided to use the BBM as it was clearly understood, robust, and adaptable to the peculiar characteristics of the Ganga. These characteristics include the large scale of the river, the seasonal precipitation fluctuations, the history of human modification, and the importance of cultural and spiritual issues.

Figure 10.6 Ritual bathing in the Ganga River. Source: Melissa Schalke/Adobe Stock.

Himalayan rivers in this region have high spring flow due to snow melt, low flows during the winter months, and flood pulses from monsoon rains. The Ganga river supports a diverse ecosystem in its upper reaches, including large mammals such as river dolphins and the endangered *Gharial* crocodile (Figure 10.7), and local villagers are beginning to appreciate these creatures as part of the local ecosystem (Indian 2018). The river's basin is home to industries, especially sugar processing, and is attractive to tourists who camp, fish, and kayak in its higher quality waters.

Ideally, the reference conditions of a river are the unaltered, natural conditions. Because of many years of water diversions, disposal of waste, and disconnection from the surrounding floodplain, the study reaches of the Ganga river had been modified so greatly that natural conditions were no longer apparent. Thus, it was decided to use the present-day conditions upstream of Narora as a realistic set of reference conditions. A six-level classification system was used to classify each segment of the river according to unmodified/natural (Class A) to critically modified with near-complete loss of natural habitat (Class F). Environmental flow assessments were designed to establish levels of the top three classes (Classes A–C) in all reaches. Because historical data was not available, the Soil–Water Assessment Tool (SWAT) model was used to model the river without the barrage dams that have been put in place.

Each of five study groups conducted specific analyses:

- The fluvial geomorphology group estimated the flow velocities and depths required to move, sort, and deposit different sizes of sediment, so as to maintain or restore channel size and other important channel features (such as multiple channels and bars).
- The biodiversity group concentrated on the habitat characteristics (e.g. depth, flow velocity) required for the life cycle of important flow-dependent species such as the river dolphin, selected fish species, macro-invertebrates, and floodplain vegetation.

Figure 10.7 The unique Gharial crocodile has a *ghara* (pot) on the tip of its snout. Source: Lazar/ Adobe Stock.

- The livelihood group focused on depth, water quality, and river width required to maintain certain livelihood activities (such as ferrying or rafting).
- The spiritual/cultural group had to ascertain the depth and water-quality issues that would affect religious and cultural activities (such as ritual bathing).
- The water-quality group responded to the recommendations of the other groups, estimating the effects that the recommended flows would have in mitigating pollution or other water-quality issues (O'Keeffe et al. 2012).

Table 10.7 gives the results of the environmental flow assessment for one of the reaches. The recommended flow for January, in the dry season, would be 330 m^3/second, as this would meet the biodiversity needs of this stretch while simultaneously meeting the other needs. In August, the recommended flow is 3303 m^3/second, as this flow is required to meet the cultural/spiritual needs of the community.

This example of an IWRM approach to environmental flow assessment utilized the expertise of specialists to represent various societal and environmental needs.

Table 10.7 Results of environmental flow assessment (EFA) for one study site.

	January			August		
	Flow (m^3/second)	Depth (m)	Velocity (m/second)	Flow (m^3/second)	Depth (m)	Velocity (m/second)
Biodiversity	330	3.5	1.03	1986	6.7	1.7
Fluvial geomorphology	330	3.5	1.03	1986	6.7	1.7
Cultural and spiritual	98.8	1 m at bottom step	0.73	3303	8.08	1.2
Livelihoods	330	3.81	1.03	1500	6.13	1.5

Source: Recommended flows for the Upper Ganga River. Adapted from O'Keeffe et al. (2012).

10.7 Conclusion

In developed countries, a growing awareness of ecosystem importance has led to new regulations aimed at protecting these or, at least, attempting to redress the loss by balance or replacement. For example, the US Department of Agriculture's Wetland Reserve Program encourages landowners to voluntarily protect, restore, and enhance wetlands on their own private property. This program has been effective in stemming the rate of wetlands loss, which has declined by some 80% in the 1990s over previous decades. But the United States still loses more than 50 000 wetland acres per year, according to the US Fish & Wildlife Service (Earthtalk 2008). Globally, 171 governments have now signed on to the 1971 Ramsar Convention on Wetlands, a framework for cooperation in the conservation and wise use of wetlands. (Note: the Ramsar Convention is named after the city in Iran where the agreement was first signed.) Some 2416 wetland sites – totaling almost 255 million hectares – have been protected as "Wetlands of International Importance" under the terms of the treaty (Ramsar 2021). Such agreements and programs strive to ensure that wetlands good and services are maintained for future generations.

Regarding populations in need, we have noted elsewhere that economic water scarcity may be alleviated by the construction of water infrastructure, most notably dams and reservoirs for flood control, water supply, and irrigation. These constructed works deliver goods to the benefiting populations and become ever more important in a rapidly changing climate. However, they should be built in a manner that also recognizes the benefits of downstream ecosystems, which rely upon natural flows of water, both baseflow and flooding events. Even the reuse of wastewater (discussed in Chapter 12) has a local benefit to an urban community but may be to the detriment of downstream ecosystems as a substantial portion of stream baseflow is now being diverted and reused. Wise water management will incorporate these benefits into infrastructure design and practice to the betterment of long-term water security.

*Foundations: Biomagnification, Ecological Risk, and Determination of Total Maximum Daily Loads

Biomagnification and ecological risk. When aquatic organisms live in water that is tainted with exogenous chemicals, they may consume and accumulate these in their organs or other tissues. Such a process is called bioconcentration. If the accumulation of a chemical also comes from the food source of an organism, the composite process is called bioaccumulation (Hemond and Fechner 2014).

In some cases, the organism may metabolize and excrete a chemical. But if the organism is itself a food source, such as fish, there is much concern that the chemical will bioaccumulate as it moves up the food chain. For example, indigenous peoples of Canada were found with high blood levels of highly toxic methylmercury after consuming fish from a reservoir that was formed over flooded land for a hydropower installation. The bioaccumulation of the pesticide DDT in birds of prey resulted in the premature collapse of bird eggshells, a phenomenon reported in the influential book by Rachel Carson, *Silent Spring* (Hemond and Fechner 2014).

A partitioning constant called a bioconcentration factor, *BCF*, can be defined as the ratio of a chemical's concentration in an organism to that chemical's concentration in the surrounding water. This factor may be based on observations for various chemicals and species, or it may be the prediction of a portioning model. Because the fish tissue has an organic percentage of oil and fat (called the lipid content), the various partitioning constants of the chemical may be used in estimating the *BCF*. For example, the K_{ow} is the octanol–water partitioning coefficient, which is a ratio of a chemical's equilibrium concentration in octanol ($C_7H_{15}CH_2OH$) to its concentration in water. Developed by the pharmaceutical industry, this coefficient approximates the portioning between body fat and water, a very useful piece of knowledge when prescribing medicinal doses. Large K_{ow} values indicate a greater likelihood of sorption into other phases, such as tissue fat, by virtue of being less soluble in water. Similarly, K_{oc} is a partitioning constant ratio of a chemical's concentration sorbed to organic carbon to its concentration in water at equilibrium.

An example regression equation for the observed bioconcentration in the fish species of rainbow trout, fathead minnow, and a few others is (Hemond and Fechner 2014):

$$\text{Log } BCF = 0.76 \times \log K_{ow} - 0.23 \tag{10.2}$$

For naphthalene ($C_{10}H_8$), a common polycyclic aromatic hydrocarbon (PAH), K_{ow}, is 3.36:

$$\text{Thus, } BCF = \log^{-1}\,[0.76 \times \log(3.36) - 0.23] = 1.48$$

When the concentration of the chemical naphthalene is known in the dilute aqueous habitat, then the chemical concentration in the fish species can be estimated from this ratio. This result can be used to estimate human health risk based on certain exposure scenarios. (See Chapter 6.)

Ecological risk assessment adopts the perspective of the aquatic species but may use similar tools. The goals of an assessment may be narrowly focused, such as to protect brook trout from a shortage of DO, or may be broader in focus, such as the restoration and maintenance of the diversity and abundance of fish species in an estuary bay. The assessment endpoint may be breeding success, fry survival rate, or adult return rate as in the case of spawning species who return to an upstream place of origin.

Although there are other physical stressors, water quality criteria have been developed by the US EPA to protect aquatic life from chemical stressors. A laboratory LC_{50} – the concentration that results in lethality for 50% of the sample tested – for at least eight species and a chronic no-observed-adverse-effect level (NOAEL) for at least three species will be used to record both acute and chronic effects of a chemical (US EPA 1998).

Uptake in aquatic species is unique to the situation. Uptake is usually estimated using the soluble concentration, its bioavailability by partitioning, and the internal dose calculation. This latter can be estimated using a pharmacokinetic model or by measuring standard biomarkers or tissue residue (US EPA 1998).

An ecological risk assessment adopts a pattern of source identification, primary and secondary (resultant) stressors, and primary and secondary effects on aquatic organisms. Figure 10.8 shows an assessment of a logging operation, which results in stream siltation and smothering of the habitat of benthic insects, necessary for aquatic species. The end result is the indirect effect of loss of insectivorous fish (US EPA 1998).

Determination of total maximum daily loads (TMDLs). The US EPA recognizes that surface waterbodies differ by the ways in which they have historically been used, from drinking water to recreation to environmental habitat alone, or to some combination of these. Clean water legislation is intended to preserve the water quality of each stream, river, lake, or wetlands for the designated use, as determined by the local governing body, usually a state or tribe. Water-quality standards have three components: (i) the designated use, (ii) water-quality

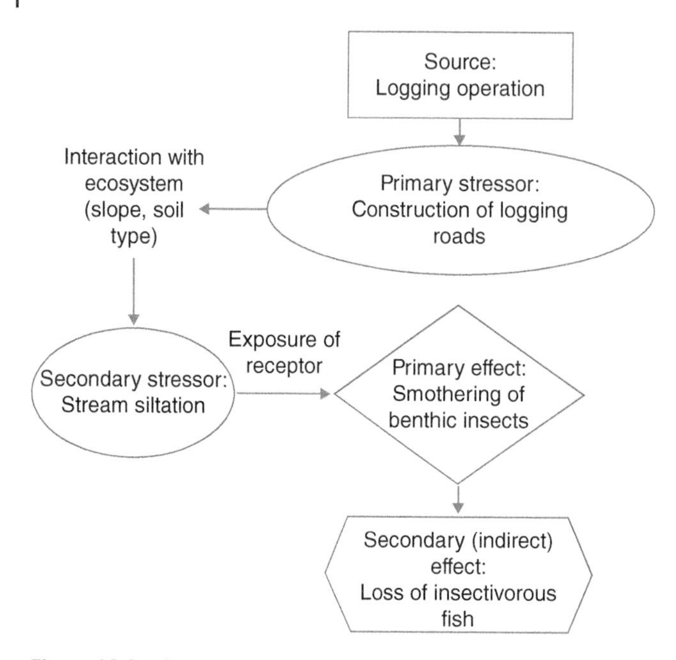

Figure 10.8 Conceptual model for ecological risk assessment using a logging operation as an example. Source: US EPA (1998).

criteria, and (iii) an antidegradation policy. These are normally expressed in concentration units, such as "mg/l," or special units, such as "NTU" for turbidity. Water-quality criteria work to protect the waterbody's designated use.

To this end, the Clean Water Act of 1972 (with its various amendments) requires the states and tribes to:

- Identify portions of surface waters that are impaired for their intended use,
- Prioritize these waters, and
- Develop TMDLs to restore the designated use for impaired waterbodies.

Once a designated use has been set for a waterbody, an assessment is done to determine whether the body is meeting its water-quality criteria for that intended use. If not, it is placed on an "action" list, also called a 303d listing (Figure 10.9).

The next step is to determine a **total maximum daily load** (TMDL) for various pollutants that have caused that waterbody to be so listed. Simply put, a TMDL is the maximum daily amount of a pollutant that a waterbody can receive and still meet water-quality standards for its intended use. It represents a regulatory requirement and will indicate the amount of pollution reduction needed for that particular pollutant (US EPA 2015b).

Pollutant sources are characterized as either:

- Point sources that receive a waste load allocation (WLA), such as an allowed source release from an National Pollutant Discharge Elimination System (NPDES) permit or
- Nonpoint sources that receive a load allocation (LA). These are generally caused from septic systems, forestry activities, and stormwater runoff across agricultural settings.

Permitted point sources may include an industrial site, wastewater-treatment facilities, or a concentrated animal feeding operation (CAFO). Nonpoint sources include all remaining

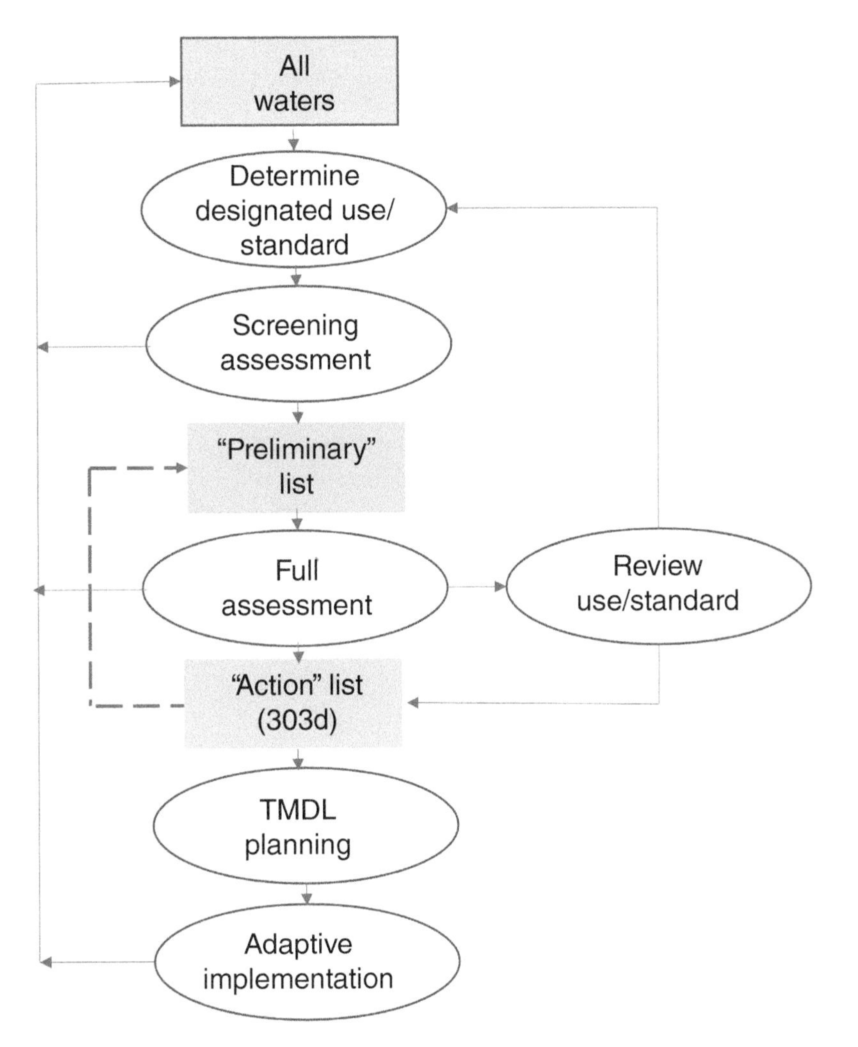

Figure 10.9 Steps of the designation, listing, and TMDL planning process. Source: NRC (2001). Permission given by National Academies Press: https://www.nap.edu/content/request-rights-and-permissions.

non-discreet sources of the pollutant in addition to natural background sources. A margin of safety (MOS) is added to provide a conservative cushion to cover fluctuating flow regimes as well as the uncertainty as to the effectiveness of the pollutant reductions in meeting the water-quality standards.

In mathematical terms:

$$\text{TMDL} = \Sigma\text{WLA} + \Sigma\text{LA} + \text{MOS} \tag{10.3}$$

In this equation, TMDL is the total maximum daily load (mass/time) for a particular contaminant, WLA is the sum of all waste load allocations (point sources), LA is the sum of load allocations (nonpoint sources and background concentrations), and MOS is the margin of safety.

In practice, TMDLs may be developed using a variety of techniques, from simple models (e.g. mass-balance calculations) to more complex modeling approaches that involve computer-driven software and extensive databases. The degree of analysis often depends upon the complexity of flow conditions, the waterbody type, and the type of pollutant that is causing impairment. For example, the determination of a TMDL for a lake with secure riparian filtration and receiving only point sources may be a simple spreadsheet process.

There are many mid-range and complex watershed-scale loading models available, including:

- GWLF (Generalized Watershed Loading Functions; Cornell University),
- SWMM (Storm Water Management Model; U.S. EPA), and
- STORM (Storage, Treatment, Overflow, Runoff Model; U.S. Army Corps of Engineers) (US EPA 1997).

These models will utilize land-use knowledge, soil characteristics, chemical fate and transport characteristics, and precipitation data to simulate storm runoff over the watershed.

It may seem obvious that the most difficult pollutant sources to quantify are those that are nonpoint and diffuse. We present here the "Simple Method" developed by the Metropolitan Washington Council of Governments (MW-COG) as a first-level quantification for diffuse storm runoff. The model is intended for small (less than 1 square mile) watersheds that have recently been stabilized through suburban development. The method is useful for estimating loading of phosphorus, nitrogen, COD, BOD, and most metals (US EPA 1997).

The following equation solves for pollutant loading, L, in pounds per year (lb/year):

$$L_i = P \times P_j \times R_v \times C \times A \times \frac{2.72}{12} \tag{10.4}$$

where:

L_i = pollutant loading (lb/year)
P = average annual rainfall (inches)
P_j = unitless correction factor to account for fraction of storms that produce runoff versus no runoff
R_v = runoff coefficient (dimensionless) = $0.05 + 0.009 \times \text{PI}$
PI = percent imperviousness
C = flow-weighted mean pollutant concentration (mg/l), also known as the event mean concentration (EMC)
A = area of development (acres)

For example, in the Washington, DC, metro a P_j correction factor is 0.9, indicating that 90% of rain events will produce runoff. The area receives an average of 43 in. of rain per year. If the impervious percentage is 50%, then the runoff coefficient is calculated as 0.0545. For a 600-acre watershed and a mean sediment concentration of 40 mg/l,

$$L_i = 43\frac{\text{in.}}{\text{year}} \times 0.9 \times 0.0545 \times 40\frac{\text{mg}}{\text{l}} \times 600 \times \frac{2.72}{12} = 11{,}470\ \text{lb/year}$$

The result is more than 5 tonnes of sediment over the course of a year. Note that the method is only accurate for small watersheds and the final conversion factor (2.72/12) assumes the use of proper units as given. This calculation is for nonpoint source loading only (LA), so any additional point source loadings (WLA) and an MOS would need to be included for a full TMDL calculation.

TMDL case study – Lake Thunderbird, Oklahoma. The designated uses of this 6000-acre lake in central Oklahoma are flood control, municipal water supply, recreation, and fish and wildlife propagation (OK DEQ 2013). Three major cities have stormwater permits to discharge into the lake via various streams, one of which enjoys the lake as a primary water supply, and about 99 600 people live in the lake's watershed (OK DEQ 2013). The state's Department of Environmental Quality (DEQ) had determined that the lake was not supporting two of its designated uses:

- Public water supply – excessive levels of *chlorophyll-a*
- Fish and wildlife propagation – excessive levels of turbidity and low DO

The Hydrological Simulation Program – FORTRAN (HSPF) model was used to estimate flow and pollutant loading for a one-year period and the Environmental Fluid Dynamics Code (EDFC) model was used to simulate water-quality conditions in the Lake for sediments, nitrogen, phosphorus, organic matter, DO, and *chlorophyll-a*. This lake model was calibrated with observed water-quality data from the same year. The linked calibrated lake model was then used to determine the water-quality response to various reduction scenarios in the watershed. The final set of TMDLs, thus, were the set of permit reductions for each of the three major municipal sources that would result in acceptable levels of turbidity, *chlorophyll-a*, and DO. An implicit MOS was applied by decreasing the water-quality targets for *chlorophyll-a* and turbidity by a factor of 10% (Table 10.8).

In order to reach the target goals, a municipality might employ one or more of the following structural and nonstructural best management practices (BMPs) (Table 10.9).

Table 10.8 TMDLs established for Lake Thunderbird, Oklahoma.

			WLA (total and individual source allocations)				
Water-quality constituent	**TMDL (kg/day)**	**LA (kg/day)**	**Total (kg/day)**	**City 1 (kg/day)**	**City 2 (kg/day)**	**City 3 (kg/day)**	**MOS (kg/day)**
Total nitrogen (TN)	807.7	21.3	786.4	205.1	319.4	261.8	Implicit
Total phosphorus (TP)	158.4	4.4	154.0	44.5	60.1	49.4	Implicit
CBOD	2480.8	57.4	2423.4	781.3	955.6	686.5	Implicit
Suspended solids (TSS)	76 951	2069	74 882	16 236	31 596	27 050	Implicit

Source: OK DEQ (2013).

Table 10.9 Listing of structural and nonstructural best management practices (BMPs) for waterbody improvement.

Structural BMPs include:	Nonstructural BMPs include:
• On-site retention basins	• Critical area plantings
• Grassed channels and waterways	• Xeriscaping (requiring little or no irrigation water)
• More-frequent street sweeping	• Constructed wetlands
• Porous pavement	• "Stoop and scoop" ordinances
• Diversion channels and berms	• Storm drain marketing and education
• Wetlands rehabilitation	• Mulching in new construction areas

The implementation of these practices should result in a "delisting" of the waterbody over time, restoring it to its designated uses.

End-of-Chapter Questions/Problems

10.1 Pick one example of an ecosystem service from the "Provision of Needs" category in Table 10.1. Do an Internet search and write a short essay (max. 500 words) about a recent example of a threat to this service and a response to protect it.

10.2 Pick one example of an ecosystem service from the "Regulation of Threats" category in Table 10.1. Do an Internet search and write a short essay (max. 500 words) about a recent example of a threat to this service and a response to protect it.

10.3 Pick one example of an ecosystem service from the "Support of Habitat" category in Table 10.1. Do an Internet search and write a short essay (max. 500 words) about a recent example of a threat to this service and a response to protect it.

10.4 Pick one example of an ecosystem service from the "Enhancement of Culture" category in Table 10.1. Do an Internet search and write a short essay (max. 500 words) about a recent example of a threat to this service and a response to protect it.

10.5 Why is it necessary and important to distinguish between surface water and GDEs? Give two reasons.

10.6 Imagine yourself as a rural villager in 1 of the 10 countries of Southeast Asia, e.g. Cambodia, Vietnam, or Thailand. Write about a typical day in your life and describe how your activities are dependent upon various ecosystems and their services. Give at least three examples.

10.7 Find one example of dam construction, which adversely affected the ecosystem and livelihoods of residents who lived downstream of the dam. Write a short essay (max. 500 words) describing this situation and the response (or needed response).

10.8 The Everglades wetlands of Florida are one of the national treasures of the United States. List three things that are being done now to protect these vital wetlands.

10.9 Describe the purpose of an Environmental Impact Statement (EIS). How is it used to ensure protection of ecosystem services? Give an example.

10.10 Show mathematically that this statement is true: "For very wide rivers, the hydraulic radius (R) can be approximated as the average depth of water."

10.11 Use Manning's equation (above) to estimate the average velocity for each monthly flow rate given in the table below. Use the following values: $n = 0.025$, $s = 0.12$, channel bottom width = 18 m with side slopes of 20° angles.

Average Monthly Flowrates (m^3/second) for the Clinch River, Tennessee

JAN	FEB	MAR	APR	MAY	JUN	JUL	AUG	SEP	OCT	NOV	DEC
64.3	74.5	82.7	80.4	47.6	32.3	23.6	21.0	16.7	16.4	32.0	63.4

Further Reading

Daily, G.C. (1997). *Nature's Services: Societal Dependence on Natural Ecosystems*. Island Press.
DeBarry, P.A. (2004). *Watersheds: Processes, Assessment and Management*, 1e. Hoboken, N.J: Wiley.
US EPA (2002). Guidelines for Reviewing TMDLs under Existing Regulations Issued in 1992.
Wescoat, J.L. and White, G.E. (2003). *Water for Life*, 1e. Cambridge; New York: Cambridge University Press.

References

Acreman, M. (2001). Ethical aspects of water and ecosystems. *Water Policy* 3 (3): 257–265. https://doi.org/10.1016/S1366-7017(01)00009-5.
Arthington, A.H., Bhaduri, A., Bunn, S.E. et al. (2018). The Brisbane declaration and global action agenda on environmental flows (2018). *Frontiers in Environmental Science* 6: https://doi.org/10.3389/fenvs.2018.00045.
Assessment, M.E. (2005). *Ecosystems and Human Well-Being: Opportunities and Challenges for Business and Industry*. Washington, D.C: World Resources Institute.
Black, M. (2016). *The Atlas of Water: Mapping the World's Most Critical Resource*, 3e. University of California Press.
Brander, L. and Schuyt, K. (2010). Benefits Transfer: The Economic Value of World's Wetlands. *TEEBcase*. 3.
Daily, G.C. (1997). *Nature's Services: Societal Dependence on Natural Ecosystems*. Island Press.
Darling, S.B. and Snyder, S.W. (2018). *Water Is … The Indispensability of Water in Society and Life*. World Scientific Publishing Co.
Davie, T. and Quinn, N.W. (2019). *Fundamentals of Hydrology*, 3e. London, New York: Routledge.
DeBarry, P.A. (2004). *Watersheds: Processes, Assessment and Management*, 1e. Hoboken, N.J: Wiley.
Earthtalk (2008). Wetlands Update--Has Preservation Had an Impact?. Scientific American https://www.scientificamerican.com/article/wetlands-update.
Gillilan, D.M. and Brown, T.C. (1997). *Instream Flow Protection: Seeking a Balance in Western Water Use*, 2e. Washington, D.C: Island Press.
Hemond, H.F. and Fechner, E.J. (2014). *Chemical Fate and Transport in the Environment*, 3e. Elsevier https://smile.amazon.com/Chemical-Transport-Environment-Harold-Hemond/dp/0123982561/ref=sr_1_1?dchild=1&keywords=Hemond+Chemical+Fate+and+Transport&qid=1610379244&s=books&sr=1-1.
Hornberger, G.M. and Perrone, D. (2019). *Water Resources: Science and Society*. Baltimore: Johns Hopkins University Press.

Indian, The Logical (2018). Good News: Population Of Ganga River Dolphins Rises To 33 From 22 In 2015 (October 20, 2018) https://thelogicalindian.com/news/ganga-river-dolphin.

Jowett, I. (1997). Instream flow methods: a comparison of approaches. *Regulated Rivers-Research & Management - REGUL RIVER* 13 (March): 115–127. https://doi.org/10.1002/(SICI)1099-1646(199703)13:23.0.CO;2-6.

Kreamer, D.K., Stevens, L.E., and Ledbetter, J.D. (2015). Groundwater dependent ecosystems: science, challenges and policy directions. In: *Groundwater: Hydrogeochemistry, Environmental Impacts, and Management Practices* (ed. S. Adelana) Water Resource Polanning, Developmant and Management, 205–230. New York: Nova Publishers.

NRC (2001). *Assessing the TMDL Approach to Water Quality Management*. National Academies Press.

OECD (1982). *Eutrophication of Waters; Monitoring, Assessment and Control*. https://agris.fao.org/agris-search/search.do?recordID=XF19830847706.

OK DEQ (2013). Lake Thunderbird Report for Nutrient, Turbidity, and Dissolved Oxygen TMDLs. Oklahoma Department of Environmental Quality.

O'Keeffe, J. and Le Quesne, T. (2009). Keeping Rivers Alive: A Primer on Environmental Flows and Their Assessment. WWF.

O'Keeffe, J, Kaushal, N., Bharati, L., and Smakhtin, V. (2012). Assessment of environmental flows for the upper ganga basin. WWF-India.

Ramsar (2021). Homepage | Ramsar. Ramsar.Org. https://www.ramsar.org.

Shore, R. (1999). Water in crisis - spotlight democratic republic of congo. The Water Project. https://thewaterproject.org/water-crisis/water-in-crisis-congo.

Tennant, D.L. (1976). Instream flow regimens for fish, wildlife, recreation and related environmental resources. *Fisheries* 1 (4): 6–10. https://doi.org/10.1577/1548-8446(1976)001<0006:IFRFFW>2.0.CO;2.

US EPA (1997). Compendium of Tools for Watershed Assessment and TMDL Development. EPA841-B-97-006. US Environmental Protection Agency.

US EPA (1998). Guidelines for Ecological Risk Assessment. EPA/630/R-95/002F.

US EPA, OW (2015a). What Is a Wetland? Overviews and Factsheets. US EPA (September 18, 2015). https://www.epa.gov/wetlands/what-wetland.

US EPA, OW (2015b). Overview of Total Maximum Daily Loads (TMDLs). Overviews and Factsheets. US EPA. September 29, 2015. https://www.epa.gov/tmdl/overview-total-maximum-daily-loads-tmdls.

US Forest Service (2021). Ecoregions of the United States | Rocky Mountain Research Station. https://www.fs.usda.gov/rmrs/ecoregions-united-states.

Wescoat, J.L. and White, G.E. (2003). *Water for Life*, 1e. Cambridge; New York: Cambridge University Press.

10a The Practice of Water Security: Start with the Children

Eric Stowe is the founder and CEO of Splash, an international nonprofit focused on water, sanitation, and health (WaSH) programs for children in urban settings (Figure 10a.1). After four years working as a relief coordinator for an international aid organization, Eric became frustrated at the lack of quality, transparency, and sustainability of the WaSH projects that were being implemented. He realized that he could not promise his donors that the infrastructure built with their money would still be functional after even one year of service.

Figure 10a.1 The work of Eric Stowe is aimed toward the happiness and well-being of children in urban settings. Source: Courtesy of Eric Stowe.

Along with his despair came the realization that there was a better way of doing development work. His WaSH approach is based on these primary elements:

- Target the urban populations. The density of urban populations makes it feasible to reach a lot of people (big impact) with less money. In addition, future global populations will mostly be living in big cities.
- Replicate the model of successful business. Why reinvent the wheel? Instead, Eric learned about clean water from multinational businesses such as McDonald's restaurants and Hilton Hotels. By working directly with a water filter manufacturer, he could access the same supply chains and use economies of scale to minimize costs.
- Succeed and depart. As Eric and his Splash team bring clean water and sanitation to schools and orphanages in major cities, they also prepare for the team's exit by handing over operations to local personnel. Indeed, Eric's motto is "Kill your

charity" by making its presence superfluous in the long run.

One of the key ingredients to the sustainability of Splash's work is its long-term commitment. Eric and his team commit to providing clean water at each site for 10 years during which time all maintenance, training, and replacement parts are provided. By the end of this 10-year period, the local ownership has developed relationships with local organizations and government such that Splash is no longer needed.

Splash is currently managing over 1500 projects that deliver clean water to children in eight countries (China, Cambodia, Bangladesh, Ethiopia, India, Nepal, Thailand, Vietnam), with plans to reach 100% of government schools in two major growth cities: Addis Ababa, Ethiopia, and Kolkata, India. Clean water alone will not prevent children from bacterial pathogens. Because they must also learn good hygiene habits, handwashing stations are an integral part of each Splash installation (Figure 10a.2). Water security for children in the urban setting comes from a nexus of clean drinking water, improved sanitation, and a regular practice of handwashing.

Figure 10a.2 Dr. Jim Chamberlain stands in front of a Splash hand-washing station in Addis Ababa. Source: Courtesy of author JFC.

Splash is proud of its transparency and offers a regular look at both its successes and failures. As one colleague stated: "If Splash can't prove it, they won't claim it. It's that simple. This measure of uncompromising integrity, spearheaded by Eric, is unfortunately somewhat uncommon in the charity world." (Breslin 2016). But of course, it all starts with the children.

Eric Stowe is the recipient of the 2017 OU International Water Prize

Reference

Breslin (ed.) (2016). *Nomination packet for Eric Stowe*. OU International Water Prize.

Part IV

Sustainable Responses and Solution

11

Conservation and Water-Use Efficiency

As the global population continues to rapidly increase, and as associated energy, agriculture, industrial, commercial, and institutional entities do likewise, our overtaxed water supply and infrastructure systems will require expensive water acquisitions and/or infrastructure upgrades. These upgrades will generate collateral impacts as well, including competition with the food and energy sectors, impacts on aquatic ecosystems, and increased greenhouse gas emissions. Conservation and water-use efficiency can help avoid the large capital costs of expanded water resource and supply systems while helping to mitigate the collateral impact of these systems. Further, water conservation can save money for both consumers and industry by improving efficiency and decreasing costs with payback periods of months to years. In this chapter, we discuss conservation and water-use efficiency in the residential, industrial–commercial–institutional (ICI), agricultural, and energy sectors. We will see that conservation and water-use efficiency should be highly prized and promoted in homes, businesses, industry, and beyond, to save money in both the near and long term while at the same time helping to protect and preserve our environment and society as we have come to know them.

Learning Objectives

Upon completion of this chapter, the student will be able to:

1. Understand the role of conservation and water-use efficiency in water security.
2. Be familiar with residential conservation and water-use efficiency measures.
3. Recognize industrial, commercial, and institutional conservation and water-use efficiency measures.
4. Be familiar with agricultural conservation and water-use efficiency measures.
5. Recognize water-saving approaches in the energy sector.
6. *Quantify the benefits of conservation and water-use efficiency.

11.1 Introduction

With population increase and industrial expansion, two main approaches exist to providing enough water – either increase the supply or decrease the demand. Historically,

Fundamentals of Water Security: Quantity, Quality, and Equity in a Changing Climate, First Edition.
Jim F. Chamberlain and David A. Sabatini.

Companion website: www.wiley.com/go/chamberlain/fundamentalsofwatersecurity

supply increase was the preferred approach. Increasing water supply often involves greater withdrawals from surface water or groundwater, dam and reservoir construction, or implementation of desalination or water reclamation plants, all of which are very costly. Thus, prior to embarking on such expensive projects, water utilities are increasingly assessing the management of their current water resources. Beyond economic benefits, improved efficiency can minimize impacts on aquatic resources, providing greater ecosystem protection, while liberating current water uses to meet additional needs.

As identifying additional water sources becomes more challenging and costly, curbing demand has become an increasingly attractive option. In response, the US EPA developed a best practices manual to assist water utilities and governmental agencies in evaluating water conservation and efficiency measures as an alternative to water-supply expansion. Entitled "Best Practices to Consider When Evaluating Water Conservation and Efficiency as an Alternative for Water Supply Expansion," the manual describes six major practices and includes suggested metrics to evaluate performance throughout the process:

1. The first step in the process is to conduct a *water audit*, which accounts for all inflows, outflows, incorporation into products, losses of water, etc. The report suggests use of American Water Works Association (AWWA) Free Water Audit Software, which facilitates conducting a water balance as well as generating performance indicators to aid in water-management decisions.
2. Recognizing that leakage accounts for major water loss, the second practice assesses water-loss minimization through *leakage control*. This assessment focuses on determining leakage relative to system characteristics, identifying the economic value of the water loss, and identifying measures to monitor and mitigate the loss.
3. *Water metering*, the third step, makes it possible to accurately close the water balance while helping to identify unrecognized uses and leaks. Water metering also helps to inform the water user of water-loss economics and thereby incentivize more efficient water-use practices.
4. Evaluating the viability of *water rate structures* to promote water conservation is a fourth step. It is imperative that cost structures incorporate future life cycle expenses of operating and maintaining a water utility while encouraging and rewarding conservation and water efficiency.
5. *End-use conservation and water efficiency analysis*, the fifth step, considers customer demographics (e.g. single- or multi-family residences, industrial, commercial, and institutional [ICI] facilities, etc.) to identify demand drivers as well as conservation and water-efficiency opportunities that can incentivize user adoption.
6. The final step is to develop a written *water plan* based on definitive and measurable goals for optimizing system performance and ensuring efficient water use, with timelines for implementation, assessment, and adaptation, as warranted (US EPA 2016).

Such measures can be both beneficial to the end user (via lower water costs) as well as the utility (via lessening the need for expensive water resource acquisition and treatment). At the same time, decreasing water use reduces utility revenue. While part of the revenue loss is offset by reduced production costs (chemical and electrical costs to treat the water), a portion of the revenue covers previous capital expenditures. This points to the common practice of water utilities to charge both a fixed fee, which helps recoup capital expenditure costs, and a per volume fee (see Chapters 15 and 16 for further discussion of rate structures).

Conservation and water efficiency involve both hardware (more-efficient appliances) and software (behavior modification) approaches, both of which will be discussed throughout this chapter. In a national online survey on water conservation, most of the 1000 respondents preferred curtailment or software measures (e.g. taking shorter showers, turning off the water while brushing teeth) rather than hardware improvements (e.g. replacing toilets, retrofitting washers). By answering questions on 17 different activities, these participants underestimated water use by a factor of two overall, with the underestimates being largest for high water-use activities such as toilets and washing machines (Attari 2014). Thus, it is important to consider both hardware and software when pursuing conservation and water-use efficiency.

With this background, we will now look at water-use efficiency as it relates to residential, ICI, agricultural, and energy sectors. Commonalities and overlap do exist between these sectors. For example, bathrooms occur in residential as well as ICI facilities, landscape irrigation happens in agricultural and residential settings, and energy production occurs in some industrial units. Nonetheless, unique aspects occur in some sectors, such as unique hospital or industrial equipment. Similarly, incentives for conservation and water-use efficiency adoption have common endpoints and motivations – cost savings, sustainability, and reduced environmental impacts.

11.2 Residential Settings in Developed Countries

By one estimate, the average US household uses approximately 310 liters (82 gal) per person per day (gpcd) for indoor and outdoor uses (Dieter et al. 2018). As a comparison, the World Health Organization estimates that 50–100 l (13–26 gal) per day are needed to provide adequate water for drinking, cooking, cleaning, and hygiene in households across the globe (Howard and Bartram 2003). The reason for the disparities become evident as we consider the various water-using appliances that are common in industrialized households that have ready access to abundant water.

11.2.1 Interior Home Water Usage

Residential efforts to increase conservation and water-use efficiency should be targeted to major water uses within the home. Water conservation is especially important in developed countries where household water use is significant. In rural villages of developing countries, the opposite challenge may be encountered. Given that villagers often have to walk a kilometer or more multiple times a day to retrieve and carry water in five-gallon containers, they are so oriented to conserving water that when it becomes more available the challenge is to encourage the use of this precious commodity for personal hygiene, such as hand washing.

Figure 11.1 shows that in the developed world toilets, faucets, showerheads, and the clothes washers are the top four users of water in an average household, together accounting for 80% of indoor household water use in the United States (Deoreo 2016). The remaining water is accounted for in leaks (13%) with baths, dishwasher, and other accounting for the remainder of household use. This average household "water audit" helps identify where conservation and water-use efficiency efforts can have the greatest impact within the home.

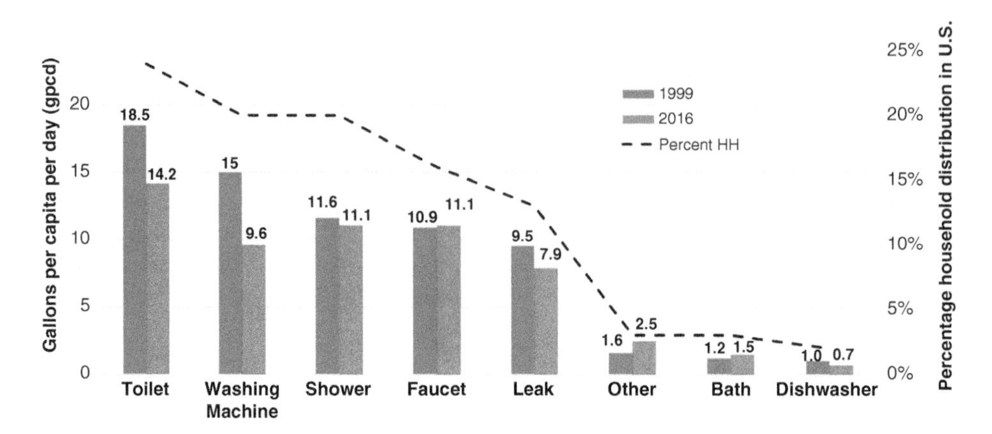

Figure 11.1 US household water-use distribution in 1999 (quantity) and 2016 (quantity and percentage). gpcd, gallons per capita-day. Source: Original, adapted from Deoreo (2016).

The Water Research Foundation (WRF) has documented progress in reducing water use in many of these eight categories (Deoreo 2016). Figure 11.1 shows the reductions in per capita water use for household water use from 1999 to 2016, demonstrating that toilets and clothes washing account for the greatest reductions. The report also documented that overall household water use decreased by 22% between 1999 and 2016 (Deoreo 2016).

Clothes washing machines. One might ask "How did the clothes washer category achieve such a significant decrease?" Much of this improvement was due to the increased use of front loading (25–30 gal per load) versus top loading (30–40 gal per load) clothes washers. (Vickers 2001). This reduced water use results from the agitation and water-contact approaches in the two types of washers. In the traditional top-loading machine, the clothes are submerged in the water and the agitation occurs through a vertical shaft agitator, while in the front-loading machine the agitation results from the clothes tumbling as the drum rotates around a horizontal axis and the water sloshes through the clothes as they tumble (Figure 11.2a). Thus, maintaining adequate water/clothes contact requires less water in the front-loading, horizontal-axis than the top-loading, vertical-axis machines. Further, manufacturers maintain that the greater agitation and water contact in the front-loading machines improves the cleaning process (Vickers 2001).

On a household level, it may be easier to transition to more water-efficient devices when equipment replacement is necessary. The purchase of a new clothes washer is a "hardware" (machine) approach to improving water-use efficiency. There are additional "software" (human behavior) approaches to promote water conservation in clothes washing. For example, the user should operate the washer with full loads, choose settings to promote load-appropriate water use (when the machine allows), pretreat stains to avoid rewashing, use shorter wash cycles for lightly soiled clothes, and check hoses regularly for potential leaks (Vickers 2001).

Toilets. Increased toilet efficiency accounted for the next largest improvement; between 1999 and 2016, the average toilet flush volume decreased by 29%, from 3.65 gal per flush (gpf) to 2.6 gpf (Deoreo 2016). While the number of toilet flushes per day remained the same (5.0), the use of low-flush toilets increased from just 5% in 1999 to 37% in 2016. This improvement can be attributed to regulatory requirements for new and replacement construction to use dual-flush toilets that improve efficiency by allowing low water flush for liquids (urine) and

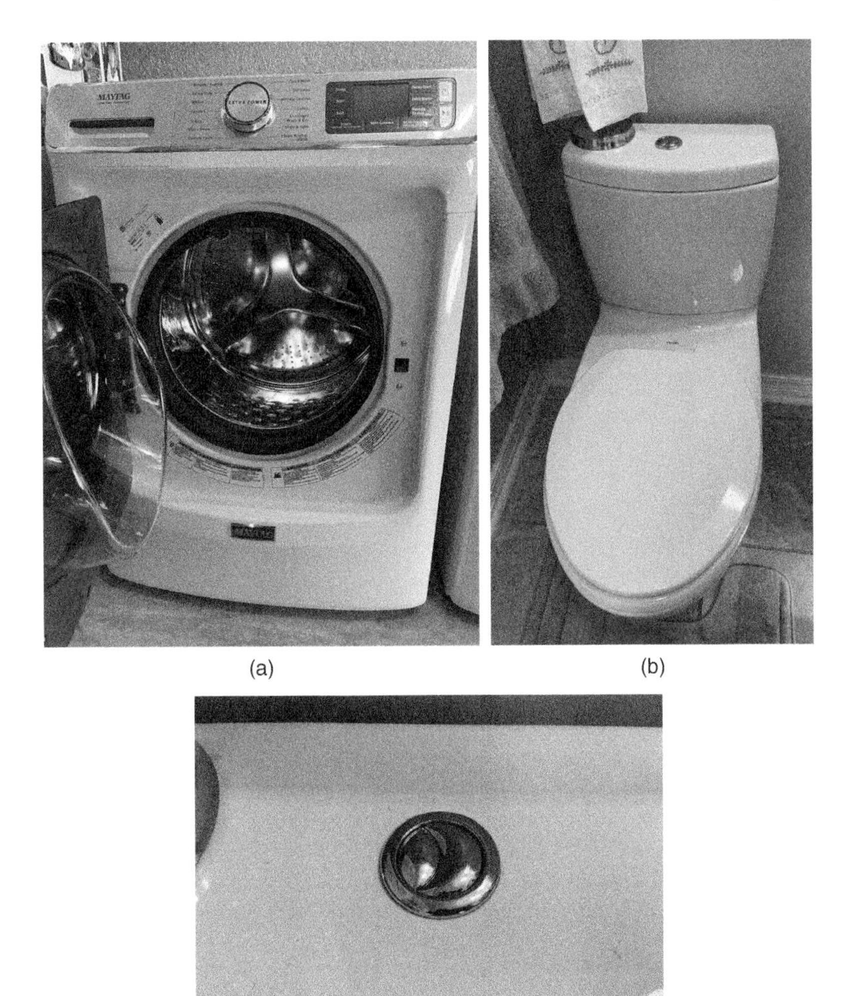

(a) (b) (c)

Figure 11.2 (a) New front-loading washer, (b) dual-flush toilet, and (c) dual-flush mechanism (left side only for liquid and both for solid flushing). Source: Photos courtesy of author, David Sabatini.

high-volume flush for solid feces (Figure 11.2b, c). Operational suggestions for improving toilet water efficiency are to fix leaks immediately and to not use the toilet for waste disposal (e.g. flushing of coffee grounds).

Showers and faucets. Looking again at Figure 11.1, an obvious question might be "why was there not more improvement in showers and faucets in this time period?" These devices are less expensive and are replaced more frequently than toilets and washing machines and likely had been more widely updated by the study period. Certain communities have even provided complementary water-saving showerheads to their customers as the cost was more than offset by not having to expand their water supply (Vickers 2001). US standards for water-conserving household devices were established in 1994 – mandating decreased flow faucets/aerators and showerheads (2.2 gpm at 60 psi) for new construction and purchases. The traveler notices a difference in showerheads when traveling outside the United States and Europe to places where conservation showerheads are yet to be adopted and shower water flow is much higher.

In addition to updating with low-flow devices (hardware), water use can be improved by operational changes (software), for example:

- Take shorter showers or taking a shower in place of a bath. An eight-minute shower at 2.5 gpm uses 20 gal of water while even a small bathtub can use 30–40 gal (Vickers 2001).
- Turn off faucet when not in use.
- Check for drips.

Dishwashers. The WRF study reported that 84% of the study homes had automatic dishwashers in 2016 and that the average water volume per dishwasher load had decreased by 39% from 1999 to 2016 (Deoreo 2016). For example, considering a sink with 4 gal of wash water and a limited five minutes of rinse water at 2.5 gpm (12.5 gal), washing the dishes in the sink uses over twice to almost three times more water than an automatic dishwasher. My wife and I (DAS) like this calculation!

The following operational (software) recommendations can also improve efficiency:

- Only operate the automatic dishwasher when full,
- Scrape off food rather than water rinsing,
- Limit prerinsing before loading dishwasher,
- Use shorter wash cycles for less-soiled dishes, and
- When hand dish washing, use a water-filled sink for washing and limit rinse water time.

Leaks. Figure 11.1 shows that the average daily per capita leakage decreased 17%, a sign of definite progress. Different techniques for managing leaks include active leakage control, pipeline and asset management, speed and quality of repairs, and pressure management. Pressure management recognizes the fact that leakage in the distribution system is proportional to the pressure in the system. Thus, the water utility may reduce water pressures at night when there are fewer users, thereby reducing system losses during this time. A clever solution!

Bathtubs. The increase of water in bathtub use may simply reflect habits or demographic shifts in the study. The fact that this normalized value is so much lower than showers demonstrates that showers are more common than baths. Hotels are increasingly providing rooms with a shower only, thereby responding to personal preferences and helping to conserve water.

Other household water uses. This final category accounts for other uses such as evaporative cooling, humidification, water softening (regeneration water), and other uncategorized indoor uses. Evaporative cooling is a cooling method that takes advantage of the heat of vaporization of water. As water evaporates, it cools (and humidifies) the surrounding air. Thus, in arid regions an evaporative cooler can be used by blowing air through a porous pad saturated with water. This process is less efficient, and less comfortable, in regions with high humidity (Vickers 2001). While this section focuses on residential applications, the same concept applies for ICI applications. From a hotel window in Bangkok, Thailand, one can see the evaporative coolers on top of tall adjacent buildings. In humid climates such as Thailand, engineers address the humidified air issue by cooling the air first to a lower temperature, thereby causing humidity to condense, and then raising to the target temperature with an acceptable relative humidity. This approach expands the applicability of evaporative cooling to areas of higher humidity for ICI entities and is widely used due to its cost savings (Vickers 2001).

11.2.2 Outdoor Home Water Usage

In certain geographic locations, and at certain times of the year, residential water use outside the home can far exceed that inside the home. Using a rule of thumb to provide 1 in. of irrigation water per week for exterior home landscaping, even a moderate size yard (e.g. 6000 ft^2) would require water use comparable to indoor use by a family of five or more, and this ratio increases for larger yards in drier climates. Manual watering often results in lower water use as individuals are less patient and less consistent than automated systems.

Lawn irrigation obviously varies significantly with climate. Vickers reports lawn irrigation values ranging from 25 gpcd (Seattle, Washington) to 100–180 gpcd in several cities in California and Arizona. The higher outdoor water uses in California and Arizona reflect the desire to have lush, green lawns in arid regions, which require higher irrigation rates, a consequence of high infiltration and evapotranspiration rates.

The US fascination with lush green lawns may be attributed to the gardens of early European immigrants and as an evolution from the post-Civil War period when home and landscape development exploded. The first US patent for a lawn mower was in 1868 and for a lawn sprinkler in 1871. The book "The Art of Beautifying Suburban Home Grounds" was first published in 1870, extolling the virtues of the lawn as the "home's velvet robe" (Vickers 2001). All this occurred in the shadows of the US Civil War and provided a welcome respite from the tolls of that tragic period.

Given that the home lawn is often interwoven into the fabric of American society, how does one promote conservation and water efficiency in lawn irrigation? Incorporation of water-wise landscaping principles is paramount. Vickers (2001) recommends these seven basic steps to aid in water-efficient irrigation:

- Lawn care *planning* includes an orientation that makes the best use of rainfall and sunlight.
- It is important to understand the *soil analysis* so that you can plan accordingly. The natural soil may require significant soil amendment (sand and compost) to improve it for gardening.
- Proper *plant selection* can reduce not only the water bill but also one's frustration level! This leads to the concept of *xeriscape*, which, contrary to misconceptions, does not imply "zero landscaping" but rather comes from the Greek word "*xeros*," meaning "dry." The xeriscape concept promotes use of climate-appropriate plants and grasses, often native to the region.
- A number of water utilities have even incentivized *turf reduction* by the incorporation of water-wise vegetation and improved water systems and schedules. Utilities find that they can save money in the long run by subsidizing the adoption of native ground cover ("cash for grass"), thereby avoiding the need to make expensive water acquisition and/or infrastructure expansion. While the installation of native landscaping may be capital (cost and labor) intensive, it leads to maintenance savings in terms of watering and mowing.
- *Efficient irrigation* relates to both hardware and software. Drip irrigation applies water directly to the plant's roots and shutoff rain gauges deactivate the watering system during rain events. Software examples include efficient water scheduling, turning off the system after a recent rain, checking and repairing leaks, minimizing loss of water on driveway or street, and the restriction of watering during early morning when evaporation is minimized. In addition, the usage of rainwater and gray water (non-kitchen sink, shower, wash water) where permitted can offset water use from the distribution system.
- Proper use of *mulches* can help minimize irrigation by reducing evaporation, cooling the soil, and controlling weeds, which compete with desirable plants for water.

- Finally, *appropriate maintenance* is needed in order to ensure the functions are all operating most water-efficiently.

The examples in Table 11.1 illustrate some of the water-conservation measures discussed above and provide a basis for future water-conservation programs at the community level.

Table 11.1 Examples of water-conservation measures for municipalities and select industries in the United States.

City	Water-conservation measures
Albuquerque, NM*	Water surcharge for water use >200% of winter rate, public education, landscaping guidelines for new construction, promotion of Xeriscape *Result: Overall water use decreased by 20%, peak use by 14%*
Ashland, Oregon*	To avoid expensive new water source, promoted leak detection/mitigation, conservation-based water rates, high-efficiency showerheads, and toilet retrofit/replacement *Result: Water savings of approximately 395 000 gal per day (16% of winter usage) as well as a reduction in wastewater volume*
Cary, NC*	In light of rapid population growth (doubling in 10 years), implemented public education program, landscape and irrigation codes, toilet flapper rebates, residential audits, conservation-motivating rate structure, landscape water budget, and water reclamation facility *Result: Estimated 4.6 MGD (million gallons per day) reduced water demand with a savings of approximately 16% in water production, allowing delay in two water plant expansions*
Houston, TX*	Conservation program that included education program, plumbing retrofits, audits, leak detection and repair, and increasing-block rate structure *Result: Predicted 7.3% reduction in water demand and $260 million in savings*
Miami, FL**	In anticipation of a water supply–demand gap, the Miami–Dade Water and Sewer Department (WASD) instituted a program focused on water-loss reduction, a water-conservation plan, and a county ordinance establishing a permanent two-day-a-week landscape irrigation restriction ordinance *Result: These initiatives reduced the water demand by 40 million gallons per day and eliminated the anticipated supply shortage and thus several costly near-term alternative water-supply projects*
Greensboro, NC**	Greensboro shifted from a (i) decreasing block rate structure to a flat rate for nonresidential customers and (ii) residential rate structure to an increasing rate structure *Result: Water consumption by the top 10 nonresidential customers shrank by 31%, or 429 million gallons per year in response to the rate structure change. For residential customers the water usage decreased by over 20%*
Rockland County, NY**	The Environmental Committee of Rockland County Legislature required County facilities to use WaterSense labeled fixtures when available and established a comprehensive water policy that includes water conservation, leak detection, a summer/winter water rate structure, conducting water audits, and implementing a rebate program for customers who install water-saving appliances and irrigation tools *Result: The construction of a proposed $130 million, 7.5 MGD desalination plant was no longer needed*

Source: Developed based on information in US EPA (2002)* and US EPA (2016)**.

11.3 Industrial, Commercial, and Institutional (ICI) Facilities

While water use trends are somewhat predictable across the domestic sector, much greater variability exists within the ICI sector. This is especially true across the industrial sector, where various industries have different water demands. Such variations can also occur within a given industry. For example, within the automotive industry, an engine plant will have different water demands from a fabrication plant and an assembly plant. The same can be true for the institution sector – e.g. two schools may differ greatly in water use depending on whether they use evaporative cooling, have a swimming pool, etc.

Further, while domestic water use does vary temporally (with usage peaks in the morning and early evening) and seasonally (summer water use may be double that in the winter), even greater temporal fluctuations exist within industries due to variations in production cycles, shift work, etc. Industries may also have dramatic seasonal variations based on input availability and/or product demand. Thus, with increased demands and variability comes increased opportunity for water conservation and efficiency.

Discussing water conservation in the ICI sector requires an understanding of the range of water use across the sector. While the *industrial* sector includes a wide range of processing and manufacturing facilities (e.g. chemicals, food, beverages, paper, steel, electronics, petroleum refining), water use can be summarized into four primary functions: heat transfer (cooling and heating), materials transfer (industrial processing), washing, and as an ingredient itself (Vickers 2001). For US industries, water for cooling, both ambient air and process/equipment cooling, has typically accounted for the largest single water use – accounting for as much as one-half of overall industrial water demand (Vickers 2001). While service and manufacturing facilities use significant amounts of water for washing and processing, food processing, and especially beverage facilities, use significant amounts of water by incorporation into the product (Vickers 2001).

Commercial or business customers typically provide a product or service, e.g. stores, commercial businesses, office buildings, restaurants, hotels, laundries, food stores, golf courses, car washes, and amusement parks (Vickers 2001). Commercial water use can be for bathrooms, employee kitchens, landscape irrigation, heating, and cooling. Cleaning and sanitation use are especially significant for hotels and motels (Vickers 2001).

Institutional water customers include public and government buildings, hospitals and health care facilities, prisons, military installations, passenger terminals, sports arenas, and places of worship. Institutional water customers often use water for cooling and heating, domestic purposes (e.g. restrooms), and landscape irrigation (Vickers 2001).

Figure 11.3 shows water-use distribution for five different types of commercial and institutional facilities, demonstrating that domestic/restroom use is dominant across all five facility types, while cooling, heating, and landscaping (irrigation) are significant for all but restaurants. Conversely, certain water uses are unique to individual facility types – kitchen/dishwashing for restaurants, laundry for hotels and hospitals, and medical equipment for hospitals. Knowing the dominant water uses for individual commercial and institutional entities is valuable when exploring water-conservation approaches.

WaterSense is an EPA program established to improve water conservation and efficiency (US EPA OW 2012). Table 11.2 summarizes the program's recommendations for water conservation and efficiency for the five types of facilities presented in Figure 11.3. While certain approaches are germane to all five facility types, other approaches are unique to individual sectors (e.g. water-efficient medical equipment for hospitals and water-efficient

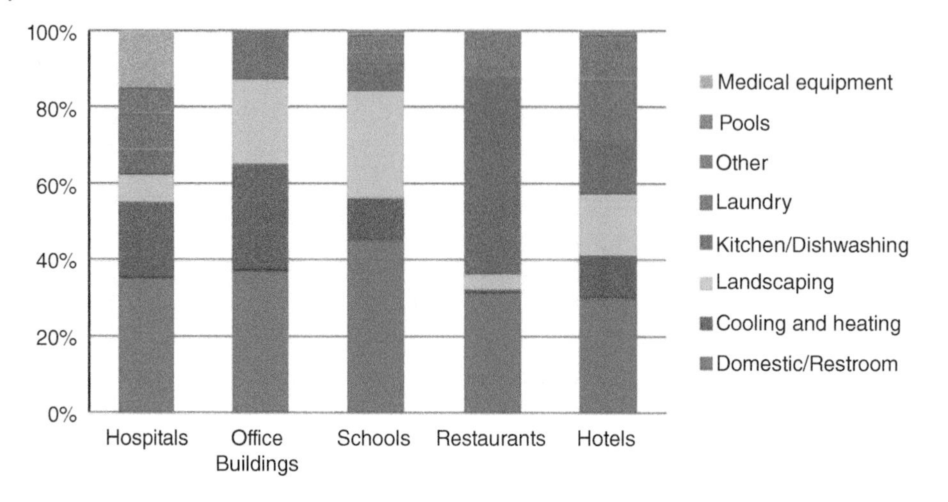

Figure 11.3 Water-use categories for select commercial and institutional facilities. Source: US EPA OW (2012).

Table 11.2 EPA's WaterSense recommended conservation measures for different commercial and institutional facilities.

Item/Facility	Office	School	Hotels	Restaurants	Hospitals
Water-management plan	x	x	x	x	x
Identify/reduce major water uses	x	x	x	x	x
Leak detection and mitigation	x	x	x	x	x
Upgrade bathroom fixtures/sensors	x	x	x	x	x
Water smart landscaping/irrigation	x	x	x		
Upgrade cafeteria/food service	x	x	x		x
Optimize water-based cooling system	x	x			x
Update laundry service equipment			x		x
Improve water efficiency of pools			x		
Upgrade dishwasher/ice machine/steam cookers				x	
Install water-efficient kitchen sprayers				x	
Update high water use medical equipment					x

Source: Developed based on information in US EPA (2017).

dishwashers/spraying systems for restaurant kitchens). The first step is to develop a water-management plan, which begins with a water audit by conducting a water balance around the facility and identifying uses and potential losses. Knowing the major uses and losses in a facility provides insight into where to focus water-conservation and -efficiency measures.

Having looked at the commercial and institutional portion of the ICI, we now turn to the industrial sector. Industries are much more diverse than commercial and institutional entities, and thus more difficult to simplify to average values. As such, we look at several examples to illustrate specific industries and potential water-saving approaches, as shown in Table 11.3.

Table 11.3 Examples of water-conservation measures for select industries including examples.

Industry	Potential water-conservation measures/Examples
Car wash	Water recycling, reduce number/flow in spray nozzles, automatic shutoffs, reduce rinse cycle, use high-pressure wands rather than hoses, monitor leaks
	Example: Seattle car wash reduces water use by 93% via *reclaimed wash water – payback period less than one year*
Hospital	Water-efficient sterilizing autoclaves – upgrade, application-appropriate size and operational time, automatic shutoff, water reuse, off-cycle shutdown
	Example: Massachusetts hospital recycles sterilizer cooling water, reducing water use by four million gallons per year (mgy) – resulting in 8% reduction in hospital water use – payback less than one year
Process wash/ rinse	Process water optimization, water reuse, rightsizing to application, install flow meters, timer-controlled shutoffs, recirculating wash, wash water treatment for reuse, upgrade nozzles, recycle cooling/heating water, avoid unnecessary dilution, pressure/flow reducers where appropriate
	Example: Chip manufacturer reduces water demand from 10 to 4 mgd using water efficiency and water recovery (RO [reverse osmosis]) techniques. NutraSweet plant uses seawater rather than freshwater for air scrubbers, reducing freshwater used by 150 mgy and $1 M per year savings from 20% reduction in water use. Metal finishing plant reuses rinse water and implements other water efficiency steps that produce 29% decrease in water use with a one-month payback period
Pulp, paper, and packaging	Recycling effluent water from pulp mills and paper-production machines (higher level treatment required for high quality paper/products), use of reclaimed water, water blending
	Example: Paperboard industry reduces water use by over 70% (savings of 1.3 mgd) by recycling and reuse of process water and installation of improved solid–liquid separation clarifier
Industrial cooling tower	Eliminate once-through cooling systems, use air-cooled systems where appropriate, install meters to monitor water use, reduce amount of makeup/bleed-off water required (e.g. continuous turbidity measurement), use advanced water treatment to enhance recycle
	Example: Ink manufacturer reduced water use by 80% by replacing once-through cooling system with a cooling tower – five-month payback; hospital replaces once-through water cooling system with air-cooled system – five-month payback; beverage plant reduces cooling tower makeup water by 75% by installing conductivity controller, saves 1.6 mgy – three-month payback

Source: Based on information in Vickers (2001).

Now let us consider two specific industries that have implemented water conservation and efficiency programs: Coca-Cola and Levi Strauss.

Coca-Cola established a goal of significantly reducing their water use in making their soda. As shown in Figure 11.4, since 2004 Coca-Cola's overall water use has decreased from 2.62 l per liter of product to 1.92 l system wide, with some of their bottling plants operating at as low as 1.4 l of water per liter of product. This progress makes Coca-Cola a leading water-conservation company in the beverage industry (Coca-Cola 2021) and has achieved significant water savings. This progress has been realized by making significant investments in new technologies and operating procedures that replace or reduce water use in their

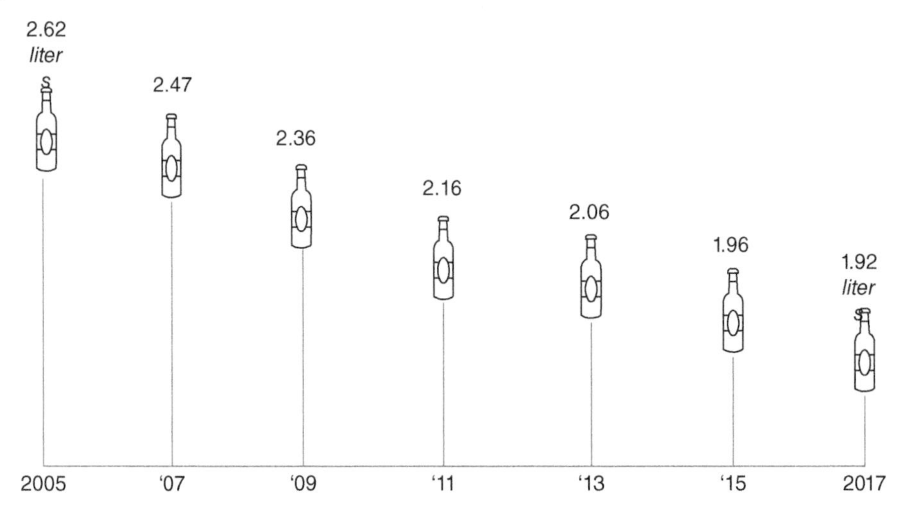

Figure 11.4 Water use per liter of Coca-Cola product (Coca-Cola 2018). Source: Original drawing – JFC.

manufacturing operations based on a system-wide water budget. Looking beyond their facilities, Coca-Cola determined that 80% of their total product water footprint comes from their agricultural ingredient supply chain, leading them to engage with their agricultural ingredient suppliers to initiate more water-sustainable agricultural practices with an initial focus on beet sugar for Coca-Cola and orange juice production (Coca-Cola 2021).

Levi Strauss' water-use distribution for the life cycle of their products is as follows: fiber (68%), consumer care (23%), fabric production (6%), packaging (2%), and cut, sew and finish (1%). Recognizing that over two-thirds of their water footprint comes from their input of cotton fiber, they were motivated to join the Better Cotton Initiative (BCI), which educates cotton farmers on how they can optimize their yields per unit water use. Beyond BCI, the Levi's® Wellthread™ collection further improves Levi's water footprint by testing sources of recycle fiber content and other innovative materials (Levi Strauss & Co. 2019a).

Further, recognizing that nearly one quarter of their product's life cycle impact occurs during the consumer care phase, Levi began implementing a Water Less Care Tag program. The company began sewing into every pair of their jeans a Care Tag for the Planet that carries a simple water-conservation message to encourage consumers to reduce the amount of water used to care for their Levi's. The tag reads "*Care for our planet: wash less, wash cold, line dry, donate or recycle.*" (Levi Strauss & Co. 2019a). Having addressed the two largest water uses in their product life cycle, Levi next focused on improving their manufacturing process at all stages of fabric development and garment finishing. The Water<Less program, launched in 2011, focused on implementing a series of technical innovations that saved water including the use of low-water washing machines and water-recycling systems (Levi Strauss & Co. 2019a).

What is the result? Currently, Levi uses more than 20 water-saving finishing techniques and shares their approach with other apparel companies to inspire industry-wide progress. Levi's Water<Less process can reduce up to 96% of the water normally used in denim finishing, the final stage in making of a pair of jeans. So far, Levi reports having saved more than three billion liters and recycled more than 1.5 billion liters of water through their Water<Less program, with more than 2/3 of all Levi's products being made with Water<Less techniques

(Levi Strauss & Co. 2019b). Levi now shares their water-saving approaches and markets their products under this Water<Less brand promoting their corporate focus on water conservation, efficiency, and sustainability (Figure 11.5).

Since a water-management plan is germane to all water-conservation approaches, it is helpful to provide a general outline for conducting a water audit. In her book, Vickers (2001) suggests a seven-step approach to developing/monitoring a water-management plan as summarized in Table 11.4.

Relative to estimating the payback period for water-conservation expenditures (Step 5), Vickers suggests that most ICI water users are willing to implement water-efficiency measures with a simple payback period of two years or less. Longer payback periods may still

Figure 11.5 *Water<Less* Levi's at a local clothing store. Source: Photos courtesy of author David Sabatini.

Table 11.4 Seven steps to developing, implementing, and monitoring an ICI water-management plan.

Category	Description
1. Buy-in	Management support is critical to success of the process
2. On-site inventory	Quantify water uses throughout facility, close the water balance
3. Water-related costs	Determine all water-related costs – purchase, treatment, disposal
4. Water efficiency	Identify all potential water-saving devices/approaches for facility
5. Payback periods	Determine simple (capital cost/savings per year) and life cycle (time value of money) payback periods
6. Implement action plan	Prepare written plan with goals, methods, projected savings, and monitoring plan
7. Evaluate progress	Monitor results and report on progress/recommended changes

Source: Based on information in Vickers (2001).

be attractive but require use of life cycle approaches incorporating time value of money. It is important that a written plan be developed with specific objectives as well as methods for monitoring progress and updating the plan accordingly.

11.4 Agricultural Sector

As discussed throughout the book, agriculture is the single greatest water user, accounting for over 70% of US consumptive water use. Seventeen western US states account for three-quarters of irrigated cropland (Figure 11.6), with 52% of irrigation water coming from surface water and 48% from local and regional aquifers. Increasing water security challenges are placing greater pressure on agricultural irrigation. Improved irrigation practices are critical as irrigated farms account for 40% of total US agricultural production, are three times more productive than dryland farming, and account for 55% of all crop products sold in the United States (Schaible and Aillery 2012).

While the use of gravity irrigation is declining in the United States (Figure 11.7), at least half of irrigated cropland still lacks water-management practices capable of conserving the most water (Schaible and Aillery 2012). Globally, 85–90% of irrigation systems are gravity-based. Thus, improved irrigation systems can increase water-use efficiencies (percent of applied water that directly benefits crop growth) by directing more of the water to the crop. For example, traditional gravity irrigation systems, such as field flooding, have been reported to be only 40–50% efficient but can be improved to 55–70% with use of furrows. In contrast, center pivot pressurized sprinkler systems are 70–85% efficient and drip irrigation is 80–95% efficient (Vickers 2001). Of course, there is a trade-off in increased energy usage.

Adopting more-efficient irrigation techniques is hindered by the fact that most on-farm irrigation investment is financed privately. Less than 10% of farmers report access to public financial assistance programs promoting use of higher efficiency systems. In addition, less than 10% of US irrigated farms used soil- or plant-moisture sensing devices or computer-based crop-growth simulation models leave significant room for improved water-use efficiencies (Schaible and Aillery 2012). High-efficiency systems are most

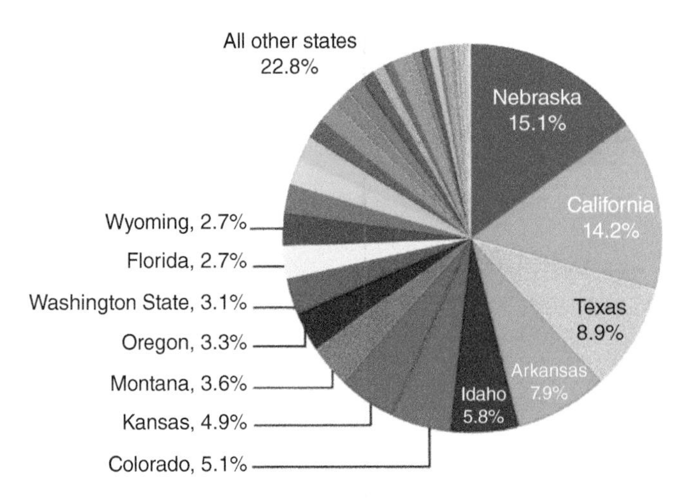

Figure 11.6 State shares of total US irrigated acres, 2007. Source: Schaible and Aillery (2012).

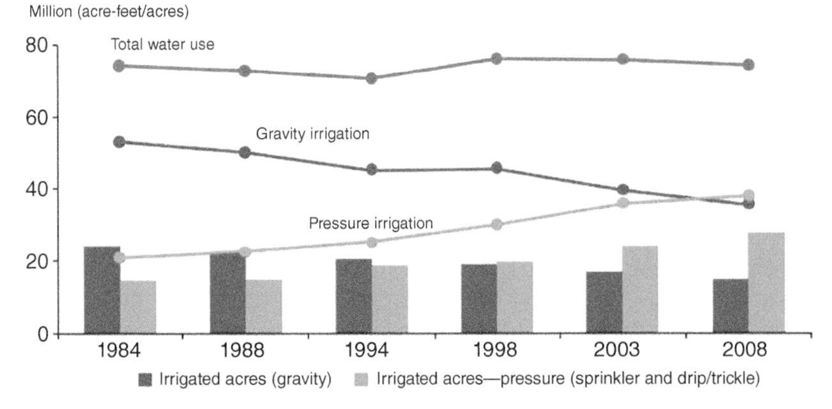

Figure 11.7 Trends in irrigated acres and water applied, for Western States, 1984–2008. Source: Schaible and Aillery (2012).

commonly applied to high-value crops such as fruits and vegetables, to areas requiring high rates of water application (e.g. sandy soils, uneven terrain) and in regions with higher water costs (Vickers 2001).

There are a range of methods for improving agricultural irrigation efficiency and crop productivity (per unit water), as summarized in Table 11.5. These approaches range from institutional to educational and financial in an effort to motivate practices and technology that will improve crop yield per unit water applied.

In addition to water-management improvements offered in Chapter 7, the following techniques can also be considered. The producer should:

- Minimize weeds and other non-cropped biomass that increases water demand,
- Minimize need for irrigation by improved rainfall capture and moisture retention (e.g. land grading, snow fences, use of mulch/crop residue to retain moisture),
- Capture field runoff and recycle, especially prior to entering saline water bodies rendering it unsuitable for downgradient agriculture use, and
- Optimize water-application rates to maximize economic return even if below maximum crop yield (factor in water costs to overall economic model) (Schaible and Aillery 2012).

Table 11.5 Options for improving agricultural irrigation efficiency.

Category	Options
Institutional	Conservation plan, conservation coordinator, policies, and penalties
Educational	Water audits, workshops, training materials, newsletters, internet
Financial	Conservation-oriented pricing, water marketing, low-interest loans, rebates
Managerial	Water metering, soil-moisture monitoring, irrigation scheduling, conservation tillage, tailwater reuse
Technical	Laser-graded land leveling, minimization of water loss – center pivot and drip irrigation
Agronomical	Rainwater harvesting, crop residue/conservation tillage, sequencing crops, use of drought-tolerant crops

Source: Based on Vickers (2001).

Being an early adopter of a new technique or technology involves both risk and opportunity. One way to reduce the risk and encourage adoption is to provide demonstration plots that illustrate the advantages of a new approach. Farmers are familiar with this concept as it is used to promote a new seed line. A demonstration field plot showing high yields can reduce the risk associated with trying this new seed. This approach, along with monetary incentives, may be the nudge needed to increase the adoption of water-efficient systems.

11.5 Energy Production

Water-use efficiency in the energy sector falls under the water-energy nexus (as well as the food–energy–water (FEW) nexus). Since water for thermoelectric cooling is the single largest water withdrawal in the United States, water efficiency in the cooling process is of utmost importance. As noted above, the usage of recirculating cooling towers and dry (air) cooling towers is more water-efficient than the traditional once-through cooling tower approach (Rodriguez et al. 2013).

Additional methods for improving water efficiency in the energy sector are discussed in Chapter 8. Building on these methods, a later report set out five specific energy–water security "Grand Challenges" to be achieved by 2030, as shown in Table 11.6 (US DOE 2020).

Progress in meeting these challenges will go a long way in the progress toward water security.

11.6 Conclusion

Given the continuing population increase and associated industrial expansion, water conservation and efficiency are increasingly important to ward off expensive water supply and infrastructure expenses. Further, water conservation can save money for consumers, institutions, and industries by improving efficiency and decreasing costs.

In this chapter, we have discussed conservation and water-use efficiency in the residential, ICI, agricultural, and energy sectors. Beyond the economic benefits, the efficient use of current water sources can have the added benefit of minimizing impacts on aquatic resources, providing greater ecosystem protection, and liberating current water uses to meet additional needs. Further, wise water conservation can reduce energy use needed to transport, treat, and utilize water. Thus, conservation and water-use efficiency should be a continuing priority in homes, businesses, institutions, and industry in the industrialized nations.

Table 11.6 Five Department of Energy (DOE) Water Security Grand Challenge goals to be achieved by 2030.

Goal
Goal 1: Launch desalination technologies that deliver cost-competitive clean water
Goal 2: Transform the energy sector's produced water from a waste to a resource
Goal 3: Achieve near-zero water impact for new thermoelectric power plants and significantly lower freshwater-use intensity within the existing fleet
Goal 4: Double resource recovery from municipal wastewater
Goal 5: Develop small, modular energy–water systems for urban, rural, tribal, national security, and disaster response settings

Source: US DOE (2020).

*Foundations: Water-Conservation Calculations

In this section, we cover some basic calculations to identify water and cost savings associated with select household water-conservation steps, namely toilets, showerheads, faucets, clothes washers, and lawn irrigation. While these calculations will focus on household conservation, the calculation approaches are germane to other sectors.

We will begin with *toilets*. Vickers (2001) reports toilet flush volumes for three time periods as shown in Table 11.7. Using an average of 5.1 flushes per day per person and 2.64 people per household, water usage in the post-90s period can be calculated:

$$\begin{aligned}\text{Gallons/day} &= 1.6\ \text{gal/flush} \times 5.1\ \text{flushes/day} - \text{person} \times 2.64\ \text{people/house}\\ &= 21.5\ \text{gallons/day}\end{aligned}$$

This same calculation is used for the higher-water-use toilets with results summarized in Table 11.7. The final column in the table shows water savings from the low-water use (1.6 gal/flush) to the 3.5 and 5.0 gal per flush models, respectively. At a rate of $4/kgal, that would correspond to $37–$67 in water savings for the household. Further, Vickers (2001) reports that up to one-fourth of toilets reportedly leak, with an average rate of 9.5 gpcd (gallons per person per day), or 25 gal/day and losing $36/year (see below), and in some cases as much as 100 gal/day:

$$\begin{aligned}\text{Leakage} &= 9.5\ \text{gpcd} \times 2.64\ \text{people/house} = 25\ \text{gpd} = 9000\ \text{gal/year@\$4/kgal}\\ &= \$36/\text{year}\end{aligned}$$

$$\text{Leakage} = 100\ \text{gal/day} = 36\,500\ \text{gal/year@\$4/kgal} = \$146/\text{year}$$

Thus, using a low-water toilet and immediately fixing leaks can not only conserve water but also save money!

Now turning our attention to *showerheads*, Table 11.8 shows that over time showerhead flowrates have decreased from 5 to 2.5 gpm resulting in reduced water use and cost savings. The calculation for 5.0 gpm and cost savings between 5 and 2.5 gpm are shown below and summarized in Table 11.8.

$$\begin{aligned}\text{Gal/year} &= 5.0\ \text{gpm} \times 5.3\ \text{min /person} \times 2.64\ \text{people/house}\\ &= 70.0\ \text{gal/day} \times 365\ \text{day/year} = 25\,540\ \text{gal/year}\end{aligned}$$

$$\begin{aligned}\text{Savings/year (2.5 versus 5 gpm)} &= 25.54 - 12.77\ \text{kgal/year}\\ &= 12.7\ \text{kgal/year} \times \$4/\text{kgal} = \$51.10/\text{year}\end{aligned}$$

Table 11.7 Water use for low-flush versus traditional toilets.

Time period	Gallons per flush	Flush/day	People/HH	Gallons/day	Gallons/year	Savings (gal/HH)
Post-1990s	1.6	5.1	2.64	21.5	7848	—
1980s–1990s	3.5	5.1	2.64	47.1	17 170	9322
1950s–1970s	5.0	5.1	2.64	67.3	24 570	16 722

HH, household.
Source: Using data from Vickers (2001).

Table 11.8 Water use for low-flow versus traditional showerheads. Assumption of 5.3 minutes per shower and 2.64 persons per HH.

Time	gal/min	gal/shwr	gal/day	svgs – gpd	gal/year	svgs – gpy	$ ($4/ 1 kgal)	svgs $/yr
Post late-1990s	2.5	13.25	35.0	—	12 770	—	51.1	—
1980s–1990s	3	15.9	42.0	7.0	15 320	2550	61.3	10.2
Pre-1980s	5	26.5	70.0	35.0	25 540	12 770	102	51.1

Source: Using data from Vickers (2001).

Given these savings in water costs, an older home with an original showerhead could recoup its cost in one year by replacing with a new lower flow showerhead as well as reducing water use!

Table 11.9 shows average water use per load in *clothes washing machines* over time. The calculations follow the same format as shown above. Once again, the water savings of low-water-use washing machines are evident.

Finally, let us look at *household lawn irrigation*. Nationwide the average lawn size is around 10 000 ft^2 with statewide averages ranging from a low of 6000 to >50 000 ft^2 (HomeAdvisor 2018). Assuming an average lawn size of 10 000 sq. ft and 1 in. of irrigation water per week, we calculate the gallons per day and week and kilogallons per month used. We will also determine the person equivalents, assuming 100 gpcd and dividing water use by this factor. This gives us a means of easy comparison as how many people would use that same amount of water in a day (see below). Thus, from Table 11.10 we see that irrigating an average size lawn at 1 in./week (27 kgal/month) would require the equivalent water use to that of 8.9 people. From a monetary perspective, at $4/kgal, this water use would cost $108/month. However, for communities with an increasing rate structure, this water may all fall in a higher fee category, say $6/kgal and thus cost $162/month, demonstrating the ability of the tiered rate approach to promote water conservation. Likewise, if the lawn is irrigated at a higher rate – say 1.5 in./week (either in a very dry climate or period, or due to watering at higher than 1 in./week), the water use would be 40.5 kgal/month, with costs of $162/month at $4/kgal versus $243/month at $6/kgal, further illustrating the utility of tiered water structure to incentivize water conservation:

$$\text{Gal week} = 10\,000\ \text{ft}^2 \times 1\ \text{in.} \times 1\ \text{ft}/12\ \text{in.} \times 7.48\ \text{gal}/\text{ft}^3 = 6230\ \text{gal}/\text{week} = 891\ \text{gal}/\text{day}$$

$$\text{Person equivalent} = (891\ \text{gal}/\text{day})/(100\ \text{gpcd}) = 8.9\ \text{people}$$

Table 11.9 Water use for low-water versus traditional clothes washing machines.

Time	gal/load	Loads/ cap-d	gpcd	svgs – gpd	gal/ month	svgs – gal/ month	gpy	svgs – gpy
Post late-1990s	27	0.37	26.4	—	804	—	9630	—
1990s	40	0.37	39.1	12.7	1192	387	14 260	4630
1980s	51	0.37	49.8	23.4	1519	715	18 180	8550
Pre-80s	56	0.37	54.7	28.3	1668	864	19 970	10 340

Source: Using data from Vickers (2001).

Table 11.10 Water use to apply 1 in. of irrigation water for different-sized lawns.

Area (ft^2)	gal/week	kgal/month	gal/day	Person equivalent
6000	3740	16.2	534	5.3
8600	5360	23.2	766	7.7
10 000	6230	27.0	891	8.9
15 000	9350	40.5	1340	13.4

As an aside, Tables 11.7–11.10 are spreadsheet outputs – demonstrating the benefit of using a spreadsheet over hand calculations. At the same time, hand calculations are beneficial and recommended in setting up the equations to be used in a spreadsheet. This will allow double-checking to make sure the equations were entered correctly.

End-of-Chapter Questions/Problems

11.1 What are three different methods you might consider for promoting water conservation in each of the following: (a) a domestic setting and (b) an industrial, commercial, and institutional setting.

11.2 Do a web search and pick a community in (a) the northeastern United States (higher rainfall) and (b) the southwestern United States (semiarid to arid) and discuss three to four household and industrial conservation measures that might be implemented.

11.3 Do a web search and discuss two industrial/manufacturing plants in terms of water-conservation activities they have implemented and resulting benefits.

11.4 Discuss four different approaches (e.g. in financial, educational, or policy areas) that could encourage farmers to use more water-efficient irrigation systems.

11.5 Do a web search and discuss two commercial/institutional facilities in terms of their water-conservation activities they have implemented and resulting benefits.

11.6 Develop a three-to-four-page water-conservation plan for a local industry, commercial or institutional facility.

11.7 Pick a community of 400 000 people or more and discuss a recent project they have implemented to greatly expand their water supply/infrastructure. (a) Discuss the relative costs of this expansion and how a conservation plan could help mitigate the need for this expansion. (b) Discuss why some communities can afford to provide monetary incentives to promote water conservation (e.g. free showerheads, free rain barrels, rebates for Xeriscape/turf replacement).

11.8 Consider a household with four family members. Estimate the water cost savings possible by updating from pre-80s to modern toilets, faucets, and showerheads. Assume a cost for water of $4 per thousand gallons (kgal).

11.9 Consider a city of 40 000 households, where 40% of the households still have low-water-efficiency toilets, showerheads, and clothes washing machines. Estimate the total amount of water that could be saved per day and per year by upgrading these to modern devices.

11.10 (a) Estimate the amount of lawn irrigation water saved each week if 40% of a 10 000 ft^2 lawn was converted to nonirrigated ground cover (assume 1 in. of irrigation water per week). (b) Estimate the yearly amount of water that could be saved if the irrigation system for the 10 000 ft^2 lawn was turned off 30% of the time between April 1 and September 30 (high irrigation months) due to sufficient rain in that period. (c) Estimate the cost savings of each assuming (i) non-tiered ($4/kgal) and (ii) tiered ($6/kgal) rates – assume the base water use (nonirrigation water) puts all of the irrigation water in the tiered rate of $6/kgal.

Further Reading

US EPA (2016). "Best Practices to Consider When Evaluating Water Conservation and Efficiency as an Alternative for Water Supply Expansion."

US EPA OW (2012). *WaterSense at Work: Best Management Practices for Commercial and Institutional Facilities*. US Environmental Protection Agency.

Vickers, A. (2001). *Handbook of Water Use and Conservation: Homes, Landscapes, Industries, Businesses, Farms*, 1e. United States: WaterPlow Press.

References

Attari, S.Z. (2014). Perceptions of water use. *Proceedings of the National Academy of Sciences* 111 (14): 5129–5134. https://doi.org/10.1073/pnas.1316402111.

Coca-Cola (2021). "Improving Our Water Efficiency – News & Articles." http://www.coca-colacompany.com/news/improving-our-water-efficiency.

Deoreo, W. (2016). *Residential End Uses of Water, Version 2*. Water Research Foundation https://www.waterrf.org/research/projects/residential-end-uses-water-version-2.

Dieter, Cheryl A., Maupin, M.A., Caldwell, R.R. et al. (2018). "Estimated Use of Water in the United States in 2015." USGS Numbered Series 1441. Circular. Reston, VA: U.S. Geological Survey. http://pubs.er.usgs.gov/publication/cir1441.

HomeAdvisor (2018). "The United States Ranked by Yard Size." *Home Improvement Tips & Advice from HomeAdvisor* (blog). September 25, 2018. https://www.homeadvisor.com/r/average-yard-size-by-state.

Howard, G. and Bartram, J. (2003). "WHO | Domestic Water Quantity, Service Level and Health." WHO/SDE/WSH/03.02. http://www.who.int/water_sanitation_health/diseases/wsh0302/en/index.html.

Levi Strauss & Co (2019a). "2025 Water Action Strategy."

Levi Strauss & Co (2019b). "How Levi's® Is Saving Water." Levi Strauss & Co. March 25, 2019. https://www.levistrauss.com/2019/03/25/world-water-day-2019-saving-h2o.

Rodriguez, D.J., Delgado, A., DeLaquil, P., and Sohns, A.A. (2013). *Thirsty Energy. Water Papers of the World Bank*. World Bank.

Schaible, G. and Aillery, M. (2012). "Water Conservation in Irrigated Agriculture: Trends and Challenges in the Face of Emerging Demands." USDA. http://www.ers.usda.gov/publications/pub-details/?pubid=44699.

US DOE (2020). "Water Security Grand Challenge." Energy.Gov. 2020. https://www.energy.gov/water-security-grand-challenge/water-security-grand-challenge.

US EPA (2002). "Cases in Water Conservation: How Efficiency Programs Help Water Utilities Save Water and Avoid Costs."

US EPA (2016). "Best Practices to Consider When Evaluating Water Conservation and Efficiency as an Alternative for Water Supply Expansion."

US EPA OW (2012). *WaterSense at Work: Best Management Practices for Commercial and Institutional Facilities*. US Environmental Protection Agency.

US EPA OW (2017). "Types of Facilities." Overviews and Factsheets. US EPA. January 16, 2017. https://www.epa.gov/watersense/types-facilities.

Vickers, A. (2001). *Handbook of Water Use and Conservation: Homes, Landscapes, Industries, Businesses, Farms*, 1e. Estados Unidos: WaterPlow Press.

12

Desalination and Water Reclamation/Reuse

When water demand exceeds supply, we often seek to increase the water supply. When new freshwater supplies are scarce or unavailable, two additional options are desalination and reclamation/reuse of wastewater. Given that over 97% of earth's water is saltwater, brackish or seawater desalination as a water supply is an attractive option. While initially thermal desalination was more common, recent advances in membrane desalination have improved the economics of this approach. Although coastal applications easily return the brine to the sea from which it came, inland desalination is often limited by brine management. Given the costs and brine challenges of desalination, water reclamation and reuse are receiving increased attention. Water reuse employs advanced treatment processes to further purify treated wastewater to an acceptable water quality for various nonpotable and potable uses, thereby providing a reliable, year-round resource that can sustain a community needing increased water security. Nonpotable water reuse, such as irrigation and industrial cooling, is widely practiced in many parts of the world. Potable water reuse requires a higher level of treatment and faces additional challenges but is increasingly being adopted in water-stressed regions. Despite challenges, water reuse continues to emerge as an important component of a water-security portfolio.

Learning Objectives

After completing this chapter, the student will be able to:

1. Describe the water security benefits of desalination and water reuse.
2. Understand the water characteristics critical to desalination and water-reuse systems.
3. Define the common types of desalination and water-reuse technologies.
4. Be familiar with the range of water-reuse applications and "fit-for-purpose" water-quality demands.
5. Understand the range of existing desalination and water-reuse systems and the associated costs associated with these systems.
6. Describe the technical and nontechnical challenges to desalination and water reuse.
7. *Perform design calculations for simple reverse osmosis systems.

Fundamentals of Water Security: Quantity, Quality, and Equity in a Changing Climate, First Edition.
Jim F. Chamberlain and David A. Sabatini.

Companion website: www.wiley.com/go/chamberlain/fundamentalsofwatersecurity

12.1 Introduction

Regions that are water stressed relative to freshwater may be water abundant with salty or brackish water. Given that over 97% of Earth's water is saltwater and recognizing that coastal areas and the Middle East are highly populated, water desalination is increasingly attractive for increasing water security. The Ancient Mariner lamented the reality of "water, water everywhere, nor any drop to drink" as ancient seafaring ships encountered the challenge of being unable to carry sufficient drinking water for their journeys. Distillation was thus practiced and refined by both sailors and land-based communities needing this source of drinking water.

Beyond saltwater, water-stressed regions can utilize wastewater reuse to expand their water supply. The United States alone discharges 34 billion gallons of wastewater per day into the environment (US EPA 2020). By displacing freshwater withdrawals, water reuse can greatly improve water security. Indeed, we may have reached the point where "society no longer has the luxury of using water only once" (Asano et al. 2007). Just as air is "recycled" locally, it may be time for water to be considered in a similar way. Where water reuse has been implemented, reclaimed water provides a dependable and versatile supply that is relatively drought tolerant. Improved technologies, an accumulation of industry experience, and the ongoing migration of global populations to large cities all point to reclaimed water as a potential water supply.

In this chapter, we provide a brief history of desalination and water reuse, define different types of source waters, explain the methods of desalination and advanced wastewater treatment, describe the challenges of each, and consider case studies of both methods of water-supply augmentation. In the Foundations section, we introduce the design of one important method of desalination – **reverse osmosis** (RO).

12.2 History and Modern Trends

Both desalination and water reuse have historical precedence and modern developments, which may contribute greatly to water security. Specific technologies introduced in this section will be discussed further in the chapter.

12.2.1 Desalination

Early desalination efforts were based on distillation. In 350 BCE, Aristotle wrote in *Meteorologica* that "Saltwater when it turns into vapour becomes sweet and the vapour does not form saltwater again when it condenses." (Kumar et al. 2017). This is the very definition of distillation. Heating water induces evaporation with the subsequent condensation captured as distilled water. This practice, referred to as **thermal distillation**, was used on ancient ships that could not afford to carry sufficient freshwater for extended voyages. With the advent of boiler-propelled ships, distillation of seawater was utilized to generate makeup water for the boilers. Then in the early 1900s commercial desalination plants were installed, with the first municipal seawater distillation plants installed in the late 1950s (Greenlee et al. 2009). The most common thermal distillation technologies (discussed further in the chapter) involve water evaporation by heat addition and, in select cases, are vacuum assisted.

The water vapor is then condensed to recover distilled water, free of salt and most impurities (Kumar et al. 2017).

Membrane technologies for desalination were introduced in the mid-twentieth century. While prior to the 1980s thermal technologies accounted for over 80% of the global desalination production, by the year 2000 thermal and membrane-based technologies held an equal share (Figure 12.1) (Jones et al. 2019). RO uses a membrane system under pressure that produces a desalinated fresh water. Membrane advances reduced energy requirements by an order of magnitude leading to increased adoption of RO (Kumar et al. 2017). In addition, the footprint (land area) of RO systems significantly decreased with the introduction of spiral-wound and hollow fiber technologies (mid-1960s), composite materials (1970s), and low-pressure nanofiltration (NF) systems (1980s). Currently, RO systems account for 69% of desalination systems, thermal distillation for 25%, and NF/**electrodialysis** (ED) for 5% (Jones et al. 2019).

While still technically viable, as reflected in numerous full-scale implementations, thermal desalination is energy intensive. Figure 12.1 also shows the energy consumption per unit volume of water for thermal distillation technologies as well as RO-based desalination. These relative economics demonstrate why RO systems are now eclipsing thermal distillation in terms of new installations. Currently, there are nearly 16 000 operating desalination facilities worldwide with a combined production capacity of over 95 million m^3/day (Jones et al. 2019). Figure 12.2 provides a global overview of where these facilities are located including their feedwater, technology, and capacity.

Not surprisingly, seawater facilities are located predominantly along the coast while inland plants tend to treat brackish, river or waste (reuse) water. Since 69% of all facilities

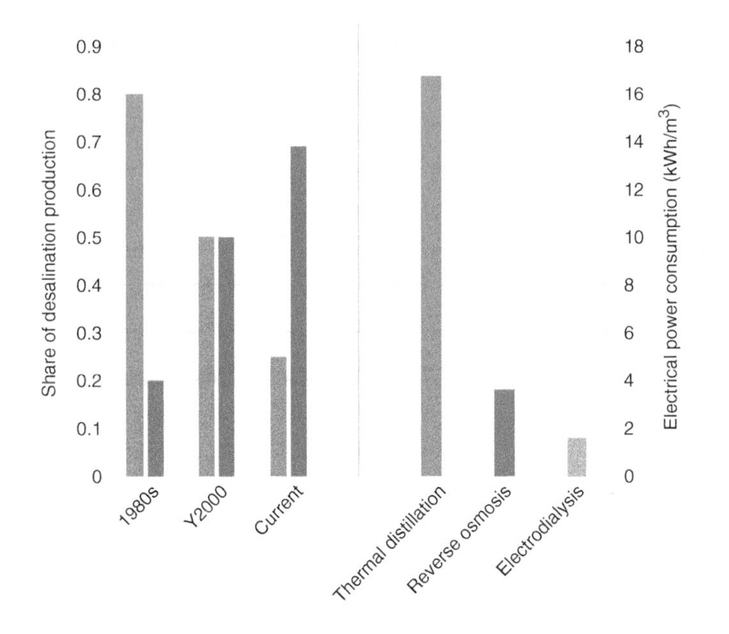

Figure 12.1 (Left) Share of global desalination production over time and (right) mean of electrical power consumption for thermal distillation (orange), reverse osmosis (blue), and electrodialysis desalination (gray) strategies. Source: Adapted from Kumar et al. (2017) and Al-Karaghouli and Kazmerski (2013).

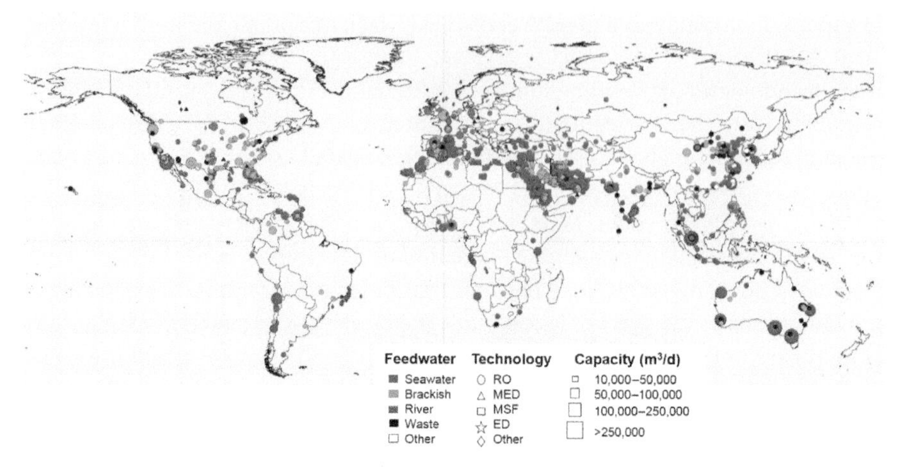

Figure 12.2 Global distribution of desalination plants based on feedwater, technology, and capacity for plants >10 000 m^3/d. Source: Jones et al. (2019)/with permission of Elsevier.

are RO plants, these dominate the landscape, while thermal facilities exist predominantly in the Middle East and the United States. Spatially, we see that every continent is host to desalination facilities, with water-stressed regions (e.g. the Middle East) having more and larger facilities. The Middle East – North Africa (MENA) region accounts for 47.5% of the global desalination capacity, followed by East Asia and the Pacific (18%) and North America (12%). The high costs of desalination facilities are prohibitive for lower income countries while high-income countries account for more than 70% of global capacity (Jones et al. 2019).

12.2.2 Water Reclamation and Reuse

In some form, water has been reused for the last 5000 years (Angelakis et al. 2018). Already in prehistoric times, domestic sewage was deemed suitable for irrigation and aquaculture. Around larger cities, such as Athens and Rome, ancient rural dwellers found a ready source of nutrient-rich water for fertilizing and watering their crops. Sewage that was thrown into the streets of Rome was washed into large sewer trenches that also collected stormwater before discharging outside the city walls (Angelakis et al. 2018). These practices received very little improvement during the Middle Ages when the emphasis was on "war, not civilization or sanitation" (Schladweiler 2020). In the western hemisphere, the *Chinampas* of Meso-America were a kind of "floating gardens" suspended over wetlands, shallow lakes, manure, and compost, demonstrating the recognized value of such materials (Angelakis et al. 2018). With new water-supply challenges, water reclamation is being rediscovered and improved upon.

Water may be characterized by levels of water quality (Figure 12.3). Potable water is defined as water suitable for direct human consumption as drinking water. But apart from the manufacturing of electronics, many commercial and domestic applications do not require such high-quality water. For example, power plants can be cooled with water of a higher salinity content than acceptable for drinking water or other applications (Webber 2016). Nonedible farm crops can be irrigated with water that is nutrient-rich and farm

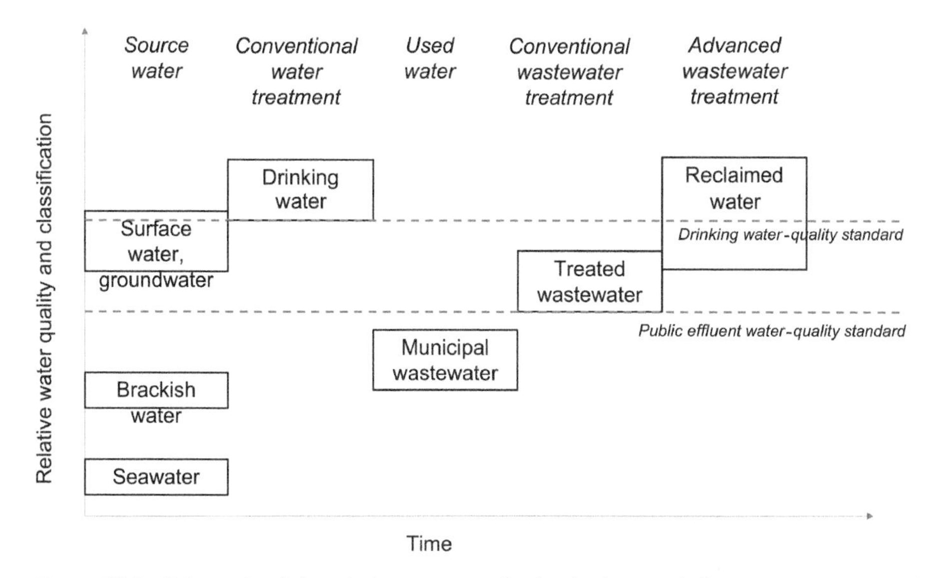

Figure 12.3 Schematic of the relative water quality levels that result for water treatment and conventional wastewater treatment. Source: Modified from Asano et al. (2007).

animals can drink water of a higher total dissolved solids (TDS) content than acceptable for human consumption (Artiola et al. 2014). Figure 12.3 is a schematic of relative water-quality levels that are reached in the various stages of water source, treatment, disposal, and retreatment. While conventional water treatment produces water that is suitable for drinking, conventional wastewater treatment produces water that can be safely released back into the environment. Reclaimed water is wastewater that undergoes more advanced treatment. This water will range widely in quality, depending on its intended use. But it will always be of equal or higher quality than conventionally treated wastewater in the United States, and for certain applications will be higher in quality than conventionally treated drinking water.

As we use the terms here, **water reclamation** is the appropriate treatment of wastewater (from municipal or industrial treatment plants) to the quality required for its intended use or discharge, while **water reuse** is the actual reappropriation of the reclaimed water for nonpotable or potable use (National Research Council [NRC] 2012). As cities grew in the age of European urbanization, water withdrawals from rivers increasingly included upstream discharges of wastewater. This **de facto** (unplanned) **water reuse** has been happening for centuries with little consideration or repulsion.

In addition to de facto water reuse, we will also discuss *nonpotable* and two types of *potable* reuse. Because the end uses of nonpotable water reuse vary widely, the desired quality and level of treatment will also vary. In a few cases, such as the usage of domestic wastewater for agriculture and landscape irrigation, the treatment may be less intense in order to take advantage of helpful nutrients in the wastewater. Several different approaches exist to implement potable reuse with the goal of safe, dependable, and desirable drinking water.

Windhoek, the capital of Namibia (population: 300 000+), began potable water reuse in 1968. The local annual rainfall is approximately 370 mm, the evaporation rate is nearly 10 times this amount, and long periods of dry spells are common (Lahnsteiner and

Lempert 2007). This water-stressed situation led Windhoek to install the world's first potable water-reuse plant to further treat domestic wastewater before blending with source water and distribution to the public. Several plant upgrades have occurred over the years, with the newest plant designed to treat a 50 : 50 mix of wastewater effluent and surface water, although the current mix is mostly effluent (90%) because the lake water quality has deteriorated dramatically (Lahnsteiner and Lempert 2007).

As expected, current water-reuse systems are more common in water-stressed regions. While countries such as Ireland, Finland, and Austria practice little or no water reuse in their water-rich landscapes, Italy and Spain are expanding their water-reuse practices, and Israel leads the way with over 85% of its domestic sewage being reused (Siegel 2017). Israel has shown that the challenges of water stress can be met with a combination of appropriate water pricing, encouragement of a water-conserving culture, and extensive water reuse.

Water reclamation and reuse have several advantages in the portfolio of water-supply options, including the following:

- Water reuse is already happening on a de facto basis.
- Reclaimed water quality may already be sufficient for many nonpotable applications, such as crop and landscape irrigation and power plant cooling and cleaning.
- Nutrient-rich reclaimed water may replace fertilizer inputs when applied to crops.
- When treatment is tailored to the end user, it will not be overtreated, thus saving valuable energy and resources.
- Municipal wastewater is relatively consistent and reliable and more drought-tolerant than most water sources.
- Water reuse may be economically attractive versus water transfer or other options, especially when considering the real costs of freshwater acquisition (UNESCO 2017).

Based on these advantages, we will further discuss current water-reuse practices and regulations, and discuss potable and nonpotable water-reuse applications, including appropriate technologies. Finally, we consider costs and challenges in implementing water reclamation and reuse.

12.3 Water-Quality Characteristics and Standards

In this section, we highlight the water-quality characteristics and standards that are relevant to the practices of desalination and water reuse.

12.3.1 Desalination

Source water composition is important when considering desalination. Table 12.1 lists several categories of water based on their TDS and energy requirements associated with membrane treatment. Whereas seawater has a TDS range of 15 000–50 000 mg/l (35 000 mg/l is often listed), brackish water is less concentrated, in the range of 1500–15 000 mg/l. Brine, often encountered in oil production, is even more concentrated than seawater, often >100 000 mg/l. The range in seawater concentrations reflects global variability. The RO energy requirement

Table 12.1 Summary of water source, salinity (total dissolved salts), and separation energy.

Water source	Concentration of total dissolved solids (mg/l)	Energy for RO separation[a)] (kWh/1000 m^3)	Osmotic pressure[b)] (psi @ 25 °C)
Produced brine	>100 000		1000
Seawater	15 000–50 000	670	400
Tampa Bay	24 500		
Atlantic Ocean/Mediterranean Sea	39 000		
Persian Gulf	43 500		
Dead Sea (no outlet)	275 000		
Brackish water	1500–15 000	170	35
River water	500–3000	40	15
Pristine water	<500	<10	<0.5
Wastewater (untreated)	250–1000	10	5
Wastewater (treated)	500–700	10	5

a) Based on average TDS of the range.
b) An approximate mid-range value.
Source: Kumar et al. (2017) and Greenlee et al. (2009).

increases with increasing salinity due to the increasing work required to separate water and salts, denoted by the osmotic pressure. Treated wastewater is included in Table 12.1 as RO and other advanced treatment processes are increasingly utilized in water-reuse applications.

Osmotic pressure is the force that causes water to move across a membrane as it seeks to balance the concentration of solutes (such as salts) in solution. As will be discussed below, reverse osmosis requires application of a pressure greater than the osmotic pressure. The osmotic pressure of seawater with an NaCl level of 35 000 mg/l is nearly 400 psi. Thus, RO treatment requires an applied pressure much greater than this to force water to flow in the opposite direction (thus the term "reverse" osmosis). An approximate rule of thumb is that every 100 mg/l of TDS produces 1 psi of osmotic pressure (Montgomery 1985). Brackish water with 3500 mg/l TDS is thus estimated to have an osmotic pressure of 35 psi – an order of magnitude lower than seawater! Thus, RO energy requirements are much lower for brackish water than for seawater, as indicated in Table 12.1. While TDS is a valuable water property, the knowledge of individual TDS components is also important. For example, sucrose (sugar), unlike salt, is nonionic and thus has minimal osmotic pressure.

The presence of salts not only impacts osmotic pressure but also changes the boiling point of water, and thus the thermal distillation process. Boiling point elevation results from water's "colligative" property, i.e. the concentration and ratio of solutes in solution. By their very presence in solution, dissolved salts reduce the ability of water molecules to leave the solution phase and go into the gaseous phase. Thus, the evaporation of saltwater requires more energy than fresh water, thereby impacting the economics of thermal distillation.

12.3.2 Water Reuse

Water quality is an especially important consideration in potable reuse applications. Although regulations vary from state to state, some generalizations may be made about the treatment levels for various nonpotable reuse categories. In cases where unrestricted public exposure is expected, such as urban irrigation of city parks, a higher quality of water is required. Where public exposure is not likely, such as in industrial cooling applications, a lower level of quality is acceptable. Treatment goals seek to protect human health by requiring a certain removal level of bacterial, viral, or protozoan pathogens (US EPA 2012). Turbidity may harbor pathogens making turbidity removal equally important. The product of the disinfectant residual chlorine and the contact time (discussed below) may also be given a specified minimum to assure pathogen disinfection. As more knowledge and experience become available, these regulations and standards will further be refined with specific targets for desired water usage.

A water-quality standard is a guideline, whereas a regulation has the force of law. As of 2017, 17 US states had potable water-reuse standards. A sample of these standards is given in Table 12.2, demonstrating both similarities and differences in state requirements. Inactivation of *Escherichia coli* and other pathogens is typically expressed in terms of "log reduction" of pathogens. So, a result of $N/N_0 = 10/100$ would express a 90% removal rate, or 1-log (10^{-1} remaining) removal, where N_0 is the original number of pathogens and N is the resulting number. Removal rate of 99% is a 2-log (10^{-2} remaining) removal, 99.9% is 3-log (10^{-3} remaining) removal, etc.

Some US states have had to issue project-specific permits for direct potable reuse (DPR) as an emergency solution. Wichita Falls, TX, received such a permit after an extended drought greatly reduced water levels in its water-supply lake. The city placed 12 miles of 32-in. diameter high-density polyethylene (HDPE) pipe on city right-of-way to carry treated wastewater to a holding pond at the water-treatment plant (Figure 12.4). The water then passed through microfiltration (MF) and RO units before being blended 50–50 with lake water before entering the water-treatment plant with its multiple disinfection processes (Force 2016).

Table 12.2 Sample of potable water-reuse standards for three US states in 2017.

	North Carolina	Virginia	Florida
Escherichia coli	≥log 6 reduction	≥11 colonies/100 ml (monthly geometric mean)	No detectable total coliforms/100 ml
BOD_5	≤5 mg/l (monthly avg)	≤10 mg/l (monthly avg)	$CBOD_5$ < 20 mg/l
TSS (total suspended solids)	≤5 mg/l (monthly avg)	—	< 5 mg/l
Turbidity	≤5 NTU	≤2 NTU	—
Other	—	Total residual chlorine <1 mg/l	Primary and secondary drinking water standards

Source: Adapted from US EPA (2018).

Figure 12.4 Wichita Falls (Texas) pipeline is laid on the ground to provide reclaimed water to residents during a time of drought. Source: Photo courtesy of DAS.

12.4 Fundamentals of Water Treatment

Water-treatment technologies are paramount to successful implementation of both desalination and water reuse.

12.4.1 Desalination Technologies

As we have noted, desalination techniques include both thermal and membrane-based desalination processes. Various thermal distillation processes seek to optimize energy input requirements while membrane processes take advantage of advanced membrane materials to decrease capital and operational costs thereby improving the economics. These processes are now described in greater detail.

Thermal desalination. As described above, thermal desalination induces water vaporization and subsequent condensation and collection of desalinated water. Thermal desalination systems are more common in the Middle East due to (i) dual-purpose cogeneration facilities that integrate steam from power generation along with the water distillation and (ii) low-energy costs in the Middle East (NRC et al. 2008). In the United States, thermal processes are commonly found in industrial applications requiring high-quality product water ($\leq$25 ppm TDS).

Three major thermal processes have been commercialized: multistage flash (MSF), multiple-effect distillation (MED), and mechanical vapor compression (MVC) (NRC et al. 2008). MSF and MED processes rely on both thermal energy (typically steam) and electrical energy, making their co-location with power generation attractive. To further improve the economics, thermal processes are configured to capture and reuse the substantial amount of energy required – 2300 kJ/kg of water under normal conditions – to evaporate water (NRC et al. 2008).

- *MSF distillation*, a forced circulation process, is the most common of the desalination technologies. MSF uses a series of chambers (stages) with successively lower temperature and pressure to enhance water evaporation from the feedwater. The water vapor is condensed using feedwater-filled tubes, which concurrently warms the feedwater (energy recovery from the heat of condensation) and reduces the overall energy requirement (NRC et al. 2008). With an increasing number of MSF stages the energy recovery improves, with the optimal design existing at the intersection of capital costs (constructing more stages) and operation costs (increasing energy recovery decreases the operating expenses).
- *MED* is a thin-film evaporation approach in which the vapor produced by one chamber condenses in the next chamber, which has a lower temperature and pressure, thereby improving condensation. The MED system reduces pumping requirements and thus power consumption compared to MSF.
- *MVC* is an evaporative process in which vapor is mechanically compressed with released heat subsequently used to help evaporate feedwater. MVC plants tend to be small (<3000 m^3/day) and are most frequently used when cooling water and low-cost steam are not readily available. Two attractive attributes of MVC systems are seen in that they (i) can operate at very high salt concentrations and (ii) do not require cooling water (NRC et al. 2008).

Membrane-based desalination. Membrane desalination utilizes a semipermeable membrane (permeable to water but not ions) and elevated pressures to desalinate the water. The water is forced through the membranes of various pore sizes, which allow the water to pass while impeding the flow of various contaminants (Figure 12.5). The four

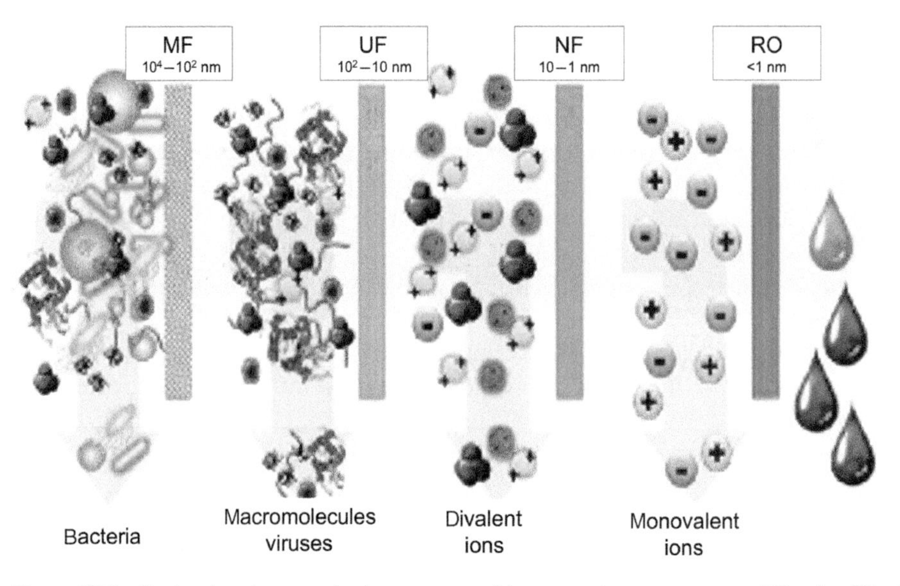

Figure 12.5 Contaminant removal using pressure-driven membrane processes. MF, microfiltration; UF, ultrafiltration; NF, nanofiltration; RO, reverse osmosis. Source: Adapted from National Research Council (NRC) (2012).

basic types of membrane systems are shown – MF, ultrafiltration (UF), NF, and RO, in order of decreasing pore size. MF targets particles passing sand filtration, including bacteria, UF removes viruses and macromolecules (e.g. large proteins), NF removes divalent ions and smaller molecules (e.g. Ca, Mg, sugars) while RO targets monovalent ions such as sodium and chloride. Thus, regarding desalination, RO is often the membrane of choice, while NF can help remove divalent ions as a precursor to RO (Greenlee et al. 2009).

RO is unique among membranes in its reliance on osmosis. Fick's law of diffusion states that more-concentrated solutions diffuse to less-concentrated regions until concentrations are equalized. In Figure 12.6a, a semipermeable membrane separates the seawater solutes from the deionized (DI) water. Without the membrane, seawater ions would diffuse into the DI water until the concentration becomes uniform. Since the semipermeable membrane prevents ions from diffusing in this manner, water molecules migrate from the dilute to concentrated salt side to reduce the concentration gradient (difference) across the membrane. This process is known as *osmosis*. Since Figure 12.6a is a closed system, the migrating water increases the pressure on the left side of the membrane resulting in osmotic pressure (ΔP).

What then is *reverse* osmosis? In Figure 12.6b, a pressure greater than the osmotic pressure ($P + \Delta P$) is applied to the concentrated salt solution side of the membrane. This total pressure counteracts the osmotic pressure and adds additional pressure, causing water to migrate from the salty side to the fresh side of the membrane. Thus, RO requires a pressure much greater than the osmotic pressure, explaining why seawater RO, with a great osmotic pressure than brackish water (Table 12.2), requires much higher application pressures than brackish water RO.

Electrodialysis is the other common form of membrane-based desalination. ED combines ion-selective membranes and an electrical potential gradient to separate ionic species from water. The anionic charged membranes contain functional groups that repulse anions migrating toward it while allowing cations to flow through. The cationic charged membrane does the opposite. In this way, every other chamber becomes depleted in ions while the alternate chambers are concentrated in ions (Montgomery 1985). As compared to RO and thermal

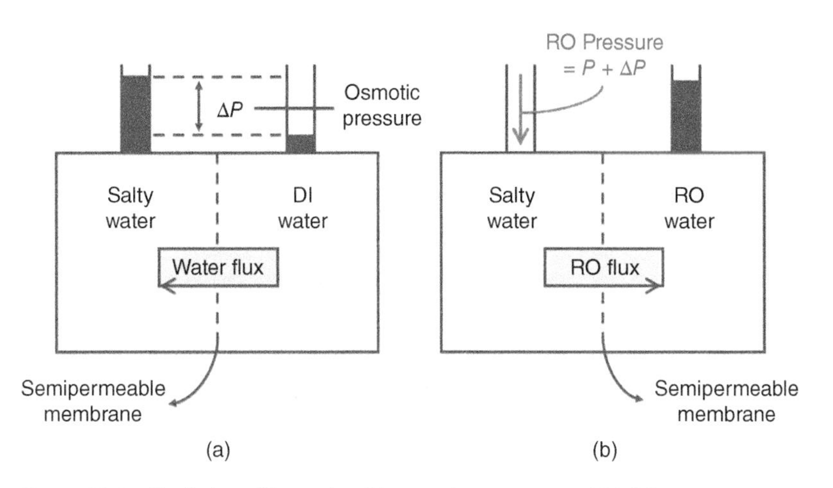

Figure 12.6 Depictions illustrating (a) osmotic pressure and (b) RO concepts.

processes, ED is limited to removal of ionic species but is more energy-efficient when applied to brackish water. Further, ED membranes tend to be chlorine-resistant, making them more robust for processing feedwaters with high organic matter levels that would otherwise foul RO membranes (Kumar et al. 2017).

Brine (concentrate) management. A major challenge for any desalination process is management and disposal of the concentrate stream, or brine. Depending on the source water, the brine salt concentration can be 2–10 times that of the feedwater. For example, seawater desalination may have a water-production efficiency of around 50%. In this case, half of the feedwater becomes product water (permeate) with virtually all the salt concentrated in the brine (retentate), thus doubling in concentration. Brackish water desalination efficiency can be 80–95%. For 90% efficiency, the concentrate concentration would be 10 times the concentration of the feedwater since the feedwater salt is retained and concentrated in 1/10 of the water.

The disposal options for this concentrated brine vary according to geography. Coastal seawater desalination has the luxury of returning the brine to the sea. Since almost half of current desalination operations are within 1 km of a coastline, and an additional 30% between 1 and 10 km of a coastline, direct disposal in the sea is a common choice (Jones et al. 2019). At the point of discharge, the brine is roughly twice the concentration of the seawater and can have a negative impact on the local ecosystem. Thus, brine introduction through a series of perforated pipes reduces shock concentrations and is preferred to end-of-pipe discharge (NRC et al. 2008; Greenlee et al. 2009).

For inland applications, potential brine management approaches are given in Table 12.3. Surface water discharges are limited to water bodies with elevated salinity values, where available. Sewer discharge requires that the salinity discharge will not negatively impact the sewer or treatment facility. Injection wells dispose of the brine in impaired subsurface formations but suffer from potential leakage into pristine zones overlying the target formation and, as is well known in Oklahoma, can produce increased seismicity. Evaporation ponds are prone to leakage and, along with land disposal, can severely damage local vegetation and greatly increase erosion. Relative costs for these disposal options demonstrate why surface water and sewer disposal are so attractive when viable. The brine concentrator (ZLD, or zero liquid discharge) in Table 12.3 utilizes evaporation and crystallization to virtually eliminate liquid brine discharge but at a high cost (see Morillo et al. 2014).

Table 12.3 Brine (concentrate) management options, costs, and critical considerations.

Disposal option	Cost ($/m^3)	Considerations
Direct discharge to surface water	0.03–0.30	Piping, pumping, outfall construction, proximity to ocean or suitable surface water
Discharge to sewer	0.30–0.66	Disposal rate, salinity, sewer capacity, fees
Deep well injection	0.33–2.64	Tubing diameter/depth, injection rate
Evaporation pond	1.18–10.04	Pond size/depth, evaporation rate, salt level
Brine concentrator (zero liquid discharge)	0.66–26.41	Disposal rate, energy costs, salinity

Source: Greenlee et al. (2009) and NRC et al. (2008).

12.4.2 Water-Reuse Technologies

Primary (settling) and secondary (biological) treatment of wastewater are the norm in almost 99% of wastewater treatment plants (WWTPs) in the United States, with about half also receiving more advanced treatment (NRC 2012). Most of the biodegradable organic matter and suspended solids are removed with conventional treatment, and indicator microorganisms are reduced by a 1–3 log removal (90–99.9% removal) (US EPA 2012). Primary and secondary treatment methods are discussed in any standard wastewater treatment text, such as Davis (2010).

Water reuse generally requires advanced treatment in addition to primary and secondary wastewater treatment. Table 12.4 illustrates various treatment processes used for a range of end uses and human exposure. For example, if water is to be reused for industrial cooling, groundwater recharge, or for nonfood crop irrigation, then secondary treatment and disinfection may be sufficient. Filtration and disinfection are required for water that has greater likelihood for human exposure. And advanced treatment – e.g. RO, advanced oxidation – is often required in the cases of indirect potable reuse (IPR) and DPR. In this section, we present a summary of advanced treatment technologies followed by a discussion of disinfection options, engineered natural processes, and treatment trains.

Table 12.4 A table of general levels and processes of wastewater treatment based on end use and human exposure.

	Increasing levels of treatment > > >			
Treatment level:	**Primary**	**Secondary**	**Filtration and disinfection**	**Advanced**
Processes:	Sedimentation	Biological oxidation and disinfection	Chemical coagulation, biological or chemical nutrient removal, filtration, and disinfection	Activated carbon, reverse osmosis, advanced oxidation processes, and soil aquifer treatment
End Use:	No uses recommended	Surface irrigation of orchards and vineyards	Landscape and golf course irrigation	Indirect potable reuse: • Groundwater recharge • Surface water augmentation
		Nonfood crop irrigation	Toilet flushing	Direct potable reuse
		Restricted landscape impoundments	Vehicle washing	
		Groundwater recharge of potable aquifer	Food crop irrigation	
		Wetlands, wildlife habitat, stream augmentation	Unrestricted recreational impoundment	
		Industrial cooling processes	Industrial systems	
Human exposure:	**Increasing levels of acceptable human exposure > > >**			
Cost:	**Increasing levels of cost > > >**			

Source: Adapted from US EPA (2012).

Advanced treatment technologies. Advanced technologies can be added to primary and secondary treatment to further reduce total suspended dissolved solids, remove nutrients and trace organic compounds, and provide further barriers for pathogen protection (NRC 2012) as follows:

- TDS reduction using ED, NF, and/or RO
- Nutrient removal with nitrification/denitrification, gas stripping
- Organic compound removal using carbon adsorption, advanced oxidation (ozone, advanced oxidation process [AOP]), NF, and RO
- Further pathogen protection using advanced oxidation

More detail on these technologies can be found in a standard water-treatment engineering text, such as Asano et al. (2007) or Crittenden et al. (2012). A brief description of relevant technologies is given here:

MF and UF are used in either a pressurized or submerged configuration. The key operational parameter is flux, or the rate of water flow volume per area of membrane. These membranes are effective at removing 1–5 logs of bacteria and protozoa (90–99.999%) and 0–2 logs of viruses (0–99%) (National Research Council 2012). The combination of membranes should impede the movement of all pathogens into the finished water, in the absence of leaks or membrane irregularities. The result is a high-quality water that can be used either directly or indirectly in a potable water system. But these membranes do generate a residual brine waste component, very high in TDS, which must be safely managed and disposed, as discussed for desalination.

AOPs can be used to mitigate a variety of natural and synthetic organic chemicals that may remain after the above treatment processes. Highly oxidizing hydroxyl radicals ($OH^{\cdot}$) can transform organic compounds into primary constituents of CO_2, water, and mineral acids, such as HCl (Asano et al. 2007). AOPs degrade molecules rather than simply transfer them from one phase to another as in other conventional water-treatment processes. AOPs generally combine ozone, ultraviolet (UV) light, and/or hydrogen peroxide (H_2O_2) (Asano et al. 2007).

Biological activated carbon (BAC) filtration, a modification of granular activated carbon (GAC), targets some naturally occurring organic compounds from plant and animal matter. If these compounds are allowed to remain, chlorine or ozone disinfection may produce harmful disinfection by-products. Microorganism growth is encouraged on GAC using the target organic matter as a food source. A pretreatment step of ozonation or advanced oxidation can break complex organic compounds into simpler compounds that are more readily bioavailable.

Disinfection. As part of the overall water-reuse scheme, *disinfection* is critical as it targets biological pathogens. Disinfection may be applied at multiple places in the water-treatment scheme to both disinfect the water and provide a residual for the distribution system. Common disinfection options are described in Table 12.5.

The effectiveness of a disinfectant can be found in the relationship between microbial inactivation and a quantifiable product known as the "C-t product" (concentration times chemical contact time) or "I-t product" (UV intensity times detention time) in the case of UV disinfection. For example, California requires a *Ct* of 450 mg-min/l when using chlorination as a disinfectant and Florida requires 120 mg-min/l for fecal coliform for bacterial counts >10 000/100 ml (US EPA 2012). California also has standards of 1 mg-min/l for ozone usage and 100 mJ/cm^2 for UV disinfection followed by sand filtration (US EPA 2012).

Table 12.5 Common disinfection options for drinking water treatment.

Disinfectant	Description	Advantages	Disadvantages
Chlorine gas, Cl_2	Most effective on bacteria, followed by viruses and protozoa	Well-developed; produces a residual disinfectant in the distribution system	Requires long contact time; not effective with *Giardia* and *Cryptosporidium*; may produce harmful by-products
Ozone	Generated on-site; more commonly used in Europe; strong oxidant and virucide	Effective in removal of CECs and color; may be combined with H_2O_2 to reduce bromate formation	Harmful by-products possible when bromide >0.10 mg/l; more complex, higher O&M costs; no residual in system
UV	Generated on-site; recent advances make this option more cost-effective	Short contact time; no known disinfection by-products; effective against *Giardia* and *Cryptosporidium*	Particles can shade target microbes, reducing effectiveness; no residual in distribution system

Source: Adapted from Davis (2010) and National Research Council (NRC) (2012).

Engineered natural processes. Natural processes can be mimicked and used as a primary treatment step or for polishing the treated water. Example processes include soil-aquifer treatment, riverbank filtration, and natural/artificial (constructed) wetlands (Figure 12.7). The processes can help in removal of metals, dissolved organic carbon (DOC), trace organic chemicals, and particulate and suspended solids. While these are often passive systems – with limited pumping and chemical addition – they frequently require a large land footprint.

Treatment trains: three levels of advanced treatment. Frequently, these advanced technologies are combined in "treatment trains" to achieve a treatment goal (Chamberlain et al. 2020). A full-advanced treatment ("FAT") is considered the "gold standard" of wastewater treatment for water reuse (Schimmoller and Kealy 2014). The FAT includes

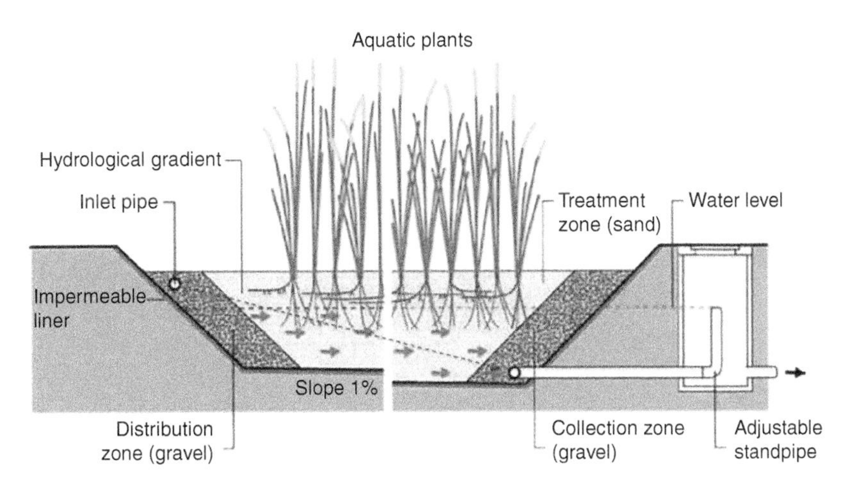

Figure 12.7 Schematic of a constructed wetlands for wastewater treatment. Source: Wang et al. (2017)/with permission of Elsevier.

low-pressure membrane filtration (e.g. NF), high-pressure filtration (RO), and advanced oxidation (UV/H_2O_2). This level of treatment achieves a log-removal credit that reaches or exceeds 12/10/10 for viruses/*Giardia*/*Cryptosporidium* and is the treatment schema at the Orange County GWRS (groundwater replenishment system) in California (Raucher and Tchobanoglous 2014).

A second treatment approach utilizes BAC and GAC in place of RO for removal of organic and dissolved constituents. The scheme will typically also include ozone pretreatment (O_3), membrane filtration (MF), and UV for disinfection. This treatment scheme is about 60% of the cost of FAT and is being utilized in eastern and southeastern US states, e.g. the Gwinnett County, GA, plant and the Upper Occoquan Service Authority (UOSA) plant in Centreville, VA (US EPA 2012).

A third, simpler treatment scheme uses only a combination of ozone and BAC technology. The ozone is used to oxidize compounds and improve bioavailability in the BAC system, which also includes adsorption for final removal. This schema is effective in removal of constituents of emerging concern (CECs) found in municipal wastewater (Sundaram 2017).

12.5 "Fit-for-Purpose" Water

In a constrained world, water security will increasingly depend upon matching an appropriate level of treatment with the intended use. The treatment level is best considered with regard to environmental, social, and economic concerns, i.e. the "triple bottom line" (TBL) approach (Schimmoller and Kealy 2014). A higher or more costly level of treatment than what is necessary for the water's purpose may result in unintended negative consequences and is therefore undesirable and unsustainable. Water that is treated appropriately for its intended use is called "**fit-for-purpose**" water.

Table 12.6 provides a summary of reuse applications, producing water fit for a given application, along with associated constraints and concerns.

The quality of water suitable for these purposes is described in the following sections.

12.5.1 Nonpotable Water Reuse

Reclaimed water can be used for a wide range of nonpotable (not for drinking) purposes (Table 12.6). Such applications include irrigation for public landscapes and golf courses, cooling water for power plants, mixing of concrete, groundwater recharge, hydraulic fracturing (HF), and recreational lakes. In locations of public access, these systems typically run in a separate (or dual) piping system, which is often colored purple to make clear the distinction (Figure 12.8).

Irrigation is the easiest and most readily accepted reuse of domestic wastewater, whether it be for farms or city landscapes. As early as the sixteenth century, European wastewater was applied to "sewage farms" for the dual purpose of wastewater disposal and nutrient addition. In the late-nineteenth century, wastewater began to be used for irrigation in US cities such as Boulder, CO, San Antonio, TX, and Vineland, NJ (Angelakis et al. 2018). In 2005, over 1600 parks nationwide and 525 golf courses in Florida alone were irrigated with reclaimed water (National Research Council 2012). Reclaimed and disinfected water can be used for landscape irrigation in parks, golf courses, and other public spaces with due consideration of proximity to the public.

Table 12.6 Water-reuse categories and associated constraints and concerns.

Category	Typical applications	Major constraints and concerns
Agricultural irrigation	Crop irrigation, plant nurseries	Seasonal demand; need for winter storage
Landscape irrigation	Parks, freeway medians, golf courses	Point-of-use often far away from water reclamation facility
Industrial recycling and reuse	Cooling water, boiler feed water, process water, high-quality water for electronics manufacturing	Constant demand, but quality is site- and use-specific
Environmental and recreational uses	Lakes and ponds, stream flow augmentation, snow-making for ski resorts	Site-specific, seasonal demand
Nonpotable urban uses	Fire protection, toilet flushing, street cleaning, car washing, water for cooling	Requirement for dual-piping, cost
Groundwater recharge	Groundwater replenishment, seawater intrusion barrier	Requires suitable aquifer between point of reclamation and reuse
Planned potable reuse– direct/indirect	Augmentation of drinking water supplies	Environmental buffer, social acceptance, health risks, cost of engineered buffers, lack of water-quality regulations

Source: Adapted from Angelakis et al. (2018).

Figure 12.8 Purple pipe at the South Bay Advanced Recycled Water Treatment Facility, San Jose, California. Source: Used with permission from Nathan Kuhnert, US Bureau of Reclamation.

Reclaimed water can also be used as *cooling water and as boiler feedwater* for steam-driven power plants. When used as boiler feedwater, inorganic constituent (e.g. metals, salts) must be removed to prevent scaling and damage to the boilers. Bicarbonate alkalinity can also lead to the release of carbon dioxide, which can cause equipment corrosion, limiting its use as feedwater but not as cooling water. In 2012, there were more than 40 US power plants that used reclaimed water for cooling, often with inclusion of a biocide to prevent biological fouling (National Research Council 2012).

Reclaimed water can be used as process water in *other industrial uses*, such as concrete manufacturing, with no additional postsecondary treatment. Advanced treatment may be needed for uses such as carpet dyeing, electronics manufacturing, and high-quality paper and pulp industries. In chemical industries, TDS and salt content may be limiting in certain applications (National Research Council 2012).

Groundwater recharge is becoming more widely practiced as a reuse application. **Managed aquifer recharge** (MAR) is the umbrella term given to the replenishment of underground aquifers for one or more of three purposes:

- Replenishment of groundwater basins in terms of both quantity and quality (e.g. decreasing the salinity)
- Creation of hydraulic barriers to prevent saltwater intrusion
- Mitigation of ground subsidence caused by the over-withdrawal of groundwater below the surface

California has been recharging its Central and West Coast Groundwater Basins (Los Angeles) for over 50 years (Land and Reichard 2016). Mexico is using reclaimed water to mitigate a subsidence rate that reaches 16 in./year in some locales (US EPA 2012). Communities in western Washington are using reclaimed water to replenish their drinking water aquifers while reducing the nutrient load to the Puget Sound (Hansen 2017).

One special case of water reuse is the recycling of water produced from *HF* operations, thereby closing the loop on water usage in this sector. The westernmost oil plays in the United States, the largest global oil and natural gas producer in 2018, produced much higher volumes of produced water than was needed for the fracking operation. Currently, produced water is mostly reinjected into the subsurface, inducing seismic earthquakes in some plays in Oklahoma. But less-stringent chemical requirements for suitable fracking water and more accommodating state regulations may result in more water being recycled back into the HF process (Scanlon et al. 2020).

Reclaimed water can also be preserved in wetlands or impoundments for both *recreational and educational* purposes, as in the Orlando Wetlands Park (FL) and Santee Parks (CA). These impoundments are popular for non-bodily contact activities such as boating, hiking, fishing, photography, and bird-watching. A facility at the Sidwell Friends School in Washington, DC, shows children how wastewater can be treated with a constructed wetlands (US EPA 2012). Snowmaking with reclaimed water can help extend the skiing season in colder climates when adequate filtration and disinfection are used to ensure no detectable fecal coliform in the final product (US EPA 2012).

12.5.2 Potable Water Reuse

Intentional potable water reuse has been practiced in some form globally and in the United States for over 50 years (National Research Council 2012). Three common ways of implementing potable reuse are illustrated in Figure 12.9. De facto water reuse is the unplanned reuse of wastewater discharged upstream, often by a downstream community. Intentional (planned) potable reuse can be either direct or indirect. In DPR, water is treated to an advanced level before being sent to a conventional water-treatment plant and then to the consumer. With IPR, the advanced-treated water is sent to an environmental buffer (lake, river, wetland, aquifer) for storage prior to water treatment and distribution.

De facto *water reuse*. Most urban residents are unaware of the extent to which de facto water reuse is already a portion of their drinking water supply. The majority of municipal WWTPs, if not located in a coastal area, discharge their effluent to a waterbody that has a downstream recipient (National Research Council 2012). A prime example of de facto reuse is the Trinity River in north Texas. The Trinity River drains much of the Dallas–Fort Worth metropolitan area, with its baseflow dependent almost entirely upon wastewater effluent from this densely populated region (National Research Council 2012). Treatment plants in the metro area use tertiary processes to remove nutrients before release into the river channel. Without much inflow for dilution, the river makes its way southward to Lake Livingston, one of Houston's main drinking-water reservoirs. With a travel time of two weeks, and a baseflow cross-section of 2 m in depth and 50 m in width, a detailed study showed attenuation of wastewater-derived trace organic contaminants (e.g. pharmaceuticals), which was attributed to biotransformation and photolysis (Fono et al. 2006).

DPR. Planned direct potable water reuse requires an additional level of treatment beyond secondary wastewater treatment. DPR describes the addition of advanced purified reclaimed water to the raw water supply of a drinking water treatment plant (DWTP). The water may be retained temporarily in engineered storage (e.g. a storage tank, transmission pipeline) to allow monitoring of water-quality parameters before introduction into the DWTP (Raucher and Tchobanoglous 2014).

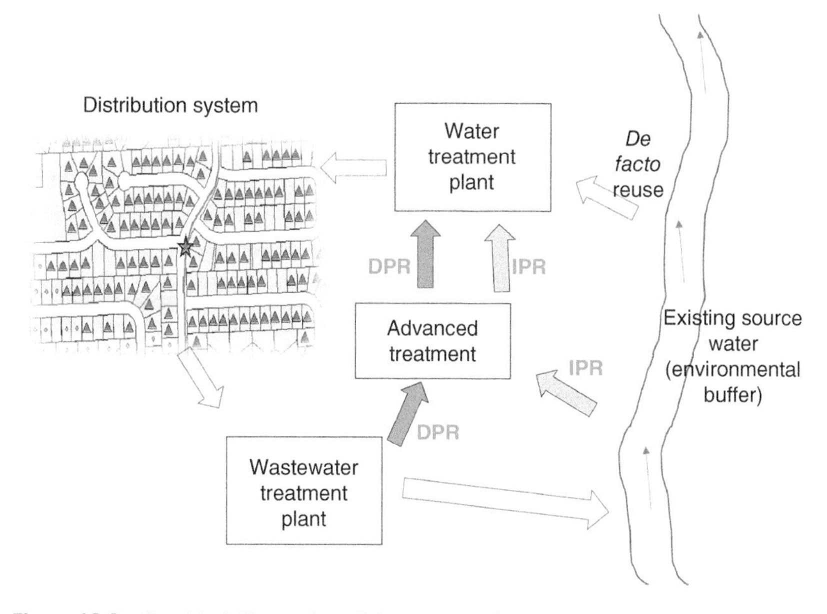

Figure 12.9 Graphical illustration of three types of potable water-reuse schemas – de facto (white), direct potable reuse (DPR) (green), and indirect potable reuse (IPR) (yellow).

IPR. For IPR, municipal wastewater receives advanced treatment before it is released into an environmental buffer, such as a lake or aquifer. These environmental buffers provide three critical elements of protection: (i) retention time, especially in the case of wastewater treatment malfunction; (ii) natural attenuation of contaminants; and (iii) dilution (blending) of wastewater with raw source water (National Research Council 2012). Hydrological data can be used to determine both the dilution ratio and average residence time that water is receiving in a buffer. Utilities often prefer this method rather than DPR because the presence of the environmental buffer improves public acceptance. It is possible that the buffer itself will be of lower water quality than the advanced-treated wastewater being added. Such considerations are important on a case-by-case basis as this method of reuse is employed.

Having discussed types and technologies for potable reuse, Table 12.7 lists several example systems with combined advanced treatment systems for water-reuse applications.

Water reuse comes with its own unique challenges:

- The distance between the advanced water treatment/reclamation plant and the point of use may be cost-prohibitive for implementation,
- Seasonal demand will vary with some applications requiring additional storage, and
- Social acceptance is critical for planned potable reuse, requiring accurate monitoring data and the support of public officials for success.

Public acceptance can be increased by highlighting the facts that (i) de facto water reuse is a common occurrence and (ii) planned water reuse has even tighter monitoring and controls, along with an increasing history of successful implementations in a safe and sustainable manner.

Having reviewed the many methods and uses of water reuse as well as its challenges, we can now discuss another important aspect – associated costs of desalination and water reuse.

Table 12.7 Examples of potable water-reuse projects across the globe.

Project	Country	Year	Size (MGD)	Reuse type	Technologies
Windhoek	Namibia	1968/2002	5.5	DPR/Blending	PAC-O_3-Clarification-DAF-Filtration-O_3/AOP, BAC/GAC
NEWater, Bedok	Singapore	2003	23	IPR – Surface water augmentation	UF-RO-UV
Essex and Suffolk	UK	2003	8	IPR – Surface water augmentation	Biological filtration – UV disinfection
Queensland	Australia	2008	61	IPR – Surface water augmentation	UF-RO-UV/AOP
NEWater, Changi	Singapore	2010/2017	122	IPR – Surface water augmentation	UF-RO-UV
Perth	Australia	2016	10	IPR – direct inject recharge	UF-RO-UV

AOP, advanced oxidation processes; BAC, biological activated carbon; Cl, chlorination; DAF, dissolved air flotation; DPR, direct potable reuse; GAC, granular activated carbon; IPR, indirect potable reuse; MF, microfiltration; O_3, ozone disinfection; GAC, granular activated carbon; RO, reverse osmosis; UF, ultrafiltration; UV, ultraviolet radiation.
Source: Based on US EPA (2018).

12.6 Costs of Desalination and Water-Reuse Systems

12.6.1 Costs of Desalination

Table 12.8 provides a summary of 10 large RO plants dating from 1989. The table includes information on production capacity and cost of providing the water. Older plants were smaller in size, as expected for adoption of a new technology. In general, more recent plants have lower costs reflecting advances in membrane technology and economies of scale although this is not always true.

Water costs shown include amortized capital (fixed construction) and operational (energy, labor, chemicals, membrane replacement, miscellaneous) cost breakdown for both seawater and brackish water applications. The Ashkelon (Israel) seawater plant reports that capital costs and energy are the two largest cost factors, accounting for 60% and 25% of the amortized costs ($1.17 and $0.49/kgal), respectively. The capital costs of the seawater plants are generally about five times greater than a brackish water plant of similar size, a fact that can be attributed to the more-extensive pretreatment and larger pumping and piping needed to move the seawater RO concentrate. Likewise, the energy costs of the seawater RO plant are over four times greater than those of the brackish water RO plant because of the higher pressures and lower recovery in the seawater plant. While the chemical costs are similar, the membrane-replacement costs are greater for the seawater RO system as they require more frequent replacement (Greenlee et al. 2009).

Officially opened in 2018, the Tuas Desalination Plant (TDP) was the first Singapore system to use advanced pretreatment technology (dissolved air-flotation and UF) combined with RO. As shown in Figure 12.10, the TDP treatment train includes intake screens, fine screens, dissolved air-flotation, RO, and post-treatment. The post-treatment includes remineralization as RO water is largely devoid of electrolytes. (Athletes know well that electrolytes are a critical component of healthy hydration.) The TDP produces up to 30 million gallons a day at a cost to the customer of $2.95/kgal. With the new plant, 30% of Singapore's water demand can

Table 12.8 Desalination plants using reverse osmosis (RO) technology.

Location (country)	Capacity (MGD)	Online date	Water source	Water cost ($/kgal)
Southmost (USA)	7.5	2004	BW	$2.40
El Paso (USA)	27.5	2007	BW	$5.26
San Antonio (USA)	30	2026 (expected)	BW	$3.10
Galder-Agaete (Spain)	0.9	1989	SW	$7.35
Ashkelon (Israel)	86	2001	SW	$2.00
Perth Kwinana (Australia)	38	2006	SW	$3.60
Tianjin (China)	40	2007	SW	$3.60
Sydney (Australia)	66	2008	SW	$5.60
Tampa Bay (USA)	25	2008	SW	$2.50
Tuas Desalination Plant (Singapore)	30	2018	SW	$2.95

BW, brackish water; SW, seawater.
Source: Greenlee et al. (2009) and Ziolkowska (2015).

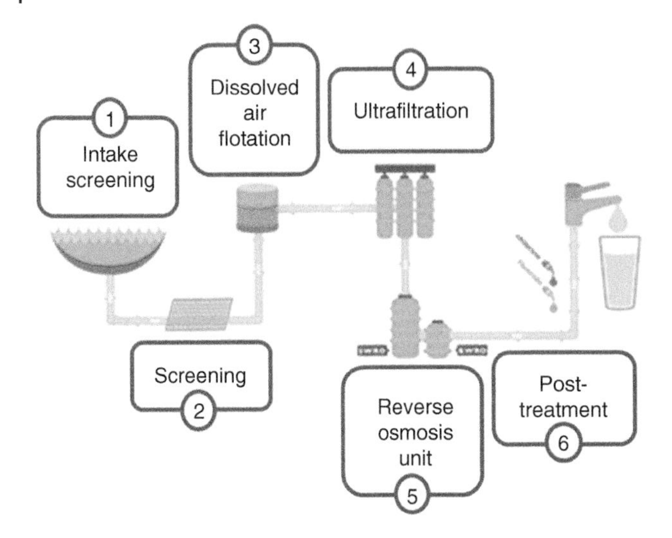

Figure 12.10 Singapore's Tuas seawater RO desalination plant. Source: Adapted from PUB-Singapore Water Agency (2020).

now be met by desalination. Further, more than half of TDP's roofs are covered with solar panels generating 1.4 million kWh a year.

The Kay Bailey Hutchison El Paso Desalination Plant (Texas) came on-line in 2007 and is the world's largest inland desalination plant, daily producing clean drinking water from a brackish ground water with a salinity of 4300 mg/l. The 27.5 MGD of drinking water consists of 15 MGD of RO water blended with 12.5 MGD of source water. Approximately 83% of the water is recovered while the remainder is produced as a concentrate. The permeate, or desalted water, is piped to a storage tank for distribution and the concentrate is disposed through deep-well injection (El Paso Water 2018). The amortized capital cost is $3.76 and production costs are $1.50 per 1000 gal. A review of this plant and five other brackish groundwater desalination plants in Texas found the overall total water production costs varied from $3.58 to $8.81 per 1000 gal (Arroyo and Shirazi 2012).

12.6.2 Costs of Water Reuse

The decision to reclaim and reuse wastewater often hinges on cost comparison with alternative water-supply augmentation. Unless the city or utility is in dire water stress, it will likely compare the costs of water reuse with existing water source supply, treatment, or interbasin transfer of a new source, and/or water conservation and efficiency.

The costs of both potable and nonpotable water reuse include the additional costs of advanced wastewater treatment, storage, and transfer of water from the WWTP to end user. The total costs will vary widely as they depend upon many factors, including size and location of the plant, regulatory requirements for level of treatment, quality of the wastewater to be reclaimed, proximity to the end user, energy cost, and brine disposal (when needed). In the United States, these costs can range from ~$1.00 to $2.50 per thousand gallons for nonpotable reuse (Table 12.9) and up to $5.00 per thousand gallons for potable reuse (Table 12.10). The variation in pipeline costs for the nonpotable applications shows the importance of proximity of the reclamation plant to the end user. In most cases, this was the most significant single cost item.

Table 12.9 Costs of nonpotable reclamation and reuse at four locations.

	Las Vegas, NV	Trinity River, TX	Denver, CO	Inland Empire, CA
Capacity (MGD)	70	10	4.4	13
Reclaimed water use	Landscape irrigation	Landscape irrigation, amenity reservoirs	Landscape irrigation, industrial cooling, zoo	Landscape irrigation, industrial cooling, paper processing
Treatment	AS, SF, UV	AS	BAF, FL, SED, SF, DIF	AS, BNR, SF, CL
Year completed	2004	1987	2000	2010
Annualized capital ($/kgal)	0.37	0.1	1.18	0.85
O&M ($/kgal)	0.68	0.05	1.06	1.18
Total annualized ($/kgal)	$1.05	$0.15	$2.24	$2.35

AS, activated sludge, SF, sand filter, UV, ultraviolet, BAF, biologically active filter, FL, flocculation, SED, sedimentation, DIF, disinfection, BNR, biological nutrient removal, CL, chlorination.
Source: Adapted from National Research Council (NRC) (2012).

Table 12.10 Costs of potable reclamation and reuse at four locations.

	Orange County, CA	El Paso, TX	Shoal Creek, GA	West Basin, CA
Capacity (MGD)	70	10	4.4	13
Treatment	Primary, AS, TF, MF, RO, UV/$H_2 0_2$	AS w/DN, SF, 0_3, BAC, DIF	AS w/NR, UV, Trtmt Wetlands	MF, RO, UV/$H_2 0_2$
Year completed	2008	1984	2003	2006
Annualized capital ($/kgal)	1.74	2.05	0.48	2.68
O&M ($/kgal)	1.16	0.33	0.31	2.38
Total annualized ($/kgal)	$2.90	$2.40	$0.79	$5.10

AS, activated sludge; TF, trickling filter; MF, microfiltration; RO, reverse osmosis; UV, ultraviolet; $H_2 0_2$, hydrogen peroxide; DN, denitrification; SF, sand filter; 0_3, ozone; BAC, biologically activated carbon; DIF, disinfection; NR, nutrient removal.
Source: Adapted from National Research Council (NRC) (2012).

In the case of potable water reuse, treatment facility costs become the highest cost item as would be expected for higher quality water (Table 12.10). In addition, one can see economies of scale by comparing the two California plants that both use MF, RO, and advanced oxidation. The West Basin plant (13 MGD) has a total annualized cost of $5.06/kgal while the larger plant in Orange County has a total annualized cost of $2.90/kgal. Potable water-reuse costs are frequently lower than desalination costs, which is why Singapore operates its reuse plant continuously and reserves its desalination plants for extreme water demand periods (PUB-Singapore Water Agency 2020).

In the United States, while Florida produces the most reclaimed water, significant production comes from the states of Colorado, California, Texas, and Arizona (Table 12.11), four states in the arid southwest subject to climate change and high population growth (National Research Council 2012).

Table 12.11 Top US states reporting of reclaimed water production and usage in 2005.

State	Reclaimed water facilities reporting data	Annual reclaimed water production (billion gallons)	Reported length of pipe in reclaimed water system (miles)
Arizona	25	18.2	221
California	85	86.9	1181
Colorado	4	5.2	65
Florida	296	218.1	5385
Texas	64	31.4	403

Source: Data from Bryck (2008).

A study comparing the cost of raw water conveyance with water reclamation/advanced treatment showed break-even distances for utilities to consider in planning (Chamberlain et al. 2020) (Table 12.12). For example, when long-term planning is conducted for a medium-sized city of about 110 000–155 000 in population (such as Norman, OK), managers would do well to consider advanced treatment and potable water reuse if an alternative source is greater than 12 miles from the city. If a higher level treatment is required (e.g. MR/RO/UVAOP), then even this level of treatment is cost-preferred if the alternative source is greater than 26 miles away. These judgments are site-specific, of course, and are provided here for comparative purposes.

Utilities in the United States often set reclaimed water rates for nonpotable use lower than potable water rates to encourage customer usage. The median rate is 80% of the potable water rate, with a range of 50–100% of the potable rate (US EPA 2012). Integrated water resources planning should ensure that providers can recoup the costs of retrofitting plants, adding treatment components, and distributing reclaimed water to the end user.

Finally, triple bottom line (TBL) accounting will consider not only economic costs but also environmental and societal costs. Economic (financial) costs have been discussed in this section but also may include cost savings to the end user as, for example, in the case of a golf course manager who needs less fertilizer inputs because of the valuable nutrients that are embedded in the reclaimed water for irrigation. Environmental costs may include the effects on ecosystem habitats in streams and rivers (see Chapter 10), the greenhouse gas and pollutant emissions that result from energy usage (see Chapter 8), and the reduction of chemical inputs for crop production when using reclaimed water. Societal costs will mostly be concerned with human health effects related to both energy usage and exposure to various reclaimed water uses.

Table 12.12 Break-even distance points for cities of three sizes in Oklahoma, based on cost of raw water conveyance by pipeline versus advanced treatment and water reuse.

City	Population Y2010/Y2060 (expected)	Water-supply demand Y2010/Y2060 (MGD)	Break-even distance point (miles) for elevation gain of 4 fpm, DPR	
			Level 1 treatment (BAC-based)	Level 3 treatment (MF/RO/UVAOP)
Small (e.g. Altus)	21 800/26400	4.4/5.3	8–10	18
Medium (e.g. Norman)	112 000/155000	15.3/21.1	12	26
Large (e.g. Oklahoma City)	565 000/673000	93.5/111.4	15	42

Source: Chamberlain et al. (2020)/with permission from John Wiley & Sons.

12.7 Challenges and Future Directions

Both desalination and water reuse in general present specific challenges that should be noted and addressed.

12.7.1 Challenges to Desalination

While desalination, and especially RO, has made significant strides in the past several decades, several challenges remain. Meeting these challenges will greatly increase potential drinking water sources, from oceans to countless brackish groundwater sources. These additional sources will significantly improve water security for numerous countries and countless people. The challenges listed below are relevant (Greenlee et al. 2009; Kumar et al. 2017):

- *Advances in desalination membranes.* While crosslinked RO membranes have been used for decades now, the fundamental nature of these membranes, and their associated separation mechanisms, is still being investigated. Further improvements in membrane chemistry and structure will enhance water flux and reduce energy requirements.
- *Concentration polarization.* Salt rejection at the RO membrane surface results in a concentrated layer at the surface that impairs mass transfer across the membrane. Reducing the thickness of this layer by enhancing the back transport of solutes will improve performance of RO-based desalination.
- *Seawater intakes and discharges.* Seawater intake and discharge provide unique challenges to seawater desalination plants. Marine organism accumulation on intake screens as well as treatment impacts on organisms passing these screens are a challenge. Further, insufficient dilution of dense brines prior to discharge can negatively impact the local marine environment.
- *Chlorine-resistant membranes.* Chlorine-based disinfection is common in water-treatment plants as is chlorine addition to reduce biofouling of membranes. Development of chlorine-resistant membranes is an important practical need to minimize membrane degradation.
- *Presence of boron.* Boron is present in seawater at an average concentration of 4.6 mg/l. The recommended level in drinking water is 0.5–2.0 mg/l. Boron removal in RO systems is limited, leading to additional treatment requirements (Wenten and Khoiruddin 2016).
- *Inland desalination brine disposal.* Processes that reduce the concentrate stream, or that identify beneficial uses of the concentrate, will improve upon current options – deep well injection, municipal sewers, and evaporation ponds. This will increase the adoption of inland desalination (Giwa et al. 2017).

12.7.2 Challenges to Water-Reuse Implementation

There are additional challenges to potable reuse implementation, both technical and non-technical in nature.

- *Trace organic compounds.* Especially in industrialized nations, trace organic compounds exist in domestic wastewater, which are not removed by conventional wastewater-treatment processes. These include pharmaceuticals and personal care products (PPCPs), and some metals, and are often grouped together under the moniker "contaminants of emerging concern". Some of these compounds act as endocrine disruptors (US EPA 2015).

While these compounds are not currently regulated, their very presence can cause public concern when considering reuse.

The World Health Organization (WHO) has studied the fate and transport of pharmaceuticals in the environment and found that concentrations in treated wastewater are generally below 0.05 μg/l (50 ng/l), far below health-based targets with conservative margins of safety. Thus, the WHO has not set guidelines nor recommended monitoring of these chemicals (WHO 2011). Australia, the first country to set national guidelines for water reuse (in 2008), suggests safety factors of 1000–10 000 relative to the pharmaceutical therapeutic dose (US EPA 2012). California has developed guidelines for several "indicator CECs" as health-based indicators or treatment unit performance indicators. Caffeine and NDMA (N-Nitrosodimethylamine) were chosen as both health- and performance-based indicators (US EPA 2012). AOPs have proven effective at reducing CEC levels to within statutory guidelines (Seibert et al. 2020).

- *Water-quality assurance and monitoring*. For water reuse to become a substantial contributor to water security, appropriate water quality must be maintained. Both nonpotable and potable reuse schemes rely upon the following four elements to ensure water quality and consumer confidence:
  1. Attenuation of contaminants using multiple barriers of treatment and disinfection,
  2. Retention in environmental or engineered reservoir to encourage natural processes and allow for response to system upsets or failures,
  3. Blending with other water sources to ensure dilution, and
  4. Monitoring at various system points and regular time intervals in order to provide clear data on efficacy of system components and achievement of water-quality target goals (National Research Council 2012).
- *Public perception*. The unfavorable public perception or "yuck factor" (envisioning the "toilet to tap" scenario) is an impediment to water reuse. Further, this perception is compounded by the presence of waste pharmaceuticals and other CECs. Routine monitoring of indicator chemicals may assure the public that these compounds are not present in concentrations of great concern.

  Public acceptance is generally greater when an environmental buffer is incorporated as in the case of IPR. However, environmental buffers have yet to be proven as more effective than engineered processes in DPR. Indeed, the highly treated water in IPR may well be of higher quality than the environmental buffer into which it is introduced (National Research Council 2012). Reminding the public that de facto water reuse has been practiced throughout history, and that multiple treatment barriers ensure safe drinking water quality, can balance these concerns. Education is the key strategy for improved public acceptance.
- *Need for regulations and standards*. Implementing water reuse is contingent upon gaining state approval for such systems. Thus, development of additional US state regulations governing water reuse will enable long-term planning and budgeting decisions for the adoption of water reuse. Regulations may need to be bolstered or adjusted by ongoing monitoring and data from actual water-reuse installations.

12.7.3 Water-Equity Considerations – A Case Study

Both desalination and water reuse depend upon advanced treatment options that are both costly and complicated – relative to conventional treatment – in their operation and maintenance. The extension of these water-supply options to rural and resource-poor communities may be accomplished but only with attentiveness to the unique challenges.

Small-scale desalination units using RO and solar power installation (ROSI) were placed in several sites in Central Australia, a remote interior region with a high proportion of indigenous peoples (Werner and Schäfer 2007). This region is characterized by low rainfall (200–300 mm/year) and 10–11 hours per day of sunlight, ideal for the harnessing of solar energy. Here, the quality of groundwater is influenced by salinity and hardness ($CaCO_3$), which make the water less palatable. ROSI is a small-scale unit that uses solar power and RO membranes to remove salts, turbidity, and trace contaminants, all without the use of batteries. The units are intended to operate on low recovery (<40%) so that the concentrate may be used for nonpotable uses such as showering, toilet flushing, animal watering, and plant irrigation.

The ROSI sites included two small indigenous communities (20–50 people), a few farms, a roadhouse, and a town of 3500 residents. Community questionnaires were used to assess the social acceptability of the ROSI units after the communities had been exposed 2–10 days. TDS levels ranged from 750 to 15 600 mg/l in the local bore hole water but was reduced in the permeate to 13–146 mg/l. The concentrate (brine) salinity was suitable for sheep stock consumption but not for grape irrigation. All respondents reported that the combination of desalination and renewable energy was "good" or "excellent." From the survey, residents expressed concern about:

- Consistency of the power supply and not having sufficient water when the sun was not shining,
- Maintenance of the unit regarding potential fouling of the membrane, and
- Very low salinity in comparison with previous water quality.

This last point is interesting as it reveals social acceptance of a change in taste that is shaped not by drinking water standards but by the community's previous experience. In other words, the community had grown used to the high hardness and salinity and were a bit skeptical about the radical improvement!

This case study demonstrates the potential of small-scale decentralized RO units that are powered by solar energy. Comparable units could find application in other similar regions of need, such as Turkey, Morocco, Jordan, and Egypt (Werner and Schäfer 2007). In order for these systems to be socially sustainable, they should (i) meet the water quantity and quality needs of the community, (ii) be within their capacity to operate and maintain, and (iii) be acceptable by the community regarding the esthetics of taste, odor, and color.

12.8 Conclusion

We close this chapter on a lighter note. Craft brewers across the country have been evaluating the use of reclaimed water in beer production. San Diego is home to more than 150 breweries, including Stone Brewery that created its own Full Circle Pale Ale from recycled wastewater (Figure 12.11) (Chappell 2017). Ten brewers in Scottsdale, AZ, also took the challenge from the local utility to produce and showcase their reclaimed-water beers in a One Water Brewing Showcase in November 2019. One brewer remarked that the recycled water is "an incredibly pure water to brew with" (Heller 2020). Stone's senior water operations manager even said that "It's higher-quality water than what we use to brew. And we actually had to add minerals

Figure 12.11 Stone Full Circle Pale Ale is a brew beer made from San Diego's recycled water. Source: Chappell (2017).

to it, to meet our brewing water quality requirements." (Chappell 2017). These examples provide credence to the suggestion that we should "judge water not by its *history*, but by its *quality*."

Thus, while water desalination taps into an abundant water source and avoids the "yuck factor" concern of water reuse, desalination suffers from higher costs and brine-disposal challenges. Overcoming public concerns about water reuse opens up another abundant water supply that is readily available, drought tolerant, and less expensive than water desalination and many water-transfer scenarios. Further development of water desalination and water reuse are thus paramount to promoting and assuring future water security for vast regions of the world.

*Foundations: RO Design Approach

RO system design is based on several factors, including: amount of water to be produced, salinity of the source water, temperature, and concentrate disposal options. The source water salinity will impact the membrane loading rate (liters of water applied per hour per square meter of membrane surface area), the water recovery (proportion of water in permeate relative to applied water), and the salt concentration (how much salt is rejected by the membrane). These factors are shown in Table 12.13 as RO permeate flux, recovery, and salt rejection, respectively. Comparing seawater and brackish water RO, the permeate flux varies more for brackish water than seawater, the same is observed for hydrostatic pressure. This can be attributed to the relative salinity ranges for the two sources. Recall from Table 12.1 that seawater has a salinity range of 15 000–50 000 mg/l (a factor of 3) while

Table 12.13 RO design parameters.

Parameter	Seawater RO	Brackish water RO
RO permeate flux (l/m^2-hour)	12–15 (open water intake) 15–17 (beach well intake)	12–45 (groundwater)
Hydrostatic pressure (kPa)	5500–8000	600–3000
Membrane replacement	20% per year	5% per year
Recovery (%)	35–45	75–90
pH	5.5–7	5.5–7
Salt rejection (%)	99.4–99.7	95–99

Note: 1 kPa = 0.145 psi.
Source: Based on Greenlee et al. (2009).

brackish water ranges from 1500–15 000 mg/l (an order of magnitude). This also explains the variation in water recovery – brackish water, with its lower salinity, has a much higher water recovery.

A seawater recovery of 50% has two important implications: (i) only half of the source water treated ends up as drinking water (and so you will need to have access to twice as much water as you need for drinking water) and (ii) half of the water ends up as concentrate (brine) with twice the feedwater salinity and requiring transport and disposal. Note that the target salinity level can be achieved for brackish water with a lower relative salt rejection than seawater. For example, 99% salt rejection of seawater (35 000 mg/l → 350 mg/l) has higher permeate salinity than 95% rejection of brackish water (4000 mg/l → 200 mg/l).

A final observation from Table 12.13 is that seawater RO membranes are "stressed" to a higher level than brackish water RO membranes (higher pressures, loading, etc.). This explains the more-frequent replacement cycles for seawater RO systems (20% per year versus 5% per year).

Table 12.14 provides actual operating parameters for three different operating seawater plants with salinities ranging from 38 000 to 41 000 mg/l. The recoveries are all in the 40–50% range, the permeate flux values are all in the range of 13 l/m^2-hour and the feed pressures are all in the 5700–7000 kPa range, all consistent with design values reported in Table 12.13.

Table 12.14 Examples of operational seawater RO desalination plants with feedwater TDS, recovery, permeate flux, and feed pressure.

Plant location	Feed TDS (mg/l)	Recovery (%)	Permeate flux (l/m^2-hour)	Feed pressure (kPa)
Eni Gela, Sicily	40 070	40–46	13.3	6200–6400
Gran Canari, Spain	38 000	42	13.0	5700
Eilat, Israel	41 000	50	13.0	6100–6300

Note: 1 kPa = 0.145 psi.
Source: Based on Greenlee et al. (2009).

To illustrate usage of the design parameters, consider that you are designing a 10 000 m^3/day RO system for seawater with a salinity of 35 000 mg/l. Further, assume that your system will achieve 50% recovery (permeate versus feedwater) and 99.7% salt rejection. Assume a permeate flux rate of 15 l/m^2-hour and that RO systems come in racks of 90 modules/rack with 15 m^2 of surface area/module. You want to determine (i) the minimum membrane surface area, (ii) the number of racks, (iii) the feedwater rate (permeate + reject [also known as retentate or concentrate]), (iv) the concentrate rate, (v) the permeate salinity, and (vi) the concentrate salinity.

(i) Calculating the membrane surface area involves dividing the treated water goal (10 000 m^3/day) by the permeate flux rate as below:

$$\text{Surface area} = \frac{10\,000\ \text{m}^3/\text{day}}{15\frac{1}{\text{m}^2-\text{hour}}}\ \frac{1\ \text{day}}{24\ \text{hours}}\ \frac{10^3\ \text{L}}{\text{m}^3} = 27\,800\ \text{m}^2$$

(ii) Knowing the minimum surface area, we can calculate the number of racks by dividing the total surface area by the surface area per rack (90 modules/rack* 15 m^2/module = 1350 m^2/rack) as below:

$$\text{Racks} = \frac{27\,800\ \text{m}^2}{1350\ \frac{\text{m}^2}{\text{rack}}} = 20.6\ \text{racks, round up to 21 racks}$$

(iii, iv) The feedwater consists of both the permeate and the retentate (concentrate), so for a 50% water recovery the feedwater rate would be the permeate divided by 0.5, or twice the permeate (1/2 of the feedwater will be permeate and 1/2 will be concentrate).

$$\text{Feedwater flowrate} = 20\,000\ \text{m}^3/\text{day}$$

$$\text{Permeate flowrate} = \text{Concentrate flowrate} = 10\,000\ \text{m}^3/\text{day}$$

(v) The permeate salinity can be estimated based on the seawater salinity and the salt-rejection percentage. Since the salt-rejection rate is 99.7%, it means that 99.7% of the salinity will be retained by the membrane in the concentrate and only 0.3% (0.003) will pass through the membrane.

$$\text{Salinity in permeate} = 0.003 \times 35\,000 = 105\ \text{mg/l}$$

(vi) The salinity that is retained by the membrane will be added to (concentrated with) the salt already in the retentate. The simple estimate would be that all the permeate salt is retained and accumulates with the retentate, so for a water recovery of 0.5 the salt level in the concentrate would be twice that in the feedwater or 70 000 mg/l. More accurately, 0.997 of the salt is added to (concentrated in) the retentate, leading to 1.997 (almost double) the feedwater salinity.

$$\text{Salinity in concentrate} = 1.997\left(35\,000\frac{\text{mg}}{\text{l}}\right) = 69\,895\ \text{mg/l}$$

Thus, for every liter of drinking water produced, a liter of concentrate is generated requiring transport and disposal. This sample calculation was for seawater; the same procedure would be followed for brackish groundwater using parameters appropriate for that water.

End-of-Chapter Questions/Problems

12.1 Do a web search and describe three historical (pre-1950) examples of desalination.

12.2 Based on a search, what would be a typical salinity value for seawater in (a) Atlantic Ocean, (b) Pacific Ocean, (c) Mediterranean Sea, (d) Persian (Arabic) Gulf, and (e) Indian Ocean. What is the significance of the range in these values relative to RO desalination?

12.3 List and define the three general types of water reuse as well as the two primary methods for intentional potable water reuse.

12.4 If a given sample of wastewater has 10 000 colonies per 100 ml of liquid, how many colonies remain after the following disinfection removal rates: 1-log, 2-log, 3-log, and 4-log?

12.5 What are three RO brine (reject/concentrate) management options and what are the advantages and disadvantages of each?

12.6 Do a search and find a recent journal article discussing recent advances in (a) membrane technologies and (b) concentrate management approaches and provide a one-page summary of each.

12.7 Why is remineralization an important step in RO-based drinking water. What are several ways of implementing remineralization?

12.8 Based on a search, provide a one-page summary each of a full-scale system for:

a. RO-based desalination.
b. Thermal-based desalination.
c. For each, discuss the nature of the source water, the treatment system, cost information, and concentrate management steps.

12.9 Much of the difficulty with nonpotable water reuse has to do with proximity, i.e. how close is the WWTP to the point of reuse. Use the Internet to look at the feasibility of using reclaimed water for a nonpotable reuse application in a large city in your home state.

a. Identify the location of major WWTPs.
b. Find at least five potential reuse applications for using reclaimed water (e.g. golf courses, parks or greenspaces, power plants).
c. Determine the approximate distance for the five sites to the nearest WWTP.
d. Using one site, determine its approximate water usage.
e. Compare the water usage in (d) to the WWTP daily output.
f. Discuss the feasibility of using reclaimed water in this city.

12.10 A very rough Water-Stress Index can be calculated for a region by determining the ratio of total water use to water availability. Use census data, precipitation data, and the United States Geological Survey (USGS) document: "Estimated Use of Water in the United States in 2015" to answer the following questions:

a. Pick three US states – one in each of eastern, middle, and western parts of the country.
b. What is total freshwater demand (withdrawals) for each state in 2015?
c. What is average volume of precipitation over each state? Use rainfall × surface area to generate a volume of rainfall.
d. What is the Water-Stress Index for each state? Comment on the results.
e. Which water demand sector – irrigation, industrial, domestic, etc. – should each state focus on for water savings, in your opinion?

12.11 Conduct the following analysis for a 3000 m^3/day RO desalination system for seawater with a salinity of 32 000 mg/l. Assume that your system will achieve 50% recovery (permeate versus feedwater) and 99.6% salt rejection. Also assume a permeate flux rate of 16 l/m^2-hour and that RO systems come in racks of 80 modules/rack with 15 m^2 of surface area/module. You want to determine:

a. The minimum membrane surface area,
b. The number of racks required,
c. The feedwater rate (permeate + reject [also known as retentate or concentrate]),
d. The concentrate rate,
e. The permeate salinity, and
f. The concentrate salinity.

12.12 Conduct the following analysis for a 5000 m^3/day RO system for brackish water with a salinity of 4000 mg/l. Assume that your system will achieve 85% recovery (permeate versus feedwater) and 98% salt rejection. Assume a permeate flux rate of 25 l/m^2-hour and that RO systems come in racks of 90 modules/rack with 15 m^2 of surface area/module. You want to determine:

a. The minimum membrane surface area,
b. The number of racks required,
c. The feedwater rate (permeate + reject [also known as retentate or concentrate]),
d. The concentrate rate,
e. The permeate salinity, and
f. The concentrate salinity.

Further Reading

Al-Karaghouli, A. and Kazmerski, L.L. (2013). Energy consumption and water production cost of conventional and renewable-energy-powered desalination processes. *Renewable and Sustainable Energy Reviews* 24 (August): 343–356. https://doi.org/10.1016/j.rser.2012.12.064.

Asano, T., Burton, F., Leverenz, H. et al. (2007). *Water Reuse: Issues, Technologies, and Applications*. New York: McGraw Hill.

NRC, Division on Earth and Life Studies, Water Science and Technology Board, and Committee on Advancing Desalination Technology (2008). *Desalination: A National Perspective*. National Academies Press.

Tchobanoglous, G. (2015). *Framework for Direct Potable Reuse*. Alexandria, VA: Water Reuse Research Foundation.

References

Al-Karaghouli, A. and Kazmerski, L.L. (2013). Energy consumption and water production cost of conventional and renewable-energy-powered desalination processes. *Renewable and Sustainable Energy Reviews* 24 (August): 343–356. https://doi.org/10.1016/j.rser.2012.12.064.

Angelakis, A.N., Asano, T., Bahri, A. et al. (2018). Water reuse: from ancient to modern times and the future. *Frontiers in Environmental Science* 6: https://doi.org/10.3389/fenvs.2018.00026.

Arroyo, J. and Shirazi, S. (2012). *Cost of Brackish Groundwater Desalination in Texas*. Texas Water Development Board.

Artiola, J.F., Hix, G., Gerba, C., and Riley, J. (2014). *An Arizona Guide to Water Quality and Uses*. University of Arizona.

Asano, T., Burton, F., Leverenz, H. et al. (2007). *Water Reuse: Issues, Technologies, and Applications*. New York: McGraw Hill.

Bryck, J. (2008). *National Database of Water Reuse Facilities Summary Report*. Water Reuse Foundation.

Chamberlain, J.F., Tromble, E., Graves, M., and Sabatini, D. (2020). Water reuse versus water conveyance for supply augmentation: cost and carbon footprint. *AWWA Water Science* 2 (1): e1170. https://doi.org/10.1002/aws2.1170.

Chappell, B. (2017). "Beer Brewers Test A Taboo, Recycling Water After It Was Used In Homes." NPR.Org. 2017. https://www.npr.org/sections/thetwo-way/2017/03/24/521388995/beer-brewers-test-a-taboo-recyling-water-after-it-was-used-in-homes.

Crittenden, J.C., Rhodes Trussell, R., Hand, D.W. et al. (2012). *MWH's Water Treatment: Principles and Design*, 3e. Wiley https://smile.amazon.com/MWHs-Water-Treatment-Principles-Design/dp/0470405392/ref=sr_1_2?dchild=1&keywords=Crittenden+Water+Treatment&qid=1601087601&s=books&sr=1-2.

Davis, M.K. (2010). *Water and Wastewater Engineering*. MacGraw-Hill https://www.amazon.com/Water-Wastewater-Engineering-Mackenzie-Davis/dp/0071713840/ref=sr_1_2?dchild=1&keywords=MacKenzie+Davis+water+and+wastewater&qid=1588104742&sr=8-2.

El Paso Water (2018). "Kay Bailey Hutchison WTP - El Paso Water." 2018. https://www.epwater.org/cms/one.aspx?portalId=6843488&pageId=7422402.

Fono, L.J., Kolodziej, E.P., and Sedlak, D.L. (2006). Attenuation of wastewater-derived contaminants in an effluent-dominated river. *Environmental Science & Technology* 40 (23): 7257–7262. https://doi.org/10.1021/es061308e.

Force, J. (2016). "How Wichita Falls Turned a Drought Into…" Text/html. Treatment Plant Operator. Treatment Plant Operator. Https://www.tpomag.com/. November 10, 2016. https://www.tpomag.com/editorial/2016/12/how_wichita_falls_turned_a_drought_into_enlightenment.

Giwa, A., Dufour, V., and Hasan, S. (2017). Brine management methods: recent innovations and current status. *Desalination* 407: 1–23.

Greenlee, L.F., Lawler, D.F., Freeman, B.D. et al. (2009). Reverse osmosis desalination: water sources, technology, and today's challenges. *Water Research* 43 (9): 2317–2348. https://doi.org/10.1016/j.watres.2009.03.010.

Hansen, J. (2017). "Aquifers to Store Reclaimed Water: Nature Provides an Elegant Solution | HDR." 2017. https://www.hdrinc.com/insights/aquifers-store-reclaimed-water-nature-provides-elegant-solution.

Heller, C. (2020). "Why You Should Drink Beer Made From 'Recycled' Water." Why You Should Drink Beer Made From 'Recycled' Water. 2020. https://oct.co/essays/recycled-water-toilet-tap-beer.

Jones, E., Qadir, M., van Vliet, M.T.H. et al. (2019). The state of desalination and brine production: a global outlook. *Science of The Total Environment* 657 (March): 1343–1356. https://doi.org/10.1016/j.scitotenv.2018.12.076.

Kumar, M., Culp, T., and Shen, Y. (2017). Water desalination: history, advances, and challenges. National Academy of Engineering. In: *Frontiers of Engineering: Reports on Leading-Edge Engineering from the 2016 Symposium*. Washington, DC: The National Academies Press. https://doi.org/10.17226/23659.

Lahnsteiner, J. and Lempert, G. (2007). Water management in Windhoek, Namibia. *Water Science and Technology* 55 (1–2): 441–448. https://doi.org/10.2166/wst.2007.022.

Land, M. and Reichard, E. (2016). "Geohydrologic Study of the Central and West Coast Basins of Los Angeles County." Scientific Investigations Report. Usgs.Gov. Scientific Investigations Report. 2016.

Montgomery, J.M. (1985). Water treatment principles and design. In: *Hardcover*, 1e (ed. J.M. Montgomery). Wiley-Interscience.

Morillo, J., Usero, J., Rosado, D. et al. (2014). Comparative study of brine management technologies for desalination plants. *Desalination* 336 (March): 32–49. https://doi.org/10.1016/j.desal.2013.12.038.

National Research Council (NRC) (2012). *Water Reuse: Potential for Expanding the Nation's Water Supply through Reuse of Municipal Wastewater*. Washington, DC: The National Academies Press https://doi.org/10.17226/13303.

NRC, Division on Earth and Life Studies, Water Science and Technology Board, and Committee on Advancing Desalination Technology (2008). *Desalination: A National Perspective*. National Academies Press.

PUB-Singapore Water Agency (2020). *Tuas Desalination Plant*. PUB, Singapore's National Water Agency www.pub.gov.sg.

Raucher, R.S. and Tchobanoglous, G. (2014). *The Opportunities and Economics of Direct Potable Reuse*. Alexandria, VA: Water Reuse Research Foundation.

Scanlon, B.R., Ikonnikova, S., Yang, Q., and Reedy, R.C. (2020). Will water issues constrain oil and gas production in the United States? *Environmental Science & Technology* 54 (6): 3510–3519. https://doi.org/10.1021/acs.est.9b06390.

Schimmoller, L. and Kealy, M.J. (2014). *Fit for Purpose Water: The Cost of Overtreating Reclaimed Water*. Alexandria, VA: WateReuse Research Foundation.

Schladweiler, J.C. (2020). "Tracking down the Roots of Our Sanitary Sewers | The History of Sanitary Sewers." Sewerhistory.Org. 2020. http://www.sewerhistory.org/time-lines/tracking-down-the-roots-of-our-sanitary-sewers/.

Seibert, D., Zorzo, C.F., Borba, F.H. et al. (2020). Occurrence, statutory guideline values and removal of contaminants of emerging concern by electrochemical advanced oxidation processes: a review. *Science of the Total Environment* 748 (December): 141527. https://doi.org/10.1016/j.scitotenv.2020.141527.

Siegel, S. (2017). *Let there be Water*. Griffin https://smile.amazon.com/Let-There-Be-Water-Water-Starved/dp/1250115566/ref=sr_1_1?crid=2SIV12GFQR65K&dchild=1&keywords=let+there+be+water+by+seth+siegel&qid=1598400233&s=books&sprefix=Siegel+let+there+%2Caps%2C177&sr=1-1.

Sundaram, V. (2017). "Need More Water? Think Ozone-BAC For 'One Water' Resolution." Water Online. August 25, 2017. https://www.wateronline.com/doc/need-more-water-think-ozone-bac-for-one-water-resolution-0001.

UNESCO (2017). "UN World Water Development Report 2017. Wastewater: An Untapped Resource." UN-Water. https://www.unwater.org/publications/world-water-development-report-2017/.

US EPA (2012). "2012 Guidelines for Water Reuse." EPA/600/R-12/618.

US EPA (2020). *National Water Reuse Action Plan*. US Environmental Protection Agency.

US EPA, OW (2015). "Contaminants of Emerging Concern Including Pharmaceuticals and Personal Care Products." Reports and Assessments. US EPA. August 18, 2015. https://www.epa.gov/wqc/contaminants-emerging-concern-including-pharmaceuticals-and-personal-care-products.

US EPA, OW (2018). "Potable Reuse Compendium 2017." Data and Tools. US EPA. https://www.epa.gov/ground-water-and-drinking-water/potable-water-reuse-and-drinking-water.

Wang, M., Zhang, D.Q., Dong, J.W., and Tan, S.K. (2017). Constructed wetlands for wastewater treatment in cold climate — a review. *Journal of Environmental Sciences* 57 (July): 293–311. https://doi.org/10.1016/j.jes.2016.12.019.

Webber, M.E. (2016). *Thirst for Power: Energy, Water, and Human Survival*. Yale University Press https://smile.amazon.com/Thirst-Power-Energy-Water-Survival/dp/0300212461/ref=tmm_hrd_swatch_0?_encoding=UTF8&qid=1596749901&sr=1-2.

Wenten, I.G. and Khoiruddin (2016). Reverse osmosis applications: prospect and challenges. *Desalination, Advances in Membrane Des: Keynotes from MEMDES 2-Singapore* 391 (August): 112–125. https://doi.org/10.1016/j.desal.2015.12.011.

Werner, M. and Schäfer, A.I. (2007). Social aspects of a solar-powered desalination unit for remote Australian communities. *Desalination*, EuroMed 2006 203 (1): 375–393. https://doi.org/10.1016/j.desal.2006.05.008.

WHO (2011). *Pharmaceuticals in Drinking Water*. World Health Organization.

Ziolkowska, J.R. (2015). Is desalination affordable?—Regional cost and price analysis. *Water Resources Management* 29 (5): 1385–1397. https://doi.org/10.1007/s11269-014-0901-y.

13

Adaptation for Drought and Flooding Resilience

In this chapter, we examine various ways to make both urban and rural communities more resilient to drought and flooding events. Modern technologies, such as remote sensing and seasonal to sub seasonal forecasts, form an integral part of early warning systems that help populations anticipate and prepare for natural hazards. Drought adaptation has a long history in the form of rainwater harvesting, both on a watershed and on a household scale. Managed aquifer recharge is an intentional form of temporary storage in an aquifer from which water can be withdrawn when needed. Flood protection has often taken the form of reservoir construction, river channel modification, and placement of dikes. Modern urban planning includes low-impact development that increases stormwater infiltration while reducing peak flow runoff and entrapped pollutants. Large dams are becoming less desirable as a default technology as the social and environmental and economic impacts are now more widely understood. Water security, as threatened by floods and drought, will necessarily include adoption of many of these forms of adaptation and resiliency.

Learning Objectives

Upon completion of this chapter, the student will be able to:

1. Understand the devastating impacts of droughts and flooding on human and environmental well-being.
2. Understand some of the remote sensing and modeling tools that can be used to form early warning systems for drought and flooding.
3. Recognize some of the common drought adaptation methods and technologies, including rainwater storage and infiltration.
4. Recognize some of the common flood adaptation methods and technologies, including low-impact development in urban areas.
5. Understand the social and environmental impact of large dams.
6. Understand some of the options for drought and flood mitigation available to urban areas through use of a case study.
7. *Be able to perform the calculations regarding household rainwater vessel sizing and retention basin design.

Fundamentals of Water Security: Quantity, Quality, and Equity in a Changing Climate, First Edition.
Jim F. Chamberlain and David A. Sabatini.

Companion website: www.wiley.com/go/chamberlain/fundamentalsofwatersecurity

13.1 Introduction

Water security is a function of water availability over a time frame of interest. Nature itself provides punctuated periods of extremes in water quantity – both droughts and flooding – which may be exacerbated or mitigated by human modifications of natural topography. In addition, these punctuations are becoming more severe in nature as a consequence of a warming climate (Bates 2008) (See Chapter 6).

Historically, extreme events were generally intermittent in any one location, giving local communities time to recover from one event and prepare for the next one. But as events become more frequent and more extreme, recovery times are often not sufficient for resource-challenged communities. Droughts can reduce irrigation potential, cause wells and ponds to run dry, devastate agricultural production, disrupt hydroelectric power, stimulate migration, and produce or exacerbate conflicts from competition over scarce resources. Floods can destroy homes, crops, workplaces, and water-treatment facilities; carry water-borne diseases; and diminish water quality to a point of diminished usability. These direct impacts are often accompanied by indirect impacts that include loss of ecosystem services, impaired transportation, and diminishment of culture and leisure. Each of these possible outcomes is magnified in communities with little or no resources for adaptation.

In the United States alone, droughts and flooding result in enormous economic costs as well as loss to human life (Table 13.1). The occurrence and severity of modern wildfires in the western states were exacerbated by drought conditions. Flooding and hurricanes take an enormous economic toll, while Hurricane Harvey alone (2017) resulted in $125 billion dollars of damage in Houston and forced 32 000 people out of their homes and into shelters (Amadeo 2020). Globally, the World Resources Institute (WRI) estimates that by the year 2030, riverine (inland) flooding will impact 132 million people and result in a loss of $535 billion each year (WRI 2020). Money spent for infrastructure adaptation would seem to be money well spent in areas that are most vulnerable.

The findings of the Fifth Intergovernmental Panel on Climate Change (IPCC) Assessment report predict an increase in annual mean precipitation at high latitudes and Pacific nations near the equator. Wet tropical regions are expected to see more frequent and intense precipitation events resulting in possible flooding. Conversely, mid-latitude and subtropical dry regions are expected to suffer severe reductions in freshwater resources. The predicted risks to flooding and increased landslides are very high in Central and South America, while drought-related water and food shortage are ranked high in Asian nations (IPCC 2014).

Table 13.1 Billion-dollar weather events that affected the United States, 1980–2017.

Disaster type	No. of events	Total losses ($USD billions)	Average event cost ($USD billions)	No. of deaths
Drought	25	236.6	9.5	2993
Wildfires	15	53.6	3.6	238
Flooding	28	119.9	4.3	540
Severe storm	91	206.1	2.3	1578
Tropical cyclone	38	850.5	22.4	3461

Source: Based on Smith (2018).

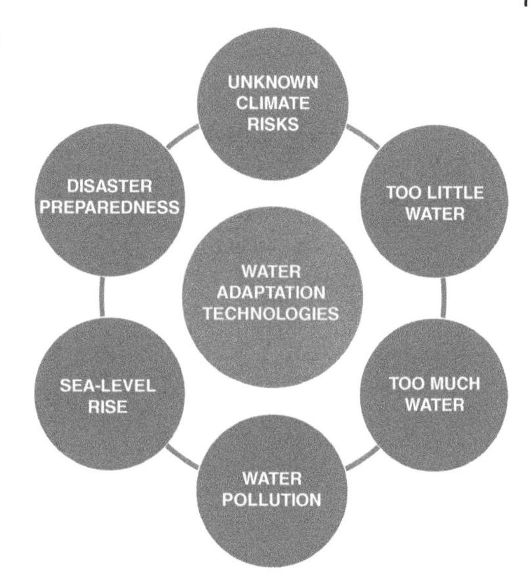

Figure 13.1 Challenges to water security from climate change. Source: Original adapted from UNEP (2017).

In the parlance of climate change, **mitigation** measures are those pre-emptive actions that are taken to reduce the flow of greenhouse gases into the atmosphere. **Adaptation** measures are those taken to lessen the risk and severe effects from climate change on people, property, and the environment (NASA 2021). The goal is to reduce vulnerability and provide added resilience. The United Nations Environmental Program (UNEP) has proposed and described over 100 water-adaptation technologies that can be used to reduce the effects of six major water challenges due to climate change (Figure 13.1).

Regarding droughts and flooding, these technologies include water efficiency, augmentation, and storage (droughts) as well as riverine flood protection and urban storm water management (flooding). In this chapter, we explore the various means by which humans can predict, prepare for, and adapt to these two extremes of water-related weather events.

13.2 Remote Sensing and Early Warning

Because drought, by definition, results in a lack of locally available water, any type of forewarning or prediction would be of great benefit in the proper allocation of current and alternative water supplies. Satellite-based remote sensing (RS) has been a proven tool for detecting and monitoring changing land conditions as an impending drought threatens. By providing real-time observations of precipitation and vegetation, RS can enable drought assessment at multiple scales and at different levels of detail (Mapedza et al. 2019).

In 2006, the United Nations developed a knowledge portal called Space-based Information for Disaster Management and Emergency Response (UN-SPIDER). Practitioners can both share and retrieve data from the latest technologies in order to aid decision-making in preparing for potential drought conditions (United Nations 2021) (Figure 13.2). Most countries now have local meteorological agencies that collect and analyze precipitation data, which can be used to produce a standardized precipitation index (SPI). The SPI is a widely used index that quantifies the observed precipitation as a departure or variation from a standardized probability function of the long-term historic precipitation (Keyantash 2018). Once the historical data is transformed into a normal (symmetric) distribution, then SPI values less than 0 are

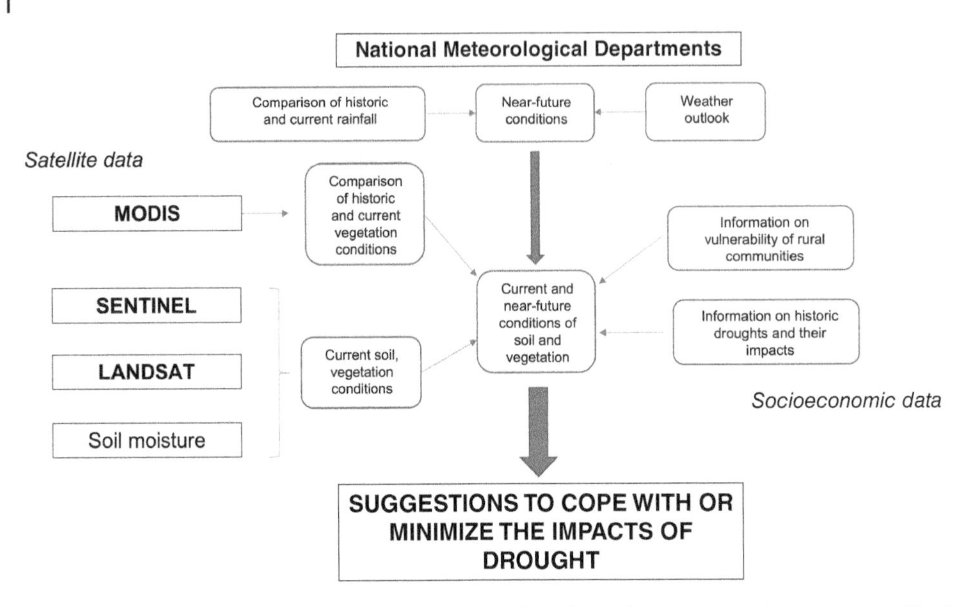

Figure 13.2 Decision support system for drought adaptation using early warning systems, utilized by UN-SPIDER. Source: Adapted from Graw et al. (2019).

below the median while those above 0 are higher than the median. The SPI can be calculated for different time spans, e.g. to match the agricultural cycle. The United States also uses the Palmer Drought Severity Index (PDSI) that includes both temperature and precipitation data in drought forecasting. Vegetation and land cover can be monitored via satellite data, such as that from MODIS, LANDSAT, and other platforms, while soil moisture is detected using infrared and passive microwave sensors (Graw et al. 2019).

The consequences of drought, as well as decisions on how to cope, are based on the exposure and impact on vulnerable populations. Socioeconomic data can also inform decision-making by providing information on historic droughts and their consequences as well as information on current vulnerable communities. Small-scale farms lacking irrigation are much more dependent on rainfall and will be affected much earlier than larger farms.

RS-based data has increasing value in providing early warning for drought conditions via the monitoring of climate-related variables, vegetation conditions, and the dynamics of surface waterbodies. When coupled with local historical meteorological and socioeconomic data and models, managers have a large suite of tools with which to minimize the impacts of drought to their communities.

Floods are, by their nature, more sudden and unexpected. Nevertheless, early warning systems can utilize data from sensors that measure water levels at strategic points throughout the watershed, using river and reservoir gages to forecast a potential flood event. The government of India has begun implementing a flood early warning system (FLEWS) in the state of Assam, one of the poorest states in India with 36% of the population living below the poverty line. Poor infrastructure, remoteness, and loss of productivity from frequent flooding have impeded development in the region. FLEWS utilizes real-time hydrometeorological data from weather radar and on-ground stations to evaluate flood risk and issue flood warnings. The system relies on the accurate modeling of catchment response to this precipitation data. Such a tool can be one element of saving lives and lifting people out of poverty (Government of India 2013).

13.3 Drought Adaptation

The water security risk of "too little water" is driven by climate change, population growth, and the per capita increase in water demand. Droughts can be categorized as one of three types. A *meteorological* drought is a situation in which precipitation levels fall well below average for a given location (Mapedza et al. 2019). In actuality, it is the easiest to measure, although rainfall may have wide variability over a single large basin, and its severity is conditioned by the timing of the drought within a wet or dry season.

Practically speaking however, the absolute amount of precipitation that falls is of less practical importance than the residual soil moisture, groundwater reserves, and water in available surface storage and access. Consequently, a *hydrological* drought is declared when water levels drop in surface waterbodies, wetlands, and aquifers. A most recent example is found in the depletion of water storage for the city of Cape Town, South Africa, in 2018. Successive years of below-average rainfall left the reservoirs nearly empty as the city imposed severe water restrictions before rains finally returned that summer.

Another useful category of drought is *vegetative* drought in which there is insufficient soil moisture for healthy plant growth and survival (Mapedza et al. 2019). Plants will self-indicate when the wilting capacity of soil has been reached, turning from green to brown as chlorophyll becomes inactive. As mentioned above, the PDSI uses temperature data to predict levels of evapotranspiration during periods of low rainfall. This index has been successful at quantifying long-term drought (NCAR 2021). A "flash drought" is a combination of high surface temperatures and declining soil moisture over a relatively short period of time (Mo and Lettenmaier 2015).

UNEP guidance presents a series of technological categories for meeting the challenge of water deficit:

- Water efficiency and demand management, such as conservation measures
- Alternative water sources, such as seawater desalination
- Equitable water allocation using modeling, forecasting, and seasonal rationing
- Water-source augmentation and storage, such as reservoirs, rainwater harvesting (RWH), managed aquifer recharge (MAR), transmission pipelines, and stormwater retention (UNEP 2017)

Each of these is discussed below:

Water efficiency and conservation are the low-hanging fruit for developed nations to reduce water demand (Chapter 11). The US EPA's WaterSense program certifies water-efficient appliances and promotes smart water use inside the house. Xeriscaping is a practice of planting native species in landscapes and gardens, reducing the need for lawn watering and promoting smart water use outside the house. Drought-resistant crops, no-till farming, and drip irrigation are all techniques to conserving water on the larger scale of agriculture (Chapter 7). Seepage and evaporative losses in canals, reservoirs, and impoundments can be reduced by liners or plastic shade balls on pond surfaces (Howard 2015).

Alternative water sources can come from urban wastewater as well as brackish and saline waters in regions of great need. Small and decentralized desalination systems are becoming more cost-effective in supplying water for drinking and other household uses. Reclaimed wastewater, with additional treatment, can provide a year-round supply of water in urbanized areas with piped networks (Chapter 11). A national water grid, much like a power grid, would transfer water from water-rich areas to water-poor areas in times of need.

The *allocation of water among users* utilizes a variety of techniques. Hydrological and economic models may be used in tandem to efficiently allocate water among competing water users. Once a total water availability is estimated, the environmental and economic impacts of each type of use are used to devise optimal water-management plans. Allocations may vary on a seasonal or areal basis with the overall goal to maintain high levels of water productivity throughout the year. Allowance should be made for both voluntary and nonvoluntary reallocations during times of particular stress, such as critical water shortages during a drought (UNEP 2017). Some archaic water laws, such as prior appropriation in the United States, preclude optimal allocations. These can be a real barrier to water security.

Water-source augmentation and *storage* form the last strategy for drought adaptation. Surface water reservoirs and aquifers provide short- and long-term storage, and transmission pipelines carry water across watersheds to points of high demand. These are major infrastructure projects requiring large financial outlays and much energy in the case of moving water, thus placing a heavy burden on low-resource communities.

Given this burden, we focus now on two of the most promising practices for the future – RWH (large and small scale) and MAR.

RWH on a large scale has been utilized for nearly as long as agriculture has been practiced. On the farm scale, this technology can be as simple as intercepting the flow of rain with berms and barriers. RWH via field-scale infiltration is also known as in situ water harvesting. It can be of various forms including the following most common:

- Terracing along a contour slows the overland sheet flow enough to moisten the soil and infiltrate into the groundwater. A small impoundment will serve the same purpose and has been used for many centuries in rural parts of India and other countries (Figure 13.3). Rajendra Singh, winner of the Stockholm Water Prize in 2015, has assembled a team to newly construct these water-harvesting structures (*johads*, in traditional jargon) across rural India. These improvements increased groundwater recharge by as much as 20%, and dry borehole wells began to refill (Postel 2017).

Figure 13.3 View of a *johad* in Thathawata village (India) built to collect and store rainwater overland flow. Source: LRBurdak / CC BY 2.5 / Wikimedia Commons.

- A recharge (soak) pit or land depression allows the rain to slowly replenish a groundwater aquifer. The pit can be filled with stones so that it is not a safety hazard. Farm ponds tend to be larger, serve the same function, and have been commonly used in the rural United States.
- Conservation tillage is any tillage practice that minimizes soil disturbance and erosion while conserving water. This can be accomplished by no- or low-till practice and leaving crop residues on the surface to increase soil moisture and rain infiltration (Table 13.2). The table shows that as more residue is left in place on the field, less rain runs off the site and the less soil is lost.

RWH on a small scale has also been practiced for centuries. On the household level, rainwater collection is still a common practice for supplemental water in rain-rich regions, such as southeast Asia and Central America, and it is gaining popularity in the United States. Rainwater is usually diverted via rain gutters off of a roof of impermeable surface, such as metal, tin, or tile. In addition to drinking water, RWH can provide a supplemental source of water for cleaning, cooking, personal hygiene, home-based gardening, and watering of livestock (Figure 13.4). Any suitable roof structure, on a dwelling, school, or community building, can be used to collect rainwater, and it can be stored in vessels of various sizes between rain events.

Table 13.2 The effects of surface residue cover on loss of rain and soil in the Midwestern United States.

Residue cover (%)	Runoff (% of rain)	Soil loss (tons/acre)
0	45	12.4
41	40	3.2
71	26	1.4
93	0.5	0.3

Source: Janssen and Hill (1994).

Figure 13.4 A rainwater-harvesting system in the Maldives. The pipe system allows for a "first flush" of rainwater during the initial minutes of a rain event. Source: Roel/Adobe Stock.

Household RWH has the following advantages:

- It is a simple technology that local people can be trained to construct, use, and modify.
- The water source is free, except for the cost of the infrastructure – guttering, pipes, storage vessels.
- After the first flush, the rainwater is generally clean and free of pathogens.
- The system reduces reliance on groundwater sources and springs.
- Rainwater will not contain geogenic contaminants, such as arsenic and fluoride, in areas where these may be prevalent in groundwater.
- The source is available in very close proximity to its end use, so there is no need for transport or piped delivery.

The disadvantages of household RWH include the following:

- The water's taste and color may not be as desirable as groundwater.
- Uncovered storage vessels may be a breeding place for mosquitos.
- It is only amenable to areas with sufficient rainfall.
- First flush water quality may be impaired by bird droppings and dust accumulation on the roof.

When properly sized, RWH systems can help households, schools, or communities through recurring periods of drought or low rainfall. Because of its low cost and adaptability, RWH will likely continue to be a drought adaptation option for many years to come.

Managed aquifer recharge. MAR is the practice of temporarily storing water underground for future water needs. The storage may be accomplished through the use of spreading basins, aquifer storage and recovery (ASR) wells, or intermittent stream beds. These practices may be used in arid regions, such as the Southwest United States or central Australia, both for storing water for times of drought and for alleviating risk from stormwater flooding. The recharge is typically done during times of high precipitation and/or low water demand.

Below-ground storage offers some advantages over above-ground storage of water which may suffer from:

- Losses from evaporation;
- Sediment accumulation, such as behind dams;
- Inability to satisfy ecosystem needs;
- High real estate and construction costs; and
- Disruptions on cultural heritage and value (IWR 2020).

Managed aquifer storage consists of five major components – water source, recharge method, storage and management, recovery method, and end use of the recovered water. Each of these is discussed briefly below:

- Source waters can range from urban stormwater runoff to treated wastewater or drinking water. Water quality may be improved as desired through the use of natural or constructed wetlands.
- The recharge method chosen will depend on the availability of suitable land, the infiltration characteristics of the soil, drilling costs, and the depth and confinement type of aquifer. Spreading (infiltration) basins and intermittent streams may be used to recharge an *unconfined* aquifer. A drilled well is required to use a *confined* aquifer (Figure 13.5).
- Management and storage include attention paid to both quality and quantity of the stored water. Quality may be improved by biodegradation of organics, inactivation of pathogens

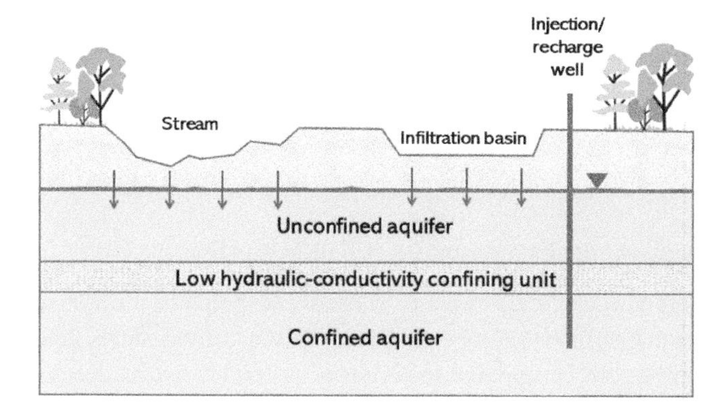

Figure 13.5 Aquifer recharge for unconfined and confined aquifers.

through soil aquifer treatment (SAT), and adsorption of metals. But it may also be reduced by increase in salinity and uptake of geogenic contaminants, such as arsenic or fluoride. Migration may occur from leakage from a confined aquifer.

- Recovery of the stored water may be through the same well as used for recharge (in the case of ASR wells) or a separate well downgradient for this purpose.
- Recovered water may be used for drinking water upon treatment or for industrial purposes depending on the resultant water quality (e.g. low total dissolved solids, low alkalinity) (IWR 2020).

Water for storage can also be obtained from increased (excess) streamflow from melting glaciers. Early runoff from global warming can be harvested. Water-rights issues may be a problem, especially in the western United States.

As part of the Comprehensive Everglades Restoration Plan (CERP), two ASR facilities were installed next to Lake Okeechobee and the Hillsboro Canal in central Florida. Their purpose was to inject treated surface water into the Floridan aquifer system for recovery during dry periods. The recovered water could provide necessary environmental flow for endangered aquatic ecosystems. After a decade of operation, the wells were found to produce 100% recovered water of satisfactory quality, although brackish water from one well will need successive cycles of treatment. Arsenic mobilization occurred in early stages but has attenuated over time (SFWMD 2018).

The US Army Corps of Engineers (USACE) is allowed to assist in developing MAR projects in the United States as long as the project cost is borne by the non-Federal owners (IWR 2020). The USACE is currently involved in over 30 projects around the United States. These projects, and others like it, show the potential benefits of this type of underground water storage. (The reader is also referred to Chapter 12, which includes a discussion of MAR as a technology for water reuse.)

13.4 Flood Resilience

Flooding events are usually assessed and compared using the common metric of total cost of damages tallied *after* the event has occurred. But a basic knowledge of hydrology can often provide a clearer picture of both the inherent vulnerability of a watershed and the steps needed to correct its fragility, or at least mitigate losses.

Strategies for flood resilience. Flood-resilience strategies can occur under four primary topical areas (Figure 13.6):

- Plan for new development in *safer areas* that are less prone to flooding.
- Conserve land in *river corridors* and discourage development in these areas.
- Protect people, buildings, and facilities in *vulnerable settlements* in or near the floodplain.
- *Manage stormwater* in the watershed to reduce stormflow (US EPA 2014).

First, *safe development areas* should be identified by cities. These areas should be promoted for development over areas that may be more prone to flooding. In the United States and developed nations, building codes should be updated to conform to flood-resistant design and construction, such as the American Society of Civil Engineers' (ASCE) Standard 13.

The US Federal Emergency Management Agency (FEMA) has tools and resources to assist local communities in managing a floodplain. Community functions include zoning, establishment of building codes, enforcement, and education. FEMA sets minimum floodplain standards, but higher standards may be adopted at a local level to provide greater protection against flooding risk (FEMA 2021b).

Second, *land in river corridors* should be conserved and protected from development. Local governments can acquire conservation easements on land that is frequently flooded. Riparian and wetland vegetation can be restored in those areas subject to erosion. For those citizens who live in riparian zones, they can be encouraged to install swales and ponds to capture

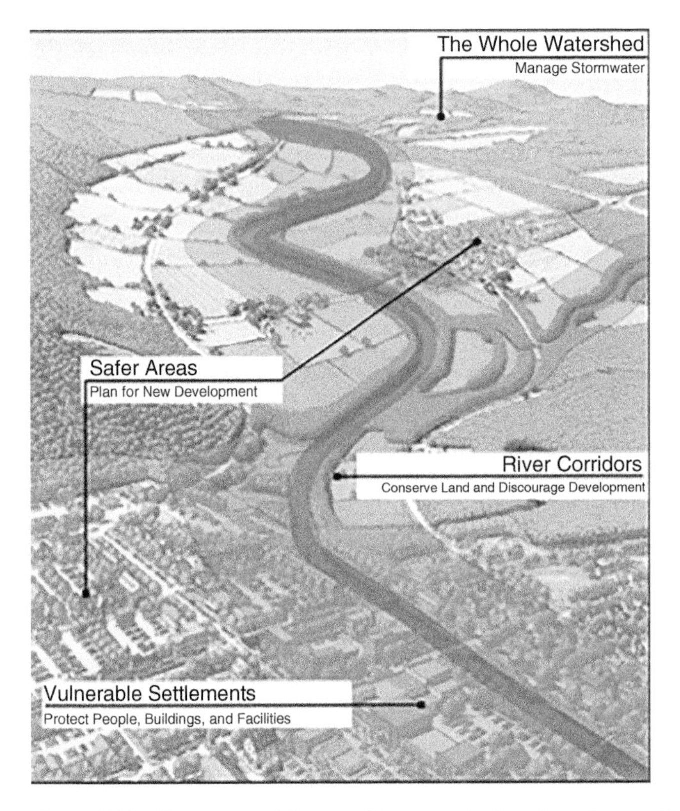

Figure 13.6 Four areas of flood resilience and adaptation. Source: US EPA (2014)/with permission from US EPA.

stormwater and alleviate flooding. Agricultural land–management practices can be used to improve the capability of soil to retain water. These practices include crop rotations, no-till farming, and the addition of water-retaining organic matter.

Third, dwelling and buildings that are already located in *vulnerable areas* should be protected as much as possible in a sustainable fashion. Damage can be ameliorated through flood-proof basements and elevation of HVAC (heating, ventilation and air conditioning) systems. Levees may be constructed to encircle or disconnect vulnerable facilities from floodwaters. New buildings should be constructed at or above the base flood elevation (BFE), which is the elevation of surface water resulting from a flood that has a 1% chance of occurrence (Federal Alliance for Safe Homes [FLASH] 2021; FEMA 2021a). Additional flood-storage capacity can be added with parks and undeveloped green spaces to amend stormwater absorption.

The Dutch have centuries of experience in flood resilience. Approximately 30% of the Netherlands (especially in western regions) is below sea level, making it extremely vulnerable to flooding (Figure 13.7). In addition, 55% of housing is in flood-prone areas (ClimateWire 2012). In 2006, the Dutch launched a "Room for the River" program to address flood mitigation. Its goals are to (i) widen rivers to make room for flooding and (ii) restore the natural landscape and biodiversity of water courses. Mitigation plans are tailor-made for each river based on existing infrastructure, spatial features, and future needs.

The centuries-old tradition of building higher and higher dikes is now being replaced with new thinking regarding the nation's waterways. Centuries of deposition in a river's flood plain had raised the land elevation of the riparian zone. These zones are now being excavated, making more room for the river's water. Groynes (bank stabilization structures) are being lowered so as not to impede water flow, and obstacles are removed or bridges raised. In

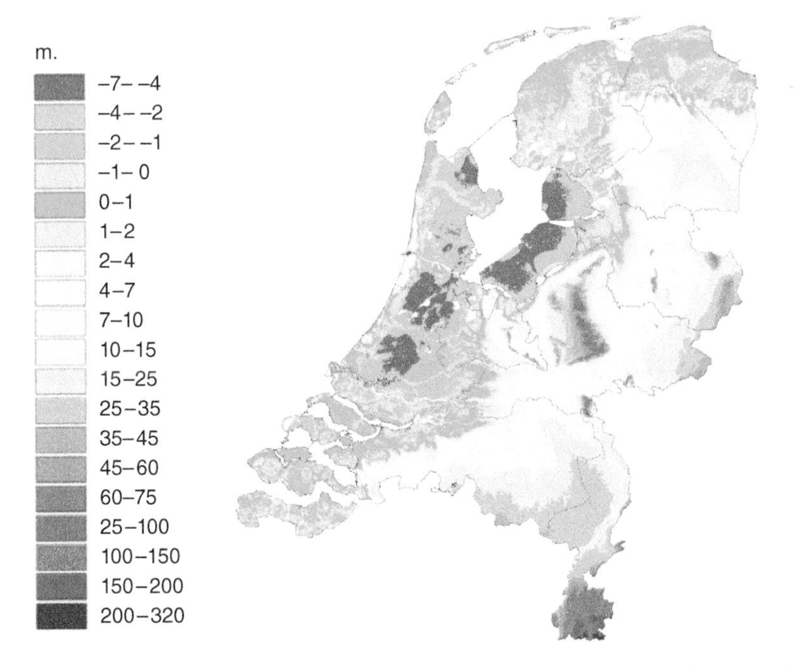

Figure 13.7 Contour map of the Netherlands showing land elevation in meters (m) above or below sea level. Source: Blom-Zandstra et al. (2009)/with permission from IOP Publishing.

addition, dikes are being relocated farther away from the river to further expand the river's carrying capacity. These concepts are not only applicable to the Netherlands but also to any place that has relied on dikes to prevent localized flooding, such as along the Mississippi River in the United States.

The fourth topical area for flood resilience is *stormwater management* within the watershed. Much that was said above about RWH will apply here as well. In rural areas, stormwater can be managed by capture and use in small earthen reservoirs and retention ponds. The ponds can serve a dual purpose for both flood control and water storage for landscape water, irrigation, and groundwater recharge.

Urban planning and best management practices (BMPs). On an urban scale, stormwater management is best employed with the techniques of low-impact development (LID). LID is the new paradigm for sustainable urban development that utilizes freshwater runoff (stormwater) as a resource that can be intercepted, stored, and reused on site (Vogel 2015). This design makes the best use of natural hydrologic processes and the creation of green infrastructure (GI) to maximize the capture and beneficial effects on both water quantity and quality. In essence, this strategy of development seeks to maintain as much as possible the original hydrology of the landscape to be developed regarding stormwater capture, infiltration, and duration and intensity of storm runoff (NAHB Research Center 2003). From the perspective of the developer/builder, additional costs in the planning stage may be offset by the reduction in conventional stormwater infrastructure, such as culverts, curbing, gutters and downspouts, detention ponds, and concrete storm channels. Instead of routing large volumes of flow away from densely populated areas, LID functions to dampen the flow and stimulate infiltration into the subsurface. This water thus replenishes aquifers that can be a source of drinking water, irrigation, or water for industrial uses. In the process of capture and infiltration, excess nutrients (nitrogen, phosphorus) and other contaminants may be removed from the surface water runoff, decreasing the concentrations of these pollutants entering downstream water bodies.

The threats to water security posed by the proliferation of impermeable surfaces and land alterations have been introduced in the chapter on hydrology (Chapter 3). Flooding has the greatest potential to do costly damage in urban areas with high-density populations and development. These areas thus have the most to gain from LID practices. The purpose of this section is to (i) review the most common practices, called Best Management Practices (BMPs), for LID; (ii) estimate the benefits that may accrue from the usage of these practices; and (iii) present a few brief LID case studies.

LID seeks to preserve and recreate natural landscape features, minimize effective imperviousness, maximize soil infiltration of stormwater, and create functional and esthetically appealing site drainage that treats stormwater as a resource rather than a waste product (US EPA 2015). BMPs used in LID can be categorized as structural and non-structural:

- Structural BMPs are those technologies that reduce runoff quantity and improve water quality before stormwater can enter the receiving stream (Omaha Stormwater 2014). Examples include:
  - Sand filters, bioswales, permeable pavement, rain gardens, rainwater collection barrels, and vegetated rooftops
- Nonstructural BMPs are those that involve proper site design or activities that reduce quantity of water and source of pollution. Examples include:
  - Redirection of rainwater into vegetated swales, disconnection of impervious cover, flattening of slopes or providing terracing, usage of native groundcover and vegetation, and maintaining natural drainage features

These practices together replace the conventional urban drainage design that relies on impermeable pavement, street curbing, and storm culverts to quickly collect and displace stormwater runoff.

Benefits of LID. Table 13.3 highlights some of the differences between conventional development and LID. LID results in an improvement of water quality as stormflow is allowed to infiltrate slowly or be retained, which allows natural processes to reduce nutrient and pollutant concentrations. The bulk of the costs in traditional land development are in the infrastructure – culverts, curbs, and channels – while the costs in LID are mostly in planning and design.

A few brief examples. Several cities in the United States have begun to implement LID design in the last two decades and are now sharing results and lessons learned.

- Seattle, Washington, is a large city in the Pacific Northwest. The city gets abundant rainfall, around 38 in. (965 mm) per year. In 2003, the city replaced 1300 linear feet of culverts and road ditches with stair-stepped pools that encourage infiltration, slow the damaging stormwater flows, and trap pollutants in slow settling pools from a 28-acre basin. The LID improvements reduced stormwater flow in these locales by 25–50%, total suspended solids by 84%, motor oil by 92%, and nutrients by 63% (EPA 2012).
- A study in the Trailwoods housing subdivision of Norman, OK, paired two halves of a five-acre development – one with conventional drainage (curbs, gutters, storm sewers) and one with an LID drainage of rain gardens, rain barrels, and permeable pavement. The results showed that the LID design reduced the peak flow of basin discharge for a 4-in. spring rain event and the total discharge volume was 57% that of the conventional design (Nairn 2016) (Figure 13.8).
- Philadelphia, PA, replaced two square miles of impervious surface with LID practices – stormwater plantings, wetlands, rain gardens, and porous paving. During a two-year period, the improvements reduced runoff by a half billion gallons, relieving pressure on the overwhelmed storm-sewer system that often overflowed and released untreated sewage into surface waters (EPA 2012).

Table 13.3 Comparison of traditional and alternative (low-impact development) practices of land development.

Element	Traditional practice of land development	Low-impact development
Overall cost	Most of the cost in infrastructure – culverts, curbs, concrete channels	Greater cost upfront in planning; lower cost in infrastructure; greater blending of infrastructure and natural topography
Cost of individual features	Impermeable paving and construction well known and less expensive	Permeable paving more expensive on a unit basis; cheaper over life cycle because of reduced maintenance
Effect on water quantity	More water leaves the site as stormflow is efficiently transported using impermeable surfaces, storm drain inlets, culverts, and concrete channels	More stormflow is detained and allowed to infiltrate into subsurface on site; detention ponds
Effect on water quality	No attempt to improve water quality of stormflow	Water quality is improved through filtration, sorption, capture by plants, and evapotranspiration

Source: Adapted from NJDEP (2021) and Mihelcic and Zimmerman (2014).

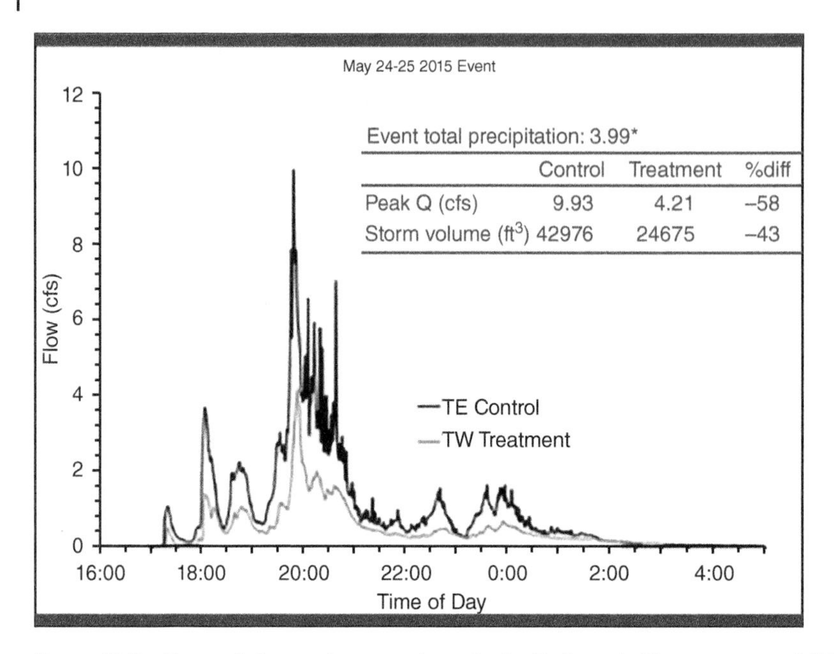

Figure 13.8 The peak flow and storm volume in the Trailwoods West treatment (LID) section were 58 and 43% lower than the control (conventional) section following a 4-in. rainfall event. Source: Nairn (2016).

The results of these case studies show that localized flooding in urban areas can be substantially ameliorated through sensible designs that increase infiltration and reduce peak stormflows.

13.5 The Global Impact of Large Dams

The last decade of the twentieth century saw increased tension and conflicts surrounding large dams and their impacts on people and the environment. The World Bank was frequently targeted as a funder for many of these large infrastructure projects (Schulz and Adams 2020). By this time, an estimated 45 000 large dams (greater than 15 m in height) had been constructed around the world, and many more were being planned in developing countries for hydropower and irrigation needs (Wescoat and White 2003).

Large dams have negative impacts that must be weighed against their positive benefits. Sediments are trapped behind the dam producing a clearer outflow that accelerates erosion and diminishes sediment deposition below the dam. Aquatic species are stressed through lack of sufficient flow downstream, while terrestrial species are lost through inundation in the reservoir. In spite of structural measures, such as fish ladders and trucking, spawning species of fish suffer declines of population. Riparian communities that depend on periodic flooding are also imperiled. On the human side, the construction of large dams often results in the displacement of persons, homes, and livestock. Research has shown that the fate of these people has only recently been considered in the deliberations. And even when resettlement plans are made, they are often not implemented in actuality and rarely are people left in a situation as favorable as prior to the relocation (Wescoat and White 2003).

A panel of experts was assembled in 1997 to comprehensively study the social, economic, and environmental impacts of global dams. The report of the World Commission on Dams (WCD), issued in 2000, proposed a new framework consisting of seven priorities that comprise a more human-centered approach to large dams (WCD 2000):

1. A new project should gain public acceptance rather than simply meet a specific need.
2. An assessment of other options for power and water supply should be conducted.
3. Existing dams need to be evaluated and decommissioning considered.
4. Impacted rivers should be recognized for their overall value in sustaining human livelihoods.
5. Displaced peoples who have suffered injustices should share in the benefits and their entitlements should be recognized.
6. More attention should be given to baseline assessments and ongoing monitoring and evaluation following project construction.
7. Rivers that cross international boundaries should be shared for peace, development, and security.

The WCD report made great strides toward inclusion of broader human values but fell short of giving entitlements to ecosystem organisms and services. A more recent protocol published by the International Hydropower Association (IHA) advocates following a series of evaluative protocols to assess the sustainability of a dam project at all life cycle stages: planning, preparation, implementation, and operation (IHA 2018). With this protocol, project effects on community livelihoods, indigenous peoples, biodiversity, and public health are considered at nearly every stage.

13.6 Case Study – San Antonio, Texas

With its beautiful river walk and diverse culture, San Antonio, TX, is both a vacation destination and a pleasant place to live (Figure 13.9). The city receives 33 in. of rain per year and sits on a climatological divide between subtropical and arid desert regions of the state. Historically, most of the city's water has been drawn from the precious Edwards Aquifer, a victim of unregulated pumping. The city has faced moderate-to-extreme drought for 17 out of the last 19 years, with 2005 and 2007 being the only drought-free years. In response, the city has developed a comprehensive set of conservation strategies to fight periods of drought.

The following actions were taken by the city to increase its drought resilience (San Antonio Water System [SAWS] 2021):

- Restrictions on watering lawns and washing impervious surfaces were enforced, depending on the water level in the Edwards Aquifer under the city.
- Rebates were given to upgrade water-efficient equipment and appliances for both residential and commercial users.
- Over 250 000 traditional toilets were replaced with low-flow models.
- A tiered water tariff structure was instituted to incentivize conservation.
- Xeriscaping was promoted to establish drought-resistant landscapes.
- Conservation easements over aquifer recharge zones were purchased.
- Direct nonpotable water reuse was begun with a capacity to deliver 31 MGD of reclaimed water for industrial use and landscaping.
- A brackish groundwater desalination plant was built with a capacity of 12 MGD.

Figure 13.9 San Antonio is a city that uses water in many capacities for both its tourists and its full-time residents. Source: City of San Antonio (2021).

One of the more successful initiatives was the "Plumbers to People" program, begun in 1994. A full one-quarter of the city's population live below the US Federal poverty level, and most live in housing that is more than 50 years old. Plumbers are employed to repair various leaks in bathtub/showers, bathrooms, kitchen, water heater, water line, and hose bibs. The SAWS pays for these repairs at no cost to the consumer. As of 2012, the program has resulted in savings of 1517 acre-feet (ac-ft; 494 million gallons) of water per year at a cost of only $80 per acre-foot. This can be compared to water-conveyance costs that range from $120 to $1250 per acre-foot (Raucher and Tchobanoglous 2014). Additional cost savings to SAWS come from a reduction of costs from bill delinquency and disconnection/reconnection of water service (SAWS 2012). Perhaps more importantly, it increases the quality of life for San Antonio residents.

On the commercial level, Frito-Lay company (based in San Antonio) spent $1.4 million on water-saving upgrades, including the recovery of steam condensate and replacement of nozzles and washers. The result was a savings of $138 000 and 43 million gallons of water a year. That is the equivalent of the water consumed by 460 city households each year (Barnett 2012).

The city is currently exploring additional drought-resilience measures, such as direct potable reuse, stormwater management to enhance runoff infiltration, and the expansion of brackish water desalination. With these measures, the city hopes to achieve an average daily water consumption of 112 gal per capita in 2025 (a reduction of 9.6% from 124 gal per capita in 2017). This consumption value is approaching other western cities in 2013 – Phoenix (111 GPCD) and Denver (100 GPCD) – but higher than New York City (61 GPCD) (Statista 2021).

As a result of these measures, while San Antonio's population has grown by 80% over the past 30 years its overall water demand has increased by only 20% (SAWS 2021). With a new desalination plant coming online and the largest groundwater-based ASR facility in the nation, the city now hopes to reduce its reliance on the Edwards Aquifer from a high of 70% in 2000 to 31% by the year 2070 (Figure 13.10) (SAWS 2017).

Figure 13.10 San Antonio's reliance on the Edwards Aquifer has been diminishing in large part to the city's conservation efforts and alternative water sources. Source: Based on SAWS (2017).

13.7 Conclusion

As we have seen previously in the water budget equation (Chapter 3), precipitation is the only major input variable into a watershed. That is to say, the only new freshwater to enter a watershed is via water from the atmosphere (neglecting some groundwater inflow). This is water that will again exit the watershed or be bound up in plants or a soil matrix unless it is captured for a time, either as surface water, in an aquifer, or in a storage vessel. Water security, then, includes the efficient capture of freshwater precipitation as needed by the human and ecological community, especially in times of drought.

The occurrence and severity of drought and flooding events have become more uncertain due to the non-stationarity of climate change (Chapter 6). However, adaptation and resilience measures have been shown to be effective in both rural and urban contexts as well as household and commercial levels. If cities are to continue to exist in water-poor or flood-prone areas, their future water security will depend in large part on the willingness of populations to design and adopt these measures.

*Foundations: Rainwater Vessel Sizing and Stormwater Detention/Retention Basin Design

In this section, we consider the design of water storage on two scales – household and small areal plot scales. In the first, rainwater collection is used to supplement household water supply. In the second, retention and detention basins are used to store and slow peak stormflow runoff, especially in urban areas of development. The first design will help carry a household through a time of drought (or low rainfall), while the second will both mitigate the effects of flooding and replenish underground aquifer storage of groundwater.

Rainwater vessel sizing. The most obvious surface upon which to collect rain is an impermeable roof, such as that of tin, sheet metal, composite shingles, or tile. A gutter system can be used to direct water into a storage container, such as a plastic tank, pjeng jars, or ferrocement vessels. A first-flush system may be used to divert the first few liters (or gallons) of rainwater away from the storage vessels, which is of low quality due to the washing of the roof of dirt, leaves, bird droppings, waterborne chemicals, etc.

A good rule of thumb for first-flush roof cleansing is to allow 40 l of wash water for every 100 m^2 of roof area (~10 gal for a 1000-ft^2 roof) (Mihelcic et al. 2009). A simple pivot assembly using PVC pipe can be used to swivel the pipe both away from and back to the storage tank.

The amount of water that can be collected from a roof (V_r) is given by the equation:

$$V_r = P \times A \times C \tag{13.1}$$

where

V_r = Volume of water collected from a roof (liters)
P = Precipitation (mm)
A = Roof area (m^2)
C = Splash coefficient, a unitless fraction of the rainwater that hits the roof and reaches the storage system, commonly 0.8–0.85 depending upon roofing material (Mihelcic et al. 2009). [Note that the unit conversion factor = 1, since "mm-m^2" × (1 m/1000 mm) × (1000 l/m^3) = liters.]

Thus, a rain event of 25 mm on an 80-m^2 roof area would collect 1600–1700 l. If the roof had a first-flush system, then ~32 l would be diverted away for cleaning the roof surfaces.

Over an annual time period, the storage needed can be calculated by doing a monthly analysis of rainfall, household demand, and first flush in order to determine the proper amount of storage needed.

The equation for volume at the end of each month would be:

$$V_t = V_{t-1} + V_r - D - (V_{ff} \times n) \tag{13.2}$$

where

V_t = Volume stored at the end of the current month
V_{t-1} = Volume stored at the end of the previous month
V_r = Volume of rain that can be collected from the roof
D = Monthly demand of the household
V_{ff} = Volume diverted as first flush per rain event
n = Average number of rain events per month

A rainwater vessel can be sized such that the household always has at least 50 l at the end of each month, accounting for some variability in demand. For example, the average rainfall in Cambodia is given in Table 13.4.

Suppose that an average Cambodian household has a roof size of 4 m by 6 m (24 m^2). There are six persons living in the house with an average demand rate of 8 lpcd. This gives a monthly demand, D, of 1340 l per month. For two rain events per month, there is a first-flush diversion of 19.2 l (40 l/100 m^2 × 24 m^2 × 2 events). We will choose a vessel size of 3600 l in order to ensure that the family does not run out of water in the dry season (December–April). Of course, a storage vessel cannot store more than its capacity. So, during the rainy season, some rainwater will not be captured as the most appropriate vessel will be as small as possible.

A sample calculation for the month of December in this setting would be (using Eq. (13.2)):

$$V_{Dec} = V_{Nov} + V_{Dec,r} - 1340 - (9.6 \times 2) = 3600 + (32.8)(24)(0.85) - 1300 - 19.2 = 2809\ \text{l}$$

Figure 13.11 shows the results over the entire year. At the end of March, the vessel storage is at its lowest, only holding 77 l.

The practitioner will discover that in many regions, the proper sizing of a RWH tank will provide adequate water during the midst of and probably a few months on either side of the rainy season. During the dry season, additional sources would likely be needed.

Table 13.4 Average rainfall in Cambodia.

Month	Jan	Feb	Mar	Apr	May	Jun	Jul	Aug	Sep	Oct	Nov	Dec
Rainfall (mm)	13.5	20.7	45.6	89.0	133.1	265.4	273.9	310.6	327.6	253.1	111.6	32.8

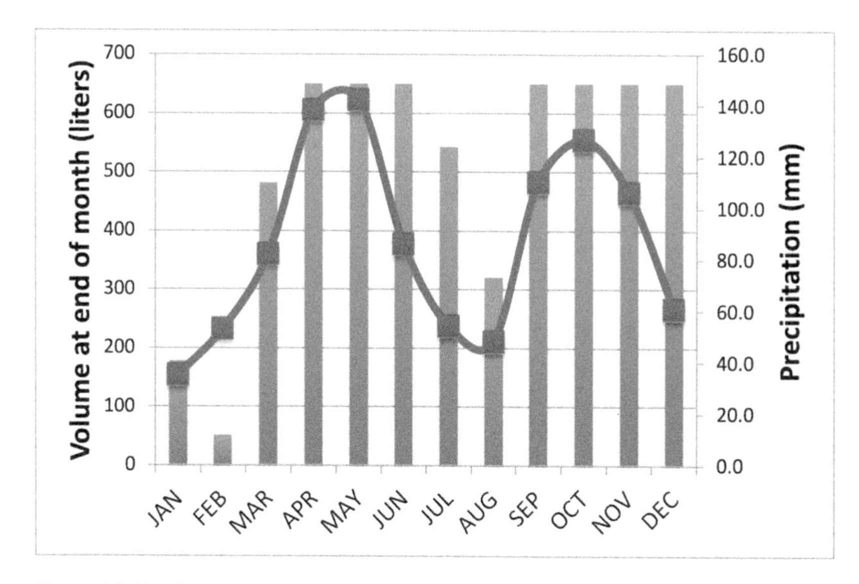

Figure 13.11 Rainwater vessel sizing for a typical household in Cambodia. Vertical bars are rainwater storage at the end of each month; solid line is average monthly precipitation.

Detention basin design. Stormwater runoff from a development with a sizeable area of impermeable surfaces can be significant. A *detention basin* collects water temporarily, reducing the peak runoff flow from a rain event (Figure 13.12). The basin may be dry in-between events as the water percolates into the subsurface or is slowly released through a low elevation outfall. A *retention basin*, as its name suggests, is similar in purpose but will retain water on a more permanent basis.

A common method used for calculating peak runoff is called the rational method, in which the peak flow runoff rate, Q_p, from an areal extent of land is given by:

$$Q_p = CiA \tag{13.3}$$

where

Q_p = Peak flow runoff rate (cfs)
C = Runoff coefficient corresponding to a land use type (dimensionless)
i = Rainfall intensity (in./hour)
A = Watershed area under study (acres)

The runoff coefficient has been estimated for various land uses and is shown in Table 3.4 of Chapter 3. As seen there, the percent impervious area in the watershed has a great influence

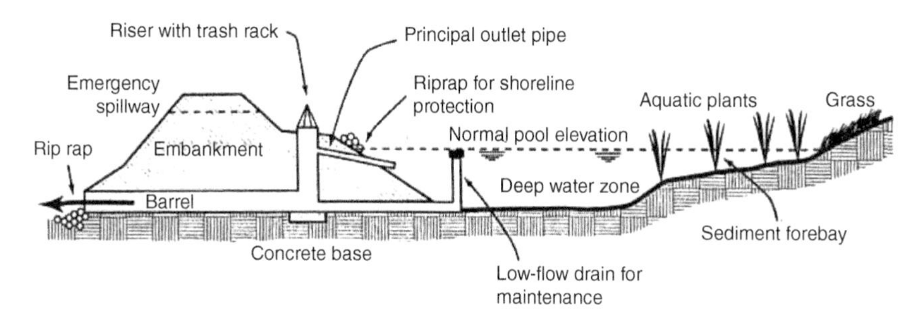

Figure 13.12 Conceptual design of a stormwater retention pond. Source: Clemson Cooperative Extension Service (2021)/with permission from Clemson University.

on the total peak flow. When a single area has multiple land uses, then the C value should be weighted according to these areas.

For example, consider an 80-acre plot of land (clayey soil, gentle slope) that is divided into three sections of commercial use (40 acres, $C = 0.72$), medium-density residential (20 acres, $C = 0.33$), and open space (20 acres, $C = 0.16$). The coefficient can be calculated as:

$$C = \frac{40}{80} \times 0.72 + \frac{20}{80} \times 0.33 + \frac{20}{80} \times 0.16 = 0.483$$

And the peak flowrate from a storm that averages 1.5 in./hour will be:

$$Q_p = 0.483 \times 1.5 \times 80 = 57.9 \text{ cfs}$$

Once a peak runoff has been determined for a particular design storm, culverts and pipes can be sized to carry this peak rate of flow. In addition, the positive effects of using LID concepts can be measured quantitatively. In the above example, if one-half of the commercial district was converted to more highly permeable surfaces, then the C would be reduced to 0.34 and the peak flow would be 41.1 cfs, a reduction of 29%.

An in-line detention basin is designed to intercept the runoff from a watershed, especially one that has undergone development with increased impermeable surfaces. This basin is often designed to ensure that the peak runoff rate after development is no greater than the runoff rate before development. For design purposes, a 2-, 10-, or 100-year storm event is often used. From published rainfall data, a runoff hydrograph can be developed for each size storm. This hydrograph relates the progression of flow over time from a certain size storm over a particular watershed. As such, the hydrograph is unique to both the storm and the watershed. Two US agencies – the Natural Resources Conservation Service (NRCS) and the USACE – have developed manuals and models that help the water scientist develop hydrographs for the area of interest.

The peak runoff rate for a developed site in an urban area will typically have a post development peak that is higher and occurs faster than the predevelopment peak (Figure 13.13).

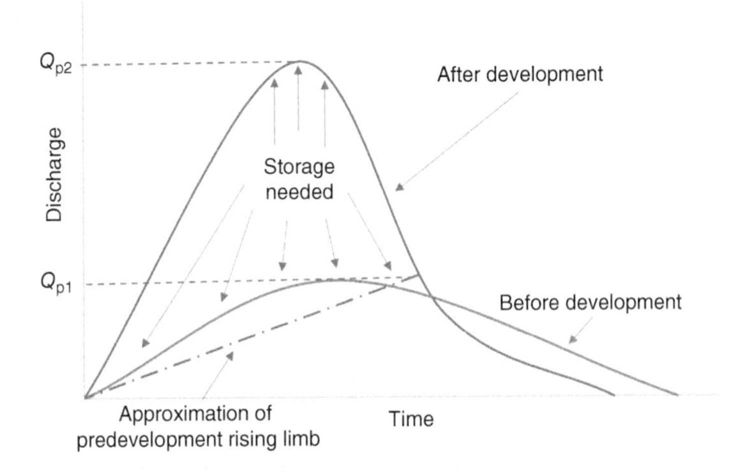

Figure 13.13 Conceptual sketch of peak flow runoff discharge rate before development (Q_{p1}) and after development (Q_{p2}).

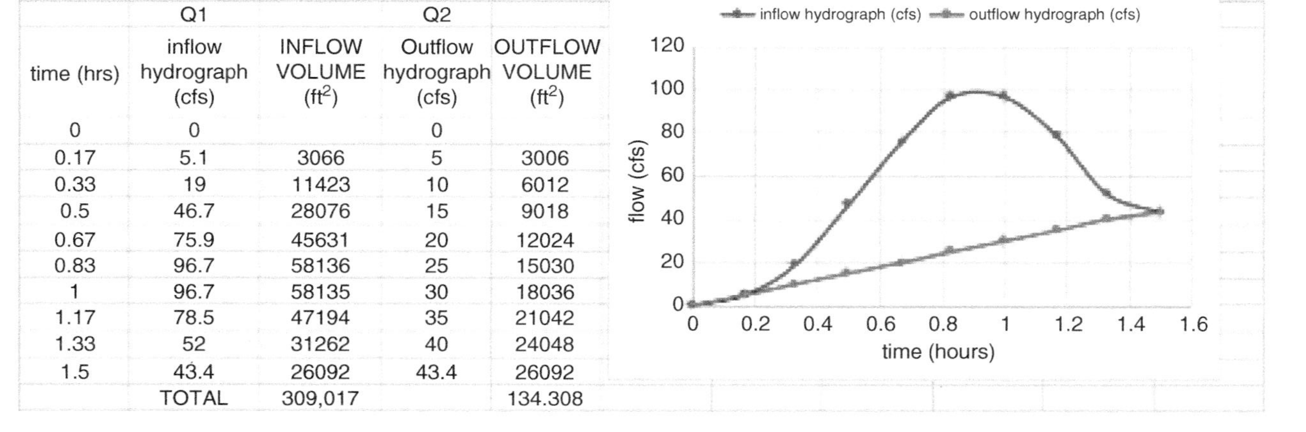

	Q1		Q2	
time (hrs)	inflow hydrograph (cfs)	INFLOW VOLUME (ft^2)	Outflow hydrograph (cfs)	OUTFLOW VOLUME (ft^2)
0	0		0	
0.17	5.1	3066	5	3006
0.33	19	11423	10	6012
0.5	46.7	28076	15	9018
0.67	75.9	45631	20	12024
0.83	96.7	58136	25	15030
1	96.7	58135	30	18036
1.17	78.5	47194	35	21042
1.33	52	31262	40	24048
1.5	43.4	26092	43.4	26092
	TOTAL	309,017		134.308

Figure 13.14 Sample design for detention basin using an inflow hydrograph and a desired outflow hydrograph at predevelopment peak. Source: Modified from Wurbs and James (2001).

Note that the area under a hydrograph represents a volume of water, calculated as the product of flowrate and time. Thus, the storage needed for a detention basin is shown in the area between the two hydrographs, as shown by the arrows. The predevelopment rising limb can be estimated using a straight line as shown in Figure 13.13 (dashed line). Note that this is merely an approximation but is adequate for a rough design.

The following example illustrates the sizing of a detention basin. (A simpler design is given in Chapter 3.) Consider a design storm with the inflow hydrograph for post development shown over a time of one and a half hours. The desired outflow hydrograph is a straight line to where the predevelopment peak, which is known to be 43.4 cfs, intersects the inflow hydrograph (Figure 13.14).

The desired detention volume storage is calculated as the difference between inflow and outflow volumes = 174 708 ft^3 = 4.0 ac-ft. A detention basin should be designed to store this volume of water in order to meet the requirement of no runoff peak greater than the predevelopment peak.

This exercise has demonstrated the first step in detention basin design. The next step would be to design the pond itself with suitable geometry of length, width, depth, and side slopes. The final step will be to choose and size an outlet structure that will release the water at appropriate flowrates during your design storms. There are software packages and design manuals available to help the manager in all these steps.

End-of-Chapter Questions/Problems

13.1 Compare and contrast the impacts of droughts and flooding.

a. Pick an LMIC (low- and middle-income country) of interest to you. Using the Internet and other sources, what have been the impacts of both drought and flooding over the past two decades? Give both economic losses and mortality/morbidity effects.

b. Pick a developed country of interest to you (European, North American, etc.). Using the Internet, what have been the impacts of both drought and flooding over the past two decades? Give both economic losses and mortality/morbidity effects.

c. Compare and contrast the results of the two nations. How are they different/the same? What variables are most significant?

d. What steps can be taken to help each country adapt to its natural hazards? Give three possible solutions.

13.2 Use the Internet to learn about the Famine Early Warning System Network (fews.net) developed to signal areas of acute food stress throughout the world.

a. List and describe three of the top areas of famine stress in the near term.

b. Which of these (if any) are experiencing famine stress due to drought conditions?

13.3 Use the UN-SPIDER web page to answer the following:

a. What services can UN-SPIDER offer to member states?

b. What are the website's current predictions regarding drought and water scarcity?

13.4 Describe how the Standard Precipitation Index (SPI) is calculated?

13.5 Research one flooding event in your hometown or region.
a. What were the impacts of this event?
b. What was the return period of the event, i.e. how extreme was it?
c. What steps could be taken to lessen future impacts of a similar-sized storm?

13.6 "Flash drought" is a term that is more commonly used today. Provide answers to and references for the following questions:
a. Define flash drought. How is it determined?
b. How does "flash drought" compare with "drought" under the definition used with the PDSI?
c. How does climate change affect the severity and frequency of flash droughts?
d. In what regions of the world are flash droughts becoming common? How do these affect agriculture and plant life in these regions?

13.7 LID may be considered to be an example of Engineering with Nature (EWN).
a. How does LID imitate the natural hydrology of a site?
b. What are three ways in which LID improves water quality?
c. What is the purpose of vegetation in LID systems?

13.8 Create a table that compares the differences of methods and costs of LID to conventional stormwater management treatment systems.

13.9 What are three alternative ways that European countries are working on reducing impervious surface coverage? Consider policy, behavioral, and technical solutions.

13.10 The Grand Ethiopian Renaissance Dam in Ethiopia has been the subject of much controversy.
a. Briefly summarize the scope and history of the project.
b. Give both pros and cons regarding the dam's construction.
c. What steps can be taken, in your opinion, to lessen the negative impacts of the dam?

13.11 Use the following monthly rainfall data to properly size a rainwater storage vessel of a family of five. Assume a roof size of 24 m^2, a demand rate of 7.5 lpcd, and a runoff coefficient of 0.8. The first flush should be 0.4 l/m^2 of roof area. Assume one flush per month. The storage vessel capacity should be given in liters and should be sufficient so as to never go below 50 l in any one month (Table 13.5).

Table 13.5 Monthly rainfall averages (mm of rain).

Jan	Feb	Mar	Apr	May	Jun	Jul	Aug	Sep	Oct	Nov	Dec
35.0	53.0	82.0	138.0	142.2	86.0	54.0	48.0	110.0	126.4	106.0	60.7

13.12 Compare the peak runoff rate for two adjacent 20-acre sites. Site A is completely undeveloped, whereas Site B is 40% commercial, 40% medium-density residential, and 20% open space. Use a rainfall intensity of 2.0 in./hour.

Further Reading

NAHB Research Center (2003). *The Practice of Low Impact Development*. Upper Marlboro, Maryland: US Housing and Urban Development.
Barnett, C. (2012). *Blue Revolution: Unmaking America's Water Crisis*. Beacon Press.
EPA (2012). *Effectiveness of Low Impact Development*. US Environmental Protection Agency.
UNEP (2017). *Climate Change Adaptation Technologies for Water*. United Nations Environmental Programme.
Wescoat, J.L. and White, G.E. (2003). *Water for Life*, 1e. Cambridge; New York: Cambridge University Press.

References

Amadeo, K. (2020). "Hurricane Harvey Shows How Climate Change Can Impact the Economy." The Balance. 2020. https://www.thebalance.com/hurricane-harvey-facts-damage-costs-4150087.
Barnett, C. (2012). *Blue Revolution: Unmaking America's Water Crisis*. Beacon Press.
Bates, B. (2008). "Climate Change and Water — IPCC." IPCC. https://www.ipcc.ch/publication/climate-change-and-water-2.
Blom-Zandstra, M., Paulissen, M., Agricola, H., and Schaap, B. (2009). How will climate change affect spatial planning in agricultural and natural environments? Examples from three Dutch case study regions. *IOP Conference Series: Earth and Environmental Science* 8 (November): 012018. https://doi.org/10.1088/1755-1315/8/1/012018.
City of San Antonio (2021). "River Walk." The City of San Antonio - Official City Website. 2021. https://www.sanantonio.gov/CCDO/DowntownEvents/ArtMID/20498/ArticleID/2516/River-Walk.
Clemson Cooperative Extension Service (2021). "Stormwater Pond Design, Construction and Sedimentation | College of Agriculture, Forestry and Life Sciences | Clemson University, South Carolina." 2021. https://www.clemson.edu/extension/water/stormwater-ponds/problem-solving/construct-repair-dredge/index.html.
ClimateWire (2012). "How the Dutch Make 'Room for the River' by Redesigning Cities." Scientific American. 2012. https://www.scientificamerican.com/article/how-the-dutch-make-room-for-the-river.
EPA (2012). *Effectiveness of Low Impact Development*. US Environmental Protection Agency.
Federal Alliance for Safe Homes (FLASH) (2021). "Floods: Flood Zone - Which One Are You In." 2021. https://flash.org/peril_inside.php?id=58.
FEMA (2021a). "Base Flood Elevation (BFE) | FEMA.Gov" 2021. https://www.fema.gov/node/404233.
FEMA (2021b). "Floodplain Management | FEMA.Gov." 2021. https://www.fema.gov/floodplain-management.
Government of India (2013). *FLEWS: Flood Early Warning System - a Warning Mechanism for Mitigating Disasters during Flood*. Government of India.
Graw, V., Dubovyk, O., Duguru, M. et al. (2019). Chapter 9 - Assessment, monitoring, and early warning of droughts: the potential for satellite remote sensing and beyond. In: *Current Directions in Water Scarcity Research*, Drought Challenges, vol. 2 (ed. E. Mapedza, D. Tsegai,

M. Bruntrup, and R. Mcleman), 115–131. Elsevier. https://doi.org/10.1016/B978-0-12-814820-4.00009-2.
Howard, B.C. (2015). "Why Did L.A. Drop 96 Million 'Shade Balls' Into Its Water?" Science. August 12, 2015. https://www.nationalgeographic.com/science/article/150812-shade-balls-los-angeles-California-drought-water-environment.
IHA (2018). *Hydropower Sustainability: Assessment Protocol.* International Hydropower Association.
IPCC (2014). "Climate Change 2014: Synthesis Report Summary for Policymakers." IPCC.
IWR (2020). "Managed Aquifer Recharge and the U.S. Army Corps of Engineers: Water Security through Resilience." US Army Corps of Engineers.
Janssen, C. and Hill, P. (1994). "Conservation Tillage - Purdue University Cooperative Extension Service." 1994. https://www.extension.purdue.edu/extmedia/ct/ct-1.html.
Keyantash, J. (2018). "Standardized Precipitation Index." *The Climate Data Guide* (blog). 2018.
Mapedza, E., Tsegai, D., Bruntrup, M., and Mcleman, R. (2019). Drought challenges: policy options for developing countries. In: *Current Directions in Water Scarcity Research*, vol. 2, (ed. E. Mapedza, D. Tsegai, M. Bruntrup, and R. Mcleman), iii. Elsevier. https://doi.org/10.1016/B978-0-12-814820-4.21001-8.
Mihelcic, J.R. and Zimmerman, J. (2014). *Environmental Engineering: Fundamentals, Sustainability, Design*, 2e. Wiley.
Mihelcic, J.R., Fry, L.M., Myre, E.A. et al. (2009). *Field Guide to Environmental Engineering for Development Workers: Water, Sanitation, and Indoor Air*. American Society of Civil Engineers.
Mo, K.C. and Lettenmaier, D.P. (2015). *Flash Droughts over the United States.* Denver, CO: NOAA.
NAHB Research Center (2003). *The Practice of Low Impact Development*. Upper Marlboro, Maryland: US Housing and Urban Development.
Nairn, R. (2016). "Optimizing the Role of Natural Infrastructure in Watershed-Based Approaches." Presented at the Big XII Water Conference, Baylor University.
NASA (2021). "Climate Change Adaptation and Mitigation." Climate Change: Vital Signs of the Planet. 2021. https://climate.nasa.gov/solutions/adaptation-mitigation.
NCAR (2021). "Palmer Drought Severity Index (PDSI)." *Climate Data Guide* (blog). 2021.
NJDEP (2021). Chapter 2: Low impact development techniques. In: *New Jersey Stormwater Best Management Practices Manual* JFC (ed.). NJDEP – New Jersey Department of Environmental Protection.
Omaha Stormwater (2014). "Omaha Regional Stormwater Design Manual." 2014. https://omahastormwater.org/orsdm.
Postel, S. (2017). *Replenish: The Virtuous Cycle of Water and Prosperity*. https://smile.amazon.com/Replenish-Virtuous-Cycle-Water-Prosperity/dp/1642830100/ref=sr_1_1?dchild=1&keywords=Replenish+Postel&qid=1593794678&s=books&sr=1-1.
Raucher, R.S. and Tchobanoglous, G. (2014). *The Opportunities and Economics of Direct Potable Reuse*. Alexandria, VA: WateReuse Research Foundation.
SAWS (San Antonio Water System) (2012). *Plumbers to People: Water Conservation Program*. San Antonio Water System.
SAWS (San Antonio Water System) (2017). "2017 Water Management Plan."
SAWS (San Antonio Water System) (2021). "Conservation - Protecting Our Water Supplies." San Antonio Water System. 2021. https://www.saws.org/conservation.
Schulz, C. and Adams, B. (2020). "The World Commission on Dams: Then and Now." The FutureDAMS Research Consortium. November 20, 2020. http://www.futuredams.org/the-world-commission-on-dams-then-and-now.

SFWMD (2018). *Aquifer Storage and Recovery: Regional Study*. South Florida Water Management District.

Smith, A. (2018). "2017 U.S. Billion-Dollar Weather and Climate Disasters: A Historic Year in Context | NOAA Climate.Gov." 2018. https://www.climate.gov/news-features/blogs/beyond-data/2017-us-billion-dollar-weather-and-climate-disasters-historic-year.

Statista (2021). "Daily per Capita Water Use in Households by Select U.S. City 2014." Statista. 2021. https://www.statista.com/statistics/886597/us-city-per-capita-water-consumption-households.

UNEP (2017). *Climate Change Adaptation Technologies for Water*. United Nations Environmental Programme.

United Nations (2021). "What Is UN-SPIDER?" UN-SPIDER Knowledge Portal. 2021. https://www.un-spider.org/about/what-is-un-spider.

US EPA (2014). "Flood Resilience Checklist | Smart Growth | US EPA." 2014. https://www.epa.gov/smartgrowth/flood-resilience-checklist.

US EPA, OW (2015). "Urban Runoff: Low Impact Development." Overviews and Factsheets. US EPA. September 22, 2015. https://www.epa.gov/nps/urban-runoff-low-impact-development.

Vogel, J. (2015). Critical review of technical questions facing low impact development and green infrastructure: a perspective from the great plains. *Water Environment Research* 87 (9): 849–862.

WCD (2000). *Dams and Development: A New Framework for Decision-Making*. World Commission on Dams.

Wescoat, J.L. and White, G.E. (2003). *Water for Life*, 1e. Cambridge; New York: Cambridge University Press.

WRI (2020). "New Data Shows Millions of People, Trillions in Property at Risk from Flooding — But Infrastructure Investments Now Can Significantly Lower Flood Risk." World Resources Institute. April 23, 2020. https://www.wri.org/news/2020/04/release-new-data-shows-millions-people-trillions-property-risk-flooding-infrastructure.

Wurbs, R. and James, W. (2001). *Water Resources Engineering*, 1e. Upper Saddle River, NJ: Pearson.

13a The Practice of Water Security: The Power of Change Agents

Martha Gebeyehu grew up in Addis Ababa, Ethiopia, a booming metropolis of over three million people and home to one of the fastest growing economies on the continent (Figure 13a.1). While in high school she developed a love of science and pursued a college degree in applied chemistry. In the big city, she did not realize the immense difficulties faced by so many of her fellow citizens.

Figure 13a.1 Martha Gebeyehu receives the OU International Water Prize for her work as a practitioner of water security Source: Photo: OU WaTER Center.

Martha joined the Ethiopian Kale Heywet Church (EKHC) as a water quality analyst in the WaSH (water, sanitation, and hygiene) department in 2010. One day, her work took her to a rural village where she saw people fetching water from a pond and using it for drinking water. She was shocked. She recounts the incident: "As I settled down to gather a water sample from the new borehole that EKHC had dug, the community members gathered around me. As I pulled out the water people around me began to clap and shout with happiness. One very old man sang out and said he had drunk river water his whole life and he was praising God that he was able to see clean water coming from a borehole in his community within his lifetime. Before that day I had never imagined that there were people who lived their whole lives without a safe water source. It was a life changing moment for me!"

Martha continues: "It is not possible for my government to reach every village with safe, piped water. But that does not mean there are no solutions. In fact, there are many. There are both technical solutions and some systems in place that we were able to use to disseminate information and training about these solutions (that use) simple, affordable technologies. We build people's capacity through training and consulting and technical support because we believe that *people are the central part of any transformation. They are the true 'change agents'*. Human development is about the real freedom ordinary people have to decide who to be, what to do, and how to live. For people to make informed decisions they need the right information and the right skills and knowledge."

The strategy of EKHC is to build upon the WaSH structure already in place at the community level in Ethiopia. Health Extension Workers (HEWs) are paid government employees working at the local community, or kebele. There are more than 42 000 government salaried female HEWs deployed in the country. However, each HEW would be servicing 1000–3000 households – they simply could not reach everyone. Based on feedback provided by HEWs, Martha's team started to build the capacity of self-help groups (SHG). SHG are community-level groups that were initially established as small-scale savings and loan and financial literacy groups that met on a regular basis. Each group has 15–20 members. The SHG model is built on the belief that people living in poverty can be agents of change rather than merely recipients of aid.

One example comes from Abera and his family, who made great progress in their home after they received trainings about WaSH (Figure 13a.2). They built a latrine, with a hand-washing station. They installed a kitchen shelf to keep their kitchenware free of contamination. And they bought a water filter to treat their water. Moreover, Abera now teaches the people in his community about WaSH. He has become an advocate and salesman for both the Biosand Filter and the Minch filter (Desert Rose LLC). Through simple WaSH education and the peer support of the SHGs, Abera has become a change agent in his community

Figure 13a.2 Martha teaches rural families about proper sanitation and hygiene practices. Source: Photo courtesy of Martha Gebeyehu.

Martha Gebeyehu is the recipient of the 2019 OU International Water Prize

Part V

Resilience, Economics, and Ethics

14

Planning for Water-Supply Security and Resilience

Public water supplies are vulnerable to both natural and human-induced hazards, both intentional and accidental. Consequences from a hazardous event could jeopardize the health and well-being of many consumers. In this chapter, we discuss formal methods for increasing water security in community water systems. Risk, resilience, and related terms are defined as well as metrics used to measure them. There are elements of planning that are consistent across various methodologies – assessment of threats and associated risk, evaluation of physical and cybersecurity against attack, methods of monitoring and detection, and options for response to a hazard or attack. A similar type of planning can occur on a larger watershed scale. Finally, case studies are presented that highlight some of the challenges and lessons learned in water planning for security and resilience.

Learning Objectives

Upon completion of this chapter, the student will be able to:

1. Define the concepts of risk, resilience, hazards, and consequences.
2. Quantify risk and resilience.
3. Understand the methodologies for water supply planning advocated by the US Environmental Protection Agency, the World Health Organization, and a collaboration of the American National Standards Institute/American Water Works Association.
4. Understand the various elements of planning that are common to various methodologies.
5. Understand planning on a larger scale, such as the scale of a watershed.
6. Understand some of the challenges and lessons learned from water security planning case studies.

14.1 Introduction

In a rapidly expanding population, water security will not happen without careful planning. And it will not be achieved overnight. Planning for water security involves the careful consideration of the best usage and protection of water resources while meeting competing demands over time and preparing for unforeseen circumstances.

Fundamentals of Water Security: Quantity, Quality, and Equity in a Changing Climate, First Edition.
Jim F. Chamberlain and David A. Sabatini.

Companion website: www.wiley.com/go/chamberlain/fundamentalsofwatersecurity

We have already seen how droughts and flooding can be major disruptors of water for food, drinking water, hygiene, and industry. These natural occurrences and others, such as high winds and tornados, are even more common in a warming climate. But accidents may also happen, such as mistakes in system operation, malfunctioning of a monitoring system, or a pump failure. Intentional attacks on water systems include both cyberattacks and deliberate contamination or attack on the physical infrastructure. All of these incidents can and have resulted in illness, disruption of society, economic losses, and loss of life (Camarillo et al. 2014). Much less quantifiable is the resulting loss of public confidence and trust in water supply, potentially leading to fear and civic unrest.

In the United States alone, there are nearly 53 000 community water systems, a small percentage of which supply clean drinking water to 75% of US citizens (Tindall and Campbell 2011). These systems are prime targets for domestic terrorism and, indeed, any disruption of service would have major consequences. In light of both the dangers and the consequences, both national and global health organizations have promulgated planning approaches and methodologies for assessing and improving community water system risks and resilience. As we will see, these plans have a dual-use function in that they also promote and improve a higher standard of water quality and reliability even in normal operating conditions.

This chapter looks closely at methods and procedures that can be used to ensure that, as far as possible, suitable water is available and sustainable for users over a long time period. This includes the adequate response to potential disruptions from both natural and human causes, and both intentional and accidental in nature.

14.2 Concepts and Measurement

We begin with definitions of important terms. Broadly considered, **risk** is the potential for loss or harm due to any unwanted event, such as tornado, earthquake, flood, drought, accidental spills, or acts of terrorism. **Likelihood of threat** is the probability of an event occurring that might impair the utility from continuing its operation. A **hazard** is any agent (physical, chemical, biological, radiological) that can cause harm to public health, and a **hazardous event** is any incident that introduces a hazard, or fails to remove it, from a water-supply system. A **consequence** is the effect of an event, incident, or occurrence. This effect can be measured in quantifiable terms, such as gallons of water, dollars of damage, or lost revenue.

Resilience is the ability of a system to withstand a natural hazard or human attack without serious performance interruption. It is often measured as a function of the time required for a return to normal system operation. The inverse of resilience is **vulnerability**, or the likelihood of damage from a threat, should it occur. It is usually a result of weaknesses in the system.

The overall goal of a risk and resilience assessment (RRA) is to identify the highest risks to water system components (e.g. a water-treatment plant or water-supply reservoir) and lay the basis for determining cost-effective measures to increase risk resilience. In planning for water security, risk assessments are completed for any component of the system that is vulnerable to compromise from an accidental, natural, or human event. Components of interest include physical infrastructure, monitoring and maintenance systems, and financial and data infrastructure.

Regarding measurement, the probability of an event is a product of likelihood of threat and vulnerability of the system. The overall risk can then be measured as a product of probability and consequence of an event. This results in the following equation (US EPA 2020a):

$$\begin{aligned}\text{Risk} &= (\text{likelihood of threat} \times \text{vulnerability}) \times \text{consequence}\\ &= \text{probability} \times \text{consequence} \qquad (14.1)\end{aligned}$$

The first two terms can be quantified in terms of a percent probability, or can be expressed qualitatively as "high," "medium," or "low." The consequence can be expressed as a dollar amount of economic damage.

Similarly, resilience can be calculated positively as the inverse of the risk as a function of system outage, expressed as a time period of inoperability (or reduced operation) (ASCE/AWWA Committee 2010). (These metrics will be discussed again later in the chapter.)

$$\text{Resilience} = (\text{likelihood} \times \text{vulnerability} \times \text{system outage})^{-1} \qquad (14.2)$$

In summation, water-supply security planning involves the evaluation of risk from all potential hazards and the promulgation of resilience measures that reduce the threat probability or consequences of the risks.

14.3 Methods of Planning for Water Security

The temporary loss of water supply or wastewater collection and treatment can be a substantial burden on a community, both from a health and an economic standpoint. Consider the following:

- A break in a main water line due to a storm event causes a school to lose its clean water supply. The children cannot drink from a fountain or use flush toilets. If parents must keep their children home, there may be a loss of income from lost days of work. The interruption of water supply to hospitals and nursing homes has even more serious consequences.
- During plant upgrades, an operator mistakenly shuts off the chlorination system in a water-treatment plant. Dozens of households ingest water tainted with *Escherichia coli* bacteria before a "Boil Advisory" order is put in place. Elderly residents are most at risk from serious health effects.
- A cyberattack causes water outages during the busy tourist season. Hotels and restaurants lose income while temporary water supplies are secured.

In this section, we present three major planning methods used and promoted by United States and global organizations for the purpose of ensuring water security against major threats. Adequate resilience and response planning can both reduce the incidence rate of hazards and mitigate the consequences.

14.3.1 US EPA – Water Security and Risk Resilience

In October 2012, New Jersey and New York were inundated with rainfall from Hurricane Sandy (Figure 14.1). Over 100 drinking water and wastewater facilities, many with outdated infrastructure, were inundated and damaged. Some residents and businesses were without water for weeks after the storm. The estimated costs were about $45 billion (US EPA

Figure 14.1 Destruction from Hurricane Sandy devastated communities and water utilities in New York, October, 2012. Source: Sergey / Adobe Stock.

2015a). Much of this financial burden could perhaps have been lessened by careful up-front emergency response planning.

The history of water supply protection began in the early twentieth century. In 1914 the US Public Health Service set water-quality standards for many contaminants that would likely be found in community water supply sources (Figure 14.2). States adopted these standards

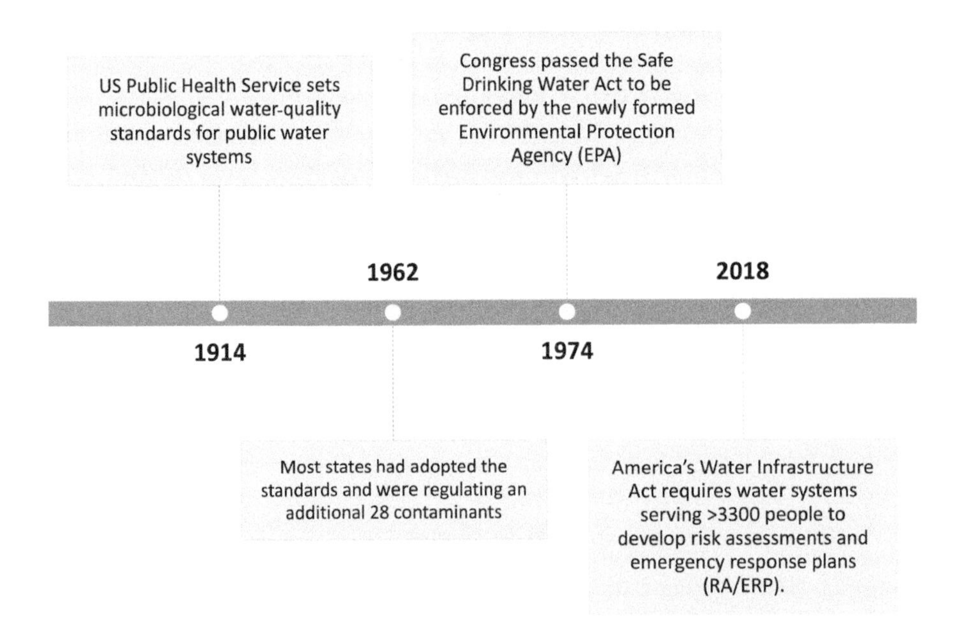

Figure 14.2 Timeline of major steps in the United States toward water-quality regulations and water security in drinking water utilities. Source: Original drawing, author JFC.

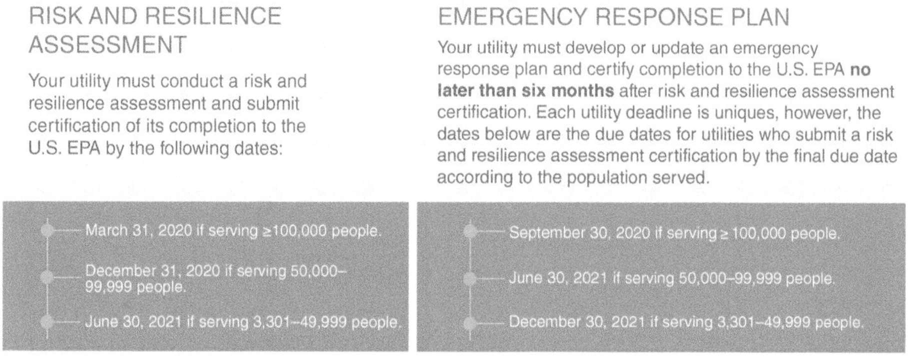

Figure 14.3 Requirements for community water systems, from the AWIA Act of 2018. Source: US EPA (2018).

as their own local guidelines. But the passage of the Safe Drinking Water Act (SDWA) in 1974 (and subsequent amendments) established more firmly a set of policies and procedures protective of human health. These policies and procedures would be the under the purview of the newly formed Environmental Protection Agency (EPA). The law mandated a three-step process in which EPA would:

1. Issue interim water-quality standards based on an update of the 1962 standards,
2. Contract with the National Research Council (NRC) and National Academy of Sciences (NAS) for a new study of drinking water contaminants that could impact human health, and
3. Revise the water-quality standards based on this new study.

In addition, water utilities must monitor their own drinking water quality and meet the standards; state agencies would provide oversight. Utilities were also required to notify customers when standards were not met or there was a lapse in monitoring. The Federal government provided financial assistance for improving water-treatment processes, replacing or repairing leaky distribution pipes, improving water supply sources, or providing other water infrastructure improvements as needed (Kimm 2020).

Following the lead and progress of the SDWA with various amendments, the US Congress passed the America's Water Infrastructure Act (AWIA) of 2018, which required all large and medium-sized water systems to develop RRA and emergency response plans (ERP) (Figure 14.3).

The process involves four discrete steps described below.

1. Step One: Conduct an RRA
   Each public water system serving a population of greater than 3300 persons must assess all of the potential risks to, and resilience of, its system.
   Such an assessment should include:
   - The risk to the system from both natural hazards and malevolent acts;
   - The resilience of the pipes and constructed conveyances, physical barriers, source water, water collection and intake, pretreatment, treatment, storage and distribution facilities, electronic, computer, or other automated systems (including the security of such systems), which are utilized by the system; and

- The financial infrastructure; the use, storage, or handling of various chemicals; and operation, maintenance, and monitoring of the system.

The assessment may also include an evaluation of capital and operational needs for risk and resilience management for the system (US EPA 2018). Using an online tool, the utility can identify the highest risks to mission-critical operations (US EPA 2015b). This assessment is required by a stated deadline and then reviewed, updated, and recertified every five years (Figure 14.3).

2. Step Two: Identify some strategies to build resilience

In order to assist utilities in this step, the EPA provides a "Resilient Strategies Guide for Water Utilities" (Guide) (US EPA 2019). Upon launching the Guide, the water manager will be asked to select priorities for his/her system. For example, the manager may decide that flooding, wildfires, and water-supply management are the biggest concerns for their system. Then, potentially vulnerable assets are chosen, including aquifers, buildings, data-acquisition systems, drinking water–treatment plant, forested lands, etc. The manager selects which of these assets are most vulnerable in his/her system. The Guide than presents a list of strategies based on the highest priorities of the water system. Some of the resilience strategies given are to:

- Implement a proactive approach to community alerts for flooding and wildfires. Notify customers of any anticipated or potential disruption of service.
- Integrate flood management and modeling into land-use planning. High-risk flood areas should be avoided when planning new infrastructure.
- Increase raw water storage capacity by increasing a dam height, practicing aquifer storage and recovery, removing accumulated sediment in reservoirs, or lowering water-intake elevation (US EPA 2019).

The Guide then lists possible funding sources based on the selected strategies and gives links to US case studies in which these strategies have been adopted.

3. Step Three: Develop an ERP

Pursuant to the AWIA, a utility must then complete an ERP within six months of certification from its RRA. This plan describes the utility's strategies to prepare for and respond to incidents, whether natural or human-induced. These may be as small as a main line break or localized flooding. Or they may be larger, such as a hurricane, power outage, or a cyberattack.

The ERP includes:

- Strategies and resources to improve resilience, including physical security and cybersecurity;
- Plans and procedures for responding to a natural hazard or malevolent act that threatens safe drinking water;
- Actions and equipment to lessen the impact of a malevolent act or natural hazard, including alternative water sources, relocating intakes, and flood-protection barriers; and
- Strategies to detect malevolent acts or natural hazards that threaten the system (US EPA 2018).

A template for developing this plan is given on the EPA website (US EPA 2018). Once the ERP is complete, all relevant personnel can be trained on its implementation, including individual roles and responsibilities.

4. Step Four: Monitor the system
   Finally, monitoring activities should focus on both water-quality surveillance and security monitoring.
   - Water-quality surveillance is done at the treatment plants and in distribution systems. Monitoring is done at various locations and for various parameters. The ERP will highlight the key parameters.
   - Enhanced security monitoring uses advanced security equipment and communication systems.

   In order to assist utilities in this important task, the US EPA offers a Vulnerability Self-Assessment Tool (VSAT), which can be used for assessing risk and resilience at drinking water and wastewater systems (US EPA 2020b). The VSAT user will first assess threat, vulnerability, and consequences for an asset/threat pair considering the utility's existing countermeasures. Then she will have the option to conduct a Countermeasure Analysis in which the threat, vulnerability, and consequences are re-evaluated by incorporating the capabilities of potential countermeasures. This allows the user to determine the cost-effectiveness of additional countermeasures for reducing risk and enhancing resilience at the utility (US EPA 2020a).

14.3.2 Risk Analysis and Management for Critical Asset Protection

A similar approach to utility security is provided by following the *Risk Analysis and Management for Critical Asset Protection* (RAMCAP) standard developed by the American National Standards Institute (ANSI) and the American Water Works Association (AWWA). The original RAMCAP Framework was developed by the American Society of Mechanical Engineers (ASME) after the 9/11 terrorist attacks (of 2001) to protect the nation's infrastructure and economy.

The seven steps to meet this standard, with related questions, are shown in Figure 14.4.

Within the overall process, the standards require a utility to ask more detailed follow-on questions with regard to its risk and resilience. For example,

- *High winds and intense storms* are common and may cause structural damage in my area. Which facilities are most likely to be damaged and how might I better protect them?
- *Information systems connectivity* is a vital part of modern public utilities. But "greater connectivity equals greater exposure" (Morley 2018). Thus, all relevant IT network(s) – for both process control systems (treatment processes, telecommunications, HVAC, etc.) and enterprise systems (payroll, billing, etc.) – must be examined and assessed for vulnerability.
- *Emergency power* is a key limiting factor in system recovery. Where is the power going to come from? Is there enough generator capacity to sustain critical customers, or do we have a partnership with the local power provider to give priority to key customers?
- Finally, *emergency drinking water supply* must also be considered, especially to homes and hospitals. Who is taking care of supply and distribution, and how much will be needed in terms of quantity?

In summation, the RAMCAP standards help a utility prepare for the *effect* of a threat without attempting to manage the *causes*, which are often out of their control (Morley 2018).

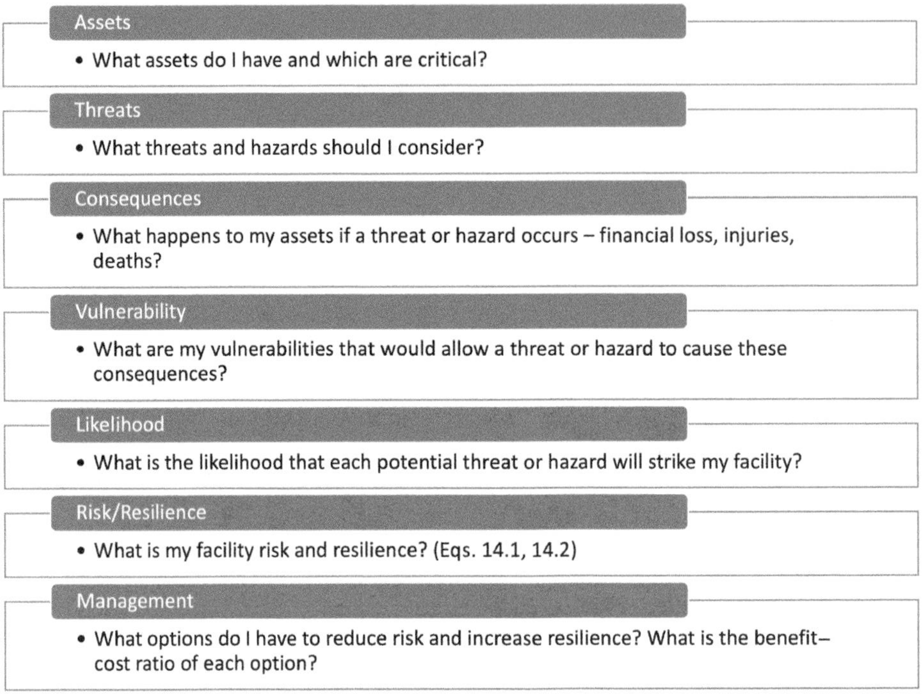

Assets
- What assets do I have and which are critical?

Threats
- What threats and hazards should I consider?

Consequences
- What happens to my assets if a threat or hazard occurs – financial loss, injuries, deaths?

Vulnerability
- What are my vulnerabilities that would allow a threat or hazard to cause these consequences?

Likelihood
- What is the likelihood that each potential threat or hazard will strike my facility?

Risk/Resilience
- What is my facility risk and resilience? (Eqs. 14.1, 14.2)

Management
- What options do I have to reduce risk and increase resilience? What is the benefit–cost ratio of each option?

Figure 14.4 The seven steps of the RAMCAP process for risk and resilience analysis. Source: Original, adapted from ASCE/AWWA Committee (2010).

14.3.3 Water Safety Plans of the World Health Organization

The Water Safety Plan (WSP) approach of the World Health Organization (WHO) is a comprehensive risk assessment and management plan that encompasses all steps in water supply "from catchment to consumer" (WHO and IWA 2009). The preliminary work is to form a WSP team that will carry out the tasks that follow. The size and composition of this team is up to the water provider, but it should include members from both within and outside the provider in order to get a broad perspective and not be committed to any one approach.

Once the team is in place, there are seven discrete tasks that comprise the bulk of the assessment and follow-on monitoring and surveillance actions (Table 14.1). Although more costly and sometimes difficult, site visits are critical for most tasks in forming an accurate plan.

Task 1 – Description of water-supply system. The first task is to describe the water-supply system, from raw water source to final customers (Figure 14.5). This exercise will help managers to pinpoint the locations and assets of greatest risk and vulnerability, whether that be the water source, a reservoir, a pump, or a treatment unit.

Task 2 – Identification of hazards. Hazardous events that can disrupt the normal operation can happen near the raw water source, in the catchment area, in the treatment plant, anywhere in the distribution network, or at or near the point of consumption (Table 14.2). The WSP team should have enough experience and knowledge to be able to assess the potential hazard risks in each of these subject areas.

Task 3 – Assess the risk using a risk matrix. The next step is to evaluate the risk represented by each hazard using a risk matrix. Using either a quantitative or semiquantitative format, this matrix incorporates both the likelihood and consequence of each hazard.

Table 14.1 Discrete tasks to be completed in the formation of the water safety plan.

Task	Site visit?	Desk exercise?
1 – Describe the water supply system	✓	✓
2 – Identify hazards that can affect safety of water supply from catchment to consumer (point of use)	✓	
3 – Assess the risk presented by each hazard using a risk matrix	✓	✓
4 – Determine if control measures (mitigating measures or barriers) are in place for each hazard and reassess the risks	✓	✓
5 – Develop and implement an improvement plan	✓	✓
6 – Define monitoring of control measures		✓
7 – Regularly review hazards, risks, and controls	✓	

Source: Modified from WHO and IWA (2009).

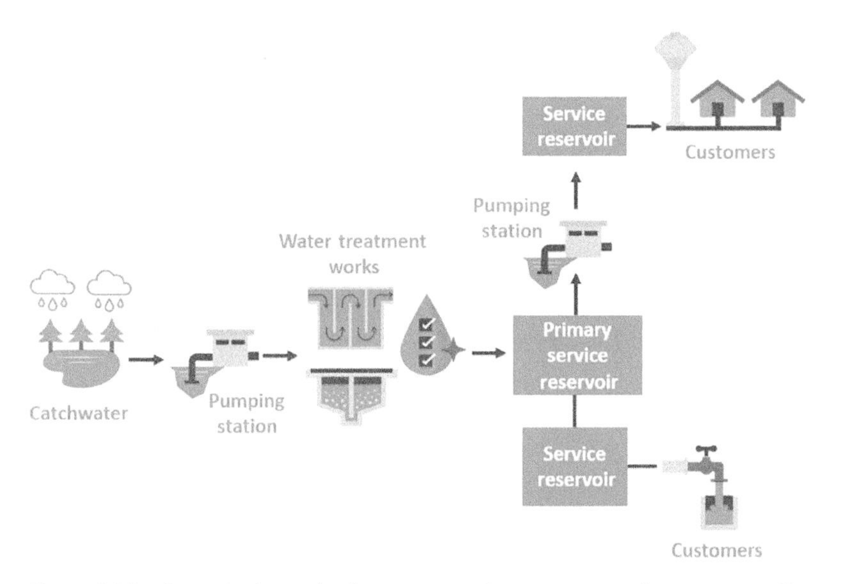

Figure 14.5 General schematic of a water-supply system to service customers. Source: Original, adapted from Hong Kong WSD (2020).

Similar to Eq. (14.1), the risk rating is a product of the likelihood (frequency) of threat and the severity (consequence):

$$\text{Risk rating} = \text{Likelihood} \times \text{severity} \tag{14.3}$$

The panel shown in Figure 14.6 is an example matrix that assigns numerical values to each level of likelihood and consequence. (In this case, vulnerabilities of all components are considered equivalent.)

For example, consider the introduction of pathogens due to the loss of network integrity. The severity of this hazard is given a "5" rating due to its public health impact, which includes disease and potential death. The likelihood of the occurrence is given a "2" since there are plumbing controls in place. The total risk score $= 5 \times 2 = 10$, which represents a "High"

Table 14.2 Examples of hazardous events in each water-supply step, from catchment to consumer.

Water-supply component	Hazardous event
Catchment	Flooding and rapid changes in source water quality
	Pesticides and nitrate from agricultural runoff
	Contamination from mining waste
	Contamination from household septic systems
	Toxins from algal blooms
Treatment plant	Interrupted treatment due to power loss
	Contamination due to vandalism
	Inadequate treatment due to lack of monitoring
	Inadequate treatment due to design failure (insufficient disinfection contact time, excessive head loss in filters, etc.)
Distribution network	Recontamination due to mains burst, corroded pipes, or defective valves
	Vandalism of storage reservoirs
Consumer	Recontamination in lead pipes
	Recontamination from leaky septic tank

Source: Adapted from WHO and IWA (2009).

		Severity or consequence				
		Insignificant or no impact - Rating: I	Minor compliance impact - Rating: 2	Moderate esthetic impact - Rating: 3	Major regulatory impact - Rating: 4	Catastrophic public health impact - Rating: 5
Like hood or frequency	Almost certain/ Once a day - Rating: 5	5	10	15	20	25
	Likely/Once a week - Rating: 4	4	8	12	16	20
	Moderate/Once a month - Rating: 3	3	6	9	12	15
	Unlikely/Once a year - Rating: 2	2	4	6	8	10
	Rare/Once every 5 years - Rating: 1	1	2	3	4	5

Risk score / Risk rating	<6 Low	6–9 Medium	10–15 High	>15 Very high

Figure 14.6 An example of a semiquantitative risk matrix, with level of severity given in columns and likelihood (frequency) given a score by rows. Source: WHO and IWA (2009).

risk score rating (WHO and IWA 2009). Risk score ratings can be used to prioritize potential events according to their ratings.

One caution in performing a quantitative risk evaluation is that, when comparing risk scores alone, one is not able to determine control measures unless the two factors making up the evaluation are known. For example, water managers in Australia considered two scenarios with the same overall risk score, but one scenario had a high likelihood but low severity in consequence, while the other had a low likelihood but high severity in consequence. The first scenario was common sporadic complaints about dirty water, which had minimal health consequences. The second scenario was about rarely occurring inadequate disinfection due to dirty water leading to severe health impacts (WHO and IWA 2009). Nonetheless, the risk-evaluation process is an important first step in prioritizing without losing sight of the supporting data.

Task 4 – Consider control measures and reassess the risk. Once a risk rating has been established for each potential hazard, control measures can be considered that will mitigate against the risk and reduce the potential consequences. This allows for a reassessment of the risk after the control has been put in place. For example, an inadequate disinfection process could result in serious health impacts if pathogens are able to survive water treatment and enter the distribution system. Two possible short-term control measures would be to (i) increase the contact time for the disinfectant and (ii) install alarms to trigger a low disinfectant level. These measures would reduce this hazard risk from "High" to a "Low" status, assuming appropriate and consistent monitoring.

Occasionally, the process of considering control measures might bring to the forefront a different prioritization of hazards. For example, in a case study from the Caribbean islands, the theft of chlorine tanks was scored "Low" in the quantitative assessment because of its low likelihood of occurrence. However, the control measure was simple – install lock boxes to prevent theft – and managers had a "gut feeling" that this simple protective measure was justified and that the lock boxes should be installed (WHO and IWA 2009).

Task 5 – Develop and implement an improvement plan. The result of Task 4 deliberations may reveal a finding that some control measures are not sufficient in their current status and measures need to be added or upgraded. The next step then is to develop and implement an improvement plan what will address specific control deficiencies. The list of improvements or upgrades should include a responsible party and a due date for implementation (Table 14.3).

For example, *Cryptosporidium* is a serious health hazard that has an increased likelihood due to lack of fencing around wellhead where cattle are allowed to graze. Two possible improvements would be to (i) install fencing around the wellhead area and (ii) include UV disinfection (or ozone) in order to deactivate the oocysts that cause cryptosporidiosis. If the latter option is chosen, validation of the disinfection technique was done by comparing theoretical treatment performance with that required in this particular setting to meet water-quality standards.

Table 14.3 Examples of drinking water system recommended improvements/upgrades with accountabilities and due dates.

Action	Arising from	Specific improvement plan	Responsible party/Due date
Control risk from *Cryptosporidium*	Cattle defecation in vicinity of unfenced wellhead	Install ultraviolet (UV) disinfection treatment	Engineer/03/2023
Control risk from agricultural pesticides	RRA identified a suite of pesticides used in area of source water intake	Install ozone and granular activated carbon filtration in treatment plant	Engineer/05/2023
Reduce risk from viral and protozoan contamination	Potential risk from nearby sewage systems	Develop additional disinfection and downstream water treatment	Water quality officer/09/2023

Source: Based on WHO and IWA (2009).

Task 6 – Define monitoring of control measures. Monitoring is the operational measurement and/or observation of the control measures to ensure that these are performing as expected. Some control effectiveness can be directly measured (e.g. pH, turbidity, chlorine concentration) while others are measured by observation (e.g. integrity of fencing, functioning of security lighting).

The task of establishing a monitoring program includes asking the following questions:

- Who will do the monitoring?
- How frequently will the monitoring be done?
- Who will analyze the samples?
- Who will interpret the results?
- Can the results be easily interpreted at the time of monitoring or observation?
- Can corrective actions be implemented in response to the detected deviations?
- Has the list of hazardous events and hazards been checked against monitoring or other appropriate criteria to ensure that all significant risks can be controlled? (WHO and IWA 2009)

For example, the chlorine residual in treated water should in the range of 0.5–1.5 mg/l. To ensure this proper concentration, a Water Quality Officer is assigned to measure the residual at various points in the distribution on a weekly basis. In the case of a deviation (chlorine level too high or too low), the Officer activates a noncompliance protocol on which she has been trained.

The sample monitoring plan shown in Figure 14.7 is from a medium-sized city in Uganda. Operational monitoring for pH, chlorine, and turbidity happens on a daily or weekly basis. Verification monitoring, which samples for the indicator organisms of *E. coli* and *Enterococci*, happens less often – weekly or monthly. The verification monitoring is done to ensure that the functioning system is meeting health-based target levels for microorganisms.

Task 7 – Regularly review hazards, risks, and controls. In order to ensure that control measures are adequate and monitoring is proceeding properly, the following verification and audit activities are necessary. The water managers should:

- Monitor compliance of treated water with chemical and microbiological water-quality standards,
- Initiate an audit of implementation activities including treatment procedures, employee training, ERP, etc., and
- Verify customer satisfaction with water-supply quality and service.

Unit process	Operational monitoring (see Module 6)			Verification monitoring		
	What	When	Who	What	When	Who
Treatment works	On-line measurement –pH –Chlorine	Daily	Water treatment operators/Analyst	*Escherichia coli*	Weekly	Analyst
				Enterococci	Weekly	
				Record audit	Monthly	
	Jar testing records	Weekly				
	Turbidity	Daily				
	Dosing records	Monthly				
Distribution system	pH	Weekly		*Escherichia coli*	Monthly	
	Turbidity	Weekly				
	Chlorine	Weekly		Turbidity	Monthly	
	Sanitary inspection	Weekly		Enterococci	Monthly	

Figure 14.7 A sample monitoring plan from water-supply system in Jinja, Uganda, with a population of ~71 000. Source: WHO and IWA (2009).

The WSP should undergo periodic review and revision, and it should be reviewed immediately following an emergency. The following questions should be asked following a significant incident (WHO and IWA 2009):

- What was the cause of the problem?
- Was the cause related to a hazard already identified in the WSP risk assessment?
- How was the problem first identified or recognized?
- What were the most essential actions required and were they carried out?
- If relevant, was appropriate and timely action taken to warn consumers and protect their health?
- What communication problems arose and how were they addressed?
- What were the immediate and longer term consequences of the emergency?
- How can risk assessment/procedures/training/communications be improved?
- How well did the ERP function?

The WSP approach has been adopted in many global regions, with both successes and lessons learned. The *Water Safety Plan Manual* offers composite case studies of lessons learned in three regions – Australia, Latin America, and Caribbean region (LAC) and the United Kingdom (WHO and IWA 2009). The insights gained here will likely apply to other regions across the world.

14.4 Discrete Elements of Planning

In this section, we look more closely at the common discrete elements of planning: (i) identification of threats and vulnerabilities, (ii) provision of physical and cybersecurity, (iii) monitoring and detection of contaminants, and (iv) countermeasures and threat evaluation and response. These elements are useful in both planning for events and improving the operational quality and efficiency of the water system.

14.4.1 Identification of Threats and Vulnerabilities

For water systems of any size or complexity, this element is crucial for generating awareness of the components (assets) of a system as well as the potential threats to each asset. Table 14.2 lists some of the hazardous events that might occur in each major component grouping – catchment/source water, treatment system, distribution system, and consumer endpoint. The specific threats from contamination may be of the following types:

- Biological contaminants can cause illness and death in an exposed population. Not all infections will produce symptoms, and the median infectious dose, ID-50, is the dose that causes infection in half of the exposed population. Similarly, LD-50 is the lethality dose that causes death in half of the exposed population. Some biological agents (biotoxins) are extremely toxic at low doses and may be used in a bioterrorism attack (Table 14.4). For example, botulinum toxins cause adverse health effects at the extremely low dose of 0.0004 ppb (μg/l) and are thus considered an extreme threat as a biological weapon (Camarillo et al. 2014).
- Chemical contaminants pose a danger that is relative to their toxicity and dose–response relationship relative to a receptor. Many commonly used, off-the-shelf chemicals can be

Table 14.4 Biological contaminants that represent a medium to high water threat.

Category	Subcategory	Examples	Associated disease
Microbiological	Bacteria	*Bacillus anthracis* *Brucella* spp. *Clostridium perfringens* *Shigella* spp., *Vibrio cholerae* *Yersinia pestis*	Anthrax Brucellosis Clostridium Shigellosis Cholera Plague
	Viruses	Enteroviruses	Enteric viruses
	Parasites	*Cryptosporidium parvum*	Cryptosporidiosis
Biotoxins	Ricin, saxitoxin, botulinum toxins, T-2 mycotoxins, microcystins		

Source: Adapted from US EPA (2015c, p. 1).

used in a malignant attack if large quantities are employed (Table 14.5). But these same chemicals may also appear from time to time in public water supply. The US EPA publishes a drinking water Contaminant Candidate List (CCL), which is a list of contaminants that are currently not subject to any proposed or promulgated national primary drinking water regulations but are known or anticipated to occur in public water systems. Contaminants listed on the CCL may eventually require future regulation under the SDWA (US EPA 2014).

- Radiological contaminants may be of concern if they are accessible to belligerent actors. Cesium-137 (30.2 years half-life), cobalt-60 (5.3 years half-life), and strontium-90 (28.8 years half-life) have been used for radiation therapy in the treatment of cancers and could be made available in a water-soluble powdery salt that is easy to disperse. Even low doses of these contaminants can result in long-term health effects (Camarillo et al. 2014).

Table 14.5 Examples of chemical contaminants found in the US EPA response planning guidelines.

Category	Subcategory	Examples
Inorganic chemicals	Corrosives and caustics	Toilet bowl cleaners, drain cleaners
	Cyanide salts	Sodium cyanide, potassium cyanide
	Metals	Mercury, lead – their salts and complexes
Organic chemicals	Hydrocarbons and derivatives	Paint thinners, gasoline, kerosene, alcohols
	Insecticides	Organophosphates, chlorinated organics
	Organics, water-miscible	Acetone, methanol, ethylene glycol, detergents
	Pesticides	Herbicides (e.g. atrazine), rodenticides
	Pharmaceuticals	Cardiac glycosides, anticoagulants, illicit drugs (LSD, PCP, heroin)
Schedule 1 chemical warfare agents		Organophosphate nerve agents (e.g. sarin, tabun, VX), vesicants, nitrogen and sulfur mustards, Lewisite

Source: Adapted from US EPA (2015c, p. 1).

14.4.2 Physical Protection and Cybersecurity

"Water system hardening" is used to denote the usage of physical barriers to deny access and prevent tampering. Physical protection systems should include system redundancy so that functionality can continue even if one barrier fails or is breached. Prevention barriers include alarms, fencing, security lighting, multiple locks, vehicular barriers, and live guards. These also include backflow prevention and emergency generators as power backup.

Water-supply networks have become dependent upon systems control and data acquisition (SCADA) systems that are used to manage automated physical processes essential to both water treatment and distribution systems (Clark et al. 2017). However, these systems are vulnerable to malicious cyberattacks, and in one year alone (2015), the US Department of Homeland Security responded to incidents in both the water sector (25 incidents) and the energy sector (46 incidents) (Clark et al. 2017).

Ransomware is any type of malicious software (aka, malware) that is designed to deny access to a computer system or data until a ransom is paid. Ransomware typically spreads through fraudulent ("phishing") emails or by a victim unknowingly visiting an infected website (CISA 2020). The malicious actor holds the data or systems hostage until a ransom is paid.

The following are some essential topics that a water-supply system should be addressing regarding ransomware and other cyberattacks:

- Backups: Do we regularly backup all critical information? Are the backups stored offline? Have we tested our ability to revert to backups during an incident?
- Risk Analysis: Have we conducted a cybersecurity risk analysis of the organization?
- Staff Training: Have we trained staff on cybersecurity best practices?
- Vulnerability Patching: Have we implemented appropriate patching of known system vulnerabilities?
- Application Whitelisting: Do we allow only approved programs to run on our networks?
- Incident Response: Do we have an incident response plan and have we exercised it?
- Business Continuity: Are we able to sustain business operations without access to certain systems in the event of a malware attack? For how long? Have we tested this?
- Penetration Testing: Have we attempted to hack into our own systems to test the security of our systems and our ability to defend against attacks? (CISA 2020)

14.4.3 Monitoring and Detection

All public water systems have monitoring strategies designed to meet regulatory water-quality requirements. These monitoring schedules were based on expected contaminants in the source water supply. They were not specifically designed to provide water security in the event of deliberate or accidental water contamination. This latter function requires an expanded array of monitoring and detection techniques.

Real-time detection of contaminants is needed to provide a quick response to a perturbation, thereby reducing exposure of the consumer to contaminated water. The methods employed must have low detection limits as these contaminants of concern may be present in very low concentrations. Direct analysis of target contaminants can be achieved by using standard analytical methods, such as gas chromatography (GC), GC/mass spectrometry (GC/MS), liquid chromatography (LC), LC/MS, immunoassay test kits, and ion chromatography (Camarillo et al. 2014).

Novel detection technologies include optical systems using lasers, flow cytometry (cell measurement), immunological assays, and even engineered cells. Some techniques detect and measure perturbations in standard parameters – pH, temperature, conductivity, dissolved solids – and can alert the manager of such when these are effectively preprogrammed.

Toxicity testing can be employed when it is not necessary or expedient to test for all of the potential contaminants. Fortunately, technologies are being developed and refined in order to meet this great need that rests on the biological response that specific species, such as algae or fish, have to toxic constituents. A sample of these techniques is given in Table 14.6.

This listing reveals a wide variety of techniques, each with their own advantages and challenges. Many techniques involve a side-stream of water in which fish or algae can be exposed; some are costly and some have not been fully tested at scale.

Table 14.6 Example detection systems used for monitoring contaminants in public water systems.

Category	Sensor name	Description	Parameters monitored
Optical	Biosentry™	Online laser-based multiple-angle light scattering	Particle counts, sizes, and shapes (rods, spores, cysts)
	FlowCAM®	Online flow cytometer and microscope	Particle size distributions
Biosensors	TOXcontrol™	Real-time biosensor using luminescent bacteria	Toxicity, TOC, BOD, turbidity, NO_3-N
	Algae Toximeter	Biosensor using green algae	Toxicity, algae
	Daphnia Toximeter™	Biosensor using water fleas *Daphnia magna*	Toxicity
	ToxProtect 64	Biosensor using fish	Toxicity
Algorithm	Hach Event Monitor	Indicates a contamination event using baseline data and predetermined triggers	pH, conductivity, organic carbon, temperature, turbidity, chlorine residual, pressure
Enzyme	Eclox	Chemiluminescent oxidation –reduction reaction catalyzed by plant enzyme	Toxicity, pH, total dissolved solids, mustard gas (optional)
Bioelectric	CANARY Bioelectronic Sensor	Engineered cells that emit photos that indicate binding with pathogens of interest	Microorganisms
Biochemical	Reveal®	Immunological assay	*Escherichia coli* O157:H7
Electrochemical	Censar® Six-CENSE™	Electrochemical technology on a ceramic chip	Color, turbidity, free chlorine, dissolved oxygen, pH, conductivity, oxidation reduction potential (ORP), temperature

Source: Adapted from Camarillo et al. (2014).

14.4.4 Threat Evaluation and Response

A system threat evaluation has three stages and associated tasks:

- "Possible" threat – characterize the site, collect data, take precautions;
- "Credible" threat – notify the public, identify contaminant, sample and analyze; and
- "Confirmed" threat – begin crisis response, contain damaged water, provide alternative water source, follow response protocol (Camarillo et al. 2014).

A countermeasure is any system or practice that reduces the overall risk to the utility's assets, either proactively or in response to a threat. These might involve reducing the likelihood of threat or the vulnerability or by dampening the consequences of an event should one occur. Example countermeasures in response to water-security threats are provided in Table 14.7.

14.5 Planning on a Large Scale – Reliability, Resiliency, and Vulnerability

We have thus far been considering water systems planning from a utility's perspective. But planning can also occur on a watershed level, or for a larger water system component, such as a reservoir or piping network.

Similar to community water–supply assessment, these components have been evaluated using three criteria of large-scale (LS) reliability, resiliency, and vulnerability (Hashimoto et al. 1982).

Table 14.7 Water-contamination threat management and responses.

General response	Potential actions – examples
Proactive countermeasures	Physical security measures such as guards and fences; equipment such as backup power for critical systems; employee training and practice exercises on ERPs
Determination of nature and extent of contamination	Evaluate contaminant properties – toxicity, reactivity, flammability, persistence, vapor pressure, solubility, etc.; use database such as EPA's Water Contaminant Information Tool (WCIT) Increase monitoring and sampling Increase surveillance – customer complaints, public health data
Operational responses	Isolate and contain contaminated water Mobilize booster chlorination Maintain water pressure for firefighting (20 psi minimum)
Public notification	Promulgate advisories: "Do not drink," "Boil water order," etc. Utilize public announcement tools to keep public updated: television and radio, reverse 911 calls, Facebook, Twitter
Alternate water supply	Provide alternate sources of water: bottled water, tanker truck deliveries, water from neighboring communities, emergency treatment of alternate sources
Return to normal operation	Notify the public of return to normalcy Discontinue any alternate water supplies

Source: Adapted from Camarillo et al. (2014).

- *LS Reliability, R_{LS}*, is a measure of the general stability of the system or, inversely, a measure of how often the system fails. It can be understood as the opposite of risk. Reliability, R, can be described as the frequency or probability, α, that a system is in a satisfactory state. Conversely, the risk or probability of failure is "$1 - \alpha$."
- *LS Resiliency, $R_{S\text{-}LS}$*, is the ability of a system to recover and return to a satisfactory state following a severe disruption or failure. If T_F is defined as a length of time needed for recovery, then resiliency, R_s, can be described as the inverse of the expected value of T_F. In other words, if $R_s = 1/T_F$, then a low T_F will yield a higher value of resiliency.
- *LS Vulnerability, V_{LS}*, here is used to describe the magnitude of failure if such should occur. (In the above text, this was labeled "consequence.") Numerous historical examples can be used to show that the maximization of production can actually lead to higher vulnerability in the case of failure. For example, monolithic (single-species) agriculture with high-yielding seed varieties can increase the short-term yields. But the invasion of a pest or crop disease in such a system is much more devastating as compared to a more biodiverse system. Flood-control structures that guard against smaller floods can provide a sense of security as expensive structures are built within the previous flood plain. But the event of a levee collapse from a major storm event will result in much more devastation than would otherwise occur (Hashimoto et al. 1982).

An example application of these criteria is the analysis of reservoir management options that consider both seasonal inflows and reservoir commitments for both flood protection and irrigation water use. Optimal policies are dependent upon inflow, storage, and desired outflow for seasonal variation. If the annual protocol for both summer and winter season release is designated by a Beta variable, then the three parameters of reliability, resiliency, and vulnerability can be assessed in terms of this variable (Figure 14.8). The graph shows

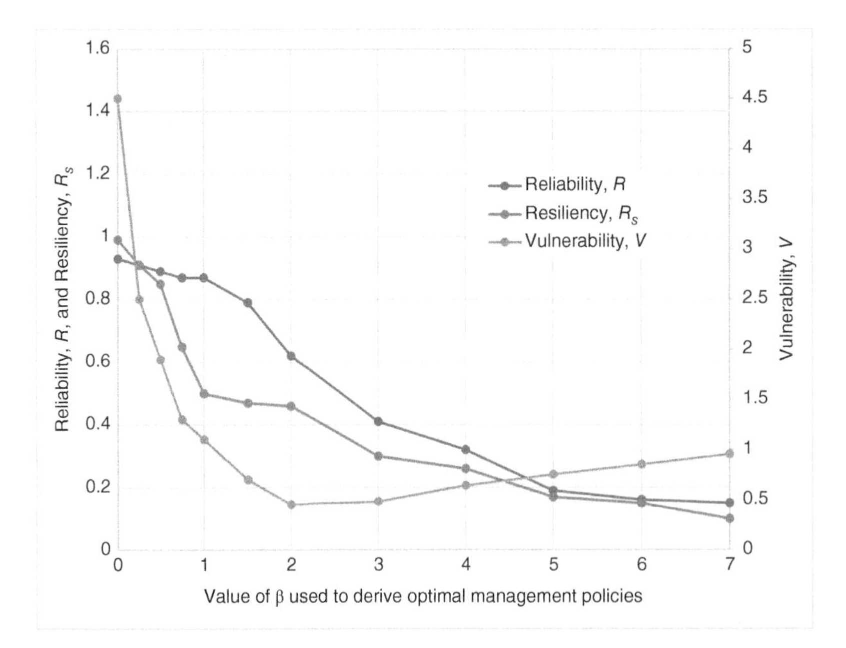

Figure 14.8 Example of reservoir management, using a function parameter, ß, to derive optimal operating policies with respect to system reliability, resiliency, and vulnerability. Source: Redrawn from Hashimoto et al. (1982).

trade-offs that ensue regarding resilience and reliability versus increased vulnerability. In this instance, a desirable management protocol would be in the range of $\beta = 1\text{–}2$, a system with low vulnerability and increased reliability and resiliency (Hashimoto et al. 1982).

14.6 Case Studies in Proactive Water Planning

The following case studies consider two urban areas and two national states in very different environments. South Korea is the most water-abundant of the examples with average rainfall 1000–1800 mm, mostly in the summer months of June–August. Hamburg and Iceland are relatively water-rich (~800 mm/year of rainfall), while Tucson is in an arid climate (~300 mm/year). Each jurisdiction has met its water challenges in slightly different ways with different methods of water-supply planning. We examine each of these now.

14.6.1 Case Study – Hamburg (Germany)

Hamburg's water utility (Hamburg Wasser) supplies water to two million people in this second-largest city in Germany. The bulk of the city's water supply comes from groundwater wells that provide spring-like water quality without the need of chlorination (Brears 2016). Water managers, however, have seen new challenges with climate change, population growth, and rising energy costs. Climate change has brought about longer dry periods and more intense storms (Figure 14.9). Stormwater collects in antiquated combined sewer pipes, which, along with the wastewater-treatment plant, get overwhelmed during storm events. In order to treat this additional volume of water, treatment plants have to use extra energy and purchase additional treatment chemicals.

As a measure to reduce water stress and encourage conservation, managers have required the installation of water meters in homes and apartments. They also instituted increasing

Figure 14.9 The historic Hamburg fish market is flooded during Storm Sabine in early 2020. Source: Westend61 / Adobe Stock.

block tariff pricing based on the water meter size. The utility has also begun inspecting the 5500-km drinking water network for leaks at a rate of 1000 km of pipeline each year. For public institutions, continuous flow controllers were installed on all new or existing public buildings and water from rainwater harvesting systems is used to flush toilets in sports arenas (Brears 2016).

The Hamburg Water Cycle (HWC) is a method of closed-loop wastewater management, which separates wastewater sources based on potential uses (Hamburg 2020). The greywater from washings is used for gardening and stormwater is allowed to soak into the ground and replenish aquifers. Flush toilets are converted into an energy source by being vacuumed with much less water than a conventional toilet. The concentrated waste is then sent for anaerobic treatment before being used for production of methane energy on site. The HWC system is being tested on an old military barracks converted into a mock residential site (the "Jenfelder Au" neighborhood).

A SWOT analysis for the Hamburg water utility revealed the following strengths, weaknesses, opportunities, and threats:

- Strengths – water is priced to ensure full cost recovery; every private consumer has an installed water meter; the system as a very low leakage rate of 4% water loss; it has a pipe replacement program in which old cast iron pipes are systematically replaced
- Weaknesses – because of the increasing block tariff system, the utility is not financially interested in a steep reduction in water consumption; new innovations require cooperation from a large number of stakeholders and need significant political will; existing infrastructure is not suitable for meeting new challenges of increased stormwater runoff
- Opportunities – automatic meter readers would help consumers monitor their own water usage; the sale of water-efficient appliances and devices could be promoted through the use of a labeling scheme
- Threats – climate change has resulted in frequent heavy storm events; population growth means greater water consumption and increased impervious surface areas; higher energy costs are putting a greater financial strain on the water utility (Brears 2016).

14.6.2 Case Study – Tucson (Arizona, US)

Tucson, AZ, has a metropolitan population of 1 million people and is located in the Santa Cruz River Basin, a desert watershed that is rich in ecological biodiversity and culture (Zuniga-Teran and Staddon 2019). Surface water for a growing population has always been scarce in this region that sees an average 300 mm (11.8 in.) of rain per year. The invention of air conditioning and the attraction of copper, cotton, cattle, citrus, and climate (the "five C's") spawned a burgeoning population through the early twentieth century. Excessive groundwater pumping began in the late 1940s and within a few decades, the Upper Santa Cruz aquifer was overdrawn and land subsidence became a reality (Figure 14.10).

The City took several actions to both increase the *supply* of water and reduce its *demand*. Regarding supply, in 1992 the City began to import water from the Colorado River via the Central Arizona Project (CAP). This water, along with treated wastewater, would become the two primary sources of water. The imported Colorado River water had pH and water-quality issues with the distribution system, however, leading to public disapproval of the project. As one citizen put it: "The trust of the public … had been destroyed by smelly, yellow, orange, brown, and red water pouring out of home faucets." (McGuire and Pearthree 2020). This

Figure 14.10 Geological fissure caused by land subsidence in south central Arizona. Source: Photo courtesy of Joe Cook, Arizona Geological Survey (AZGS) (2017).

led to an innovative approach in which the Colorado River water would be "banked," i.e. recharged into an aquifer where it would be blended with groundwater. This approach both saved the water allocation from CAP and improved the resulting water quality.

Water-conservation measures were introduced to reduce the household and commercial demand for fresh water. Utilities are allowed to extract only a certain amount of groundwater based on a predetermined natural recharge rate. Any groundwater that is extracted above the incidental recharge must be offset with an equivalent amount of renewable water from another source (Zuniga-Teran and Staddon 2019). The *Beat the Peak* program, launched in 1977, was strongly embraced by citizens who would voluntarily use less water during peak demand periods. Increasing block tariffs were used to incentivize conservation on a household level, and xeriscaping – planting of drought-tolerant plants in lawns – was regulated. The City would now "require the use of drought-tolerant plants from a published list and limits non-drought tolerant vegetation to small 'oasis' areas. Multifamily facilities may develop 5% of a site as an oasis area; commercial facilities are restricted to 2.5% of a site. Any water features or turf must be confined to the oasis areas. Canopy trees are required within all buffer yards, along street frontages, and within parking lots with one tree for every 15 spaces." (City of Tucson 2014) Notably, this ordinance also increased the number of shaded areas as well. The result of all these measures was a drop in water use per capita over the past decade even with an increasing population (Figure 14.11).

14.6.3 Case Study – Iceland

Iceland was one of the earliest nations to be proactive in water-supply planning (Figure 14.12). In 1995, the national government mandated water utilities to implement the Hazard Analysis and Critical Control Point (HACCP) approach to ensure water safety. This approach had already been proven in the food industry, which now included the water sector by law and regulation. The HACCP process was found to be complicated for smaller utilities and a simpler five-step model was subsequently approved for utilities serving 500–5000 people (Gunnarsdóttir et al. 2012).

By the year 2008, 31 water utilities serving 81% of the population were reported to have WSPs. Of these, 14 used the five-step model for smaller systems while 17 had adapted the

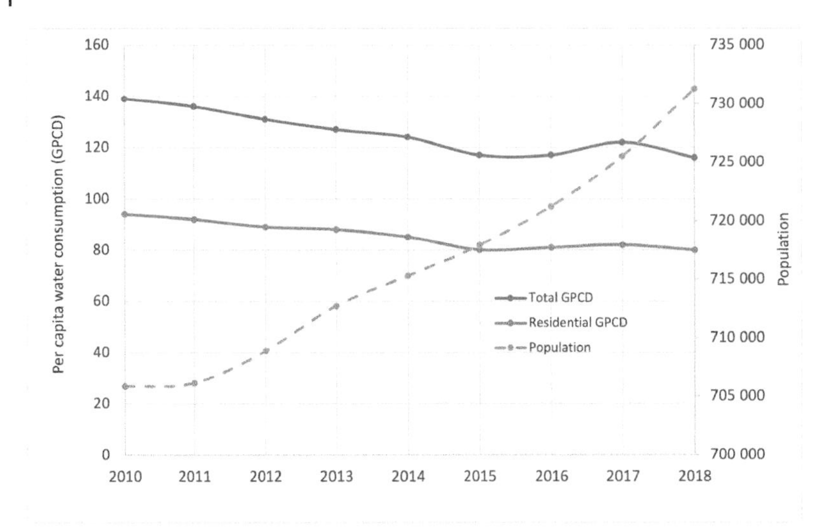

Figure 14.11 Even as Tucson's population has risen, the total and residential per capita water demand has fallen slightly (GPCD, gallons per capita per day). Source: Data from City of Tucson (2020).

Figure 14.12 Iceland, with its colorful cities, was one of the earliest adopters of formal water-supply planning. Source: WHO and IWA (2009). Photo: boyloso/Adobe Stock.

HACCP model. A study was conducted on 16 of these utilities with WSPs and the findings reported (Gunnarsdóttir et al. 2012). When asked to give their top three priorities for implementing the WSP process, the top three reasons were to provide safe water (100% of utilities), fulfill the regulatory requirements (88%), and improve service to their customers (88%). Only 12% of utilities conducted the process in order to decrease complaints from consumers. A correlation among factors analysis revealed that the utilities with a formal WSP committee (or steering group) had higher overall WSP rating and were more likely to carry out internal

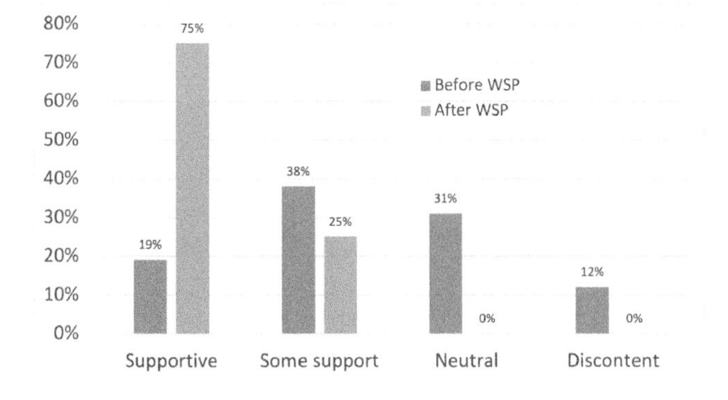

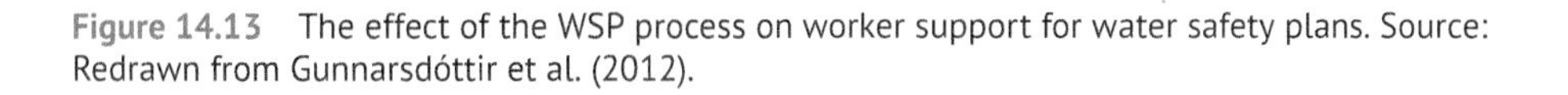

Figure 14.13 The effect of the WSP process on worker support for water safety plans. Source: Redrawn from Gunnarsdóttir et al. (2012).

audits and put emphasis on regular cleaning. The presence of an active training program was also correlated with frequent audits and a high overall WSP rating. One of the most striking revelations was how staff attitudes changed after the WSP process. Before implementation, 57% of the workers were supportive or somewhat supportive of the WSP process. After implementation, that number rose to 100% (Figure 14.13).

The study revealed a few shortcomings in the lack of documentation, especially after an incident, and the need for more staff training. When the benefits of better management increased reliability and financial gain were stressed to employees, the WSP plans and implementation provided their greatest benefit (Gunnarsdóttir et al. 2012).

14.6.4 Case Study – South Korea

As part of its "Green New Deal," South Korea has plans to reinvigorate and renew its four major river systems (Figure 14.14). The planning included a wide range of stakeholders: government ministers from water resources, water quality, ecology and environment, landscape, and culture and tourism as well as advisory groups consisting of professors, specialists, academics, and local representatives. These groups also heard from NGOs such

Figure 14.14 Sunrise over Han River, Seoul, South Korea. Source: pimplub / Adobe Stock.

as religious groups, environmental groups, and local citizen associations on a regular basis (Cha et al. 2011).

The Four Major Rivers Restoration Project includes the following activities to achieve its five major objectives:

- *Water storage.* Sixteen new weirs will secure 800 million cubic meters of water and an additional 500 million cubic meters will be collected in higher water levels for 96 agricultural reservoirs and 3 multipurpose dams. This added infrastructure will enable the storage of water needed during the dry season. The additional dredging of riverbeds will decrease the flood-water level and significantly reduce annual floods and the damage they cause.
- *Flood control.* An expansion of the water gates of tributaries will allow a quick water level decline and fast draining of flood waters. In addition, flood control capacity will be expanded up to 920 million cubic meters of water.
- *Water quality and ecological restoration.* By 2012, the average water quality of the main stems will be improved to not greater than 3 ppm BOD by the expansion of sewage treatment plants and establishment of green algae–reduction facilities. Moreover, plans are in place to restore ecological rivers, create wetlands, and relocate farmlands in the rivers to rehabilitate the river ecosystem.
- *Creation of multipurpose recreational spaces for local residents.* To create the riverfront as a multipurpose recreational area, 1728 km of bicycle lanes will be developed and walkways and sports facilities will be expanded.
- *River-oriented community development.* Local communities are expected to benefit by the improved infrastructure and scenery. A few example slogans are "Our major rivers that flow with culture" and "Creating a vivid land of beautiful scenery" (Cha et al. 2011).

The total cost of the project was expected to be almost 20 billion USD (Lah et al. 2015). This example shows how significant change can be accomplished when all the important actors are involved in the planning.

14.7 Conclusion

In this chapter, we explored several different methodologies for water-supply planning on a utility scale for a community water system. Many of the elements in these approaches are common to one another and serve a dual purpose for both water delivery improvement and planning for accidents, natural disasters, or malignant acts. Several case studies have been reviewed to demonstrate the impact of water-supply planning and coordination among managers and stakeholders. The lesson is that, with good planning, water security is attainable.

US President and former General Dwight D. Eisenhower once remarked that "in preparing for battle I have always found that plans are useless, but planning is indispensable." The formulation of a water-supply security plan has perhaps its greatest benefit in the *process* (the planning) rather than the *product* itself (the plan). When the process of planning is conducted with rigor and thoroughness, then the utility and its customers will be much better off. In addition, expected events rarely go as planned! The act of careful planning results in both a set of contingency actions and an increase in confidence that emergencies can be safely managed.

End-of-Chapter Questions/Problems

14.1 Hurricane Sandy did major damage to cities along the east coast of the United States. Name and describe three planning activities that could have been done up-front to mitigate some of the damage from this storm. Be very specific and use information from city discussions that will support your recommendations.

14.2 Find an example of a community water system that has implemented the EPA's risk and resilience planning method (RRA, ERP, etc.). Describe the plans that have been made.

14.3 Find an example of a water system that has implemented the RAMCAP approach to water-supply planning. Describe the plans that have been made.

14.4 Find an example of a water system that has prepared and promulgated a Water Safety Plan (WSP) according to the WHO method of water supply planning. Describe the plans that have been made.

14.5 Consider your own place of residence. If your water utility were to prepare a Water Supply Plan, which two tasks (of the seven tasks) would be the most difficult to implement? Use Table 15.1 as a guide. Why?

14.6 Pick one of the contaminant detection methods given in Table 14.6. Find and summarize a case study in which the method has been, or is being, used.

14.7 Find a recent study on the HWC. Summarize the study and comment on challenges and progress being made.

14.8 Compare the water-resilience strategies of *two* of these major southwestern US cities: Phoenix (AZ), Tucson (AZ), Los Angeles (CA), and Las Vegas (NV).
a. How was a water-planning process involved in each city?
b. What elements of resilience are common to both cities? Which elements are different?

14.9 Compare Iceland's per capita water usage with one other Nordic country – Denmark, Finland, Norway, or Sweden.
a. Why are the usage rates similar or different?
b. Compare the long-range water-security planning methods of both countries.

14.10 Find a recent study that shows the progress of South Korea's river restoration project. Summarize the progress that has been made and the challenges being faced.

14.11 Download the VSAT application from the US EPA website. Do a rough mock assessment of the water-supply system in your home city. Give reasonable estimates for data that is unavailable. The purpose is to simply learn how to use the tool. Summarize your results.

Table 14.8 Results of a water-supply system analysis.

Component	Threat	Likelihood	Vulnerability	Consequence	Outage time
Rusty roof over chlorine shed	High winds/tornado	Medium	Medium	\$8000	1 month
Shortage of household backflow preventers	Backflow contamination from sanitary lines	Low	High	\$20 000	1 month
Outdated and vulnerable software	Ransomware attack	Low	Medium	\$10 000	1 week

14.12 The analysis of a water-supply system resulted in Table 14.8. Use Eqs. (14.1) and (14.2) to answer the following questions:

a. What is the relative probability for each of the listed threats?
b. What is the relative risk for each of the threats?
c. What is the system resilience for each of the threats?
d. As a water manager, how would you prioritize these threats?

Further Reading

Camarillo, M.K., Stringfellow, W.T., and Jain, R. (2014). *Drinking Water Security for Engineers, Planners, and Managers*, 1e. Amsterdam: Butterworth-Heinemann.

Tindall, J.A. and Campbell, A.A. (2011). *Water Security: Conflicts, Threats, Policies*. Chapter 14.

WHO & IWA (2009). Water Safety Plan Manual: Step-by-Step Risk Management for Drinking-Water Suppliers. *World Health* Organization.

References

ASCE/AWWA Committee (2010). *J100 Risk and Resilience Management of Water and Wastewater Systems: AWWA Standard*. Washington, D.C.; Denver: American Water Works Assn.

AZGS (Arizona Geological Survey) (2017). "Earth Fissures & Ground Subsidence." AZGS. July 12, 2017. https://azgs.arizona.edu/center-natural-hazards/earth-fissures-ground-subsidence.

Brears, R.C. (2016). *Urban Water Security*, 1e. Southern Gate, Chichester, West Sussex, UK; Hoboken, NJ: Wiley.

Camarillo, M.K., Stringfellow, W.T., and Jain, R. (2014). *Drinking Water Security for Engineers, Planners, and Managers*, 1e. Amsterdam: Butterworth-Heinemann.

Cha, Y.J., Shim, M.-P., and Kim, S.K. (2011). The four major rivers restoration project. In: *Proceedings of The Event Water in the Green Economy in Practice*, 10. Zaragoza, Spain.

CISA (2020). "Ransomware | Cybersecurity & Infrastructure Security Agency." 2020. https://www.cisa.gov/ransomware.

City of Tucson (2014). "Xeriscape Landscaping and Screening Regulations – Ordinance 7522." July 10, 2014. https://www.tucsonaz.gov/water/ord-7522.

City of Tucson (2020). "Tucson Water Conservation Program FY 2018–2019 Annual Report." Tucson, AZ.

Clark, R.M., Panguluri, S., Nelson, T.D., and Wyman, R.P. (2017). Protecting drinking water utilities from cyberthreats. *Journal of the American Water Works Association* 109 (February): https://doi.org/10.5942/jawwa.2017.109.0021.

Gunnarsdóttir, M.J., Gardarsson, S.M., and Bartram, J. (2012). Icelandic experience with water safety plans. *Water Science and Technology* 65 (2): 277–288. https://doi.org/10.2166/wst.2012.801.

Hamburg (2020). "Hamburg Water Cycle." 2020. https://www.hamburgwatercycle.de/en/home.

Hashimoto, T., Stedinger, J.R., and Loucks, D.P. (1982). Reliability, resiliency, and vulnerability criteria for water resource system performance evaluation. *Water Resources Research* 18 (1): 14–20. https://doi.org/10.1029/WR018i001p00014.

Hong Kong WSD (2020). "Hong Kong Water Supplies Department-Building Water Safety Plan." 2020. www.wsd.gov.hk/tc/water-safety/water-safety-in-buildings/index.html.

Kimm, V.J. (2020). *Drinking Water: A Half Century of Progress*. EPA Alumni Association.

Lah, T.J., Park, Y., and Cho, Y.J. (2015). The four major rivers restoration project of South Korea: an assessment of its process, program, and political dimensions. *The Journal of Environment & Development* 24 (4): 375–394.

McGuire, M.J. and Pearthree, M.S. (2020). *Tucson Water Turnaround: Crisis to Success*. AWWA.

Morley, K. (2018). "Water Sector Approach to Manage Risk & Resilience." Presented at the Intelligent Water Networks Summit (AWWA).

Tindall, J.A. and Campbell, A.A. (2011). *Water Security: Conflicts, Threats, Policies*.

US EPA (2020a). "Risk and Resilience Assessment Basics Tutorial."

US EPA (2020b). "VSAT Web 2.0." 2020. https://vsat.epa.gov/vsat.

US EPA, OW (2014). "Basic Information on the CCL and Regulatory Determination." Other Policies and Guidance. US EPA. May 5, 2014. https://www.epa.gov/ccl/basic-information-ccl-and-regulatory-determination.

US EPA, OW (2015a). "Basics of Water Resilience." Overviews and Factsheets. US EPA. February 5, 2015. https://www.epa.gov/waterresilience/basics-water-resilience.

US EPA, OW (2015b). "Conduct a Drinking Water or Wastewater Utility Risk Assessment." Data and Tools. US EPA. April 16, 2015. https://www.epa.gov/waterriskassessment/conduct-drinking-water-or-wastewater-utility-risk-assessment.

US EPA, OW (2015c). "Module 1: Water Utility Planning Guide for Drinking Water Utilities." Other Policies and Guidance. US EPA. June 2, 2015. https://www.epa.gov/waterutilityresponse/module-1-water-utility-planning-guide-drinking-water-utilities.

US EPA, OW (2018). "America's Water Infrastructure Act: Risk Assessments and Emergency Response Plans." Other Policies and Guidance. US EPA. November 15, 2018. https://www.epa.gov/waterresilience/americas-water-infrastructure-act-risk-assessments-and-emergency-response-plans.

US EPA, OW (2019). "Resilient Strategies Guide for Water Utilities." Data and Tools. US EPA. February 25, 2019. https://www.epa.gov/crwu/resilient-strategies-guide-water-utilities.

WHO & IWA (2009). "Water Safety Plan Manual: Step-by-Step Risk Management for Drinking-Water Suppliers." World Health Organization. http://www.who.int/water_sanitation_health/publications/publication_9789241562638/en.

Zuniga-Teran, A. and Staddon, C. (2019). Tucson Arizona – a story of 'Water Resilience' through diversifying water sources, demand management, and ecosystem restoration (chapter 13). In: *Resilient Water Services and Systems: The Foundation of Well-Being*. IWA Publishing.

15

The Economics of Water Security

In this chapter, we examine the economic aspects of water security. We describe the types and characteristics of economic goods, especially natural resources that are available for public consumption. We consider the time value of money and the ways in which water pricing is applied in a manner appropriate for its use. Then we describe the process of cost–benefit analysis, a widely used approach for assessing large water infrastructure projects. These projects often address the need to augment (increase) the supply of water. Conversely, decreasing water demand is the other side of water security and can be addressed through rate structures and pricing strategies that may result in greater water-use efficiency or conservation.

Finally, ecosystem services may be directly or indirectly valued using methods borrowed from the social sciences.

Learning Objectives

Upon completion of this chapter, the student will be able to:

1. Define the various types and characteristics of economic goods.
2. Explain and calculate the time value of money using a discount rate factor.
3. Perform a cost–benefit analysis (CBA) in order to assess the economic desirability of one or more water infrastructure projects.
4. Explain the advantages and shortcomings of a CBA.
5. Critique various commonly used water rate structures.
6. Understand the price and income elasticity of water demand.

15.1 Introduction

The last decades of the twentieth century have seen the rise of very large water infrastructure projects in Africa – the Grand Ethiopian Renaissance Dam (GERD), the world's largest irrigation project in Libya, and the Northern Collector Tunnel project bringing water to Kenya's capital, Nairobi (Ali 2017; Blomkvist and Nilsson 2017). While *upstream* activities may be used to secure new water sources, such as in the building of a dam and reservoir, *downstream*

Fundamentals of Water Security: Quantity, Quality, and Equity in a Changing Climate, First Edition.
Jim F. Chamberlain and David A. Sabatini.

Companion website: www.wiley.com/go/chamberlain/fundamentalsofwatersecurity

projects are built to distribute and/or treat the water for the end users (Blomkvist and Nilsson 2017). Both upstream and downstream projects are concerned with increasing the *supply* of water, while measures that deal with efficiency, conservation, and reuse are matters of decreasing the *demand* for new water sources. Both supply and demand interventions depend in large part on financing and economics for their success.

Economics has been called the "science of scarcity." That is, it studies why and how people make decisions in the face of limited resources. In this sense, it is a "descriptive" science, describing what can be observed both now and historically. But it can also be used in a "prescriptive" sense. Economic theory can be action-oriented, helping managers and guiding decision-making on many levels. Water resource economics can be both a descriptive and a prescriptive tool in that knowledge of past projects with their financial impacts can be used to optimize future interventions regarding this important resource.

Water is a natural resource whose scarcity, and thus its value, depend upon time and place. Farmers in southern India may be inundated with water during the monsoon season, and months later find themselves plowing a dry, cracked earth. In periods of abundant rainfall, water may not seem to have much economic value at all – or even a negative value if flooding occurs. But during times of drought, the value may be more than people are willing or able to pay. A large dam project may be able to store water for various purposes during the dry season as well as release water slowly during times of heavy rainfall, thereby decreasing the risk of downstream flooding (Figure 15.1). Such an infrastructure project serves to dampen the contrasting effects of flood versus drought. Economics has a role to play in alleviating both seasonal and temporal disparities. Cost–benefit analysis (CBA) can be used to compare and select large infrastructure (*supply*) projects. Appropriate pricing of water can help incentivize greater efficiencies and conservation while also helping to more equitably allocate water resources among multiple users (*demand*).

The ultimate goal of water-resource economics is recognizing and stimulating the most *efficient* choices given a particular set of demands along with supply constraints (Griffin 2006). Because access to a suitable water supply is a basic human need, the traditional option to "let the market run its course" is not always appropriate. For our purposes, the most efficient choice is the one that provides the greatest availability of water of suitable quality for the most people. This chapter is intended to help the reader understand the applications of resource economics theory that help create and bolster the conditions needed for choices leading to greater water security.

Figure 15.1 The Cherokee Dam on the Holston River in East Tennessee provides hydroelectric power, flood control, and recreation, all under the management of the Tennessee Valley Authority. Source: TVA (2021).

15.2 Initial Economic Considerations

Water becomes an economic commodity when it enters the market as an improved good, because it is here that someone is willing to provide it as desired (*producers*) and someone else is willing to purchase it (*consumers*).

This section presents a brief consideration of some economic principles and terms that are relevant to water security – private and public projects, types of economic goods, the time value of money, and the benefits of pricing. These terms and principles will underpin the discussion in the remainder of the chapter.

15.2.1 Private and Public Projects

The field of economics has widespread application in market-based economies, where players and firms strive toward economic optimization in which both suppliers and consumers are relatively satisfied. As opposed to private, profit-seeking ventures, water projects are often public projects that are authorized, financed, and maintained by public entities, such as governments or public utilities. In addition to being typically larger, more expensive, and built for a longer lifetime, public projects differ in other ways from private projects (Table 15.1).

15.2.2 Types of Economic Goods

Even though water is generally the same entity in all cases, it has a variety of uses, including agricultural, industrial, recreational, transportation, ecological, and human health and

Table 15.1 Basic differences between privately owned economic projects and public projects.

	Private projects	Public projects
General purpose	To provide goods or services at a maximum profit (examples: manufactured goods, cellular telephone, package delivery)	To protect health and property; provide services, jobs, water/environmental quality; provide jobs; improve well-being of general public (examples: reservoirs for flood control/hydropower, water treatment, irrigation canals, constructed wetlands)
Beneficiaries	Entity undertaking the project; shareholders; venture capitalists	General public (sometimes local to the project)
Sources of capital (financing)	Private investors and lenders; partnerships, corporations	Taxpayers, low or no-interest loans, bonds, subsidies
Project life	Usually relatively short (5–20 years)	Usually relatively long (20–60 years)
Nature of benefits	Monetary rewards, return on financial investment	Real but often intangible and nonmonetary, difficult to monetize
Conflict of purposes and/or interests	Moderate	Common (examples: water for irrigation versus hydropower versus recreation)
Quantitative assessment of efficiency	Internal rate of return (IRR)	Cost–benefit analysis (CBA)

Source: Modified from Sullivan et al. (1999).

Table 15.2 Examples of types of goods associated with usage of water resources.

	Rival	Nonrival
Excludable	Private goods: Water consumed by industry, agriculture, or domestic use	Club goods: Instream fishing rights by fee Protection of aquatic species in stream
Nonexcludable	Common goods: Riparian managed surface water withdrawal for irrigation	Public goods: Instream water usage, such as transportation Minimum flow for ecosystem health, groundwater withdrawals (common pool)

Source: Adapted from Young and Loomis (2014).

sustenance. As we have noted earlier, water is *withdrawn* if it is taken from the water source and then returned to the local hydrologic system. It is *consumed*, however, if it is embodied in a product, lost to evapotranspiration, or if its quality becomes so degraded that it is rendered unusable to local consumers. *Rival* uses are those in which one person's use reduces the amount available to another person. *Excludable* uses are those in which one entity (or person) can easily exclude another entity from using the water.

A private use of water for irrigation or industry, for example, is a rival, excludable usage. Water is withdrawn and a portion is consumed to the detriment of other potential uses. Other uses are non-consumptive uses that happen instream and do not involve a substantial loss of water. These uses include recreational (boating, fishing), transportation, ecosystem stability and hydroelectric power generation (to a great extent). These are public goods if they are nonexcludable and are private (club) goods if a fee is charged for usage. Table 15.2 offers a few examples.

15.2.3 Time Value of Money

Public water infrastructure projects tend to require large outlays of money, both upfront and over time, and have long project lives (Table 15.1). In addition, since water is a bulky commodity, it is expensive to extract, transport, and treat. Cash outlays may include capital costs at or near the beginning of the project, equipment-replacement costs at various anticipated times, and operation and maintenance costs that occur throughout the project lifetime. Money that is thus allocated changes value over time and cannot be used for other things. A dollar today would be worth more in the future because it could be invested and gain interest (a benefit for allowing others to have access to your money now).

Most people are familiar with the interest that can be gained by depositing money in a bank savings account. Compound interest on a present investment (or deposit), P, over a period of n years with an interest rate d can result in a future value, F, according to the formula:

$$F = P(1 + d)^n \tag{15.1}$$

Likewise, a future expense (or benefit) has a present value by solving for P. This is how much money you would need to have in the bank today (P) to have a certain amount of money (F) in a future year (n) for a given interest rate (d):

$$P = \frac{F}{(1 + d)^n} \tag{15.2}$$

To evaluate large projects, or to compare alternative projects, economists use this *present value* of both the costs and the benefits associated with a project. By mathematically quantifying the time value of money, each cost outlay or benefit can be *discounted* (brought back) to the present, thus extrapolating all future dollars to the equivalent dollars in the present. The present value of a benefit that occurs in year t (B_t), with a discount rate, d, is calculated and summed over year $t = 0$ to n using the following formula:

$$\text{Net Present Value of Benefits} = \sum_{t=0}^{n} \frac{B_t}{(1+d)^t} \tag{15.3}$$

A similar formula is used for net present value (NPV) of all costs:

$$\text{Net Present Value of Costs} = \sum_{t=0}^{n} \frac{C_t}{(1+d)^t} \tag{15.4}$$

Many projects can have a relatively constant annual cost, e.g. operation and maintenance, or a steady stream of revenue, such as payment for hydropower electricity. If a steady-value cost or benefit is recurring annually, A, a slightly more complicated formula can be used to calculate the present value:

$$\text{Net Present Value of Annual Costs (or Benefits)} = A\left[\frac{(1+d)^n - 1}{d(1+d)^n}\right] \tag{15.5}$$

From the above formulas, it can be seen that a higher discount rate will make the *initial* capital costs figure weigh more heavily in the ratio of benefits to cost, since future costs and benefits will be discounted (reduced) more heavily to the present-day value (you will need less money in the bank today to make a future payment if the interest rate is higher) (Tietenberg 2002). Conversely, a lower discount rate will shift the overall weight to *future* costs and benefits. For this reason, the choice of discount rate can alter the choice between options in time-weighted monetary decision-making (Griffin 2006).

Traditionally, economists have used the interest rates on long-term government bonds in economic CBA. Private agents may use the *market rate* for borrowing funds. For Federal US projects, a mandated discount rate is often applied that is generally closer to a social discount rate. The *social discount rate* incorporates both the opportunity cost (lost benefits) of investing public funds in the market, while also considering the perceived risks associated with the project, it's possible reduction of future utility, and the changing needs of the citizen population. This rate may also be adjusted downward to give more weight to the well-being of future generations, as ethically desired.

The sensitivity of economic analysis on the discount rate can make a project seem favorable or not. For example, both Canada and the United States did an analysis on a tidal power project to be built in the waters shared by New Brunswick and Maine. The project would have heavy initial costs but low operating costs. Using the exact same data, the two nations came up with different conclusions regarding the favorability. Using a discount rate of 4.125%, Canada determined that the project should not be built, while the United States, using a rate of 2.5%, argued that it should be built (Tietenberg 2002). The US annual inflation rate in the last two decades (2000–2020) was around 2.5%, whereas that of developing nations can often be higher. For example, the appropriate social discount rate in Bangladesh was 9–11% during a period with an annual inflation rate of 6.0% and long-term government bond rate of 9.17% (Jalil 2010).

An understanding of the time value of money and its usage in CBA is a crucial step in the economic analysis of water security and equity.

15.2.4 Economic Value of Water

Because water falls freely from the sky, many assume that it "ought to be free." It is true. People can collect rainwater in a bucket or a cistern and they will have collected free water. But if they want an ample supply of water that is conveniently accessed, of a certain quality and is available to them all day long (or at least when they need it), then there may justifiably be some price to pay. It becomes then an economic commodity that has a relative cost and benefit.

When a market is functioning properly, the equilibrium market price is the price that consumers are willing to pay at the current supply of the gadget, good, or service, i.e. at the intersection of the supply and demand curves (Figure 15.2). The market intervention of government subsidies, practiced almost universally, is intended to increase access. But it may lead to water pricing that is not reflective of the true cost. In a typical scenario, if a government intervention increases the supply (from S_1 to S_2), the quantity demanded will almost certainly increase and the price paid by water consumers will be lower. Likewise, if the government provides a price subsidy (from P_1 to P_2, following the demand curve only), the water demand will increase as if there were an increased supply. This latter condition could potentially lead to increased water scarcity.

These observations are true when water is treated as an economic commodity. But it is then also subject to the phenomenon of elasticity as will be discussed below.

15.3 Supply Enhancement – Cost–Benefit Analysis of Water Projects

Public water projects may render many services, including improvement of water quality, reduction of nutrient runoff, mitigation of floods and drought, maintenance of biodiversity, recreation, power generation, or some combination of the above. Because water has many potential uses and users, water projects are often designed on a grand and expensive scale. And so the idea of a new water project invariably raises critical questions: "What are the costs of building this dam and reservoir, and who will receive the benefits?," "What are the benefits of restoring this urban stream to its natural channel?," "What are the economic implications of diverting a portion of this river water for agricultural usage?."

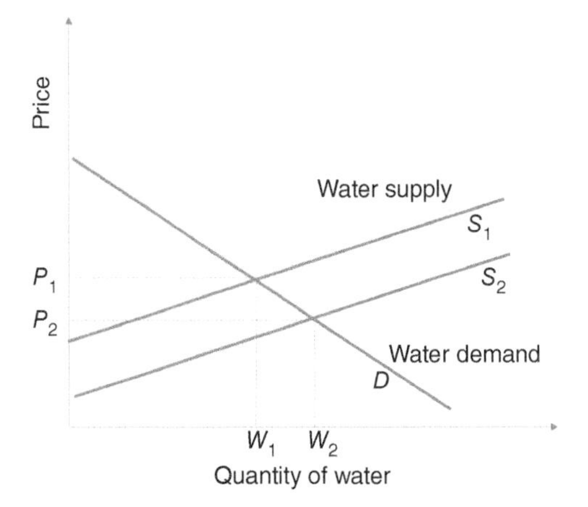

Figure 15.2 Water supply and demand are both reflective of the price of water, and price fluctuations can shift the equilibrium point.

How then should a proposed water project be assessed? Conventional economic theory focuses on the decision-making of individual consumers or owners of private enterprises. A private enterprise may rely completely upon access to public water, thereby making a claim upon this commodity as a rival, excludable good (Table 15.2). The economics of a *public good*, such as water, is more often focused on economic decisions that affect the economic interests of many users. This is sometimes called *welfare economics*, as it describes the resulting welfare of various groups of people who may be affected differently by major water economic decisions (Griffin 1998). For example, public water projects, such as dams, can be very expensive and require a commitment of public funds. The benefits, however, may be enjoyed by a small local segment of the population, albeit for a long time period.

15.3.1 Description and Methodology

CBA is often mandated for water works paid for by the US Federal government. A CBA (or benefit–cost analysis, BCA) entails the comparison of both costs and benefits using a single metric, usually a US dollar. The "Principles and Guidelines" (P&G) published in the Federal Register (originally in 1983 and updated from time to time) specify the Federal projects to be evaluated and provide methodologies for assessment of these. (Note: Agency-specific procedures are referred to as PR&G, "Principles, Requirements, and Guidelines" [US CEQ 2014]).

The central objective of the practice of CBA was summarized in the US Flood Control Act of 1936: projects are deemed economically acceptable "if the benefits to whomsoever they accrue are in excess of the estimated costs" (Griffin 1998). Of course, this does not mean that the project will be funded and built but simply that it is deemed economically acceptable. Note that there is no attention paid to the distribution of costs or benefits to various public groups; simply that the total benefits must be in excess of the total costs.

However, the latest update to the process was done in 2013 and specifies that the following six principles must apply to Federal water-resource projects, although in no particular order or ranking:

- *Healthy and resilient ecosystems* must be protected and restored so that their functions may be preserved. Their resilience to climate change must be supported.
- *Sustainable economic development* should result and improve the well-being of both current and future generations.
- *Floodplains* connect land and water ecosystems and support high levels of biodiversity and productivity. These should not be adversely impacted.
- *Public safety* must be ensured by using both structural and nonstructural measures to reduce and mitigate human risks.
- *Environmental justice* requires that Federal actions identify any disproportionate burdens on minority, tribal, or low-income populations.
- *Watershed approach* is necessary to accumulate a comprehensive incorporation of relevant stakeholders and risks (US CEQ 2013).

CBA determines a cost–benefit ratio (*CBR*) over total project time, n, using the following equation:

$$\mathrm{CBR} = \frac{\mathrm{NPV}_{\mathrm{Benefits}}}{\mathrm{NPV}_{\mathrm{Costs}}} = \frac{\sum_{t=0}^{n} \frac{B_t}{(1+d)^t}}{\sum_{t=0}^{n} \frac{C_t}{(1+d)^t}} \tag{15.6}$$

The discount rate, d, is generally set by the assessing institution or entity, as discussed above. Expressed thusly, if the CBR > 1.0, then the project is deemed economically beneficial.

15.3.2 Cost–Benefit Analysis – Three Examples

Consider a hypothetical new dam to be constructed in Kenya for both hydropower (initially) and irrigation (later). The following analysis presents both costs and benefits over a 20-year period (Table 15.3). Costs include the construction of a dam and reservoir over a four-year period, followed by annual operation and maintenance costs. In year 9, an irrigation canal is added to divert some of the water in the reservoir for irrigation. Income is generated from sales of both electricity and irrigation water.

By discounting all of the costs and benefits to the present value, these can be compared as "apples to apples." The analysis shows that the ratio of benefits to cost (CBR) is

Table 15.3 Sample cost–benefit analysis for a proposed reservoir and irrigation canal. Discount rate is 5%.

Year	Cost	NPV_{cost}	Benefit	$NPV_{benefit}$	Notes
0	$250	$250	$0	$0	Construction of dam/reservoir/hydro plant over 4 years ($1 million)
1	$250	$238	$0	$0	
2	$250	$227	$0	$0	
3	$250	$216	$0	$0	
4	$50	$41	$95	$78	Annual income from electricity ($95 000); annual O&M costs ($50 000)
5	$50	$39	$95	$74	
6	$50	$37	$95	$71	
7	$50	$36	$95	$68	
8	$50	$34	$95	$64	
9	$550	$355	$95	$61	Addition of irrigation canal ($500 000)
10	$50	$31	$105	$64	Annual income from sales of irrigation water ($10 000)
11	$50	$29	$105	$61	
12	$50	$28	$105	$58	
13	$50	$27	$105	$56	
14	$50	$25	$105	$53	
15	$50	$24	$105	$51	
16	$50	$23	$105	$48	
17	$50	$22	$105	$46	
18	$50	$21	$105	$44	
19	$50	$20	$105	$42	
20	$50	$19	$105	$40	
	Totals:	$1740		$979	

Note: All dollar values are in $ thousand.

979/1740 = 0.56, less than 1.0. Thus, the project does not seem desirable from an economic standpoint. If external funding is obtained to subsidize some of the initial costs, or if the benefit can somehow be increased, then the project may be viewed more favorably.

A second example considers only the costs of water treatment to be compared later with the health benefits that accrue. The maximum contaminant level (MCL) is a health-based target for contaminants that may be found in public drinking water supply systems (US Environmental Protection Agency [US EPA] 2020). Arsenic is of serious concern in some US communities where the metalloid is naturally occurring in groundwater sources. The US MCL for arsenic in drinking water is 10 μg/l (US EPA 2015a). According to EPA analysis, the estimated CBR to reach this standard is 0.9–1.25 for a 3% discount rate and 1.0–1.4 for a 7% discount rate (US EPA 2000).

The great majority (82%) of US public water systems serve communities of less than 3300 persons (US EPA 2009a). Because of the economies of scale associated with unit treatment costs, large community water systems (CWS) can treat water at a lower annual cost on the household level (Table 15.4). But for small rural populations (<3300 residents), the annual costs can be significant. In such cases, the health benefits accrued to the small community may not be sufficient to justify the costs. From both an equity and economic standpoint, subsidized household arsenic removal units would be a less-expensive alternative on a per household basis (Cho et al. 2010).

Although CBA is widely used, it is only objectively accurate to the extent that both cost and benefit data are quantifiable and accurate. Thus, practitioners of water security should be aware of the potential shortcomings and subjectivities. This third example is used to illustrate several of these.

The Platte River drains the eastern slopes of Wyoming and Colorado before it flows through the state of Nebraska and joins the Missouri River (Figure 15.3). It has a reputation for being wide, muddy, and shallow. The common description was that the Platte was "a mile wide at the mouth, but only six inches deep." Other pioneers moaned that it was "too thick to drink, but too thin to plow." It is, however, part of the Central Flyway, the major migratory

Table 15.4 Mean annual costs per household of meeting the arsenic MCL standard (10 μg/l) for drinking water.

CWS size category (population served)	EPA – estimated annual cost per household to meet arsenic standard
25–100	$407
101–500	$202
501–1000	$88
1001–3300	$72
3301–10 000	$47
10 001–50 000	$40
50 001–100 000	$31
100 001–1 million	$25
More than 1 million	$1
Weighted average across all size categories	**$39**

Source: Data from US EPA (2000).

Figure 15.3 The Platte River flows eastward across the state of Nebraska. Source: MarekPhotoDesign.com/Adobe Stock.

north–south route for hundreds of thousands of birds, including sandhill cranes, who inhabit the sandbars at night and feed on local fields during the day.

In the mid-1940s, farmers thought that land on the north side of the river could be developed agriculturally with irrigation. The Bureau of Reclamation planned to build a series of canals and reservoirs using public monies and water contracts with local farmers. At that time, there were only around 1300 irrigation wells. But as the project was debated and stalled in Congress, more and more irrigation wells were drilled, and more land was being irrigated. Thirty years later, the project still had the endorsement of the Bureau of Reclamation who claimed that the groundwater was being depleted by excessive pumping. But it was now opposed by both farmers who said they no longer needed the extra water (and expense) and the public who was concerned about the loss of baseflow in the river for fish and ecosystem health. Most of the remaining water would be diverted under the new project (Hanke and Walker 1974).

The original project scope was a composite of a large diversion dam, a main supply canal, 23 interconnected reservoirs, four distribution canals and networking system, large pumps to supply extra water during peak demands, and a Federal wildlife refuge on three of the proposed reservoirs (USFWS 1957). CBA analyses were made and adjusted through various iterations as the scope reduced in size and appurtenances.

The CBA performed for the original scope, in the original context of water demand in 1967, resulted in a benefit-to-cost ratio (BCR) of 1.24, sufficient to justify the economic expenditures. However, a later, more conservative third-party analysis resulted in a BCR of 0.23–0.87. This new analysis claimed three modifications: (i) a revised, higher discount rate (5.375%) from the original rate (3.125%), which was deemed inappropriately low; (ii) a reevaluation of flood control and fish and wildlife benefits, which were originally overstated; and (iii) a removal of the assumption that 44 000 acres of "new lands" from the project would be converted from dry land to irrigated farming and yield significant benefits.

From this example, it is clear that economics alone cannot produce infallible answers to complicated questions. When the projects involve large outlays of public monies, it behooves decision-makers to utilize public input as much as possible, including both experts and affected stakeholders.

15.3.3 Considerations and Limitations

The CBA method as currently practiced has these considerations and limitations:

- Costs and benefits are often pre-evaluated using "willingness to pay (WTP)" of the beneficiaries and "willingness to accept (WTA)" of compensation for the losers. Especially in pre-project conditions, these dollar amounts are subjective and/or not easily obtained.
- Costs may more severely affect groups of people that are already disadvantaged, thereby increasing their misery – for example, poor tenant farmers who are displaced by inundation of their land for a reservoir. The costs may also bring harm to a large pool of taxpayers over a long term. These cost inequities are not automatically accounted for in CBA.
- Welfare benefits typically are only considered for the human species, with no regard for other species or ecosystems, who have an intrinsic value of their own apart from the benefits they give to humans. The importance of endangered species when conducting an Environmental Impact Assessment is an attempt to mitigate this shortcoming.
- Prices used in assessing costs may not coincide with social opportunity costs, which are more appropriate for public welfare projects. For example, the hiring of day laborers has a higher social value in times of extreme unemployment. Social costs also include *externalities*, such as the carbon emissions and air pollution associated with burning fossil fuels. These are often not included in the market price.
- As discussed above, benefits and costs must consider the time value of money, so that alternatives may be compared as "apples to apples" in present worth dollars. Although individuals and businesses may have different "rates of time preference," the CBA process uses a social discount factor that is deemed suitable, although critics think that future generations are being shorted (Griffin 1998).
- Intangible, non-monetized benefits and costs are difficult to include in the CBA process. Cultural practices and environmental benefits may be affected by a dam – reservoir project, but these are not easily assigned an economic value. For example, a hydroelectric project may grant an independent African nation, such as Ethiopia, a measure of political independence from its neighbors. But how is this to be valued in real dollars? Conversely, a dam on a free-flowing river in the northwest United States may disrupt the traditional fish harvest of indigenous people during the seasonal spawning runs. The food source may be adequately compensated by technology such as fish ladders, but the tribe's cultural identity suffers from the loss of ritual.

These considerations do not imply that the CBA is not a useful tool, but only that it is only one metric that should be augmented with other criteria for project evaluation. Other methodologies, such as the Community Capitals Framework, incorporate economic capital along with other community resources, such as natural, cultural, human, social, political, and built capital (Beaulieu 2014).

15.4 Demand Management Through Water Pricing

Public water supply institutions can be considered as natural monopolies. That is, high startup costs and economies of scale usually exclude multiple competitors for the same public water supply. The public utility usually sets prices that do not include a profit but that allow operation and maintenance costs to be met on an annual basis and capital costs to be recovered over time resulting in prices that match the "full marginal cost". Because capital costs are so high for infrastructure projects, this second requirement may be difficult to meet – especially for small communities. The marginal cost for raw water itself is often not included, although access to the raw water source may be expensive depending on location, political ownership, built infrastructure, and purchases of land needed.

Financing in the United States has shifted in the past 40 years. In the middle of the twentieth century, Federal grant funding could often be tapped to provide capital expenditures for large reservoirs, treatment plants, and pipelines, while both capital and operating costs were recouped through monthly utility bills to the consumers. In 1988, the government stopped providing grants and initiated a "revolving fund" that could provide low-interest loans for large infrastructure projects. As these loans were paid off, the fund was replenished to provide more loans (US EPA 2015b).

As systems and pipelines age and degrade, these need to be repaired and replaced. Thus, the average utility bill in many developed countries has increased higher than, and sometimes double, the rate of inflation. Even so, only about 80–90% of the funds needed are provided by consumer payments. This large national funding gap has been the focus of much political debate over the last 20 years and will continue to be so even as local communities try to reduce expenditures by delaying maintenance needs, possibly resulting in property damage, greater wastage, dangers to human health, and loss of consumer confidence (Sedlak 2015).

In this section, we discuss various options for water pricing intended to reach water security through equity, conservation, and more efficient financing.

15.4.1 Water Rate Structures

Water rates can be structured to help promote greater equity and sustainability through efficiency (US EPA 2016). In economic terms, water is a necessity rather than a luxury good. Thus, low-income households will experience more hardship when water rates are raised as their water fees are typically a greater percentage of their monthly income. Water rate structures are the applications of price to volumes of water used. (The following is taken from US EPA, "Pricing and Affordability of Water Services" [US EPA 2016].)

The following price structures are possible:

- Price structures that provide equitable minimum access
  - "Lifeline rate" – households are charged lower rates on a threshold, nondiscretionary water consumption. In the United States, this amount is 6000 gal minimum required per month per household for sanitation and hygiene (US EPA 2016). Water that is consumed beyond this amount is charged at a higher rate.
- Price structures that encourage conservation
  - Increasing block rates – Using block rates or tiered pricing that increase with water usage. The per-unit charges for water increase as the amount of water used increases. The first block is charged at one rate, the next block is charged at a higher rate, and so on.
  - Time of day pricing – Charging higher prices for water used during a utility's peak demand periods, mirroring an approach used in electricity rates.
  - Water surcharges – Charging a higher rate for "excessive" water use (i.e. water consumption that exceeds the local or regional average).
  - Seasonal rates – Water prices that rise or fall according to weather conditions (such as drought) and/or the corresponding demand for water.

- Price structures that are less effective in encouraging conservation
  - Uniform rate structures – A uniform rate charges the same price-per-unit for water usage beyond the fixed customer charge, which covers some fixed costs. The rate sends a price signal to the customer because the water bill will vary by usage. Uniform rates by class charge the same price-per-unit for all customers within a customer class (e.g. residential or nonresidential).
  - Flat fee rates – Flat fee rates that do not vary by customer characteristics or water usage.
  - Declining block rate – Tiered pricing that decreases with water usage, seeking to encourage consumption when the commodity is in excess of demand.

Table 15.5 gives a snapshot of the usage of various rate structures for public water supply systems in the United States up to 2021. Both the uniform rate charge and increasing block rate remain popular, while some utilities use a tiered rate structure that also acts as a seasonal structure. For example, Denver Water has a seasonal structure for its irrigation customers. Most customers reach the higher tiers only during the summer period (Cristiano 2021).

Figure 15.4 offers an increasing block rate structure conceptual design that sets a low unit price (P_2) for all households up to a total volume of water needed to meet the basic needs of

Table 15.5 US public water-supply system rate structures for four years.

Rate structure	Percent of sample utilities (residential)			
	1997[a]	2014	2019	2021
Flat fee, regardless of volume	2	1	1	1
Uniform rate charge	33	29	36	39
Declining block rate	34	16	11	11
Increasing block rate	31	50	47	46
Seasonal rate	0	4	5	3
Total	100	100	100	100

a) This year includes nonresidential rate structures.
Source: Cavanagh et al. (2002), US EPA (2009b), and Raftelis (2021).

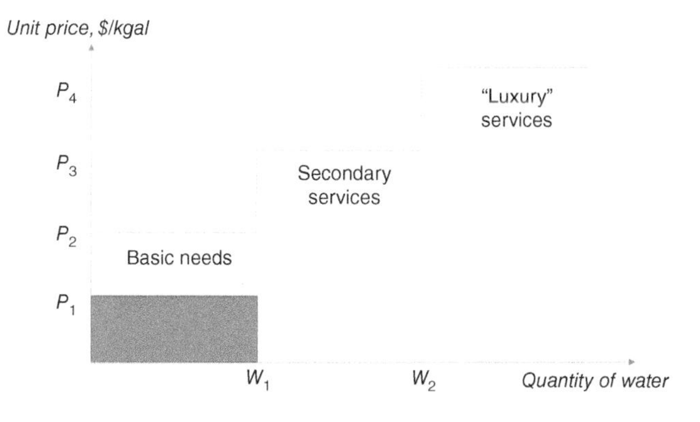

Figure 15.4 Conceptual model of a proposed increasing block structure for social equity and conservation. Source: JFC original.

cooking, drinking, toilet flushing, and cleaning. As stated above, this volume (W_1) is 6000 gal per household per month in the United States. This unit price may be added to a uniform flat fee (shaded block, $P_1 \times W_1$) that will recoup capital costs over a long time period. Then the unit price increases to P_3 for additional volumes of water ($>W_1$ but $<W_2$). These may include typical secondary services such as landscape watering. A final jump at W_2 to price, P_4, would be for water that is used for nonessential "luxury" services, such as washing vehicles and filling a swimming pool. Those who can afford this extra water will purchase it, while others will simply not use it.

Each community would need to do its own analysis in order to determine fair rates and average household demands for the various step quantities. Such a rate structure would promote greater access to basic needs while reducing the wastage of extraneous water.

15.4.2 Price and Income Elasticity

A final concept that is relevant to water economics is that of elasticity. *Elasticity* of demand refers to the change in demand (positive or negative) in response to a variable factor, such as the price or household income. This is calculated as the percentage change in quantity divided by the percentage change in the second variable. For example, in the case of price elasticity,

$$\text{Elasticity, €} = \frac{\%\text{ change in demand}}{\%\text{ change in price}} \tag{15.7}$$

If the demand for water does not change significantly with price (i.e. the absolute value of € is less than 1), then the demand is said to be "inelastic" with respect to price. Elasticity is thus a measure of the *sensitivity* of water demand with respect to a variable such as price. In the example for Israel (discussed below), the elasticity could be estimated by € = −16%/40% = −0.4, since the household demand dropped 16% following a price increase of 40%. Thus, in this case, household demand is said to be slightly elastic with respect to the price of water.

Some goods are highly elastic because there are suitable substitutes. For example, if the price of sugar goes up significantly, consumers can switch to alternative sweeteners and the demand for sugar decreases dramatically. Since there is no substitute for water, its demand tends to be inelastic. Research supports the intuition that price has some limited effect on the consumer's demand and that it has less effect on individuals (or households) in the higher income bracket since the monthly water bill in this case is a less-significant portion of total household expenses (Renwick and Archibald 1998). Research on 11 urban areas in the United States and Canada found that water demand was more sensitive to water rate *structure* than to the rate itself (Cavanagh et al. 2002).

Bottled water represents a modern phenomenon that has ancient roots in the association of some spring waters with healing powers. Natural sulfates and bicarbonates would treat gastrointestinal illnesses, while calcium strengthened bones and teeth and lithium is soothing to people with depression (Salzman 2017). People began to demand this water even when they were unable to personally travel to the source. Bottled water in the United States and Europe became popular in the 1970s first with Perrier and later with brands launched by

PepsiCo (Aquafina) and Coca-Cola Company (Dasani). These products have an extremely high markup with the retail price being 240–10 000 times the actual cost of producing the water (Salzman 2017). In addition, bottled water in the United States is not as closely regulated and monitored for impurities as is water from a community water-supply system. Nonetheless, consumers will prefer this product over tap water because of its marketing mystique, convenience, taste, and association with fitness and health (Salzman 2017). Perhaps it is no small irony that "Evian" spelled backward is "naïve." Such a commodified water is relatively price elastic since a drastic price increase in one brand of water would likely cause a consumer to jump to tap water or an alternative brand signaling a drop in quantity demanded for the more expensive brand.

In order to understand the variation between income quintiles, it may be helpful to parse out household water consumption micro-activities. Figure 15.5 shows the results of a survey of over 400 households in Duhok, a medium-sized city in northwestern Iraqi Kurdistan (Hussien et al. 2016). In addition to domestic wells, water is supplied from a central treatment plant to each household, three to four times a week for six hours at a time. The water is stored in overhead tanks and used for drinking and other activities. The average survey household occupancy was seven persons.

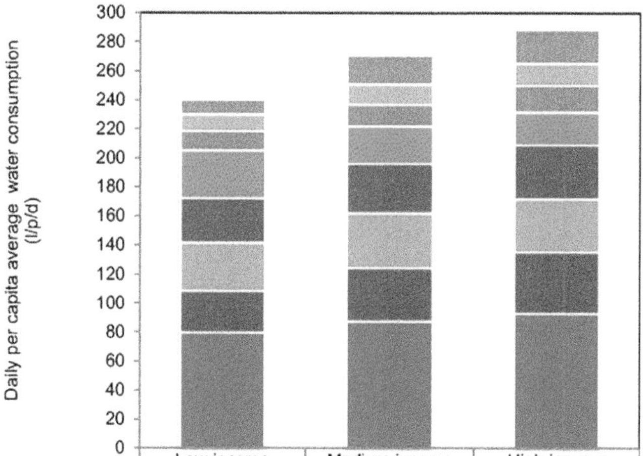

Daily per capita average water consumption (l/p/d)

	Low income	Medium income	High income
Swimming pool	0	0	0.19
Bath	0	0	1.36
Vehicle washing	1.4	1.65	0.53
Garden watering	10.38	20.09	23.3
House washing	11.18	14.23	15.41
Cooking and drinking	13.2	14.85	18.33
Toilet flushing	32.99	25.45	22.51
Laundry	30.91	33.99	37.14
Dishwashing	32.98	37.98	36.69
Shower	28.74	36.67	42.31
Taps	79.43	87.27	92.59

Figure 15.5 Impact of per capita monthly income on various water-consuming activities in Duhok, Iraq (population 295 000). Source: Hussien et al. (2016).

In this analysis, the average per capita water consumption increases with the per capita annual income – low income (<\$1500/241 lpcd), medium income (\$1500–3000/272 lpcd), and high income (>\$3000/290 lpcd). Higher income families have a few additional amenities – e.g. a bath and swimming pool. Medium-income households use more water for vehicle washing because they are likely to have more vehicles, whereas high-income families will use an outside car-washing service. Toilet flushing decreases in medium and higher income households who are more likely to have low-flush toilets. Although the frequency of showers was higher in high-income households, the shower heads were newer and likely to be more water efficient. The highest water usage in each income group is water used in the category of "taps," or hand-washing basins, as the observance of Islamic faith includes ablutions before each prayer time.

15.4.3 Water Pricing – A Case Study

For the people of the modern nation of Israel, the word "water" conjures up more than just a scarce commodity. It appears 600 times in the Hebrew Bible, and there are several words for "rain," denoting its importance and its frequency of occurrence. Average annual rainfall varies from the wetter north (1100 mm/year) to the arid south (<100 mm/year), for a country-wide average of ~450 mm/year. In contrast to private water rights in the United States, water rights in Israel belong to and are managed by the government on behalf of all its people (Siegel 2017).

In 1964, the Israeli government completed the National Water Carrier, a 130-km-long pipeline that carries water from the Sea of Galilee in the north to the Negev Desert region in the south (Kantor 2008) (Figure 15.6). Originally allocated at 80% for agriculture and 20% for drinking water, the allocation has reversed so that the majority goes toward drinking water for the growing populations of people and higher standard of living.

Figure 15.6 Israel's National Water Carrier is a concerted effort to "make the desert bloom" through increased irrigation. Source: Ariel Palmon / Wikimedia Commons / Public domain.

In 2008, the Israel Water Authority, charged with managing the country's water supply, established a national water fee that would cover the real cost of water. This real cost would cover not only the pumping costs, as was customary, but also the existing and new planned infrastructure. Household water prices rose by 40%, while farmers' prices rose too with the promise that they would be guaranteed a regular supply, even in times of drought. Almost immediately the usage of household water dropped by 16%, as families learned how to recycle water within their own homes, such as by using dirty bathwater to water plants. Farmers, too, managed to conserve water even before they switched to new crops that were less water-intensive. Consumer education on the value of water had been tried previously, but water pricing was a much more effective incentive (Siegel 2017).

The following assertions can be drawn with respect to price and income elasticity:

- Water demand is generally *price* inelastic. A meta-analysis of 314 price elasticity estimates resulted in a sample mean of −0.41 and a median of −0.35 (SD = 0.86) (Dalhuisen et al. 2003). A separate analysis had similar results, with a mean of −0.51, and about 75% of the estimates were between −0.02 and −0.75 (Espey et al. 1997). Water is a basic need for all households, and it has no substitute, but households can learn how to conserve and reuse water, as in Israel.
- Water demand is also relatively *income* inelastic but varies in a positive direction since higher income quintiles tend to use more appliances that consume more water. A meta-analysis of 162 income elasticity estimates resulted in a sample mean of 0.43 and a median of 0.24 (SD = 0.79) (Dalhuisen et al. 2003). This would suggest that the water demand will rise at about half the rate of the rise of income.
- Conservation and efficiency may result from a price increase but mostly over a longer term rather than the short term. This is because most consumers do not replace appliances (dishwashers, clothes washers, etc.) immediately in response to a price hike but will do so at the end of the product's lifetime. At that time, they will likely choose a more efficient unit as available, such as a low-flush toilet and more-efficient clothes washer. Manufacturers of these products will respond likewise as long as there is some economic certainty for the price hike. Unfortunately, this latter shift is not revealed in elasticity studies (Achttienribbe 1998).
- Bottled drinking water is considered a luxury good and is much more expensive on a unit basis than tap water. It follows that bottled water is more price elastic as well, since the consumer can choose a different brand of bottled water, choose to use less-expensive sachet water (water in a sealed plastic bag), or consume tap water as a response to a price jump.
- Water demand in the urban setting is subject to availability. In most cultures if it were more readily available and affordable, the patterns of use would change. The present observed patterns are determined by the accessibility and affordability in major cities.

The science of economics provides tools that can lead to a more complete analysis of water security as well as recommendations for a more suitable price structure. Water has value, and its value is expressed in both the supply and consumption costs.

15.5 Economic Valuation of Ecosystem Goods and Services

The CBA method we have discussed is most applicable when costs and benefits are discrete and easily monetized. But when considering water-based or water-related public goods, the method is not so straightforward. For example, consider a decision to be made whether to remove a dam on the Columbia River, returning the stream to its natural flow patterns and ecology. How do the benefits of protection of fish populations, esthetic appeal to recreational tourists, and the cultural importance to native tribes measure up against the loss of hydropower and water storage for irrigation? Can the costs and benefits of this proposal be compared using a common metric, the US dollar?

Or consider the diversion of water to grow a bioenergy crop, such as corn or soybean, in order to reduce the consumption of fossil-based fuels. How does the loss of a delicate river ecosystem now compare with the benefit of reduced global warming for future generations? Economists speak of *externalities* as those side effects, typically negative, which are the result of economic decisions. These effects are not included in the market price of a good (i.e. external to the price) and thus go unnoticed and uncounted by neither the producer nor the consumer but are nonetheless borne by the society at large.

In the last several decades, CBA has given way to a broader assessment, sometimes called "resource evaluation," and this large group of intangible goods are now considered "assets" to be evaluated (Smith 1996). In the case of water, these intangible goods often are "indirect services" that water provides to humanity, such as recreation; protection of wildlife, cultural, and spiritual beauty; preservation of biodiversity and endangered species; and the transport and dispersion of wastes (Figure 15.7). Or, in a different sense, some values are *intrinsic* values of the water, e.g. its beauty in a shimmering lake or its replenishment function as drinking water. Other values are *instrumental* values in that the water is a means to another useful end, such as its use in industry or in growing food or removing wastes. In this section, we present several ways in which these resources can be evaluated and thus included in an overall decision assessment.

A water-rich acre of wetland might have an inferior economic value when commercial development is a primary criterion. But its value is quite different when considered from the point of view such as wildlife habitat, a means of flood control, purification of surface runoff, or recharge to the groundwater aquifer. The economic value of these water-related functions is distinct from a market value, since these functions are not bought or sold in markets (Freeman 2014). The economic value must then be determined indirectly using tools that owe their formation in the social sciences.

For example, the metrics of WTP (Willingness-to-Pay) and WTA (Willingness-to-Accept) are important tools in the valuation of intangibles. WTP is the maximum sum of money that an individual is willing to pay for an increase of an environmental good, while WTA is the minimum amount of money an individual would willingly accept for the loss of an environmental good. In principle, the two measures should be equivalent, but WTP is often constrained by an individual's income, whereas there is no limit to a WTA (Freeman 2014).

Valuation methods generally fall into two broad categories: revealed preference (RP) and stated preference (SP) methods. An RP, as its name implies, is "revealed" in an actual

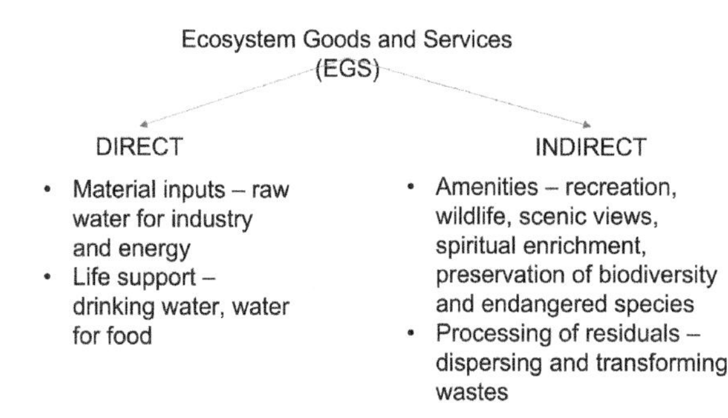

Figure 15.7 Ecosystem goods and services fall into two broad categories – direct and indirect services. Source: Adapted from Freeman (2014) (Also discussed in Chapter 10).

expenditure for an environmental good, such as the amount of money paid for a fishing license. An SP is based on a "what if" scenario that is presented to the consumer in a consideration of proposed or hypothetical changes in ecosystem goods and services (Young and Loomis 2014). Examples of both types of methodologies are given in Table 15.6 along with a third, more over-arching method, that of "benefit transfer."

The most commonly used SP method is contingent valuation. This method is best illustrated with an example. Suppose the public authority sees the future need to protect a vital watershed in southern Ontario. They might pose the following questionnaire to all residents living in the watershed (Brox et al. 1996):

> *Imagine a situation whereby commercial development in the Grand River Watershed begins to threaten fish, waterfowl, songbirds, and other creatures in marshes and woodlands (Figure 15.8). Conserving these areas might require the government to acquire properties and/or pass regulations restricting land use.*
> *If funds are required to purchase land, or compensate owners for further restrictions on land use, would you be willing to pay an additional annual tax exclusively for this purpose?*
> ____ *Yes, I would be willing to pay an additional annual tax to a maximum of:* ____________

Table 15.6 Description of various revealed preference and stated preference environmental evaluation methods.

Valuation method	Description
Revealed preference methods:	
Travel cost	Value is based on the frequency of trips to an outdoor recreational site; includes the opportunity cost of the consumer's time
Hedonic property value	Value is based on the incremental value given to properties with environmental benefits, such as a lake view and riverfront
Defensive behavior	Value is based on the willingness of a participant to avoid an environmental pollutant, such as the purchase of bottled water to avoid drinking contaminated water
Damage cost	Value is based on damages caused by an environmental change, such as the damages from a flood
Stated preference methods	
Contingent valuation	Consumer is asked what they would be willing to pay contingent on some hypothetical change. For example, how much would you be willing to pay for access to a nearby whitewater river for rafting and kayaking?
Choice modeling	Consumer is asked to choose between a suite of options, thereby revealing his most preferred alternative
Benefit transfer method	
Benefit transfer	Applies numerical benefits obtained from primary research at one location/environmental service ("study site") to another location/environmental service of interest ("policy site")

Source: Based on Young and Loomis (2014).

Figure 15.8 The Grand River watershed in Ontario, Canada, home to the Six Nations of the Grand River Territory, is currently 70% agricultural farmland but under pressure for commercial development. Source: Sara J. Martin / Wikipedia / Pubic Domain.

In the Grand River case, the survey results suggested a WTP of $4.56 per household for a water project described in one of the scenarios. When multiplied by ~259 000 households and using a discount rate of 5%, this gave the investigators a capital value of $284 million with which to protect the watershed. Because this number exceeds the estimated cost of protection, the perceived social benefit of improved water quality was greater than the cost and the project can be deemed economically desirable.

Both RP and SP surveys can be used to establish economic value of intangible environmental goods. But government requirements for CBA analyses are often confounded by low budgets and time constraints such that primary valuations are not feasible (Johnston and Rosenberger 2010). An alternative method is employed in the case of benefit transfer in which data collected from one site (or ecosystem service) are applied to a second site (or service). This method is often used in cases where the second "policy" site is small or study resources are limited. The USDA Forest Service has used this method to estimate the consumer surplus, or net WTP, for various recreational activities in its National Recreation Areas. From this estimate, the nationwide average economic value of fishing is $82.37 per person per day, while the value of just enjoying nature is $70.99 per person per day (Rosenberger et al. 2017).

Similar methods were employed on a per land area scale in the classic study by Costanza and others and updated in a 2014 publication (Costanza 1997; Costanza et al. 2014). The value of wetlands services is in the range of $140 000/ha-year, while coastal estuaries provide services in the range of $28 000/ha-year. The recent update revealed a net loss of $4–20 trillion per year since that time due to land-use change. The authors remind us that these ecosystem services are common pool resources, and these numbers are not meant to be employed as market price valuations (Costanza et al. 2014).

As these systems and services become more scarce, public policy makers and businesses would be wise to be aware of the potential costs and disruptions to the vital services upon which they depend (Millennium Ecosystem Assessment 2005). Valuation tools can be useful methods for use in long-range planning.

15.6 Conclusion

In the way that most of us experience it, water, like air, is a natural resource that is widely considered available, universal, and critical for nearly all human and environmental activities. Unlike air, however, water is bulky, transportable, localized in availability, and highly variable regarding quality. No one is selling bottled air at the grocery store – yet! Thus, water is more likely to fall under the laws of economics as it can be collected, stored, piped, traded, bought, and sold.

In its Dublin Statement, the United Nations declared that "Water has an economic value in all its competing uses and should be recognized as an economic good … Managing water as an economic good is an important way of achieving efficient and equitable use, and of encouraging conservation and protection of water resources." (quoted in [Young and Loomis 2014]). Efficient pricing of water can be used to both support water-supply improvements and achieve efficient usage and reduce wastage. Major projects can be evaluated on the basis of comparison of present value costs and benefits. And the value of less-tangible benefits can be estimated with indirect methods of valuation.

In this chapter, we have reviewed some of the basic economic concepts that are useful in optimizing the management of water. The field of economics supplies tools and methodologies that can be helpful in the pursuit of water security.

End-of-Chapter Questions/Problems

15.1 Do an Internet search for the large irrigation project in Libya, described as the "world's largest irrigation project." In one paragraph, describe this project – where the water will come from, how will it be purchased, transported, distributed, etc.

a. What are the cost figures associated with the project?
b. What are the expected benefits?
c. What are some potential problems/challenges that come with this project based on its size, location, politics, etc.

15.2 Develop a good working definition of "welfare economics."

a. How does this differ from traditional economics?
b. Why does this subset of economics apply to water-resource projects?

15.3 Developing an economic value for natural ecosystem functions is an art and science unto itself. Use trusted Internet resources to set a value ($/unit) for each of the following ecosystem services. Provide a reference for each.

a. Water purification by wetlands
b. Riparian buffer zones
c. Flood and erosion protection of mangrove plantations along coastal zones
d. Coral reefs for providing biological habitat

15.4 Conduct a spreadsheet-level CBA for the following irrigation project, including dam and pond and major canal for irrigation:

a. Discount rate: 6.5%
b. Project lifetime: 20 years

c. Costs:
   i. Year 1 – capital construction of dam and reservoir – \$50 000
   ii. Year 2 – capital construction of canal and pumps – \$21 000
   iii. Year 10 – replacement of pumps – \$8000
   iv. Year 18 – replacement of pumps – \$8000
   v. Annual maintenance and operation – \$4500
d. Benefits:
   i. Annual collection of water fees, beginning in Year 3 – \$9500
e. What is the BCR of this project? Show your spreadsheet and all your work.

15.5 The Series Present Worth Factor is used to calculate the present worth of a series of uniform annual payments, or "how to calculate P when given A." The formula is expressed below, and the factor that is multiplied by A to get P is also shown:

$$P = A((1 + d)^n - 1)/(d(1 + d)^n)$$

factor $(P/A, i, n) = ((1 + d)^n - 1)/(d(1 + d)^n)$
Derive the formula for the factor, P/A, from simply the knowledge of a term, n, and a discount rate, d.

15.6 Figure 15.5 lists typical household activities in Iraq with daily water consumption for each. Find a resource that gives a similar listing for US households.
a. How does the United States list compare with the high-income household in Iraq?
b. Compare the top three most water-intensive activities for each country. Comment on the differences/similarities.

15.7 The USGS has a Benefit Transfer Toolkit that includes databases on resource evaluation, including water quality. This Toolkit is found at sciencebase.usgs.gov. Open the toolkit and find the database that has evaluations of water quality.
a. Find a study that examines a location near where you live.
b. What services were being evaluated?
c. What type of valuation method was employed?
d. What is the range of WTP values associated with this service(s)?

Further Reading

Freeman, A.M. (2014). *The Measurement of Environmental and Resource Values: Theory and Methods*, 3e. Abingdon, Oxon; New York, NY: Routledge.

Griffin, R.C. (1998). The Fundamental Principles of Cost-Benefit Analysis. *Water Resources Research* 34 (8): 2063–2071. https://doi.org/10.1029/98WR01335.

Griffin, R.C. (2006). *Water Resource Economics: The Analysis of Scarcity, Policies, and Projects*, 1e. Cambridge, Massachusetts: The MIT Press.

Young, R.A. and Loomis, J.B. (2014). *Determining the Economic Value of Water: Concepts and Methods*. RFF Press.

References

Achttienribbe, G.E. (1998). Water price, price elasticity and the demand for drinking water. *Aqua* 47 (4): 196–198. https://doi.org/10.1046/j.1365-2087.1998.00099.x.

Ali, M. (2017). "Libya's Great Man-Made River Irrigation Project: The Eighth Wonder of the World?" Qantara.de - Dialogue with the Islamic World. 2017. https://en.qantara.de/content/libyas-great-man-made-river-irrigation-project-the-eighth-wonder-of-the-world.

Beaulieu, L.J. (2014). *Promoting Community Vitality & Sustainability: The Community Capitals Framework*. Purdue University.

Blomkvist, P. and Nilsson, D. (2017). On the need for system alignment in large water infrastructure: understanding infrastructure dynamics in Nairobi, Kenya. *Water Alternatives* 10 (2): 20.

Brox, J.A., Kumar, R.C., and Stollery, K.R. (1996). Willingess to pay for water quality and supply enhancements in the Grand River watershed. *Canadian Water Resources Journal* 21 (3): 275–288.

Cavanagh, S.M., Hanemann, W.M., and Stavins, R.N. (2002). "Muffled Price Signals: Household Water Demand Under Increasing-Block Prices." SSRN Scholarly Paper ID 317924. Rochester, NY: Social Science Research Network. https://doi.org/10.2139/ssrn.317924.

Cho, Y., William Easter, K., and Konishi, Y. (2010). Economic evaluation of the new U.S. arsenic standard for drinking water: a disaggregate approach. *Water Resources Research* 46 (10): https://doi.org/10.1029/2009WR008269.

Costanza, R. (1997). The value of the World's ecosystem services and natural capital. *Nature* 387: 253–260.

Costanza, R., de Groot, R., Sutton, P. et al. (2014). Changes in the global value of ecosystem services. *Global Environmental Change* 26 (May): 152–158. https://doi.org/10.1016/j.gloenvcha.2014.04.002.

Cristiano, T. (2021). "Public Water System Rate Structures," November 5, 2021.

Dalhuisen, J.M., Florax, R.J.G.M., de Groot, H.L.F., and Nijkamp, P. (2003). Price and income elasticities of residential water demand: a meta-analysis. *Land Economics* 79 (2): 292–308. https://doi.org/10.2307/3146872.

Espey, M., Espey, J., and Shaw, W.D. (1997). Price elasticity of residential demand for water: a meta-analysis. *Water Resources Research* 33 (6): 1369–1374. https://doi.org/10.1029/97WR00571.

Freeman, A.M. (2014). *The Measurement of Environmental and Resource Values: Theory and Methods*, 3e. Abingdon, Oxon; New York, NY: Routledge.

Griffin, R.C. (1998). The fundamental principles of cost-benefit analysis. *Water Resources Research* 34 (8): 2063–2071. https://doi.org/10.1029/98WR01335.

Griffin, R.C. (2006). *Water Resource Economics: The Analysis of Scarcity, Policies, and Projects*, 1e. Cambridge, Massachusetts: The MIT Press.

Hanke, S.H. and Walker, R.A. (1974). Benefit-cost analysis reconsidered: an evaluation of the mid-state project. *Water Resources Research* 10 (5): 898–908. https://doi.org/10.1029/WR010i005p00898.

Hussien, W.'e.A., Memon, F.A., and Savic, D.A. (2016). Assessing and modelling the influence of household characteristics on per capita water consumption. *Water Resources Management* 30 (9): 2931–2955. https://doi.org/10.1007/s11269-016-1314-x.

Jalil, M.M. (2010). "Approaches to Measuring Social Discount Rate: A Bangladesh Perspective." SSRN Scholarly Paper ID 1921987. Rochester, NY: Social Science Research Network. https://doi.org/10.2139/ssrn.1921987.

Johnston, R.J. and Rosenberger, R.S. (2010). Methods, trends and controversies in contemporary benefit transfer. *Journal of Economic Surveys* 24 (3): 479–510. https://doi.org/10.1111/j.1467-6419.2009.00592.x.

Kantor, S. (2008). "The National Water Carrier." http://research.haifa.ac.il/~eshkol/kantorb.html.

Komives, K., Foster, V., Halpern, J., and Wodon, Q. (2005). *Water, Electricity, and the Poor*. World Bank.

Millennium Ecosystem Assessment (2005). *Ecosystems and Human Well-Being: Opportunities and Challenges for Business and Industry*. Washington, D.C: World Resources Institute.

Raftelis (2021). "Excerpts from AWWA Rate Surveys."

Renwick, M.E. and Archibald, S.O. (1998). Demand side management policies for residential water use: who bears the conservation burden? *Land Economics* 74 (3): 343–359.

Rosenberger, R.S., White, E.M., Kline, J.D., and Cvitanovich, C. (2017). *Recreation Economic Values for Estimating Outdoor Recreation Economic Benefits from the National Forest System.*" PNW-GTR-957. Portland, OR: USDA Forest Service.

Salzman, J. (2017). *Drinking Water: A History*. Revised and Updated edition. Overlook Duckworth.

Sedlak, D. (2015). *Water 4.0: The Past, Present, and Future of the World's Most Vital Resource*. Reprint edition. Yale University Press.

Siegel, S. (2017). *Let There Be Water*. Griffin. https://smile.amazon.com/Let-There-Be-Water-Water-Starved/dp/1250115566/ref=sr_1_1?crid=2SIV12GFQR65K&dchild=1&keywords=let+there+be+water+by+seth+siegel&qid=1598400233&s=books&sprefix=Siegel+let+there+%2Caps%2C177&sr=1-1.

Smith, V.K. (1996). *Estimating Economic Values for Nature: Methods for Non-Market Valuation*. Cheltenham, U.K.; Brookfield, VT: Edward Elgar Pub.

Sullivan, W.G., Wicks, E.M., and Bontadelli, J.A. (1999). *Engineering Economy*. Subsequent edition. Upper Saddle River, N.J: Prentice Hall.

Tietenberg, T. (2002). *Environmental and Natural Resource Economics*, 6e. Pearson Education.

TVA (2021). "Cherokee Dam on the Holston River." TVA.Com. 2021. https://www.tva.com/energy/our-power-system/hydroelectric/cherokee.

U.S. Environmental Protection Agency (US EPA) (2009a). *Community Water System Survey, 2006*. EPA 815-R-09-001. U.S. Environmental Protection Agency.

U.S. Environmental Protection Agency (US EPA) (2009b). *Community Water System Survey, 2006 - Volume II: Detailed Tables and Survey Methodology*. EPA 815-R-09-002. U.S. Environmental Protection Agency.

US CEQ (2013). *Principles and Requirements for Federal Investments in Water Resources*. U.S. White House.

US CEQ (2014). *Interagency Guidelines*. Council on Environmental Quality.

US EPA, OW (2015a). "National Primary Drinking Water Regulations." Overviews and Factsheets. US EPA. 2015. https://www.epa.gov/ground-water-and-drinking-water/national-primary-drinking-water-regulations.

US EPA, OW (2015b). "How the Drinking Water State Revolving Fund Works." Overviews and Factsheets. September 29, 2015. https://www.epa.gov/dwsrf/how-drinking-water-state-revolving-fund-works.

US EPA, OW (2016). "Pricing and Affordability of Water Services." Overviews and Factsheets. US EPA. January 25, 2016. https://www.epa.gov/sustainable-water-infrastructure/pricing-and-affordability-water-services.

US EPA, OW (2020). "SDWA Evaluation and Rulemaking Process." Overviews and Factsheets. US EPA. 2020. https://www.epa.gov/sdwa/sdwa-evaluation-and-rulemaking-process.

US EPA (2000). "Arsenic in Drinking Water Rule: Economic Analysis." EPA 815-R-00-026.

USFWS (1957). "Mid-State Project Will Help Fish and Wildlife," 1957.

Young, R.A. and Loomis, J.B. (2014). *Determining the Economic Value of Water: Concepts and Methods*. RFF Press.

16

Developing a Twenty-First Century Water Ethic

As water challenges continue to grow, competing interests will increasingly vie for limited water supplies or suffer at the hands of drought or elevated flooding. How should future decisions be made in this increasingly complex and interconnected system? While the preferred approach may seem obvious to one sector, this same choice may be suboptimal for another sector. Do we need to increase water availability to satisfy an ever-increasing thirst for more water, or should we learn to get by with what we have and what we can recycle? Should we learn to be locally water independent rather than continually look to grab water from others? How do ecosystem services and environmental flows factor into a human-centric decision-making process? How do we negotiate the competing needs while maximizing the benefit realized for both society and the environment? Further, what about the water needs of future generations? As a resource that is essential for virtually every aspect of human and nonhuman life – water has particularly complex ethical dimensions. In the twenty-first century we must encounter, and struggle with, these and many other difficult questions in a way that we never have before. We suggest that the answers must flow from a deeply rooted water ethic based on solid fundamentals. This water ethic is one that parallels and complements the land ethic and environmental ethic that our ancestors bequeathed to us. How does one develop a water ethic? Does it start with individuals, business, or government? What lessons can we learn from the past and from various perspectives?

These are issues that the reader will encounter, and wrestle with, in this chapter. While not easy, and at times unsettling, developing a twenty-first century water ethic is not a choice but an obligation.

We owe it to our forebears, whose vision and leadership have enabled us to come this far. And we owe it to our children, grandchildren, and beyond, who deserve a secure water future.

Learning Objectives

Upon completion of this chapter, the student will be able to:

1. Understand basic concepts of values and ethics.
2. Be familiar with the evolution of several other environmental ethics.
3. Understand the historical development of water ethics to date.
4. Understand how water ethics relates to various sectors of society.
5. Recognize the unique perspective that Indigenous people have on water ethics.

Fundamentals of Water Security: Quantity, Quality, and Equity in a Changing Climate, First Edition.
Jim F. Chamberlain and David A. Sabatini.

Companion website: www.wiley.com/go/chamberlain/fundamentalsofwatersecurity

6. Be familiar with the role of water and water ethics in various religions.
7. Understand factors leading to a water ethic in three countries.
8. Understand the importance of as well as pathways to a twenty-first century water ethic.

16.1 Introduction

One can speak of both "water ethics" and a "water ethic." The former describes the framework by which one makes water-related decisions among competing considerations at the community to global scale. Conversely, a water ethic is a set of principles that guide individual actions as well as collective behavior. In this chapter, we will first provide a general introduction to the field of ethics. Next, we will discuss several environmental ethics that evolved in the twentieth century – specifically, the land ethic of Aldo Leopold and the environmental ethic spawned by Rachel Carson's *Silent Spring*. Insights gained from these pivotal movements can guide us towards a twenty-first century water ethic. We will next discuss ongoing efforts to increase societal understanding of our valuable water resources as well as discuss how a water ethic intersects with various sectors of society and the environment. This water ethic should be informed by various perspectives, including individuals, business, ecosystem advocates, and unique vantage points from Indigenous peoples and various religions. After looking at three countries that are leaders in promoting and sustaining a water ethic, we close this chapter by suggesting various pathways to achieving a twenty-first century water ethic that will promote water security, both now and for future generations.

16.2 Ethics – An Introduction

The English word "ethics" is derived from the Ancient Greek words êthos (ἦθος) and ēthikós (ἠθικός), which refer to one's character and moral nature. In English, it can refer to philosophical ethics or moral philosophy to answer various questions of proper behavior. In the general sense, ethics helps us answer the question "how should we live," i.e. what guides our decision-making and behavior amid competing forces or motivations?

In this section, we draw on the work of ethical scholars to provide a brief introduction to the broad field of ethics as a basis for our discussion of water ethics. Ethics and morals are intertwined concepts. They help us distinguish right from wrong and good from evil. In the extremes, these distinctions may seem obvious and may generally hold true between cultures and across time (e.g. theft, murder, dishonesty). But for other choices, and other contextual settings, the distinctions may be much more challenging and less straightforward – for example, water for human needs versus water for ecosystem services, water privatization versus governmental control, etc.

Louis Pojman defines four normative institutions that help shape our behavior and choices – ethics, religion, law, and etiquette (Pojman 1999). He proposes that ethics distinguishes itself from law and etiquette by relying on a deeper foundation and differs from religion by being based on reason rather than a supreme authority. Ethics is necessary to prevent social chaos, and that while on one hand ethical behavior may seem to restrict our freedom, in the larger sense it leads to greater freedom and well-being.

Pojman suggests that ethics support the following positive aspects in regulating a healthy and just society: (i) to keep society from falling apart, (ii) to ameliorate human suffering, (iii) to promote human flourishing, (iv) to resolve conflicts of interest in just and orderly ways, and (v) to assign praise and blame, rewards and punishment, and guilt. Imagine what society would look like if everyone was obsessed with satisfying their own needs and desires without consideration of others! Chaos and disorder would ensue, along with increased human misery.

David Groenfeldt states that "we need ethical principles because most behavioral choices involve multiple and conflicting values, and we need to choose which values to honor and which to ignore" (Groenfeldt 2019). He suggests that ethics is the manifestation of our individual, and at times overlapping, societal values. An example of this would be the application of the popular business concept of the "triple bottom line," considering the bottom lines of profit, society, and environment.

An ethical framework may stem from any of several foundational principles, including ethical relativism, egoism, altruism, utilitarianism, and egalitarianism.

- *Ethical relativism* denies the existence of universally valid moral principles but rather suggests that ethical behavior is location and time dependent. Ethical relativism differs from moral skepticism, which suggests there are no moral principles, by virtue of suggesting the moral principles can differ between societies. For example, while stealing and bribery are considered unethical in certain cultures, they may be highly valued in other cultures (Pojman 1999).
- *Universal ethical egoism* promotes self-interest as of ultimate importance, even if it causes harm to others. The economist might cite Adam Smith as an example, who in the eighteenth century established modern economic theory as the place where self-interest motivates manufacturers to make products at the lowest possible cost thereby maximizing profits.
- *Altruism*, as viewed by purveyors of ethical egoism, is futile. The failure to pursue one's own self-interest prevents individuals, and thus society, from achieving their maximum benefit. Conversely, altruists see ethical egoists as unable to sacrifice immediate gratification for a greater good (Pojman 1999).
- The appeal of Jeremy Bentham's nineteenth-century *utilitarianism* stems from its simplicity and focus on maximizing pleasure and minimizing suffering. Who could argue with that? However, others suggest that the simplicity is illusory and that "the devil is in the details." How does one quantify pleasure and suffering? Isn't there more to life than pleasure and suffering?
- *Egalitarianism*, which comes from the French *égal*, meaning "equal," is sometimes referred to as equalitarianism. This school of thought values social equality and holds that all peoples deserve equal rights and opportunities (Pojman 1999).

The goal of this section was to briefly introduce basic concepts in the field of ethics. Admittedly, this brief overview is far from exhaustive – the interested reader is directed to volumes devoted to this subject. It is our hope that this section provides a basis for understanding water ethics as discussed in subsequent sections.

16.3 A Progression of Ethics – Land, Environmental, and Carbon

Prior to discussing water ethics, it is insightful to look at other natural resources where a transition from exploitation to stewardship occurred – land, environment, and atmospheric carbon. In his classic book *A Sand County Almanac* (Leopold 1949), Aldo Leopold, an ecologist, forester, and environmentalist, described the land around his home in Sauk County, Wisconsin. By doing so, Leopold showed that land stewardship occurs at the interface of people and the land they inhabit. Leopold's *land ethic* is captured in his statement: "A thing is right when it tends to preserve the integrity, stability, and beauty of the biotic community. It is wrong when it tends otherwise." (Leopold 1949). *A Sand County Almanac* begins by providing a look at the month-to-month seasonal changes in nature and their impact on the ecological system. The book goes on to describe how human activities disrupt this delicate ecosystem balance. Finally, the book ends with a plea for a land ethic, a wilderness ethic, and a conservation ethic, all to protect and preserve our precious natural environment. Leopold observes that the human experience and human freedom are tarnished when humans lack wild spaces for roaming. Interestingly, Schmidt suggests that Leopold's land ethic "flowed" from his deep respect for water (Schmidt 2019).

As Leopold explained in his book: "The land ethic simply enlarges the boundaries of the community to include soils, waters, plants, and animals, or collectively, the land. This sounds simple: do we not already sing of our love for and obligation to the land of the free and the home of the brave? Yes, but just what and whom do we love? Certainly not the soil, which we are sending helter-skelter down river. Certainly not the waters, which we assume have no function except to turn turbines, float barges, and carry off sewage. Certainly not the plants, of which we exterminate whole communities without batting an eye. Certainly not the animals, of which we have already extirpated many of the largest and most beautiful species. A land ethic of course cannot prevent the alteration, management, and use of these 'resources,' but it does affirm their right to continued existence, and, at least in spots, their continued existence in a natural state. In short, a land ethic changes the role of *Homo sapiens* from conqueror of the land-community to plain member and citizen of it. It implies respect for their fellow-members, and also respect for the community as such." Leopold goes on to observe, "No important change in ethics was ever accomplished without an internal change in intellectual emphasis, loyalties, affections and convictions." (Leopold 1949). Leopold's message so captivated its readers that a generation of people embraced both his land ethic and a wider interest in ecology.

Just over a dozen years later, one of those readers, marine biologist Rachel Carson, published her book *Silent Spring* (Carson 1962). Starting in the late 1950s, Carson focused her attention on *environmental conservation* and the environmental impact of industrial processes and products. *Silent Spring's* overarching theme is that humans can have a powerful, negative effect on the environment. Carson highlights the detrimental effects of pesticides on the biota, suggesting that they be termed biocides rather than pesticides because their impacts often go beyond the targeted pests. Further, bioaccumulation of these synthetic compounds can extend the range and time frame of their impacts. Carson also warned of unanticipated future consequences, such as pesticide resistance and weakened ecosystems' susceptibility to invasive species. Carson then calls for a natural, biotic approach to pest control (e.g. pest-resistant crops) as an alternative to man-made pesticides (Carson 1962).

Carson's book raised societal concerns, curbed the agricultural use of the chemical DDT, and spawned an environmental movement that ultimately resulted in the establishment of the US Environmental Protection Agency (US EPA). Prior to the establishment of US EPA, the US Department of Agriculture (USDA) was responsible for both promoting agricultural productivity, which pesticides could enhance, and regulating pesticide use, including any negative environmental and human impacts. Carson viewed this as a conflict of interest and suggested that USDA's focus on agriculture limited their ability to consider the broader human and environmental impacts of pesticide use (Carson 1962). Carson's book *Silent Spring* is thus widely credited with spawning an environmental movement and ultimately establishing an environmental ethic.

Thirty years later, newly elected Vice President Al Gore published *Earth in the Balance: Ecology and the Human Spirit* (Gore 1992), which delineated the need for a *carbon ethic*. In the book's first section on "Balance at Risk," Gore describes the world's delicate ecological balance, including land, water, air, climate, and other environmental systems. He then goes on to discuss "The Search for Balance" citing how individual and societal choices impact the balance of maintaining versus disrupting the delicate environmental balance. In the book's final section on "Striking the Balance," Gore states that "Human civilization is now so complex and diverse, so sprawling and massive, that it is difficult to see how we can respond in a coordinated, collective way to the global environmental crisis. But circumstances are forcing just such a response." Just as the post-WWII Marshall Plan helped rebuild and preserve war-torn Europe, Gore's book calls for an environmental equivalent – an "environmental Marshall Plan" to rebuild and preserve our delicate global ecosystem and maintain Earth's delicate balance (Albert Gore 1992).

In 2006, Gore published *An Inconvenient Truth: The Planetary Emergency of Global Warming and What We Can Do About It*, which appeared along with a film release of the same title (Albert Gore 2006). The book publication and film release have been credited with elevating global warming to the forefront of public awareness and spawning a carbon ethic. And, to show how one ethic can beget another, in his introductory comments on the 1994 edition of *Silent Spring*, Gore shares that while growing up his mother insisted that he and his sister read Carson's book and discuss it at the dinner table, crediting this as a source of his environmental passion. The ethic legacy is compelling – Leopold's land ethic spawned Carson's environmental ethic which spawned Gore's carbon/global warming ethic. The time is right to spawn yet another ethic – a water ethic!

16.4 Water Ethics – A Historical Look

A pivotal statement on water and sustainability was put forth at the International Conference on Water and the Environment (ICWE), held in Dublin, Ireland, on 26–31 January 1992. The resulting Dublin Statement on Water and Sustainable Development (the Dublin Principles) recognized the increasing scarcity of water, due in part to conflicting uses and overuses of water, and put forth four guiding principles, as listed in Table 16.1.

This set of principles generated much interest from the public. While certain sectors complained that Principle 4 placed economic value above the universal human right to water, others pointed to the full statement that emphasizes that "it is vital to recognize first the basic right of all human beings to have access to clean water and sanitation at an affordable price" (Table 16.1). Subsequently, on November 2002, the UN Committee on Economic, Social and

Table 16.1 Dublin statement on water and the environment – four principles and comments.

Principle No. 1:
Fresh water is a finite and vulnerable resource, essential to sustain life, development, and the environment
Since water sustains life, effective management of water resources demands a holistic approach, linking social and economic development with protection of natural ecosystems. Effective management links land and water uses across the whole of a catchment area or ground water aquifer.
Principle No. 2:
Water development and management should be based on a participatory approach, involving users, planners, and policymakers at all levels
The participatory approach involves raising awareness of the importance of water among policymakers and the general public. It means that decisions are taken at the lowest appropriate level, with full public consultation and involvement of users in the planning and implementation of water projects.
Principle No. 3:
Women play a central part in the provision, management, and safeguarding of water
This pivotal role of women as providers and users of water and guardians of the living environment has seldom been reflected in institutional arrangements for the development and management of water resources. Acceptance and implementation of this principle require positive policies to address women's specific needs and to equip and empower women to participate at all levels in water-resource programs, including decision-making and implementation, in ways defined by them.
Principle No. 4:
Water has an economic value in all its competing uses and should be recognized as an economic good
Within this principle, it is vital to recognize first the basic right of all human beings to have access to clean water and sanitation at an affordable price. Past failure to recognize the economic value of water has led to wasteful and environmentally damaging uses of the resource. Managing water as an economic good is an important way of achieving efficient and equitable use and of encouraging conservation and protection of water resources.

Source: Adopted at the UN International Conference on Water and the Environment in Dublin, Ireland, in January of 1992 (United Nations 1992).

Cultural Rights recognized water as a human right (Groenfeldt 2019). In their book *Water Politics: Governance, Justice and the Right to Water*, Sultana and Loftus probe this topic in depth (Sultana and Loftus 2019). The human right to water is further imbedded in both the Millennium Development Goals and Sustainable Development Goals (SDGs; Chapter 1).

Even prior to the Dublin Statement, Integrated Water Resources Management (IWRM) played a pivotal role in promoting water ethics. The genesis of IWRM can be traced to the UN Conference on Water held in Mar del Plata, Argentina, in 1977 (Groenfeldt 2019). IWRM served as the de facto water ethics policy for several decades prior to the Dublin conference. By virtue of incorporating basic values of social welfare, environmental sustainability, and economic cost–benefit analysis, and with its emphasis on a watershed perspective, with its transboundary implications, IWRM was an excellent foundation for subsequent discussions of water ethics. IWRM provides a holistic view of water and its many facets (rivers, lakes, aquifers, watersheds) and political realities (cities, states, countries). IWRM grew in popularity in the 1980s and blossomed in the 1990s as water professionals adopted the approach. At the World Water Forum in 2000, IWRM was further embraced by the water sector (Groenfeldt 2019). Thus, water ethics and IWRM are closely intertwined.

More specific to water security, in 1998 the UNESCO World Commission on Ethics of Scientific Knowledge and Technology (COMEST) formed a working group on freshwater use and published a report entitled *The Ethics of Freshwater Use: A Survey* (Selborne 2000). The resulting report listed three major themes: (i) a shared purpose and harmony with nature, (ii) a balance between individual values and technological innovation, and (iii) integration of the utilitarian and sacred aspects of water resources (Brown and Schmidt 2012b). COMEST then published a series of 14 reports on water security-related topics, which culminated in the summary report *Best Ethical Practice in Water Use* (Brelet and Selborne 2004).

A 2010 United Nations (UN) resolution addressed water ethics by declaring "the right to safe and clean drinking water and sanitation as a human right that is essential for the full enjoyment of life and human rights" and "Calls upon States and International Organizations to provide financial resources, capacity building and technology transfer, through international assistance and cooperation, in particular to developing countries, in order to scale up efforts to provide safe, clean, accessible and affordable drinking water and sanitation for all." (UN 2010).

In 2018, COMEST proposed a holistic approach to water ethics that considers interrelationships between different water bodies (freshwater, coastal, ocean) as well as the interdependency between humans and the ecosystem. Consistent with the SDGs, the 2018 report considers human, environmental, and ecosystem health in pursuit of a water ethic, with the resulting guiding principles summarized in Table 16.2 (COMEST 2018). The report proposes that all human and ecosystem life forms deserve access to ample water of sufficient quality and that all countries should assure sustainable use of natural resources and terrestrial and marine ecosystems. COMEST recommends that a participatory process be used in policy and decision-making, with due consideration of youth, gender, and diversity inclusion. The commission further supports the inclusion of the latest scientific advances, as informed by indigenous knowledge, ancestral practices, and cultural diversity, in the decision-making process. COMEST suggests that by incorporating these considerations, policy, decision-making, and project implementation will follow.

Groenfeldt and Schmidt discuss four different ways to contextualize global water governance (Groenfeldt and Schmidt 2013):

- A *management* perspective, designed to unite both physical and social concerns. For most of the twentieth century, water management was about human mastery of the water environment. Engineered systems were used to increase water supply while taming the water ecosystem when necessary.
- An *institutional* approach, of applied economics, political science, and law. As drawbacks of the management approach became apparent, institutional approaches were evaluated, including engaging the private sector in water management. As drawbacks of this approach came to light, it became apparent that political, economic, and legal institutions that were already engaged in the water sector needed to be recognized and integrated into sustainable water systems.
- A *sustainability* perspective, focused on social–ecological dimensions of water systems. By considering the social–ecological aspect of water projects (e.g. decreased biodiversity, forced resettlements, loss of livelihood), water-related decisions became more holistic when balanced against earlier project approaches. For example, consideration of environmental flow and ecosystem services integrates the health of the river's natural ecosystem into the decision-making process – a consideration of the rights of the river as an entity (River Rights is a topic of growing interest).

Table 16.2 Guiding principles for COMEST study on water ethics: ocean, freshwater, coastal areas.

i. **Human Dignity and Human Rights**: Clean drinking water and sanitation are essential to human rights; further, respect for other living beings and for nature should not be at odds with the concept of human dignity.
ii. **Solidarity**: Solidarity recognizes the interdependency between humankind, and between humans and ecosystems for survival and existence.
iii. **Common Good**: Water is the common good of all living beings and ecosystems; respect for common good opposes the privatization and commodification of commonly owned systems of water supply by individuals and multinational water corporations.
iv. **Frugality**: Frugality as a virtue requires the individual to restrict and simplify his/her needs to be happy; frugality, therefore, implies moderation and rationalization in the consumption and utilization of water.
v. **Sustainability**: Sustainability depends on the judicious management of water resources to meet not only the needs of the present but also those of future generations, including water quantity and biodiversity.
vi. **Justice**: Environmental justice assures that the fair distribution of environmental goods and burdens to all humans is particularly relevant to water management; water injustice includes impaired quantity and quality of water and inequity in the above.
vii. **Justice and International Transboundary Waters**: Asymmetries in state power can promote unfair international water-sharing arrangements favoring the more powerful actor, which is contrary to the concept of distributive justice.
viii. **Gender Equity**: The fact that women play a more important role in population growth and use of water is recognized but there is still a lack of support, commitment, and necessary sex-disaggregated data on water use and management.
ix. **Research Integrity**: Scientific data and technological innovation are critical for responding to the challenges of water security.
x. **Sharing Knowledge and Technology (Capacity Building)**: All countries of the world are likely to face water disasters exacerbated by climate change, population growth, and increasing urbanization as well as refugee migration. Sharing knowledge and technology in water-resources management will ensure that best practices are available to all.

Source: COMEST (2018).

- A *values-based* approach that utilizes values (or ethical principles) to guide decision-making. When considering the human and ecological benefits and detriments of a proposed water project, we recognize the value judgments that must be superimposed on such decisions. We begin to recognize the dynamic nature of these intercoupled systems. This is further highlighted when considering transboundary watersheds, where different cultures hold different values for the various components of the system. The result is a complex, dynamic system where competing values need to be balanced (e.g. between human and environmental compartments and cultural entities).

While these approaches are complementary, the values-based approach illuminates motivations guiding disparate views, thereby producing deeper dialogue that may enlighten and enable decisions based on a common water ethic. Jeremy Schmidt's more recent book provides an in-depth analysis of the philosophies/ethics underpinning modern American water management and discusses the right to water versus water as a resource dilemma (Schmidt 2019).

Having now provided a general view of motivations and ethics, the next section will provide a more detailed look at specific examples of water ethics.

16.5 Water Ethics – "Where the Water Meets the Land"

Water ethics happens at the critical junction points "where the water meets the land": in river and groundwater management, agriculture, urban/domestic water use, industry, and water governances (Groenfeldt 2019). In this section, we will briefly touch on each of these.

River management: Over the last half century, the prevailing water ethic for managing river systems has been shifting from "command-and-control" to "accommodating coexistence with nature." The art and science of river management has increased in complexity as competing factors of ecology, economics, and local preferences come into play. The combined economic and societal value of river systems highlight the importance of ethical river management.

Historically civilizations often evolved along rivers, fostering a delicate balance of ready water access with susceptibility to water extremes (e.g. flooding). In fact, community efforts to maximize the upside while mitigating the dangers suggest that "to the oft-quote mantra 'Water is life' might be added, 'Manipulating water is civilization' "(Groenfeldt 2019). From canals and levees in ancient Egypt to irrigation canals in China (built in 200 BCE and still functioning today) to canals in the northeastern United States and irrigation reservoirs in the arid western United States, man's efforts to command and control water are evidenced throughout history. Nonetheless, in the early 1900s John Muir, founder of the Sierra Club, battled unsuccessfully to preserve the natural beauty of Hetch Hetchy Valley in Yosemite National Park, losing to the water-thirsty growth of nearby San Francisco. (The US government actually withdrew the valley's protected status in order to allow construction of the dam and reservoir.) (Groenfeldt 2019).

By the latter part of the twentieth century, the era of US dam construction was over, partly because preferred sites had already been developed but also due to increasing recognition of the environmental and human costs. Biodiversity loss due to flooded ecosystems promoted a shift from a utilitarian to ecological perspective, while excessive human toll (e.g. the estimated displacement of 40–80 million people – mostly members of minority groups) led to special rights being attributed to Indigenous people (Groenfeldt 2019). Thus, while dams do have beneficial aspects – India's Prime Minister Nehru called dams "the temples of modern India" – they can also be conduits of pain and suffering. Cochiti Tribal members (New Mexico, USA) watched with excruciating pain as their peaceful village of 700 years, and their very way of life, was destroyed by the Cochiti Dam on the Rio Grande River. Further, the need to maintain environmental flow has received increasing attention – making sure there is enough river baseflow to maintain the native ecosystem, for biodiversity and social well-being (from personal enjoyment to cultural ceremonies). This has led to the incorporation of environmental flows into water laws since the 1990s (Chapter 10) (Groenfeldt 2019).

The Universal Declaration of River Rights, which promotes a river water ethic, states that all rivers shall possess, at minimum, the following fundamental rights: (i) The right to flow, (ii) The right to perform essential functions within its ecosystem, (iii) The right to be free from pollution, (iv) The right to feed and be fed by sustainable aquifers, (v) The right to native biodiversity, and (vi) The right to restoration. In 2017, four rivers were granted river rights in their local jurisdictions: the Whanganui in New Zealand, Rio Atrato in Colombia, and the Ganga and Yamuna rivers in India. These rivers thus have a right to representation by a "guardian" or "trustee" when their well-being is threatened (Tignino and Turley 2018).

Groundwater ethics: Beyond river management, it is important to consider groundwater management and ethics. Some have called for a groundwater ethic akin to Leopold's land ethic, pointing out that we tend to approach groundwater management from a solely human perspective. By focusing on safe yield and sustainability to maximize beneficial use for humans – mitigating resource depletion, degradation of water quality, land subsidence, and saltwater intrusion – we fail to consider potential ecosystem impacts of these decisions. Anderson proposes that a Leopold-inspired groundwater ethic would look beyond human economics and esthetics to consider the environmental and ecosystem impacts of groundwater-management decisions (Anderson 2007). Megdal further identifies the importance of this "invisible" groundwater resource and calls for good groundwater governance and management based on integrated water resource management tenets that include ecosystem considerations in decision-making (Megdal 2018).

Agricultural water ethic: As already mentioned, agriculture accounts for 2/3 of consumptive water use, mainly for irrigation purposes. This fact raises the question – does the agricultural sector really need all this water? More water-efficient methods for the agricultural sector have led to the mantra "more crop per drop" – whether it be more efficient irrigation systems, improved cultivation techniques, or development/adoption of more water-tolerant crops (Groenfeldt 2019). The need for water in agriculture is obvious; we all need food, and crops need water, just as humans do. But rainfall is often inadequate for agriculture, in terms of both amount and timing. One estimate found that 40% of the world's food comes from 16% of the world's irrigated land (FAO 2021). And with increasing population, the importance of irrigated croplands will continue to increase.

Are we getting the greatest return on our "agricultural water investment?" The economist would think in terms of *incremental return*, whether the beneficial return for every increment of water invested exceed the return from investing that water elsewhere. In the context of this chapter, we might ask: is it ethical to invest that increment of water in agriculture when investing it elsewhere (e.g. cities, industry, energy) would have greater societal benefit? If water were valued in this way, agricultural users would likely end up paying more for their water or adopt more expensive water-saving approaches. In either event, food prices would increase. One can thus suggest that we are subsidizing lower food costs by giving agriculture access to "cheap" or "underpriced" water.

So, how do ethics apply to water in agriculture? The challenge lies in the fact that people from different segments of society would apply differing weights and values to the water–agriculture nexus. As Groenfeldt states, "Finding agricultural solutions that meet the value preferences of diverse stakeholders is a challenge that can best be met through a process of negotiation and consensus building." (Groenfeldt 2019). This process can be enabled and informed by first exposing the underlying ethics motivating diverse stakeholders, exploring common interests and motivations, and developing mutually acceptable solutions.

Human consumption – urban/domestic water use: As stated above, the UN 2010 resolution on *The Human Right to Water and Sanitation* declares that access to safe and clean drinking water is a human right, and challenges states and international organizations to provide safe, clean, accessible, and affordable drinking water and sanitation for all (UN 2010). This resolution vividly portrays the water ethics of the United Nations General Assembly. The UN SDG 6 – to ensure availability and sustainable management of water and sanitation for all – seeks to achieve this goal by 2030 (UN-Water 2017).

Given that the rural poor in developing countries are farthest from realizing SDG 6, much attention has been focused on addressing their very real and pressing needs. But ethics are also important when considering urban water management. The top three urban water challenges are widely recognized as (i) flood protection, (ii) secure water supply, and (iii) safe wastewater management. Each of these is costly and highly technical, leading to quantitative oversight by financial and engineering professionals. But equally important are urban designers who think beyond function and consider form. Seeing the urban landscape as a canvas, the urban designer paints a picture not only of meeting the need – e.g. flood control – but also of the human experience – parks and increased green spaces to accommodate flood waters as opposed to levees and dams – thereby building a water—human connection that promotes a water ethic (Groenfeldt 2019).

Just over a decade ago, we passed the global threshold whereby more people live in cities than in rural areas, and this trend will only increase in future decades. With ever-increasing populations overtaxing the water-management infrastructure, water ethics will play a critical role in guiding leaders and decision makers to navigate this new paradigm.

Industrial water ethic: Just as water is critical to agriculture, it is likewise vital to virtually every industry, whether it be connected to raw materials, manufacturing, processing, transport, or disposal. Industrial water ethics can be captured in three ways: (i) responsible water use (the water footprint), (ii) water impacts to people and ecosystems (the water handprint), and (iii) water advocacy for policy and norms (water blueprint) (Groenfeldt 2019).

- *Responsible water use* can be informed by looking at the *water footprint* at each stage of the industrial process – from raw materials to manufacturing to delivery. By lessening the water footprint of industry, one can achieve more responsible water use and potentially save money as well. Likewise, improving the energy footprint of the manufacturing process has the "ripple" effect of reducing the water footprint associated with the energy consumption.
- By focusing on *water impacts to people and ecosystems*, industries can be better neighbors to the local community and the environment. By being good stewards of water and the local ecosystem, industry promotes public relations with the local community while also protecting and preserving a critical input to their industrial process – both of which are critical to the industry's long-term success. This is likened to a water commons – by realizing that these are shared natural resources, and by looking for ways to maximize everyone's benefit, each individual or industry also benefits (Groenfeldt 2019). Much like the adage "rising tide lifts all ships" (a win-win situation where when one benefits, everybody benefits), an ethics-driven water commons lifts all water users.
- While industry *advocacy for water policy and norms* historically involved efforts to relax requirements and improve profitability, industry has begun to recognize that what is good for water management is good for their long-term sustainability. In this, industry can help establish a *water blueprint* that enables their long-term sustainability by preserving a critical input, improves their public relations, and is the right thing to do. This harkens to the business concept of the "triple bottom line," which goes beyond the single goal of profit and suggests that business success should be based on being profitable, good for society, and good for the environment. (The push for Corporate Social Responsibility does likewise. See Chapter 9.)

In short, industry poses both the greatest threat and the greatest hope to sustainable water management, as guided by water ethics. Businesses and industry require great creativity and tenacity to become successful and to maintain that success. When mobilized by water ethics, these same skills can go a long way to addressing our looming water crisis.

Water ethic and governance – Which entity, be it local, state, national, or international, should be governing water, and at what scale? IWRM helps address this question. IWRM is defined as "a process which promotes the coordinated development and management of water, land and related resources, in order to maximize the resultant economic and social welfare in an equitable manner without compromising the sustainability of vital ecosystems" (GWP 2000). IWRM has served as water policy for local, national, and global governance issues and thus has served as a de facto ethic for water governance (Groenfeldt 2019).

But management alone is not enough. Management is often based on predefined operating rules and guidelines, and when a crisis arises existing policies are often called upon. Alternatively, water governance, while rooted in management principles, needs more room to maneuver. By assigning value to water in various sectors and society, water governance demonstrates an underlying water ethic. Water governance must wrestle with fundamental questions about water and society – questions such as (i) Is the water source inadequate or is the water demand too high? (ii) Why is water being preferentially allocated to one sector (e.g. golf courses) versus another (e.g. environmental flow)? (iii) Are certain sectors paying too little for water (e.g. agriculture) while other sectors are paying too much (e.g. financially challenged people)? (iv) Can creative leadership and good governance identify innovative solutions not apparent to competing sectors suffering from tunnel vision? (v) Should water governance codify the water values of the majority, or should it inform, and even transform, those interests to what is best for society and the environment? A strong water ethic can motivate leaders to go the "extra mile," to address the big challenges, to achieve progress not available to the faint of heart. Such an ethic would make Aldo Leopold and Rachel Carson proud.

16.6 Water Ethic – A North American Indigenous Perspective

Indigenous people have much to teach the world about water ethics, both from their deep respect for nature and from their astute understanding of the interplay of place and society. Embracing the water ethic of Indigenous people will go a long way toward navigating our global water crisis. As Groenfeldt states, "At the risk of generalizing, it could be said that Indigenous cultures, diverse though they are, feel a relationship with water. Water and water ecosystems are considered part of the extended family that includes the community, the animals, the land, and water. But Indigenous concepts of water go beyond the trope of functioning as a relative; water, and particularly bodies of water, are inextricably interwoven into the fabric of Indigenous identity." (Groenfeldt 2019). Groenfeldt goes on to describe three distinct but synergistic elements of the Indigenous water paradigm: water as metaphor, water as map, and water as responsibility.

1. *Water as metaphor*: For some Indigenous cultures, water is a metaphor for life in that everything is interconnected. Just as my life relates to yours, even so my life is connected with water. Just as water cycles through earth, even so virtues and people cycle through life. Water, in its simplicity and complexity, is a metaphor for life.
2. *Water as map*: Rivers, lakes, wetlands, springs, and even glaciers are physical features that demarcate place and, in certain instances, cultures. Cultures grow up along bodies of water. Large bodies of water separate cultures. Indigenous stories of water can have deep cultural meaning and geographical significance.

3. *Water as responsibility*: Indigenous cultures understand the importance of water, and being good stewards of water, as demonstrated by Indigenous prayers, ceremonies, and storytelling that demonstrate their high regard for water. They hold water as sacred. Their deep respect for and compulsion to protect water was evidenced by Indigenous protests surrounding an oilfield pipeline at Standing Rock in South Dakota, where they even referred to themselves as "water protectors," clearly demonstrating their deep-seeded water ethic.

In recognition of the unique role of Indigenous Peoples in society, the 2007 UN Declaration on the Rights of Indigenous People (DRIP) made the following two statements relative to water (UN General Assembly 2007):

1. Article 25 states: "Indigenous peoples have the right to maintain and strengthen their distinctive spiritual relationships with their traditionally owned or otherwise occupied and used lands, territories, waters and coastal seas and other resources and to uphold their responsibilities to future generations in this regard."
2. Article 32 states: "(1) indigenous peoples have the right to determine and develop priorities and strategies for the development or use of their lands and territories and other resources. (2) States shall consult and cooperate in good faith with the Indigenous peoples concerned through their own representative institutions in order to obtain their free and informed consent prior to the approval of any project affecting their lands or territories and other resources, particularly with the development, utilization or exploitation of mineral, water or other resources. (3) States shall provide effective mechanisms for just and fair redress for any such activities, and appropriate measures shall be taken to mitigate adverse environmental, economic, social, cultural, or spiritual impact."

It is worth noting that Indigenous rights are still routinely violated and their access to water denied or withheld. For this reason and to raise awareness for water's importance, Grandmother Josephine Mandamin of the Anishinabekwe (Ojibway) First Nation initiated the Mother Earth Water Walk to pray for the water while walking along the shores of the Great Lakes and St. Lawrence River. Grandmother Mandamin states that "when we are walking with the water, we are also collecting thoughts with that water. And in the collecting of thoughts, we are also collecting consciousness of people's minds. The minds, hopefully, will be of one, some time." Grandmother Mandamin recommends water fasting – going without water for a while – to gain appreciation for its importance and our need to be good stewards of this precious entity.

16.7 Water Ethic – Perspectives from World Religions

According to three major Western world religions – Judaism, Christianity, and Islam – life on earth begins in a garden that is nourished by a spring: "A river rises in Eden to water the garden; beyond there it divides and becomes four branches." (Gen 2:10). These four branches become rivers that go out and provide water to the known world. The source of this water is God who, in a slightly different version of creation, swept his mighty wind over the dark immensity of water, dividing it into dry land and sky and an immense basin of water that we call the ocean (Gen 1:1-10).

Water is a pervasive topic throughout the Bible, occurring over 400 times in the Old (Hebrew) and New (Christian) Testaments. In addition to appearing in the first book of the Bible, it plays a pivotal role in the last book: "Then the angel showed me the river of life-giving water, sparkling like crystal, flowing from the throne of God and of the Lamb down the middle of its street. On either side of the river grew the tree of life ... Let the one who thirsts come forward, and the one who wants it receive the gift of life-giving water." (Rev 22:1-17). This section presents ways in which water has played an essential role in the world's great religions.

16.7.1 Judaism

Not only was the world created from water, but it also seemed to be used to express God's will at various key moments in the formation of Israel. As a proof of God's power to Pharaoh, Moses turned the Nile water into blood (Ex 7:20) and parted the waters of the Red Sea for safe passage of the Jewish people in flight (Ex 14:21). During their desert wanderings, Moses struck a rock to bring forth drinking water to quench the people's thirst (Num 20:11).

The location of water wells became sacred places in the Hebrew Scriptures, providing water and serving as gathering places for people and animals. Genesis 29:2-3 describes Jacob's perspective on water and wells: "He looked, and saw a well in the field, and behold, three flocks of sheep were lying there beside it, for from that well they watered the flocks. Now the stone on the mouth of the well was large. When all the flocks were gathered there, they would then roll the stone from the mouth of the well and water the sheep and put the stone back in its place on the mouth of the well." The importance of well water quality is demonstrated in Proverbs where a polluted well is compared to a corrupt person "Like a trampled spring and a polluted well is a righteous man who gives way before the wicked" (Prov 22:26).

The Jordan River played a dominant role in the life of the Chosen People, and Naaman's leprosy was cured by washing in the Jordan (2 Kings 5:10-14). The book of Isaiah demonstrates God's concern for the human–water nexus, both in terms of sustaining life and as a potential vessel of death: "The afflicted and needy are seeking water, but there is none; their tongue is parched with thirst; I, the Lord, will answer them Myself" (Is 41:17) and "When you pass through waters, I will be with you; through rivers, you shall not be swept away." (Is. 43:2).

Finally, water is a common theme in Jewish prayer rituals. An Orthodox Jew washes hands before daily prayers, upon arising in the morning, after elimination of bodily fluids, and after being in the presence of the dead (Chamberlain 2007). An observant Jewish woman will observe seven "clean" days after each menstrual period. The *mikveh* bath immersion is a symbol to her of the promise of fruitfulness (Chamberlain 2007).

16.7.2 Christianity

Water is a meaningful symbol in Christianity as well. Jesus begins his public ministry by being baptized in the Jordan River (Mark 1:9). During his public ministry, Jesus used water to illustrate spiritual truths. In the well-known story of the woman at the well (aka Jacob's well), Jesus draws a parallel between the woman's need to continually fetch water to sustain physical life and his own ability to meet her desire for life-giving spiritual life (John 4:1-15). Interestingly, this story also touches on the point of gender equity. His disciples were amazed that Jesus was in discourse with not only a woman but a Samaritan woman as well, of an ethnic tribe that had no dealings with Jews. Further, after running a mud paste on the eyes of a blind man, Jesus instructs him to go and wash in pool of Siloam, a holy pool just outside the walls of the Old City Jerusalem (John 9:6). Jesus also draws an analogy between physical water leading to physical life and the spiritual life: "Now on the last day, the great day of the feast, Jesus stood and cried out, saying, If anyone is thirsty, let him come to Me and drink. He who believes in Me, as the Scripture said, 'From his innermost being will flow rivers of living water.' By this He spoke of the Spirit, whom those who believed in Him were to receive; for the Spirit was not yet given, because Jesus was not yet glorified" (John 7:37-39). Christian tradition continues to use water as the primary symbol of initiation through Baptism by immersion and by sprinkling. In Catholic churches, the holy water is placed in fonts and sprinkled on congregants as a reminder of their once-received Baptism.

The third-century Christian theologian, Tertullian, writes: "All natural water, because of the ancient privilege with which it was honored from the first, gains the power of sanctifying in the sacrament (Baptism). What once was used of old to heal the body now heals the soul." (quoted in Eliade and Holt [1996]). As Christianity evolved, the transcendence of God began to be emphasized resulting in an unfortunate dualism in the West: spirit versus body, humans versus nature, grace versus sin. Instead of being a wellspring of God's grace, the world and all it contains was more of a place of trial and temptation to be endured and conquered. This theology was expressed in a 1966 statement of the World Council of Churches: "The movement of history, not the structure of the setting (natural world), is central to reality. Physical creation … its timeless and cyclical character, as far as it exists, is unimportant. The physical world … does not have meaning in itself." (quoted in Chamberlain [2007]). More recently, Christian scholars are struggling to redress this dichotomy and return to the foundational sense of water as an instrument of God's power and grace – for initiating, healing, and purifying.

16.7.3 Islam

Islam begins with the Abrahamic traditions of Judaism and Christianity and extends with private revelations received by the seventh-century prophet Mohammed. A visit to Spain is not complete without a walking tour of the Mezquita-Catedral in Cordoba or the intricate fortress complex (the Alhambra) at Granada (Figure 16.1). These architectural masterpieces display not only the artistic sensibilities of the Moors, but also their appreciation for the value of water in a lived environment, for cooling, cleansing, and spiritual renewal. Gardens in the Islamic tradition were a glimpse of paradise itself, and free-flowing waters were an integral sign of the mercy of God (Allah), flowing miraculously from the earth. For it is "Allah who drives the winds that raise the clouds and … causes them to break up so that you can see the rain issuing out from the middle of them."(Quran 30:48). Water can reveal both Allah's anger in destructive storms and floods while also bestowing his kindness and mercy through gentle rains that soften the earth. As in Judaism, a water purification is mandatory before prayer, and the Prophet himself compared daily prayers to the frequent cleansings with water (Fagan 2011).

Water rights are well-defined in the Islamic religion. First, humans have a right to drinking water, to quench their thirst. Second, farm and household animals have the same right. Third, water can and should be used for irrigation. But supplies must be managed and not wasted. Indeed, Allah will punish those on the day of resurrection who have hoarded surplus water to the detriment of thirsty fellow travelers (Quran 70:18). Modern Islamic judgment allows

Figure 16.1 The Palacio de Generalife in the Alhambra, Granada, Spain. Source: Andrew Dunn / Wikimedia Commons / CC BY-SA 2.0.

wastewater to be cleaned and reused: "If water treatment restores the taste, color, and smell of unclean water to its original state, then it becomes pure and hence there is nothing wrong to use it for irrigation and other useful purposes."(quoted in Amery and Haddad [2015]). Thus, Islamic water ethics seem to be consistent with those of other world religions.

16.7.4 Water and Eastern Religions

In the ancient religions of Asia, water has also played a significant role in rituals of cleansing, healing, and initiation while rivers and lakes have been the foci of wisdom and enlightenment. The Ganges River in India is believed to be the body of the Goddess Ganga herself, and pilgrims who are close to death come to immerse themselves in this sacred water that cleanses their sins and carries their bones and ashes into eternal life. The spectacular six-week festival, the Great Kumba Fair, happens at the confluence of the Ganges and Yamuna Rivers. The Fair is an opportunity for seekers to gain wisdom from gurus, mystics, yogis, and other holy people who come to this watery juncture of divine knowledge. Practicing Buddhists in Thailand and Cambodia will sprinkle holy water on the miniature spirit houses constructed outside their houses, water that will protect their ancestral spirits by keeping away the evil spirits. The popular Dragon Boat Festival of China began over 2300 years ago with the drowning of a well-loved and righteous poet, Qu Yuan. One rower stands forward on the boat, symbolically searching for the poet's body, while the beating of drums and the dragon figures are meant to keep evil spirits at bay (Altman 2002).

The religious significance of water is both universal and particular. It is universal in its power to cleanse, heal, and grant wisdom, while the manifestations of such are particular to both time and culture.

16.7.5 Religion and Water Ethics

What is the role of religion in promoting ethics? As noted author and scientist Carl Sagan stated, "We are close to committing – many would argue that we are already committing – what in religious language is sometimes called 'Crimes against Creation.' The historical record makes clear that religious teaching, example, and leadership are powerfully able to influence personal conduct and commitment ... We understand that what is regarded as sacred is more likely to be treated with care and respect. Our planetary home should be so recognized." (Sagan 1990). Even as a critic of religion, Sagan recognizes the important role that religion-inspired morals can play in creating an environmental and water ethic. Huesemann states, "The long term protection of the environment is, therefore, not primarily a technical problem but rather a social and moral problem that can only be solved by drastically reducing the strong influence of materialistic values." (Huesemann 2001).

Pope Francis' encyclical on the environment, *Laudato Si'*, and the Vatican study "*Aqua fons vitae*: Orientations on Water" speak to a water ethic. In his encyclical, Pope Francis reminds us that "We have forgotten that we ourselves are dust of the earth (cf. Gen 2:7); our very bodies are made up of her elements, we breathe her air and we receive life and refreshment from her waters." (Pope Francis 2015). In a section devoted to water, Francis decries the wastefulness of water practices and implores the global community to honor its "grave social debt towards the poor who lack access to drinking water, because they are denied the right to a life consistent with their inalienable dignity" (Pope Francis 2015). *Aqua fons vitae* speaks of three water dimensions – first, water for human use (drinking and domestic use); second, for human activities (irrigation, energy, industry, etc.); third, "Water as Space" – highlighting the fragile nature of our oceans experiencing pollution, misuse, and loss of fisheries (Vatican Dicastery 2020). Church teachers believe that water can be an element "with which to build relational bridges among people, communities and countries. It can and should be a learning ground for solidarity and collaboration rather than a trigger of conflict." (Vatican Dicastery 2020).

To the extent that religion has an influence on the thoughts, behavior, and imagination of individuals, it can be a rich source of ethics-building in the twenty-first century. If water is considered sacred, or a symbol of the sacred, then it is not likely to be wasted, mistreated, or misused. As religion encourages people to rediscover and reclaim their religious traditions, they will likewise rediscover a water ethic that proceeds from the very actions of God Himself.

16.8 A Tale of Three Countries ... and Their Water Ethic

In this section, we will briefly describe the water ethic of three countries, two that have faced severe water shortages (Australia and Singapore) and one that has faced the opposite challenge – too much water (Holland). While the technical challenges have already been discussed, the goal here is to highlight the water ethic that resulted. If necessity is the mother of invention – then we see that while water challenges heighten awareness, and extreme challenges increase appreciation, prolonged challenges have the opportunity to generate a water ethic.

Southern Australia faced significant water-supply challenges during its Millennium drought, widely considered one of the most severe droughts on record. The Millennium drought, which began in the late 1990s and lasted until 2010, had a major impact on water availability, leading to dramatic water-conservation efforts as well as water desalination projects. On a Fall 2019 trip to Melbourne, Australia, one of the authors (DAS) experienced first-hand the resulting water ethic. Because of the drought, people routinely talk about the water level in their home rainwater catchment system and in the municipal water supply reservoirs. To increase awareness, Melbourne Water developed an iPhone app. Two screenshots from 17 December 2019 are shown in Figure 16.2. One shows the percentage of reservoir capacity filled. When the percentage falls below certain levels, rationing protocols are triggered. The second shows the average weekly per capita water usage (note that the target of 155 liters per capita per day (lpcd) [or 41 gallons per capita per day (gpcd)] is much below the US average home water consumption of 82 gpcd) (US EPA 2017). The notification that the water-use rate was 30 lpcd above the target hopefully promotes discussions between friends and neighbors about the potential danger this poses if another drought occurs. Thus, the prolonged severe drought motivated personal involvement (e.g. home rainwater harvesting, monitoring of water resource and usage) and produced a water ethic that has lived on past the drought.

In times past, Singapore has found itself in the precarious situation of relying on Malaysia for a critical portion of its water supply. Singapore took significant steps to become water independent – instituting its "Four Taps" approach to water supply: (i) domestic sources, (ii) imported water (from Malaysia), (iii) recycled water (NeWater), and (iv) water desalination. Following this roadmap, Singapore was able to greatly improve its water security, relying less and less on imported water to the point that it can become water independent. But this came with an unintended consequence. As citizens became more confident in their long-term water supply, they relaxed their conservation efforts and water-consumption rates began to increase. Two strategies – public campaigns to promote water conservation along

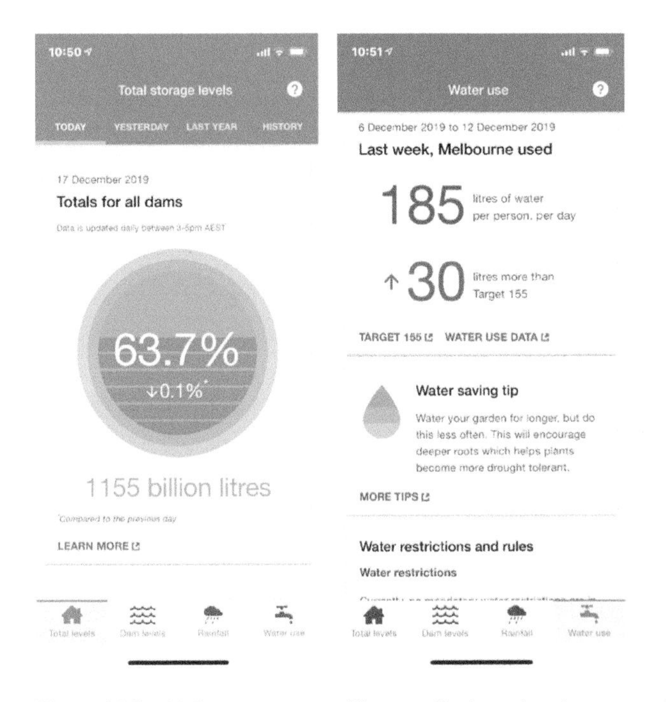

Figure 16.2 Melbourne water utility app displays showing reservoir capacity (left) and water-use rates (right). Source: Photos courtesy of DAS.

with imposing a water-conservation tax – proved unsuccessful in curbing the growing rate of water consumption. The third strategy – credited with reducing per capita consumption from 50 to 40 gal/day in 2010 – consisted of attempts to bring people "closer to water" so that they gained an appreciation for this valuable resource (Barnett 2012). The most ambitious part of this multi-prong approach was the Marina Barrage – a free waterfront public attraction designed to be inviting for the whole family, including a fountain park where adults can enjoy beautiful gardens while children in bathing suits enjoying water features designed for their entertainment. In addition, the Marina Barrage has a visitor center shaped like a seashell that houses hands-on water exhibits, a water-themed art gallery, a popular Chinese restaurant, and even a pumping station easily visible behind protective glass. The Marina Barrage has become a favorite destination for Singaporeans and visitors alike (DAS enjoyed day and nighttime visits to the park). As another example, the NeWater reuse treatment facility has an interactive, hands-on museum appropriate for interested adults and their children as well as hosting educational school tours (Figure 16.3) – all on the premises of the water-treatment facility that has numerous water features for public viewing. These two examples illustrate how Singapore went beyond traditional water-conservation methods to create a human–water connection that promoted a water ethic demonstrated in personal choices to reduce water consumption (Barnett 2012).

Holland is another example of a country with a water ethic, but in this case the challenge at hand is too much water. Barnett entitles her chapter on Holland "The Netherlands: Deluge, Dams and the Dutch Miracle" (Barnett 2012). Many will recall the legend of the little Dutch boy who saves his country by putting his finger in the hole of a leaking dike. Holland is a country surrounded by, even inundated with, water. The greatest flood in modern Dutch history was the North Sea Flood of 1953, which killed over 1800 people and 50 000 cattle. The Dutch response to this overwhelming disaster was a unified "Never again!". The resulting Delta Plan proposed to dam the tidal channels where the Rhine, Meuse, and Scheldt Rivers discharge into the North Sea as well as raise the sea and river dikes nationwide to withstand tides 16 ft above normal. Over a decade behind schedule and at 10 times the original US $600 million estimate, the project was finished in 1997. While the four-decade project is not without shortcomings and critics, these efforts demonstrate the Dutch public and national commitment

Figure 16.3 Interactive Educational Museum at NeWater Facility in Singapore. Source: Photo courtesy of DAS.

to improving the human–water interface in a way that is good for humankind and the environment. Once again, public education is demonstrated in the highly popular Dutch Water Museum, which not only informs children and adults of Dutch water history but engages them in a dialogue of the Dutch water ethic. Holland demonstrates the results of a highly effective ongoing conversation between citizens, government, and industry about water and its value – critical components of a water ethic and "the Dutch miracle" (Barnett 2012).

16.9 Conclusion: A Water Ethic for the Twenty-First Century

One could claim that the nineteenth- to twentieth-century water ethic was one of exploitation – mastering water for human benefit. And all too often the benefits accrued to individuals rather than society overall, and with little, if any, thought for the environment. As populations have dramatically increased, international trade has become more commonplace, and as individual, commercial, and industrial footprints have increased significantly, the need for a twenty-first century water ethic has become paramount. Cynthia Barnett speaks to this need in her insightful book *Blue Revolution: Unmaking America's Water Crisis* (2012). Barnett calls for a water (blue) revolution akin to the Green Revolution of the mid-twentieth century, a revolution that dramatically improved agricultural practices and productivity. Just as dire predictions of food shortages loomed on the horizon in the mid-twentieth century, even so we face a looming water crisis in the early twenty-first century. Fortunately, with visionary individual and collective actions our ancestors averted the food crisis in the twentieth century. Even so, visionary individual and collective actions are needed to avert an equally serious water crisis.

Raising public awareness and motivating lifestyle changes are certainly challenging endeavors. In the United States, public littering was more commonplace in the mid part of the twentieth century. Older readers remember fast food bags, drink cups, and plastic trash along the roadways. Fortunately, this behavior is much less common today. A Keep America Beautiful study in 1969 revealed that half of the Americans surveyed had littered in the past month; the same study replicated 40 years later (2009) showed that only 15% reported such behavior. This remarkable transformation can be attributed to widespread educational campaigns, sustained cleanup initiatives, major changes in business and industrial practices, and, ultimately, an individual and personal commitment to keep public spaces clean – that littering is wrong! We suggest that the time is right, and that time is of the essence, to mobilize a similar water movement – a movement to eliminate "water littering," i.e. the unnecessary wasting of water. To value every drop and to make every drop count (Barnett 2012).

As in the case of "de-littering," a combination of educational programs; sustained initiatives at the individual, collective, and government levels; and changes in business practices will be necessary to achieve the necessary twenty-first century water ethic. The educational programs should include advertising for adults and curricular content for school children. In addition to forming their own passions, children can share what they are learning with their parents. Children's water museums in Singapore (where they learn about water scarcity) and Holland (where they also learn about water abundance) have aided in this process. These must be replicated around the globe.

The formation of a universal water ethic will include the essential principles espoused by three important authors – David Groenfeldt (anthropologist), Sandra Postel (environmental scientist), and Cynthia Barnett (journalist). These principles are described here.

The five key values to building a water ethic are that:

1. Nature needs to be kept alive (ecological function),
2. Everyone has a right to water and sanitation (social justice),
3. Water should be used responsibly in agriculture and industries (responsible use),
4. Stakeholders should be involved in decision-making (participation), and
5. Diverse cultural identities and understandings about water should be respected (cultural respect) (Groenfeldt 2019).

An important part of building a water ethic is "make water visible in everyday experience." Some ways of doing this include: *environmental art*, which can inspire people to connect with water and the environment; *recreation*, whether it be swimming, fishing, kayaking, boating, or river rafting; *household rainwater harvesting*, which helps people identify with water in a tangible way; and *"daylighting" urban water systems*, making them visible and accessible to build water–human connections. Anything that promotes a positive human–water interaction can help create an appreciation for and value of water – critical ingredients of a water ethic, as reflected in the success of the Marina Barrage in Singapore.

Sandra Postel states that "overall we have been quick to assume rights to use water but slow to recognize obligations to preserve and protect it" (Brown and Schmidt 2012a). While economics can increase water-use efficiency, it tends to understate, if value at all, water and ecosystem services. Postel likens ecosystem protection to a solid foundation for a building – essential and critical so that what is built on it is secure. Likewise, being good stewards of water and ecosystems is the bedrock of healthy societies. Thus, we must shift away from a utilitarian approach and embrace an integrated, holistic approach that recognizes the interconnected nature of humans, water, and the environment. Postel goes on to quote Aldo Leopold's view on the extension of ethics to the natural environment as "an evolutionary possibility and an ecological necessity." The same can be said for a water ethic.

Cynthia Barnett suggests the following as evidence of a solid water ethic (Barnett 2012):

- We, as a society, value water, from appreciating local streams to pricing water right.
- We work together to use less and less, rather than fight for each other to grab more and more. We try to keep water local.
- We avoid the two big mistakes of our history: over-tapping aquifers and surface waters and over-relying on the costliest fixes that bring unintended consequences to future generations.
- We leave as much as prudently possible in nature – aquifers, wetlands, and rivers – so that our children and grandchildren, with benefit of time and evolving knowledge, can make their own decisions about water.

While achieving a water ethic will take political will, ultimately it requires individual courage to bravely step out of our comfort zone and implement practices and/or promote the public debate necessary to achieve a water-secure future, as one, and then several, individuals take steps to promote a water ethic. As we saw in the "de-littering" movement, there will be an upswell of collective action. We have seen seedlings of such a movement in water-challenged regions. We need more individuals to be inspired and motivated by these water pioneers, these visionaries, to do likewise. As more and more of us do likewise, and as this translates into public policy, we will see a twenty-first century water ethic evolve that will enable us to cherish and be good stewards of our precious water resources, thereby leaving a legacy to future generations.

End-of-Chapter Questions/Problems

16.1 Do a web search on Aldo Leopold's *A Sand County Almanac* and his resulting land ethic and provide a one-page summary of key aspects of the book and the resulting land ethic.

16.2 Do a web search on Rachel Carson's *Silent Spring* and the resulting environmental movement and provide a one-page summary of the book and the resulting environmental movement.

16.3 Discuss four to five key elements of the land ethic and environmental ethic (movement) that can be utilized in forging a water ethic.

16.4 Provide a paragraph each expounding on for four of the ten principles (four paragraphs total) in Table 16.2 "Guiding Principles for COMEST Study on Water Ethics."

16.5 How would a water ethic be similar/different for agriculture versus an automotive industry?

16.6 Identify a city, state, or country, different from the ones presented in this chapter, which has demonstrated a strong water ethic and write a one-page summary describing the evolution of this water ethic.

16.7 How can varying perspectives (e.g. Indigenous peoples, religion) enlighten and enable development of a water ethic?

16.8 What similarities/variations in approach would you anticipate if seeking to develop a water ethic in a city versus a state or a country?

16.9 Use an online version of the Quran to find all the references to "water" in the text. Summarize the various uses and importance of water according to this sacred text.

16.10 Discuss three to four factors that make developing twenty-first century water ethic different from approaching the same task at the beginning of the twentieth century.

16.11 What would you suggest are the four most important factors to developing a robust twenty-first century water ethic and why?

Further Reading

Barnett, C. (2012). *Blue Revolution: Unmaking America's Water Crisis*. Beacon Press.

Chamberlain, G. (2007). *Troubled Waters: Religion, Ethics, and the Global Water Crisis*. Lanham: Rowman & Littlefield Publishers.

Groenfeldt, D. (2019). *Water Ethics*, 2e. Abingdon, Oxon; New York, NY: Routledge.

Sultana, F. and Loftus, A. (ed.) (2019). *Water Politics: Governance, Justice, and the Right to Water*, 1e. Abingdon, Oxon; New York, NY: Routledge.

References

Altman, N. (2002). *Sacred Water: The Spiritual Source of Life*. HiddenSpring.

Amery, H. and Haddad, M. (2015). Ethical and cultural dimensions of water reuse: Islamic perspectives. In: *Urban Water Reuse Handbook*. CRC Press.

Anderson, M.P. (2007). Ground water ethics. *Ground Water* 45 (4): 389.

Barnett, C. (2012). *Blue Revolution: Unmaking America's Water Crisis*. Beacon Press.

Brelet, C. and Selborne, J. (2004). *Best Ethical Practice in Water Use*. Paris: World Commission on the Ethics of Scientific Knowledge and Technology https://unesdoc.unesco.org/ark:/48223/pf0000134430.

Brown, P.G. and Schmidt, J.J. (ed.) (2012a). Chapter 20: a missing piece: a water ethic (Sandra Postel). In: *Water Ethics: Foundational Readings for Students and Professionals*, 2e. Washington, Island: Press.

Brown, P.G. and Schmidt, J.J. (ed.) (2012b). *Water Ethics: Foundational Readings for Students and Professionals*, 2e. Island Press.

Carson, R. (1962). *Silent Spring*. Mass Paperback Edition. Crest Books.

Chamberlain, G. (2007). *Troubled Waters: Religion, Ethics, and the Global Water Crisis*. Lanham: Rowman & Littlefield Publishers.

COMEST (2018). "Report of COMEST on: Water Ethics: Ocean, Freshwater, Coastal Areas – UNESCO Digital Library." UNESCO, Paris: Commission on Ethics of Science and Technology (COMEST). https://unesdoc.unesco.org/ark:/48223/pf0000265449.

Eliade, M. and Holt, J.C. (1996). *Patterns in Comparative Religion* (trans. Rosemary Sheed). Lincoln: Bison Books. Reprint edition

Fagan, B.M. (2011). *Elixir: A Human History of Water*. London: Bloomsbury UK.

FAO (2021). "Water and Food Security." 2021. http://www.fao.org/3/x0262e/x0262e01.htm.

Francis, P. (2015). *Laudato Si': On Care for our Common Home*. Huntington, IN: Our Sunday Visitor.

Gore, A. (1992). *Earth in the Balance: Ecology and the Human Spirit*. Boston: Houghton Mifflin.

Gore, A. (2006). *An Inconvenient Truth: The Planetary Emergency of Global Warming and What we Can Do about it*. First Printing edition. Emmaus, Pa: Rodale Books.

Groenfeldt, D. (2019). *Water Ethics*, 2e. Abingdon, Oxon; New York, NY: Routledge.

Groenfeldt, D. and Schmidt, J. (2013). "Ethics and water governance." *Ecology and Society* 18 (1). doi:https://doi.org/10.5751/ES-04629-180114.

GWP (2000). *Towards Water Security: A Framework for Action*. Stockholm: Global Water Partnership.

Huesemann, M.H. (2001). Can pollution problems be effectively solved by environmental science and technology? An analysis of critical limitations. *Ecological Economics* 37 (2): 271–287.

Leopold, A. (1949). *A Sand County Almanac*. First printing edition. New York: Ballantine Books.

Megdal, S.B. (2018). "Invisible water: the importance of good groundwater governance and management." *Npj Clean Water* 1 (1): 1–5. doi:https://doi.org/10.1038/s41545-018-0015-9.

Pojman, L. (1999). *Global Environmental Ethics*, 1e. Mountain View, Calif: McGraw-Hill Humanities/Social Sciences/Languages.

Sagan, C. (1990). "Guest comment: preserving and cherishing the earth—an appeal for joint commitment in science and religion." *American Journal of Physics* 58 (7): 615–17. doi:https://doi.org/10.1119/1.16418.

Schmidt, J.J. (2019). *Water: Abundance, Scarcity, and Security in the Age of Humanity*. Reprint edition. NYU Press.

Selborne, J. (2000). "The Ethics of Freshwater Use: A Survey." Commission on Ethics of Science and Technology (COMEST). https://unesdoc.unesco.org/ark:/48223/pf0000122049.

Sultana, F. and Loftus, A. (ed.) (2019). *Water Politics: Governance, Justice, and the Right to Water*, 1e. Abingdon, Oxon; New York, NY: Routledge.

Tignino, M. and Turley, L.E. (2018). "Granting Legal Rights to Rivers: Is International Law Ready?" *The Revelator* (blog). August 6, 2018. https://therevelator.org/rivers-legal-rights.

UN General Assembly (2007). "United Nations Declaration on the Rights of Indigenous Peoples." 2007. https://www.un.org/development/desa/indigenouspeoples/declaration-on-the-rights-of-indigenous-peoples.html.

United Nations (1992). "The Dublin Statement on Water and Sustainable Development." 1992. https://www.wmo.int/pages/prog/hwrp/documents/english/icwedece.html.

United Nations (UN) (2010). "The Human Right to Water and Sanitation." A/64/L.63/Rev.1. United Nations.

UN-Water (2017). "Indicator | SDG 6 Data." Indicator 6.4.2 - Water Stress. 2017. https://sdg6data.org/indicator/6.4.2.

US EPA, OW (2017). "Water Use and Savings: Statistics and Facts." Overviews and Factsheets. January 23, 2017. https://www.epa.gov/watersense/statistics-and-facts.

Vatican Dicastery (2020). "Aqua Fons Vitae: Orientations on Water." Vatican City: Vatican Dicastery for Promoting Integral Human Development. http://www.humandevelopment.va/en/risorse/documenti/aqua-fons-vitae-the-new-document-of-the-dicastery-now-available.html.

16a The Practice of Water Security: Decolonizing Water Security

The effects of colonization often outlast one group's colonial rule. Provision of infrastructure and essential services can be highly uneven, favoring the wealthy and members of the preferred colonial (settler) class. One urban indigenous reserve in Canada – surrounded by modern cities such as Toronto – has little or no access to clean running water. As recently as February 2021, yet another boil water advisory was issued for this reserve, unthinkable in such a highly prosperous nation.

Haudenosaunee (hoe-dee-no-SHOW-nee) is a tribal name, which means "people who build a house." Located along the banks of the Grand River, Six Nations is demographically the largest First Nations reserve in Canada and is the only reserve in North America where all six Haudenosaunee nations live together (Figure 16a.1). The reservation has about 13 000 residents living on an 18 800-ha territory near Brantford, Ontario.

Figure 16a.1 The Grand River watershed is the life-giving source of indigenous First Nations culture in Ontario province, Canada. Source: Merrell-Ann Phare.

Dawn Martin-Hill is an Indigenous (Haudenosaunee) woman from Six Nations of the Grand River (Figure 16a.2). She is also a cultural anthropologist and associate professor at McMaster University and, importantly, a mother who has raised her girls in a home with no running water. She says: "Our entire way of life is governed by water – it is spiritual, it is cultural, it is our identity. When you take that away from us, you are literally taking away our culture." Such a connection between ecosystems and human culture is lacking in Western colonial ways of thinking.

Figure 16a.2 Dawn Martin-Hill dedicates her career to overturning the legacy of colonization's impact on access to clean water. Source: Dawn Martin-Hill.

Researchers have found high levels of mercury in the drinking water of the Six Nations and only 12% of the population has access to a state-of-the-art water-treatment plant, which services nearby settler populations (Phare 2021). Thus far, there has been no wastewater-treatment plant built at all. Especially during the recent pandemic, clean water is needed to maintain hygiene. To meet basic needs, indigenous communities use cistern water (often contaminated), while communities across the road are provided with safe tap water.

Dr. Martin-Hill has played a big part in the "Co-Creation of Knowledge," the integration of western science and indigenous knowledge as a core component of the Global Water Futures (GWF) research program. Co-creation of indigenous water-quality tools have been designed "*with* the community and *for* the community." She continues: "I hope that our research can show people how two ways of knowing can come together to solve problems." Her research examines the sources of water contamination on both Six Nations and the Lubicon Cree in Alberta. The three-year project will study the health impacts of water quality on people and animals that live in both communities. Her work is in water as *relation*, not *utility* alone.

Dawn is committed to understanding how water quality and security are linked to Indigenous community culture, livelihood, and health, all important in pursuit of future water security.

Dawn Martin-Hill is the recipient of the 2022 OU International Water Prize

Reference

Phare, M.-A. (2021). "OU Water Prize Nominator Support Letter for Dawn Martin-Hill," August 20, 2021.

Glossary of Important Terms

This glossary provides definitions of terms highlighted in **bold** at first use and used in multiple places throughout the text. These are terms or phrases that the authors consider to be foundational for the understanding of water security.

adaptation measures	those measures taken to lessen the risk and severe effects from climate change on people, property, and the environment
anthropogenic	environmental contamination resulting from human activities
aquatic ecosystems	systems of species that are dependent upon water present or supplied as rivers, springs, wetlands, coastal bodies of water (estuaries, lagoons) and groundwater
blue water	water that is stored in aboveground (surface) and belowground (groundwater) reservoirs
consequence	the effect of an event, incident, or occurrence; measured in quantifiable terms, such as gallons of water, dollars of damage, or lost revenue
consumptive use	removal of water from the environment for a good or service without returning the water to the local surface or groundwater source
conventional water pollutants	pollutants defined as such in the Clean Water Act (CWA) section 304(a)(4) and § 401.16; BOD_5, TSS, fecal coliform, pH, and oil and grease
corporate water stewardship	internal and external actions taken voluntarily by the private sector that ensure that water will be used in ways that are socially equitable, environmentally sustainable, and economically beneficial
disability-adjusted life years (DALYs)	the number of years lost due to illness or death as a result of exposure to an environmental contaminant
de facto water reuse	unplanned reuse of wastewater for individual or community drinking water
deficit irrigation	targeted application of water only during the critical growth stages of the plant when it is most needed

Fundamentals of Water Security: Quantity, Quality, and Equity in a Changing Climate, First Edition.
Jim F. Chamberlain and David A. Sabatini.

Companion website: www.wiley.com/go/chamberlain/fundamentalsofwatersecurity

direct potable reuse (DPR)	wastewater is treated to an advanced level before being sent directly to a conventional water-treatment plant and then to the consumer as drinking water
direct water usage	water consumed in the production of a product
drip irrigation	slow, targeted delivery of water directly to the plant roots, either from the surface or below the surface
economic water scarcity	limited access to water due to socioeconomic constraints even though water resources are sufficient to meet local demand
elasticity of demand	the change in demand (positive or negative) in response to a variable factor, such as the change in demand for water relative to price or household income
electrodialysis	a water-purification process that combines ion-selective membranes and an electrical potential gradient to separate ionic species from water
environmental flows	the quantity, timing, and quality of freshwater flows and levels necessary to sustain aquatic ecosystems, which, in turn, support human cultures, economies, sustainable livelihoods, and well-being
epidemiology	the study of the distribution and determinants of health outcomes in specified populations and the application of this study to control for health problems
evapotranspiration	the combined loss of water to the atmosphere from transmission of water through leaves and stems (transpiration) as well as volatilization from water bodies and land surfaces (evaporation)
excludable uses of goods	those uses in which one entity (or person) can easily exclude another entity from using the water
externalities	a side effect or consequence of an activity that affects other parties without being reflected in the cost of the goods or services involved
"fit for purpose" water	water that has been treated to an appropriate level - and no more - for its intended use
geogenic	environmental contamination resulting from geological processes (e.g. naturally occurring arsenic and fluoride in groundwater)
green water	water that is held in soil and available to plants; the largest freshwater resource but can only be used in situ by plants
groundwater-dependent ecosystems	systems of aquatic species that rely on springs, seeps, wetlands, aquifers, and cave ecosystems, and those gaining reaches of streams that are sustained by groundwater
guideline	a recommended but non-binding maximum allowable concentration of a contaminant in water
hazard	any physical, chemical, biological, or radiological agent that can cause harm to public health

hazardous event	any incident that introduces a hazard, or fails to remove it, from a water-supply system
hydroelectric power	power that is generated by passing water through a spinning turbine
indirect water usage	water consumed in the supply chain of a product
integrated watershed resources management (IWRM)	the consultative process of managing activities within a watershed that considers societal, economic, and environmental needs
indirect potable reuse (IPR)	advanced-treated wastewater that is sent to an environmental buffer (lake, river, wetland, or aquifer) for storage prior to withdrawal, water treatment, and distribution as drinking water
interbasin transfer (IBT)	a conveyance of water from one watershed into another in order to meet demand in the second watershed
likelihood of threat	the probability of an event occurring that might impair the utility from continuing its operation
managed aquifer recharge (MAR)	water management methods that recharge an aquifer using either surface or underground recharge techniques in order to store water for use in dry periods as needed
measures of association	the quantifiable ways in which an outcome (e.g. disease) is present in a population and its association with environmental factors
mitigation measures	preemptive actions that are taken to reduce the impact of human activities (e.g. reduce flow of greenhouse gases into the atmosphere, reduce chemical contamination of water, soil, or air)
multiple barriers of protection	the process of providing two or more preventative measures to reduce or eliminate the deterioration of water quality before the water reaches its intended use
net primary production	the difference between the energy fixed by plants and their respiration, measured as biomass per unit of land surface and time
nonconsumptive use	removal and return of water to the local surface or groundwater source from which it was withdrawn
nonstationarity	the unpredictability of the variation of natural phenomena because of the current confounding factors, especially a rapidly changing climate
physical water scarcity	regions in which more than 75% of freshwater is withdrawn for agricultural, industrial, or domestic purposes, implying a water demand that is approaching unsustainability
present value	the value on the date of evaluation of a projected cost or benefit assuming potential investment power
prior appropriation water rights	the practice of awarding access and usage of water to those who got there first in time
priority pollutants	contaminants that are regulated by the US Environmental Protection Agency (USEPA) because of the adverse health or ecological effects when released into the environment

randomized control trial	an experiment in which random groups or individuals are assigned an intervention with results being compared with those of a control group
relative species richness	ratio of the diversity of species to its areal extent of habitat
renewable energy	electrical energy derived from natural processes that are replenished within short time spans
resilience	the ability of a system to withstand a natural hazard or human attack without serious performance interruption
reverse osmosis	the pressure-driven membrane process that reverses osmotic pressure for purification of water
riparian water rights	the practice of awarding access and usage of water to those who own land adjacent to the path of the water
risk	the potential for loss or harm due to any unwanted event or threat, such as tornado, earthquake, flood, drought, accidental spills, or acts of terrorism
rival uses of goods	those uses in which one person's consumption reduces the amount available to another person
social discount rate	a discount rate used to determine the value of funds for social projects over an extended period of time
standard	a legal and enforceable maximum concentration of a contaminant in water
stationarity	the relative predictability of the variation of natural phenomena based on past occurrences
surface water ecosystems	systems of species that occupy lakes, ponds, or free-flowing streams
Sustainable Development Goals (SDGs)	the series of 17 strategies adopted by the United Nations in 2015 with the ultimate purpose of ending poverty and improving health and education, reducing inequality, spurring economic growth, and preserving our oceans and forests in the midst of a changing climate
thermal distillation	the process of heating, evaporating, and condensing water in order to cleanse it of impurities
thermoelectric power	electrical energy that is generated by the transformation of heat into power, typically through the generation of steam
total maximum daily load (TMDL)	maximum daily amount of a pollutant that a waterbody can receive and still meet water-quality standards for its intended use
virtual water	the sum total of the freshwater consumed in the growing (in food products) and production of the product on a per mass basis
vulnerability	the likelihood of damage from a threat, should it occur. It is usually a result of weaknesses in the system
water footprint	cumulative virtual water content of all goods and services consumed by one individual or by the individuals of one country
water intensity	the amount of water consumed on a unit basis (e.g. per mass, per mile traveled)

water productivity	the yield of a crop – in physical, economic, or nutritional yield – per unit of water
water reclamation	the appropriate treatment of wastewater (from municipal or industrial treatment plants) to the quality required for its intended use or discharge
water reuse	the reappropriation of reclaimed water for nonpotable or potable use
water scarcity	water supply at or below 1000 m^3 of available water per person per year
water security	the equitable availability of a suitable quantity and quality of water for health and well-being, with an acceptable level of water-related risks to people, environment, and economies
water stress	water supply at or below 1700 m^3 of available water per person per year
water transfer	the conveyance of water from a point of supply to a point of demand
watershed	a land area that channels all rainfall and snowmelt to creeks, streams, and rivers and eventually to a single outflow point (also referred to as a basin)
willingness to accept (WTA)	the minimum amount of money an individual would willingly accept for the loss of an environmental good
willingness to pay (WTP)	the maximum sum of money that an individual is willing to pay for an increase of an environmental good

Postlude

One of the earliest Greek philosophers, Thales of Miletus, believed that there exists a matter of which everything is composed and upon which everything else rests: water. It is the commonality that binds together the human and the nonhuman, the animate and the inanimate, the religious and the secular, in a watery web of life. Water endures through immediate change and persists through earth's millennial timescale.

Water's essential nature is evident in the "rule of threes" for human survival: A human person can survive no longer than three *minutes* without breathable air, three *days* without drinkable water, or three *weeks* without food. Water access thus has a timetable!

The subtitle of this book points to the many challenges involved in achieving water security: quantity, quality, equity, and a changing climate. An appropriate quantity and quality of water is needed to sustain and enrich human life, both directly (e.g. consumption) and indirectly (e.g. ecosystem services). Unlike land, water is not readily constrained by a deed or fence row. It naturally flows freely and according to the laws of physics. Nonetheless, it can be retained in a reservoir, pumped, and piped from one place to another and is often discarded with a diminished water quality. Water that is made clean and convenient may be out of reach to low-income people. Thus, inequities in water access are prevalent. And all of these are susceptible to our changing climate.

There is reason, of course, for optimism. There is enough water on earth to sustain life, both human and nonhuman, if used and managed wisely. Desalination may soon be more cost-effective by using only the sun's energy. The connections between groundwater and surface water are increasingly known and appreciated. The reclamation and recycling of wastewater are becoming more accepted in response to safe and successful implementations. Science possesses the knowledge and technology needed to purify water to its appropriate level of quality for the intended use. The fundamental human right to water is progressively being codified in national and international policies. Perhaps we will see a return to people migrating toward and protecting reliable water sources rather than pumping and piping water to them after the fact.

This book seeks to provide a general introduction to the complex and exciting field of water security. We hope that it inspires you, the reader, to further explore specific water challenges that you face in your locale, your vocation, or your field of interest.

Wonder is the prelude to gratitude. Having tasted a cool drink of water on a hot afternoon, or enjoyed a much-awaited spring rain, or stood in awe at the ocean's edge, we can pray with

Fundamentals of Water Security: Quantity, Quality, and Equity in a Changing Climate, First Edition.
Jim F. Chamberlain and David A. Sabatini.

Companion website: www.wiley.com/go/chamberlain/fundamentalsofwatersecurity

the psalmist: "You made the springs flow into channels that wind among the mountains. They give drink to every beast of the field ... Beside them the birds of heaven nest, among the branches they sing ... You bring bread from the earth and wine to gladden our hearts." (Psalm 104)

Such reflection most assuredly leads to gratitude and stewardship – gratefulness for all the blessings of water and a promise to work for its secure future.

Index

Fundamentals of Water Security: Quantity, Quality, and Equity in a Changing Climate, First Edition.
Jim F. Chamberlain and David A. Sabatini.

Companion website: www.wiley.com/go/chamberlain/fundamentalsofwatersecurity

C

i

j

k

l

r

s

x

y